13th International Conference "Modern Electrified Transport" (MET'2017)

MATEC Web of Conferences Volume 180 (2018)

Warsaw, Poland
5-7 October 2017

Editors:

A. Szelag
H. Gold
K. Karwowski
A. Zurkowski

ISBN: 978-1-5108-6803-8

Printed from e-media with permission by:

Curran Associates, Inc.
57 Morehouse Lane
Red Hook, NY 12571

Some format issues inherent in the e-media version may also appear in this print version.

Printed by Curran Associates, Inc. (2018)

For additional information, please contact EDP Sciences – Web of Conferences
at the address below.

EDP Sciences – Web of Conferences
17, Avenue du Hoggar
Parc d'Activité de Courtabœuf
BP 112
F-91944 Les Ulis Cedex A
France

Phone: +33 (0) 1 69 18 75 75
Fax: +33 (0) 1 69 28 84 91

contact-edps@webofconferences.org

Additional copies of this publication are available from:

Curran Associates, Inc.
57 Morehouse Lane
Red Hook, NY 12571 USA
Phone: 845-758-0400
Fax: 845-758-2633
Email: curran@proceedings.com
Web: www.proceedings.com

TABLE OF CONTENTS

13th International Conference
"Modern Electrified Transport"
MET'2017

Poland
Warsaw, October 5-7, 2017

ABOUT THE CONFERENCE

The series of MET – Modern Electrified Transport International Conferences (for more information visit http://ztu.ime.pw.edu.pl) is held every second year and brings together scientists, engineers, specialists, managers and representatives of institutions working for electric traction. The Conference was initiated 26 years ago by a famous researcher in the field of electric traction, Professor Jan Podoski (1904-1998), a graduate, Professor and Doctor Honoris Causa of the Warsaw University of Technology. This year MET'2017 was organised in Warsaw (more info: http://met2017.infotransport.pl) under the Honorary Patronage of the Committee on Electrical Engineering of the Polish Academy of Sciences, Warsaw University of Technology, Ministry of Infrastructure and Construction and Ministry of Energy.

Honorary Patronage

MINISTRY OF INFRASTRUCTURE AND CONSTRUCTION
MINISTRY OF ENERGY
POLISH ACADEMY OF SCIENCES, COMMITTEE ON ELECTRICAL ENGINEERING
WARSAW UNIVERSITY OF TECHNOLOGY

Organizers

PAN
Politechnika Warszawska
ZAKŁAD TRAKCJI ELEKTRYCZNEJ
University of Zagreb
tts
POLITECHNIKA GDAŃSKA
KATEDRA INŻYNIERII ELEKTRYCZNEJ TRANSPORTU
WYDZIAŁ ELEKTROTECHNIKI I AUTOMATYKI
IK INSTYTUT KOLEJNICTWA

Conference Partners

Bezpol
ALSTOM
pesa Pojazdy Szynowe PESA Bydgoszcz SA
EMENS Ingenuity for life
CORAIL
WOLTAN

Main Conference Partner

GRUPA TRAKCJA

Media Partners

AUTOBUSY Technika, Eksploatacja, Systemy Transportowe
tts

International Steering Committee

Organising Committee

Presentation of MET'2017

The participants of MET'2017, which in this year was held in Warsaw, had the opportunity for exchanging views and ideas with other specialists involved in the technical aspects of electrified transport, both rail and road modes, as well as presenting and sharing the most recent scientific, technical and as industrial developments.

In this year's edition, the topics covered were extended to include issues related to economic efficiency and safety of transport systems, planning, managing and implementation of project investments. The organisers received support from the following industry partners: main partner Grupa Trakcja, and additional partners: Bezpol, PESA Bydgoszcz S.A, CORAIL, SIEMENS, ALSTOM, Woltan. Media partners included Autobusy and TTS (Transport Technology Systems), while simultaneous interpretation was provided by Bireta – Professional Translations.

Photo 1: Welcome speech by the Vice-Rector of the Warsaw University of Technology, Prof. Stanisław Wincenciak, during an opening ceremony of MET'2017 in Mała Aula (the Main Building of WUT).

The Opening Ceremony, which gathered together approximately 90 participants and guests, took place on 5th October 2017, 10:30 a.m., in Mała Aula – the Main Building of the Warsaw University of Technology. The welcome speeches were delivered by the representatives of the main organising Institution, Warsaw University of Technology (WUT):

- Prof. Adam Szeląg – Chairman of the International Steering Committee – Warsaw University of Technology, Electric Traction Division, POLAND,
- Prof. Stanisław Wincenciak – Vice-Rector of WUT (Photo 1),
- Prof. Sylwester Robak – Director of Power Engineering Institute, Faculty of Electrical Eng., WUT,
- Prof. Marianna Jacyna – Dean of Faculty of Transport, WUT

as well as

- Prof. Josip Tomislav Mlinarić – Representative of the University of Zagreb, Vice-Dean of Faculty of Transport and Traffic Sciences (Photo 2),
- Prof. Krzysztof Karwowski – Co-Chairman, Gdańsk University of Technology (Photo 3),
- Dr Andrzej Żurkowski – Co-Chairman, Director of Railway Research Institute (Photo 4),
- Mr James Harkins – Head of Secretariat of the All Party Parliamentary Group LRT of Great Britain, Member of the Honorary Committee of MET'2017 (Photo 5),
- Mr Krzysztof Celiński, Director of Siemens.

Photo 2: Welcome speech by the Representative of the University of Zagreb, Vice-Dean of Faculty of Transport and Traffic Sciences, Prof. Josip Tomislav Mlinarić.

Photo 3: Welcome speech by the Co-Chairman, Prof. Krzysztof Karwowski, Gdańsk University of Technology.

Photo 4: Welcome speech by the Co-Chairman, Dr Andrzej Żurkowski, Director of Railway Research Institute.

Photo 5: Welcome speech by Mr James Harkins – Head of Secretariat of the All Party Parliamentary Group LRT of Great Britain.

Photo 6: Presentation made by Prof. Grzegorz Skarpetowski.

Next, during Plenary Session I, chaired by Director Andrzej Żurkowski, five invited papers were presented:

- *Life cycle assessment: assessing the environmental impact in the railway maintenance* – authors: M. Hegedić, N. Štefanić, M. Nikšić (University of Zagreb), presented by M. Nikšić,
- *Track insulation verification and measurement* – authors: J. Bongiorno (University of Genoa), A. Mariscotti (ASTM Sagl), presented by A. Mariscotti,
- *Realization of the mechanical characteristic of a series-excited DC machine in a drive with converter-fed induction machine* by G. Skarpetowski (Switzerland) (Photo 3)
- *Novel analysis methods of dynamic properties for vehicles pantographs* by A. Wilk (Gdańsk University of Technology),
- *Electrodynamics of electric power transmission and losses in devices of electric transport* by N. Kostin (Ukraine).

After a break, during Plenary Session II led by Prof. A. Wilk four papers were presented:

- *The Czech program of high speed railway development in the context international connections* by J. Ilik (Ministry of Transport, Czech Republic),
- *A mobile catenary power supply system for automotive vehicles* by T. Maciołek (Warsaw University of Technology),
- *Effectiveness of application of alternative drive vehicles in public transport* – authors: T. Dyr (Kazimierz Pulaski University of Technology and Humanities in Radom), K. Ziółkowska (University of Social Sciences), P. Misiurski (Opole University of Technology), presented by T. Dyr,
- *Traction power consumption as a component of maximum speed choice on high speed lines* by A. Żurkowski (Railway Research Institute).

After a lunch, the participants were transported to the Congress Center – Boss Hotel in Warsaw-Miedzeszyn, where a welcome dinner was organised.

On Friday, 6th of October, six thematic sessions were held:

- ***Energy effectiveness of transport / Efektywność energetyczna transportu* – A chaired by Prof. M. Nikšić**, with 7 presented papers on energy efficiency improvement in electrified transport through modernization of power supply, opportunities for charging of electric cars with solar energy, application of energy storage devices and energy consumption by modern trams.
- **Panel Session *Revitalization of urban transport – a chance for reduction environmental and transport problems in towns* chaired by Prof. A. Szeląg** with 7 papers. Particular attention should be given to the presentations delivered by J. Harkins (Head of the All Party Parliamentary Group LRT of Great Britain) and D. Giblin (Strategic Consultant of TramForward, UK All Party Parliamentary Group LRT) on the effective role of trams in tackling urban air quality (Oslo2 effect) and investments in trams in the UK. Especially interesting is how in the UK, where trams were eliminated in the mid-20th century, an increasingly strong lobby for the

improvement of environment in cities is trying to influence citizens and politicians to force return of trams to towns in the UK. The aforementioned presentations and subsequent papers dedicated to the analysis of tram power supply systems in Poland, opportunity and conditions for the use of railway track by trams, comparison of effectiveness of different means of urban transport and batteries as sources of energy for autonomous transport constituted a starting point for interesting discussions. There were no doubts that only electrified transport may be a solution to overcrowded city roads. However, slow improvement in energy storage devices technology and driving distance of autonomous vehicles is the reason why catenary supplied vehicles, especially rail ones, will still play a leading role in the nearest future. The number of trams (15) and trolleybuses (3) systems in Poland and 12,000 km of 3 kV DC electrified railways might help organising charging points for autonomous vehicles: electric cars and buses.

- **Parallel session *New technologies in electrified transport* chaired by Prof. M. Pawełczyk** with next 7 presentations concerning: SiC technology application in traction converters, new induction traction motors, methods for measurement of catenary and catenary-pantogragh co-operation, effectiveness of new 2x25 power supply system application and wayside storage devices for rail transport.
- **Parallel session IV *Automatics, control and management in electrified transport* chaired by Prof. T. J. Mlinarić** gathered 6 papers on: influence of braking control on a timetable on energy consumption by metro vehicles, reliability of supply and risk assessment of the railway traffic control system as well as introduction of new 25 kV 50 Hz systems in Poland, methods for controlling voltage influence in the 3 kV DC catenary on disturbing harmonics in a vehicle's current flowing in rails.
- **Parallel session V *Problems of construction and exploitation of transport infrastructure* chaired by Prof. P. Biczel** with 8 papers on: risk assessment of fire on-board of vehicles in tunnels, protection of pre-school children in the vicinity of suburban of railway lines, electrification criteria for railway lines and selection of high-speed rolling stock, inductive influence of 25 kV 50 Hz electrified lines, high-speed breakers in 3 kV DC traction substations, tram catenary assessment methods and implementation of new solutions in designing of railway power supply.
- **Parallel session VI *Energy effectiveness of transport- B* chaired by Prof. T. Glinka** 8 papers were presented on: LCC costs, including energy consumption as a criterion of rolling stock purchase, application of super capacitances as energy storage devices and installation of these devices in the 3 kV DC power supply system, electromagnetic processes during charging of on-board supercapacitance storage device, influence of increasing voltage in catenary on energy quality, equivalent values of currents calculations in the main circuits of a tram, application of modern technology, including SiC devices in drives and auxiliary converters of trams and electric busses.

After all-day-long sessions, the day was concluded with a social evening, which allowed the participants to discuss a number of various topics, not only related to technical and research issues.

On Saturday, 7th of October, the participants of MET'2017 visited the Alstom Workshop in Olszynka Grochowska (Warsaw), where Pendolino trains operated on Polish Railway Lines are serviced (Photo 7). The interesting solution implemented there is a movable rigid catenary (Photo 8) allowing a train to enter the service track, while later, in order to perform maintenance works on the roof, the catenary is moved away, which ensures space and safety for servicing staff.

During 3-days of MET'2017 over 120 participants and guests attended sessions, meetings and the technical visit.

The one-page abstracts of the papers were published in the Conference Proceedings, whereas the authors of the selected presentations will be asked to prepare the extended versions of the papers. Upon reviewing by the members of the International Steering Committee, these papers will be included, in an electronic form, in the Proceedings of MET'2017. The Proceedings in an open-access form will be published in 2018 on the MATEC Web of Conferences, indexed in WoS. Some papers are expected to be published in Polish in our media partner journals, i.e. TTS or Autobusy.

Photo 7: Pendolino train in the Alstom Workshop.

Statement of Peer review

In submitting conference proceedings to *Web of Conferences*, the editors of the proceedings certify to the Publisher that

1. They adhere to its **Policy on Publishing Integrity** in order to safeguard good scientific practice in publishing.
2. All articles have been subjected to peer review administered by the proceedings editors.
3. Reviews have been conducted by expert referees, who have been requested to provide unbiased and constructive comments aimed, whenever possible, at improving the work.
4. Proceedings editors have taken all reasonable steps to ensure the quality of the materials they publish and their decision to accept or reject a paper for publication has been based only on the merits of the work and the relevance to the journal.

Title, date and place of the conference

TITLE: 13th International Conference MODERN ELECTRIFIED TRANSPORT – MET'2017
DATE: October 5-7, 2017
PLACE: Warsaw, Poland
WEB-SITE: http://met2017.infotransport.pl/

Proceedings editor(s):

Editor in Chief:
Prof. Adam Szeląg - Warsaw University of Technology, Poland; adam.szelag@ee.pw.edu.pl

Board of Editors:
Krzysztof Karwowski - Gdańsk University of Technology, Poland; krzysztof.karwowski@pg.gda.pl
Hrvoje Gold - University of Zagreb, Croatia; dekan@fpz.hr
Andrzej Żurkowski - Railway institute, Warsaw, Poland; AZurkowski@ikolej.pl

Date and editor's signature

Warsaw, 29 June 2018

Adam Szeląg Krzysztof Karwowski

Hrvoje Gold Andrzej Żurkowski

https://doi.org/10.1051/matecconf/201818001001

A mobile catenary power supply system for automotive vehicles

Tadeusz Maciołek[1,*]

[1]Warsaw University of Technology, Institute of Electric Power Engineering, Electric Traction Division, ul. Koszykowa 75, 00-662, Warsaw, Poland

Abstract. Trolleybus type power supply using a two-pole overhead catenary system has been used for over one hundred years. However, energy demand in this type of road transport is the lowest. The main disadvantage of this type of transport is the lack of flexibility. Road electric vehicles equipped with batteries, which are now more and more often being put into service, allow for autonomous run on the roads without overhead contact wires. Therefore, it is possible to use them in a more flexible manner. The main problem is a limited range of autonomous run and long charge time of batteries. The paper presents a concept consisting in the application of an overhead catenary for supplying electric and hybrid vehicles (including automobiles) when a vehicle is in motion (mobile charging). Electric trucks are equipped with controlled current collectors allowing for connection with an overhead catenary during driving. Application of this solution to automobiles is rather impossible, which prevents them from being supplied directly from a catenary. The basis of the original concept presented in the paper is the creation of automobile convoys supplied via a truck equipped with current collectors. A description of the proposed system and an algorithm of its operation are presented in the paper. The discussed transport system has significant advantages in comparison to completely autonomous electric vehicles: increased range of driving with lower capacity of batteries on board and flexibility of use, longer life-time of batteries, lower energy consumption. The paper compares the proposed system with the currently used ones, which employ charging of automobiles during stops, but also underlines the advantages of the first solution.

1 Current solutions for mobile charging of automobiles

In Sweden, currently tests are being conducted for binary catenary supply of electric trucks equipped with current collectors. A 2-km section was launched in 2016. Further test sections are planned in the USA and Germany [1]. The basic difference between the trolleybus power supply system and the mentioned solution is higher speed (up to 90 km/h) and possibility of connecting pantographs during drive with the maximum speed. It is estimated that the maximum operational speed of such systems may be increased up to 140 km/h. In case of electric or hybrid automobiles with batteries, modern technical solutions for a catenary and smart current collectors would allow smooth drive at the demanded speed along sections with or without a catenary. It makes automobile driving flexible. The height of catenary suspension over a road (due to safety reasons) requires high mounting of current collectors, which are heavy and bulky. Therefore, vehicles should form a solid and stable construction. So, this solution may be used to supply busses and trucks rather than cars.

Application of mobile charging will reduce the need for stops required for charging of batteries, costs of exploitation and the number of batteries due to their lower energy capacity required (lower number of charging/discharging cycles and depth of discharging batteries, lower energy losses).

Other tested solutions include wireless inductive power supply of vehicles, including cars [2,3]. However, the main drawback of this type of power supply is high energy transfer loss (> 12%), due to additional energy transformation DC/AC at high frequency in infrastructure and AC/DC in vehicles [2,4]. Increased losses are also caused by atmospheric conditions and higher position tolerance of receiver coils in vehicles. For comparison – losses of energy transmission from a DC catenary to a vehicle are below 0.5% when the level of voltage in a catenary is above 600 V DC [5]. Due to the necessity of installing a greater number of high frequency converters, investment costs in power supply infrastructure will be a few times higher than in the case of an overhead catenary. Simultaneously, other studies are carried with the focus being put on solutions for road transport oriented towards reduction of energy consumption and increase of traffic safety. Successful tests of cooperative drive of individual cars, narrowly spaced, forming a convoy have been conducted [6,7]. Reduction of fuel used by cars in such a convoy is over a dozen percent. Equipment required for car control automation process allowing cars to drive at minimum spacing is currently being installed in serially manufactured cars. So, it is possible, using different available solutions and by modernizing them, to create a

* Corresponding author: tadeusz.maciolek@ee.pw.edu.pl

new road transport system, which could reduce fuel consumption and the number of drivers, which constitutes 70% of exploitation costs of cars [7]. Reduction of these costs is essential for operation of cars.

2 Convoy system with energy transmission between vehicles

This paper proposes a new Mobile Catenary Power Supply Convoy Cars (MCPSCC) system, with its main concept presented in Fig. 1.

a.) Connecting a new vehicle at the end of convoy

b.) driving in a convoy

c.) disconnecting from a convoy

Fig.1. A convoy in different phases of operation.

In this system, cars are composed into a road train via DC electric disconnectable connectors [8]. Connecting and disconnecting of the cars forming a convoy is possible during driving with a maximum speed. Any car in a convoy may leave it at any point of road. The convoy is headed by a Pilot – a vehicle equipped with current collectors and power connector allowing energy transfer from a catenary to cars connected to the Pilot. The Pilot is a truck or a bus with a central unit controlling the whole convoy and cars as well data transmission between the Pilot and cars, and it sends information about availability of 'space' for another car to connect the convoy. All cars in a convoy must have unified connectors in front and at the back.

The proposed MCPSCC system has the following advantages in comparison with individual traffic of cars:

- lower energy consumption,
- increased driving range,
- longer lifecycle of batteries,
- higher safety,
- higher traffic capacity of a road lane,
- decreased fatigue of drivers,
- lower unitary costs of transport,
- lower environmental load,

with flexibility of individual car transport maintained.

However, some drawbacks are evident:

- the road must have at least one lane equipped with an overhead catenary,
- equipment of cars must be unified (automatic power connectors allowing connection/disconnection at maximum drive speed),
- due to close distances between cars (<1 m) an automatic operation of vehicles during connection to/disconnection from a convoy is a must.
- central control unit in the Pilot vehicle must be used,
- continuous wireless data transmission between cars in a convoy is required,
- continuous data wireless transmission between a central control unit in the Pilot vehicle with an external data transmission network is necessary to enable localization of a specific convoy for connection.

Road convoys are assembled dynamically and the algorithm of operation may be as follows. When entering the road with a catenary, the Pilot vehicle, equipped with a double pantograph, power connector and the main computer announces, by WWW network, readiness to supply vehicles on a specific section of a road. Drivers of other vehicles (including cars) receive information about position of the Pilot and may connect the convoy, if driving in the same direction. Vehicles connect the convoy at its end part. The process of connecting an individual vehicle to the convoy, driving in the convoy and disconnecting is to be organized automatically with superior control by the Pilot's CPU. Disconnecting of any vehicle from the convoy may be done at any point of the road. The procedure requires forming gaps between the vehicle disconnecting from the convoy and the adjacent vehicles. The vehicle leaving the convoy changes a road lane and other vehicles can connect to the convoy again. When the Pilot vehicle leaves the convoy, the convoy is dissolved. At roads with low density of traffic or when a truck, which could operate as a Pilot vehicle, is excluded from traffic, it is possible to introduce special vehicles dedicated to operate as Pilot vehicles. Settlement of fees for energy consumption is to be done remotely with the use of smart energy counters installed on board of vehicles prepared to run in convoys.

3 Comparison of energy consumption by cars driving individually and in a convoy

Energy delivery to a DC supplied catenary from traction substations is associated with losses due to energy conversion and transmission, with ranges depending on the level of consumed voltage and power. The higher the voltage, the lower the losses (approx. 10%). A similar level of losses may be observed at stationary points of fast charging of electric vehicles. So, the following analysis concerns energy consumption by vehicles only.
For energy consumption to be reduced, it is required that a distance between vehicles in a convoy is below 1 m. Tests in an aerodynamic tunnel have shown [3] that aerodynamic resistances to motion could be reduced by 50% in comparison to resistances of a car driving individually (C_D/C_{D0}-Fig. 2). Automobiles driving in a convoy with a very small distances imposed by electric couplings may also ensure significant decrease of consumed energy.

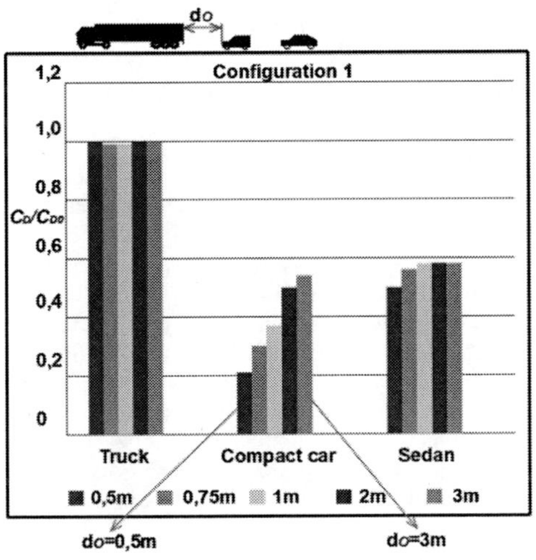

Fig.2. Dependence of aerodynamic resistance of a car driving in a convoy with a track compared with a car driving individually on a distance between driving cars. [3]

A comparison analysis for a car driving in a mix cycle, individual car and a car in a MCPSCC convoy is presented below.

A passenger car with typical parameters:
- mass 1,560 kg,
- car's frontal area [m^{2}],
- battery capacity of 36 kWh,
- mass of batteries and external connector - 200 kg,
- efficiency of a drive system - 85.5%,
- coefficient of aerodynamic resistance for a car:
- driving individually 0.35
- driving in a convoy 0.2.

Battery capacity of a convoy car was only slightly reduced, but its mass remained the same.

A UDDS urban driving cycle and a HFWET motorway driving cycle were assumed [9]. As it was expected energy consumption during a urban duty cycle (UDDS) for both types of cars (driving individually and in a convoy) are practically the same - 1.52 kWh. In a motorway driving cycle (HFWET), a convoy car uses 21% less energy – only 2.3 kWh than the second one – 2.9 kWh (Fig.4.). The savings result from lower aerodynamic resistances in a convoy. Forming a convoy of cars is required due to the operational principle of a system with a catenary power supply (MCPSCC), but a convoy of cars without catenary supply will have the same energy savings. However, when there are no catenary supply losses during frequent charging/discharging processes, it will still contribute to higher energy consumption. When catenary supply of a convoy is applied, the distance of driving under a catenary is limited by the length of catenary sections.

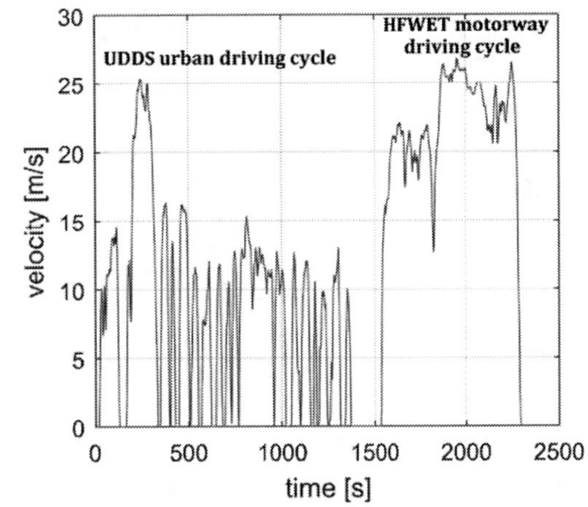

Fig.3. Speed profiles of a UDDS urban driving cycle and a HFWET motorway driving cycle used for the analysis of energy consumption by a vehicle.

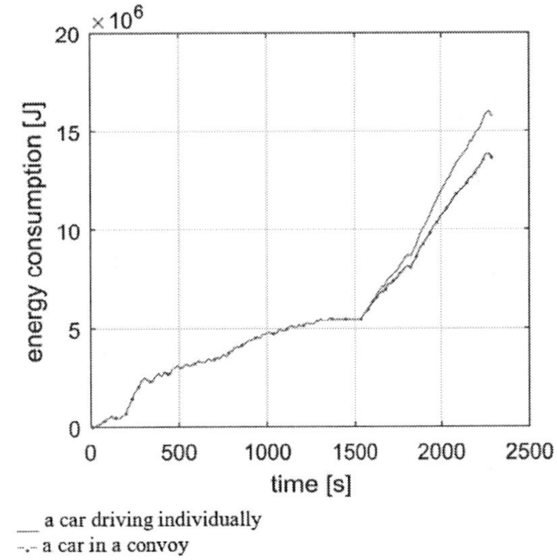

Fig.4. Energy consumption of a car driving individually and in a convoy.

Unlimited availability of power for cars during driving in a convoy makes it possible to use energy consuming auxiliary loads as air-conditions. Usage of auxiliary loads will significantly reduce SoC of battery when autonomous (no-catenary) driving is applied.

4 Comparison of mobile charging solutions

Catenary power supply for mobile charging requires construction of power supply infrastructure along a road. A schematic drawing of such a system (MCPSCC) is shown in Fig. 5. Since in a catenary there is low DC voltage present (below 1000 V), cars may be supplied directly from the catenary.

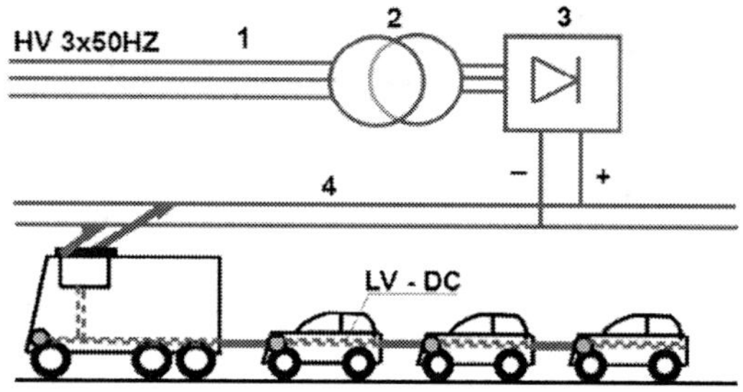

Fig. 5. A network power supply circuit for cars:
1– HV – high voltage power supply from a AC 50 Hz 3 phase
2– transformer
3– rectifier
4– low voltage (LV) DC power network

Wireless power supply of cars requires more complex infrastructure [2,4,10] (Fig. 6), including DC/HF energy conversion.

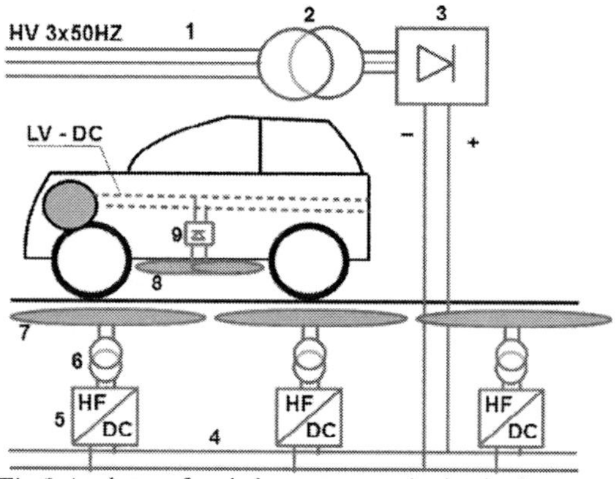

Fig.6. A scheme of a wireless power supply circuit of cars
1– HV – high voltage power supply from a AC 50 Hz 3 phase
2– transformer
3– rectifier
4– low voltage (LV) DC power network
5– HF (high frequency) generator
6– HF transformer
7– transmitter coil
8– receiver coil
9– rectifier in a car

Therefore, investment costs are much higher than in the case of network power supply infrastructure. Other advantages of catenary power supply over wireless power supply are as follows:

- higher reliability,
- lower transmission losses (only < 0.5% of losses at a contact point : pantograph-catenary at 750 V DC in comparison to 10% or more during inductive energy transfer),
- lower EMF emission,
- higher power of transfer (possible 750 kW of power delivered to a convoy of vehicles via 2 pantographs supplied by a 750 V DC catenary).

Such power allows for supply of a convoy composed of 7 trucks, each with mass of 24 Mg or a Pilot and 40 electric cars driving at a speed of 100 km/h.

When speed exceeds 100 km/h, power demand increases significantly (aerodynamic resistances are proportional to square of speed). Maximizing the number of cars in a convoy will result in all available power being used for driving and opportunity of charging of batteries will be reduced. When the number of cars is smaller, it is possible to charge their batteries as well. Cost of power supply infrastructure depends on the length of catenary sections and power demand. Increase of a length of sections supplied by one traction substation will increase losses of transmission due to low catenary voltage.

Methods of analysis and design applied in urban transport power supply systems [5] could be used to design the proposed power supply system. Since cars are equipped with batteries, it will be possible to drive on sections of a road without a catenary, where needed. In the most extreme cases only 30-40% of the length of a road for convoy driving must be equipped with a catenary. So, development of this system may be done step-by-step by increasing the length of catenary sections and the number of substations according to the demand for power dependent on the increasing traffic density.

5 Conclusion

The presented analysis has shown that catenary supplied cars in convoys used in a HFWET road cycle require 20% less energy for driving than individually driving autonomous electric cars. The use of wireless power supply will increase energy demand by another 10% comparing with the catenary supplied cars in convoys.

It was assumed that maximum speed of cars in a convoy is 100 km/h; however increase of speed to 120 km/h is assumed to be possible when automatic Pilot vehicles with pantographs driving at such a speed will be used.

Increase of speed will increase traffic capacity of the MCPSCC system.

Main advantages of the proposed MCPSCC system are as follows:

- increased range of electric car application with maintenance of flexibility and autonomy,

- reduction of energy capacity of batteries on-board of cars and a discharge level as well as the number of required cycles of charging/discharging, which increases lifetime of batteries,
- elimination of breaks required for charging of batteries and opportunity for mobile charging,
- reduction of energy consumption due to lower number of charging/discharging cycles and lower energy demand by a car driving in a convoy (in comparison to individual drive),

opportunity to develop the MCPSCC system step-by step according to traffic power demand (at the beginning only 30-40% of the road is to be equipped with a catenary). The presented system requires construction of a power supply catenary along the highways.

It also necessitates that the cars be equipped with electromechanical couplings executed as per a uniform standard. Currently, traffic of vehicles grouped in such long convoys is not permitted; therefore, changes in traffic law are required as well.

References

1. SIMENS Corporation, *eHighway – Solutions for electrified road freight transport,* https://www.siemens.com/press/ehighway
2. O.C. Onar, System Design and Experiment, in Wireless Power Transfer Systems, *IEEE Transportation Electrification Conference and Expo, :* Part 2 (2013)
3. P. Schito, F. Braghin,. Numerical and Experimental Investigation on Vehicles in Platoon , *SAE Int. J. Commer. Veh.,* 63-71, (2012), doi:10.4271/2012-01-0175
4. S. Judek, K. Karwowski, *Contactless electrical energy transfer system via magnetically coupled air coils*, (Gdańsk : Wydaw. PG, 2013. - ISBN 978-83-7348-501-3)
5. A. Szeląg, Z. Drążek, T. Maciołek, *Power supply systems of urban electric traction* (IN-W „Spatium", Radom, 2017) [in Polish: Elektroenergetyka miejskiej trakcji elektrycznej]
6. K-Y. Liang, *Fuel Efficient Heavy-Duty Vehicle Platoon Formation,* (PhD thesis, KTH Stockholm, Sweden, 2016)
7. N. Lyamin, Q. Deng, A. Vinel, Study of the Platooning Fuel Efficiency under ETSI ITS-G5 Communications, *IEEE 19th International Conference on Intelligent Transportation Systems (ITSC),* (2016) doi.org/10.1109/ITSC.2016.7795608
8. T. Maciołek, A. Szeląg, *Power supply of electric cars*, Polish patent application PL-418194, (2016) [in Polish: System zasilania samochodów elektrycznych]
9. Urban Dynamometer Driving Schedule "LA4" The UN/ECE Regulation 53 refers to the EPA UDDS, Highway Fuel Economy Driving Schedule (HWFET) www.epa.gov/vehicle-and-fuel-emissions-testing/dynamometer-drive-schedules
10. T-E. Stamati, P. Bauer, On-road charging of Electric Vehicles, *Proceedings of IEEE Transportation Electrification Conference and Expo,* 1-8 (2013)

Effectiveness of application alternative drive vehicles in public transport

Tadeusz Dyr[1,*], *Karolina* Ziółkowska[2], *Przemysław* Misiurski[3], and *Małgorzata* Kozłowska[1]

[1] University of Technology and Humanities in Radom, Faculty of Economic and Legal Sciences, Radom, Poland
[2]University of Social Sciences in Lodz, Lodz, Poland
[3]Opole University of Technology, Faculty of Economics and Management, Opole, Poland

Abstract. European transport policy for the first half the 21st century assumes creating the transport system organised with the respect of the principle of sustainable development, providing economic, social and ecological needs of the society and supporting construction of the integrated society and entirely integrated and competitive Europe. Applying alternative fuels and drives in public transport vehicles, including the development of electric-powered vehicles is an essential instrument of this policy. In the paper fuels applied in the public transport and alternative drives were described. Methodological establishments of evaluation of the investment's financial and economical effectiveness associated with acquiring public means of transport with the alternative drive were discussed. Using the method of the econometric modelling, operational and maintenance costs of use buses with the conventional and alternative drive were described. They constitute the basis of assessment of the financial effectiveness of replacing conventional vehicles with alternative propulsion vehicles. They pointed to benefits resulting from limiting the influence of public transport using alternative propulsion vehicles on the natural environment. They find expression in decreasing transport's external expenses, confirming the economic effectiveness of their application. A special attention was paid on vehicles with the electric propulsion. Their application in the urban transport is a crucial factor of improvement of the air quality in cities. Based on conducted examinations it was pointed at the purpose of applying means of transport with the alternative drive.

1 Introduction

The consistent increase in the level of urbanization effects both positive and negative. Strongly urbanised areas support the development of knowledge, learning and technology. It results above all from the great accretion of the idea on relatively small area [13], becoming a crucial factor of competitiveness and the growth in the economy of states and regions [18].

Adverse effects of the urbanization are related, among other things, to municipal transport systems. Meeting the transport needs takes place largely at using passenger cars and buses with conventional drive. Tis leads to emissions of harmful substances, including greenhouse gases, increase of the congestion and high-risk communications accidents [3]. The high share of vehicles driven with petroleum-derived fuels also causes existence of significant deficit in the trade balance of around 2.5 % GDP of the EU member states [2]. An instrument of preventing these risks is an alternative transport fuels strategy in the sector of European Union transport [17]. Such an assumption was also adopted in Polish energy policies to 2050 [15].

At the public transport in Poland, according Central Statistical Office of Poland data from December 31,2016, 11 973 buses, 3 332 trams and 223 trolleybuses were used . Enterprises of the public transport operated 454 buses with alternative drives (3.8 % of total number of buses). Buses fuelled by compressed natural gas (CNG), constituting c 70% of buses fuelled by alternative fuels have a dominating significance. Including the additional purchase of 40 hybrid buses (gas-electric) in 2015 for the purposes of the public transport in Częstochowa and 35 LNG buses in Warsaw, the participation of buses fuelled by natural gas in the total number of buses with alternative drives amounts in Poland around 95%. In 2017 a few supplies of electric buses were carried out to Polish cities, and a few cities announced public tenders for the delivery of such buses. However the participation of buses with alternative fuels must be qualified as very low. It is possible to put the following research question: why – in spite of the possibility of limiting the negative impact on the environment and for support in achieving the goals of the sustainable development – the share of vehicles with alternative drives is so low? The article constitutes the attempt of the reply to the question. Conducted examinations point, that the base of investment decisions realized by the public transport enterprises and local governments responsible for organization and financing the public transport, including the type of buses' engine types, are evaluation results of the financial and economic effectiveness.

* Corresponding author: t.dyr@uthrad.pl

2 Introduction

The possibility of using fuels and alternative drives depends on the transport kind and distance and the means of transport. The greatest possibilities of using different fuels appeare on urban areas (Table 1).

Applying fuels and alternative drives is a crucial factor of limiting the adverse impact of the transport to the environment and the counteraction for climate change. The literature review confirms that replacing buses with diesel engines with vehicles with the alternative drive generates high social benefits. And so such investments should be supported from public means.

Table 1. Coverage of transport modes and travel range by the main alternative fuels [2].

Fuel/drive		Transport mode											
		Road						Air	Rail	Water		
		Passenger			Freight					inland	short-sea	maritime
		short	medium	long	short	medium	long					
LPG		■	■	■	■	■	■			■	■	
Natural Gas	LNG	■	■	■	■	■	■		■	■	■	■
	CNG	■	■	■	■	■						
Electricity		■			■				■			
Biofuels (luiquid)		■	■	■	■	■	■	■	■	■	■	■
Hydrogen		■	■	■	■	■			■	■		

Amongst available alternative fuels a natural gas is of special importance. Using it is a crucial factor of the improvement in the quality of air in cities [10]. This fuel is recognised as the cleanest and the most hydrogen-rich hydrocarbon energy sources [6]. Using the natural gas – both in the form of compressed natural gas (CNG), as well as liquid natural gas (LNG) – is a crucial factor of achieving purposes of the sustainable development [1]. Vehicles fuelled with the natural gas produce significantly less toxic sub-stances and greenhouse gases, particularly CO_2, than vehicles with Diesel drives.

Using hydrogen is perceived as the essential instrument of the improvement in the energy security and the reduction in greenhouse gases. It may be used as the transport fuel and energy-storage agent [22]. Conducted examinations show that using this fuel in fuel cells is the most effective [23]. High costs of fuel cells and lack of refuelling infra-structure network are the main problems of applying hydrogen, as fuel to the drive of vehicles. However the permanent technological progress causes the systematic efficiency improvement of their application [11].

In recent years at the public transport electric buses have had a greater significance. In 2016 Solaris Urbino 12 electric bus, the first electric bus, was rewarded with the Bus of the Year 2017 title [19].

A major advantage of electric buses is their low-energy consumption and zero emission in the place of providing the transport services [12]. Their main defect is relatively high investments associated with the purchase of buses and loading infra-structure. Usually there needs to be higher amount of electric buses than the diesel buses to do identical transport tasks [14].

3 Methodology

Taking investment decisions, regarding the type of the drive in public transport vehicles includes results of the evaluation of the financial and economic effectiveness. After initial recognition this evaluation is conducted from investor's point of view, concentrating on the cash flow generated by the project in the transport enterprise. In the second recognition additional external costs and benefits are included [4]. They include these costs and benefits which are associated with production and consumption which are incurred by entities not-participating directly in the production, consumption or the exchange of the given good. In consequence that leads to necessity of bearing by these entities certain extra costs for which they don't receive any compensation from their perpetrator. It can also cause them benefits for which they don't pay the salary [20]. These costs and benefits determine accidental or incidental results of the entities' intentional activities. These include expressed in money negative effects of activity, including transport, to the natural environment [21].

Algorithm of financial effectiveness evaluation of investment projects, including purchase of public transport vehicles is presented in Figure 1. Financial effectiveness analysis is aimed to verify whether the project needs co-financing from public funds. And the economic effectiveness analysis should allow obtaining the answer to a question whether the project is worth co-financing. Algorithm of financial effectiveness evaluation of investment projects, including purchase of public transport vehicles is presented in chart 1, and the economic one in Figure 2.

In the evaluation of investment effectiveness incremental method is used (Sartori et al., 2014). Applying this method allows to make independent calculations from costs unrelated to vehicle's drive kind, i.e. costs which are identical for buses with different drives and aren't affecting effectivenesses evaluation results.

In incremental method it is necessary to identify cash flows diversifying compared investment variants. In the financial analysis these are investment costs, fuel costs and maintenance costs. In the economic analysis, apart from mentioned ones, external benefits are included.

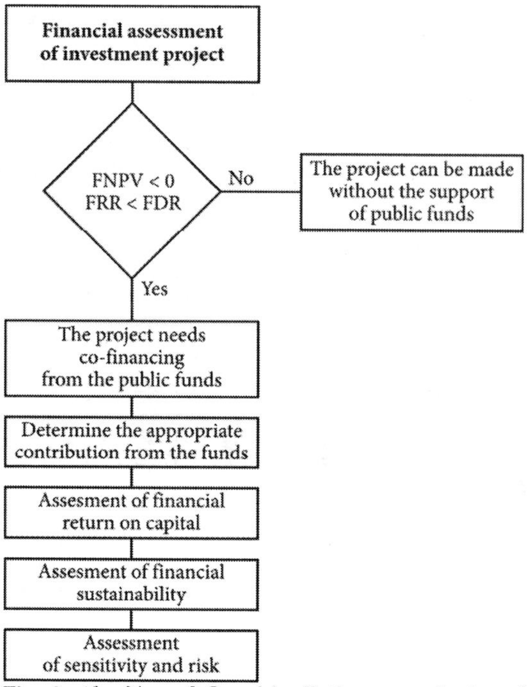

Fig. 1. Algorithm of financial effectiveness evaluation of investment projects.

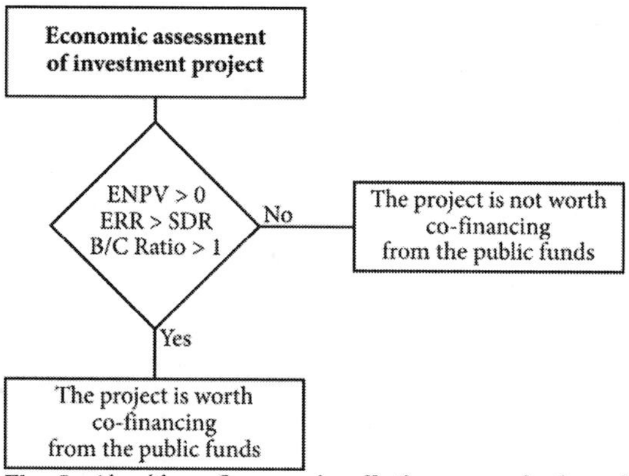

Fig. 2. Algorithm of economic effectiveness evaluation of investment projects.

4 Financial costs and benefits of purchasing buses with the alternative drive

A significant influence on the level of investments has a type of the drive. For buses with the length of 12 m they amount:

- Diesel bus – 825 thous. PLN,
- CNG Bus – 1 030 thous. PLN,
- Hybrid bus – 1 250 thous. PLN,
- Electric bus – 2 100 thous. PLN

Investment costs difference leads to diversity of depreciation amounts. However this component of costs isn't taken into account in the calculation of efficiency indicators. In this calculation only cash flows are taken into account [7]. However it has a significant influence on amount of compensation paid to the operator. According to Regulation (EC) No 1370/2007 [16] to service costs , constituting the basis of the compensation setting , include among others the depreciation costs. Application of CNG buses is a factor increasing expenses of self-government bodies on financing public transport services.

The factor that diversifies the effectiveness of replacing Diesel buses into vehicles with the alternative drive are costs of consuming fuels and the energy. Statistical data collected in examined enterprises show that the size of consuming diesel, the natural gas and the electric energy is identical in the entire life cycle of the vehicle. Hesitations in the level of the fuel consumption and the energy result mainly from weather conditions (using the air-conditioning or warming) and line load (number of passengers in the bus). In this situation for further analysis average costs of the fuel consumption of all inspected buses were taken. They amount:

- Diesel bus – 138,5 PLN/100 km (38,6 l/100 km),
- CNG Bus – 124,1 PLN/100 km (49,2 m^3/100 km),
- Hybrid bus – 110,0 PLN/100 km (31 l/100 km),
- Electric bus – 60,0 PLN/100 (200 kWh/100 km).

Maintenance costs are a crucial factor determining the effectiveness of purchasing the public transport vehicles. The relatively short operating period of hybrid and electric buses reduces precise determination of these costs in the entire life cycle. On that account the comparison was conducted for two types of buses – CNG and electric ones. For the comparison primary data from two enterprises exploiting c 130 buses were used. Conducted examinations show that the use of buses from the moment of the purchase causes gradual increase of renovations and maintenance costs. Their amount depends on the total bus mileage and the drive type. Evolution of the costs was presented in picture 3. Base of drawing the curve of costs are results of the econometric modelling. In the model a relation of unit running costs of the bus's availability being a sum of spare parts and repairs costs (dependent variable) from the accumulated mileage, reflecting the level of the vehicle's technical use (explanatory variable) was set [5].

As a consequence of faster growth of average maintenance costs of CNG buses there is an increasing difference between total maintenance costs of CNG and diesel buses (Figure 3). This component of operating cost is usually omitted in analyses. Meanwhile it has a significant influence on the financial effectiveness of purchasing CNG buses. Higher maintenance costs cause decrease of operating benefits resulting from lowest costs of fuel consumption.

Presented incremental cahs flow allows for the comparison of effectivenesses of applying buses fuelled by compressed natural gas and diesel at the public transport . Net present value indicator was used for the evaluation. For calculations of efficiency indicators discount rate in value of 4%, and time horizon – 10 years were used. It was assumed that a bus would be purchased in year 0. These values are recommended by experts of the Jaspers initiative for investment projects in the public transport carried out in Polish cities co-financed from funds of the European Union [9].

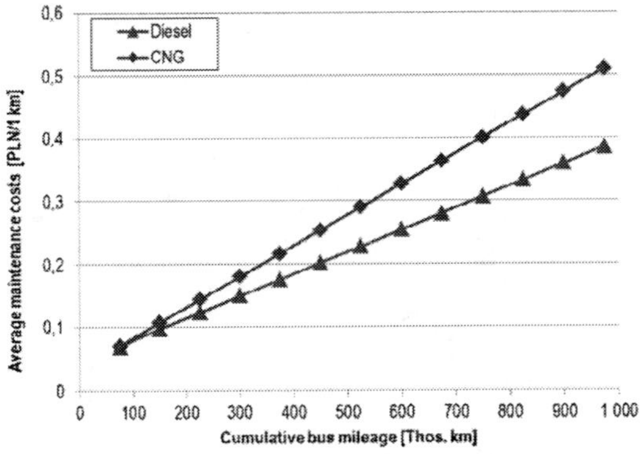

Fig. 3. Maintenance costs Diesel and CNG buses.

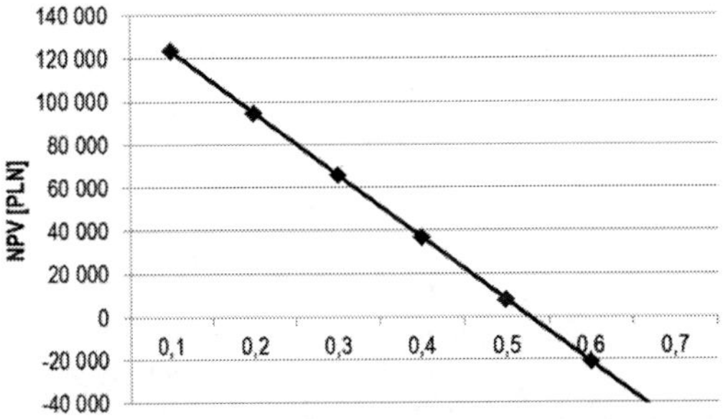

Fig. 4. Comparing the effectiveness of the purchase of the CNG and diesel bus depending on the relation of the fuels' price.

The NPV value was appointed in the function of the relation of 1 m³ CNG and 1 dm³ of diesel price. Results of calculations were presented in Figure 4. They show that the purchase of the CNG bus is more effective, if the price doesn't exceed 57.1% of price of 1 m³ CNG 1 dm³ of diesel. Otherwise for a transport enterprise purchasing a bus with the diesel drive is more profitable.

Presented analysis results assumes invariability of investment costs and maintenance cost. In reality these costs can change. Their height is determined by micro- and macroeconomic factors. Consequence of these changes can be other than appointed border value of the relation of CNG to diesel fuel's price. Sensitivity analysis reflects the assessment of such changes impact. Conducted calculations' results show that the border value of 1 m³ CNG price should not cross 52% of 1 dm³ diesel fuel. Exceeding this border is leading to the financial unprofitability of purchasing CNG buses.

Relatively little mileage of electric and hybrid buses exploited by the public transport enterprises in Poland is limiting, as already recalled, evaluation of the financial effectiveness of applying vehicles with these drives. Considering only 2 variables – investment costs and energy consumption cost a mileage, by which these costs will be levelled was appointed. They amount:

- Hybrid bus – 1,4 mln km,
- Electric bus – 1,6 mln km.

Drawn values are outside the boundary of profitable economically use of Diesel buses. Probably using hybrid and electric buses at such a course also will be unprofitable . It points out to the need of supporting investment from public means. Buses with alternative drives are characterized by a much lower negative impact on the environment. According to examinations conducted in the Motor Transport Institute in Warsaw [8], CNG application on the public transport buses causes reduction of discharges of toxic or otherwise polluting substances in the comparison from Diesel Bus with the Euro 6 engine, including CO_2 of about 25%, CO of about 50%, NOx of about 90% and PM – almost by the 100%.

In case of electric buses emission of toxic substances is entirely eliminated. In consequence in comparison with Euro 6 bus in the entire life cycle (mileage of 1 million km) reduces:

- greenhouse gas emission – for about 670 kg,
- particulates – for about 50 kg.

For CNG buses the reduction amounts:

- greenhouse gas emission – for about 380 kg,
- particulates – for about 50 kg.

Described environmental benefits confirm usefulness of applying buses with alternative drives at the public transport. These vehicles also enable a significant limitation of noise, smog and carcinogens.

5 Conclusions

Implementing the idea of the sustainable development is connected among others with the broader exploitation of alternative fuels in the transport sector, including coming from renewables. It is of special importance, how results from the conducted literature review, in cities and the urbanized area. The transport sector is there a main issuer of harmful substances, including greenhouse gases.

Barrier for the wide application of buses with alternative drive are higher investment costs than in case of purchasing diesel buses and uncertainty connected with operating costs of these vehicles, in it with costs of fuel and energy consumption. In many cases a need to adapt technical and service facilities for operating vehicles with alternative drives is also essential. In consequence applying buses with alternative drives is perceived by managers of transport enterprises, as the solution burdened with relatively high risk.

The European Union policy assuming creation of resource efficient transport system and friendly to the natural environment is a chance to increase the use of alternative fuels at the public transport. Increase in supporting investment in the communication including the purchase of low-emission vehicles is a consequence of this policy. Also there is an increase in environmental awareness of residents leading to higher interests in using environment-friendly means of transport.

References

1. Brito TLF, Moutinho dos Santos E., Galbieri R., Costa HK de M., *Qualitative Comparative Analysis of cities that introduced compressed natural gas to their urban bus fleet*, Renew Sustain Energy Rev. **71**, 502–508 (2016) doi: 10.1016/j.rser.2016.12.077
2. COM 17, *Clean Power for Transport: A European alternative fuels strategy* (2013)
3. COM 913, *Together towards competitive and resource-efficient urban mobility* (2013)
4. T. Dyr, P.R. Kozubek, *Assessment of transport infrastructure investments co-financed from European Union funds* (Spatium, Radom, Poland, 2013) [in Polish]
5. T. Dyr, P. Misiurski *Forecasting of costs of maintenace and use of a bus fleet*. Transp. Logist. Probl., **35**, 19–28 (2016) doi: 10.18276/ptl.2016.35-02 [in Polish]
6. M.J. Economides, D.A. Wood, *The state of natural gas*. J Nat Gas Sci Eng, **1**, 1–13 (2009) doi: 10.1016/j.jngse.2009.03.005
7. M. Florio, *Applied Welfare Economics: Cost-Benefit Analysis of Projects and Policies* (Routledge, New York, 2014)
8. W. Gis, E. Menes, J. Waśkiewicz. *Natural gas fuels in urban transport in Poland*, Transp Samoch (2011) [in Polish]
9. JASPERS, *Blue Book. Public Transport sector in cities, agglomerations, regions* (Warsaw, 2015)
10. G. Karavalakis, M. Hajbabaei, Y. Jiang, J. Yang, K. C. Johnson, D. R. Cocker, T. D. Durbin, *Regulated, greenhouse gas, and particulate emissions from lean-burn and stoichiometric natural gas heavy-duty vehicles on different fuel compositions,* Fuel, **175**, 146–156 (2016) doi: 10.1016/j.fuel.2016.02.034
11. A. Lajunen, *Fuel economy analysis of conventional and hybrid heavy vehicle combinations over real-world operating routes*. Transp Res Part D: Transport and Environment, **31**, 70-84 (2014) doi: 10.1016/j.trd.2014.05.023
12. A. Lajunen, *Energy consumption and cost-benefit analysis of hybrid and electric city buses*. Transp Res Part C: Emerging Technologies, **38**, 1–15 (2014) doi: 10.1016/j.trc.2013.10.008
13. V. Maráková, T. Dyr, A. Wolak-Tuzimek, *Factors of tourism's competitiveness in the European Union countries*, Ekonomika a management, **19**, 92-109 (2016) doi: 10.15240/tul/001/2016-3-007
14. J. Miles, S. Potter, *Developing a viable electric bus service: The Milton Keynes demonstration project*. Res Transp Econ, **48**, 357–363 (2014) doi: 10.1016/j.retrec.2014.09.063
15. Ministry of Energy, *The project of Poland's Energy Policy until 2050* (Warsaw, 2015)
16. OJ L 315 3.12.2007 Regulation (EC) No 1370/2007 of the European Parliament and of the Council of 23 October 2007 *on public passenger transport services by rail and by road and repealing Council Regulations (EEC) Nos 1191/69 and 1107/70* (2007)
17. K. Romejko, M. Nakano, *Portfolio analysis of alternative fuel vehicles considering technological advancement, energy security and policy*. J Clean Prod, 142, 39–49 (2017) doi: 10.1016/j.jclepro.2016.09.029
18. P.M. Romer, *Endogenous Technological Change*. J Polit Econ, **98**, 71–102 (1990) doi: 10.1086/261725
19. Z. Rusak, *Title International Bus of the Year for new Solarius Urbino electric,* Autobusy – Tech Eksploat Syst Transp, 12–20 (2016) [in Polish]
20. J.E. Stiglitz, *Economics of the Public Sector*, Third Edit. (W. W Norton & Company, New York/London, 2000)
21. D. Štreimikienė, I. Mikalauskas, *Internalization of external costs in Lithuania and Poland*, J Int Stud, **8**, 50–61 (2015) doi: 10.14254/2071-8330.2015/8-3/4
22. J. Wang, *Theory and practice of flow field designs for fuel cell scaling-up: A critical review*. Appl Energy, **157**, 640–663 (2015) doi: 10.1016/j.apenergy.2015.01.032
23. J. Wang, *Barriers of scaling-up fuel cells: Cost, durability and reliability*, Energy, **80**, 509–521 (2015) doi: 10.1016/j.energy.2014.12.007

Electrodynamics of electric power transmission and losses in devices of electric transport systems

Mykola Kostin[1,*]

[1]Dnipropetrovsk National University of Railway Transport named after Academician V. Lazaryan, Department of Electrical Engineering and Electromechanics, Lazaryana Str. 2, room 238, 49000, Dnipro, Ukraine

Abstract. For the first time, the "field" approach for explaining the processes of transmission and generation of electric power losses in devices of electric transport systems is described and theoretically substantiated on the basis of the theory of electromagnetic field. The results of the solution of the system of electromagnetic field equations show that it is energetically appropriate to design low-floor types of electric rolling stock. A qualitative view of electric power flows arriving through the air of the feeder zone from the traction substation and entering to the electric rolling stock through the roof and the front part of its body is presented. It is established that the main flow of energy enters through the roof porcelain insulator. At the same time, the electromagnetic waves partly penetrate into the metal surfaces of roof and frontal part of the body, and partially they are reflected from them creating losses of active power. The results of calculations of these losses, power factor and reactive power factor of the electric locomotive roof are shown. The relation between the standing waves, formed in the feeder zone, and the reactive power consumed by the electric rolling stock is established.

1 Introduction

Existing scientific publications, highlighting findings of electric power processes in electric transport systems, are based on the "circuit approach", i.e. on the theory of linear or nonlinear electric and magnetic circuits. Though, in essence, all electromagnetic and electric power processes at their full and detailed studying require the "field approach" based on the theory of electromagnetic field. In our opinion, it is nearly the main reason why a set of processes or existing facts in electric power engineering is not explained sufficiently. This "circuit approach" restricts and limits possibilities of scientific management of power processes. In this respect a lot of things are unexplainable. For instance, in case of "leak" (i.e. losses of the electric power in electric traction system) the meter registrations of electric power of traction substations (TS) and electric rolling stock (ERS) differ. These losses are already called "conditional losses" and they are in a directive way regulated by the Main Department of locomotive facilities of Ukrzaliznytsia (Ukrainian Railway). And though, many authors are trying to justify the reasons of occurring and change of these losses, but in our opinion these arguments are not convincing.

And, after all, we present the most important justification of the above. According to the theoretical bases of electrical engineering, electromagnetic energy (electric power) is transmitted from the source (traction substation) to the consumer (electric locomotive) not trough the wires (i.e. electric circuit), but through the dielectric (air) between (and behind) the wires and it is transmitted not by charges, but by electromagnetic field, i.e. by electromagnetic waves.

In this regard, the author of the present work covers a set of electrical power issues concerning power transmission in electric transport systems. For this purpose we will recall that, as it is well-known from the theoretical electrical engineering, the power transmission (and information as well) through dielectric is characterized by the Poynting vector $\overline{S}$ in respect of both quantity and direction [1, p. 421], [2, p. 147], [3, p. 220]:

$$\overline{S}=\left[\overline{E}\cdot\overline{H}\right], \qquad (1)$$

where $\overline{E}$ and $\overline{H}$ are vectors of electric and magnetic fields' intensities in dielectric points, i.e. the air in the feeder zone. The direction of the vector $\overline{S}$, i.e. direction of power flow in the given point, coincides with the forward movement of the right handed propeller at the rotation of vectors of $\overline{E}-\overline{H}$ from vector $\overline{E}$ to vector $\overline{H}$ towards a smaller corner between them.

2 Elements of the theory of wave reflection and refraction

As it is known from [1, pp. 590-601], [4, p. 363] if a plane electromagnetic wave, which varies sinusoidally in medium 1 with parameters ε_{a1}, μ_{a1}, γ_{a1}, meets a surface of another medium 2 with parameters ε_{a2}, μ_{a2}, γ_{a2}, and

* Corresponding author: nkostin@ukr.net

this wave falls (*incident wave* with intensities $\underline{\overline{E}}_{\text{in}}$, $\underline{\overline{H}}_{\text{in}}$) perpendicular to the border of mediums, then it is partially reflected $\underline{\overline{E}}_{\text{rf}}$, $\underline{\overline{H}}_{\text{rf}}$ with the reflection coefficient equals

$$\underline{\rho}_{\text{rf}} = \frac{\underline{Z}_2 - \underline{Z}_1}{\underline{Z}_2 + \underline{Z}_1}, \quad (2)$$

and can be described as

$$\underline{\overline{E}}_{\text{rf}} = \underline{\rho}_{\text{rf}} \cdot \underline{\overline{E}}_{\text{in}}, \quad (3)$$

$$\underline{\overline{H}}_{\text{rf}} = -\underline{\rho}_{\text{rf}} \cdot \underline{\overline{H}}_{\text{in}}. \quad (4)$$

Furthermore, this wave partially penetrates (refracts) $\underline{\vec{E}}_{\text{pn}}$, $\underline{\vec{H}}_{\text{pn}}$ into another medium 2 with the penetration coefficient (refraction) $\underline{\rho}_{\text{pn}}$, which are defined for:

1) waves of electric field intensity according to the expressions:

$$\underline{\rho}^{E}_{\text{pn}} = \frac{2\underline{Z}_2}{\underline{Z}_2 + \underline{Z}_1}, \quad (5)$$

$$\underline{\vec{E}}_{\text{pn}} = \underline{\rho}^{E}_{\text{pn}} \cdot \underline{\vec{E}}_{\text{pn}}; \quad (6)$$

2) waves of magnetic field intensity according to the expressions:

$$\underline{\rho}^{H}_{\text{pn}} = \frac{2\underline{Z}_1}{\underline{Z}_2 + \underline{Z}_1}, \quad (7)$$

$$\underline{\overline{H}}_{\text{pn}} = \underline{\rho}^{H}_{\text{pn}} \cdot \underline{\overline{H}}_{\text{in}}, \quad (8)$$

where $\underline{Z}_1$ and $\underline{Z}_2$ are normally complex wave impedances of mediums 1 and 2. For dielectric medium, the impedances have to be defined as

$$\underline{Z}_{\text{diel}} = \sqrt{\frac{\mu_{\text{a}}}{\varepsilon_{\text{a}}}}, \quad (9)$$

and they are real values. For a conducting medium, the impedances take imaginary value and should be defined according to the expression:

$$\underline{Z}_{\text{cond}} = \sqrt{\frac{\omega\mu_{\text{a}}}{\gamma}} e^{j45^0}, \quad (10)$$

where $\omega = 2\pi f$ is an angular frequency of sinusoidally varying voltage and current which produce electromagnetic field (waves).

From (2), (9) and (10) it follows that if mediums 1 and 2 have identical parameters, the reflected wave is absent. In this case, the only incident wave is in medium 1 and refracted wave takes place in medium 2.

Irrespective of the incidence angle of the incident wave on the bounder of mediums 1 and 2, the refracted wave propagates in the conducting medium, first, always in the direction of a normal to the border of mediums and, secondly, this wave dies out at the depth δ defined as:

$$\delta = \sqrt{\frac{2}{\omega\gamma_2\mu_{a2}}}. \quad (11)$$

And finally the last remark: since waves, being reflected, don't change axes of vectors but can change their sign, instead of the corresponding vectors $\underline{\overline{E}}$ and $\underline{\overline{H}}$ it is possible to consider their vector modules.

3 Power flow in the air between a catenary system and a rail

Let's consider power transmission from a DC traction substation to an electric locomotive for the case of one-sided supply and the most widespread catenary system of 2MF-100+A185 type and the rail of R65 type (Fig. 1)

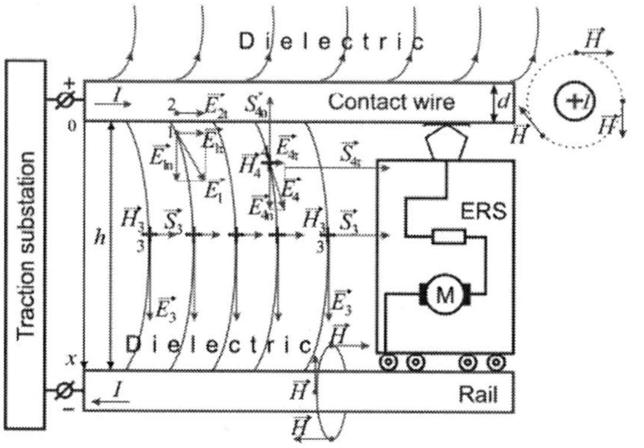

Fig. 1. A qualitative view of electricity transmission in the feeder zone.

The electromagnetic field in the dielectric (air) surrounding the catenary system is described by the system of equations [3, pp. 214-215], [5, pp. 29-30]:

$$\text{rot}\overline{H} = 0, \quad (12)$$

$$\text{rot}\overline{E} = 0, \quad (13)$$

$$\text{div}\overline{B} = 0, \quad (14)$$

$$\text{div}\overline{D} = \text{div}\varepsilon_{\text{a}}\overline{E} = 0 \quad (15)$$

$$\overline{B} = \mu_{\text{a}}\overline{H}, \quad (16)$$

$$\overline{D} = \varepsilon_{\text{a}}\overline{E}, \quad (17)$$

$$\overline{E} = -\text{grad}V. \quad (18)$$

Expression (13) demonstrates that electric field in the air is eddy-free, i.e. potential; therefore (18) is correct. Accepting that dielectric medium is homogenous, i.e. $\varepsilon_{\text{a}} = const$, and after inserting (18) in (15), we receive Laplace's equation

$$\text{divgrad}V = \frac{\partial^2 V}{\partial x^2} + \frac{\partial^2 V}{\partial y^2} + \frac{\partial^2 V}{\partial z^2} = 0, \quad (19)$$

where the potential V *(x, y, z)* is subjected in each dielectric point of the traction power system. Let's consider that in the first approximation the potential V is changing only relatively to the coordinate x, i.e. from the catenary system to the rail (Fig. 1), then (19) should to be simplified:

$$\frac{d^2V(x)}{dx^2} = 0. \quad (20)$$

The solution of (20) is possible under desired boundary conditions [5, pp. 37-39] and one of them is the equality of tangential components of electric field intensity (Fig. 1): $\vec{E}_{1t} = \vec{E}_{2t}$. The component $\vec{E}_{2t}$ in the conductor can be defined from well-known equation:

$$E_{2t} = \frac{J}{\gamma}, \quad (21)$$

where J is a current density in the catenary system, γ is electrical conductivity of material of the system.

For example, feeder current at one train, operated by the DE1 electric locomotive, doesn't exceed $I = 2000$ [A], and a cross section of the equivalent catenary system of 2MF-100+A185 type has to be found according to [6, p. 52] as:

$$s_{\text{eq}} = 2 \cdot 100 + \frac{185}{1.7} = 308.82 \text{ [mm}^2\text{]} = 308.82 \cdot 10^{-6} \text{[m}^2\text{]}.$$

Hence,

$$J = \frac{I}{s_{\text{eq}}} = \frac{2000}{308.82 \cdot 10^{-6}} = 6.48 \cdot 10^6 \text{ [A/m}^2\text{]}. \quad (22)$$

If γ for cooper equals $5.8 \cdot 10^7$ [cm/m], then

$$E_{2t} = \frac{6.48 \cdot 10^6}{5.8 \cdot 10^7} = 0.112 \text{ [V/m]}.$$

At the same time a normal intensity component $\vec{E}_{1\text{n}}$ in dielectric is $E_{1\text{n}} > \frac{U}{h}$, since the field in the dielectric is non-homogenous and the strongest one is near catenary wires. Accepting maximal value of h, the intensity is calculated as

$$E_{1\text{n}} > \frac{3300}{6{,}8} = 485.3 \text{ [V/m]}.$$

Hence, E_{1t} equals E_{2t} and is 0.112 [V/m], that is less than $E_{1\text{n}}$ in $4.3 \cdot 10^3$ times. Therefore, calculating the electric field in the air around the catenary system it is possible to ignore component $\vec{E}_{1t}$ and use boundary electrostatics condition, i.e. condition of equipotential surface of a conductor. Hence, the intensity in the air between the catenary system and the rail is defined as

$$E(x) = -\text{grad}V(x) = -\frac{dV}{dx} = \frac{U}{h}. \quad (23)$$

The intensity of magnetic field will be found using the Ampere's law (Fig. 1) as $H \cdot 2\pi r_0 = I$. Taking into account that $r_0 = x + d/2$, we get

$$H(x) = \frac{I}{2\pi(x + d/2)}, \quad (24)$$

where d is a diameter of the equivalent catenary wire.

At a further analysis we will take into account that to the strength $H(x)$, determined by (24) and produced by equivalent catenary wire, it is necessary to add the component produced by the current in rails. The equations for resultant intensity may to be found at any point (in two-wire power line), for example, according to [3, pp. 176-181]. Substitute (23) and (24) into (1), in the first approximation we get (without considering $\vec{H}$, produced by the current in a rail) the equation of the absolute value of the Poynting vector $\overline{S}$ in dielectric (air) points between the catenary system and the rail:

$$|S| = \frac{U \cdot I}{2\pi \cdot h(x + d/2)}. \quad (25)$$

In points 1, 2, 3 and 4 (Fig. 1) the vector $\vec{H}$ is directed behind a flat surface of the sheet as shown in Fig. 1. Then, $\vec{E}_4$ is divided into two components: tangential component $\vec{E}_{4\tau}$ and the normal $\vec{E}_{4n}$ to the surface of the contact wire. Taking into account the above-mentioned direction rule of the Poynting vector $\vec{S}$, it is possible to make the following conclusions.

1) In point 3, i.e. in the middle part of the space of traction network the vector $\vec{S}_3$, i.e. the energy flow, is completely directed from TS to electric locomotive.

2) In point 4 the component $\vec{E}_{4n}$ and the vector $\vec{H}_4$, according to (1), cause the formation of component of Poynting vector $\overline{S}_{4\tau}$ which also has the direction to the load (i.e. ERS).

3) In point 4 the components $\vec{E}_{4\tau}$ and $\vec{H}_4$ also form the normal component $\overline{S}_{4n}$ directed out of the dielectric (air) into the contact wire. On the border with the wire the electromagnetic wave falling from the air ($\vec{E}_{4\tau}$, $\vec{H}_4$) is partially reflected from the wire surface, and partially penetrates (refracts) into it; let's define the reflected wave.

For the air of medium 1, the wave impedance can be calculated as

$$Z_1 = \sqrt{\frac{\mu_0}{\varepsilon_0}} = \sqrt{\frac{4\pi \cdot 10^{-7}}{8.854 \cdot 10^{-12}}} = 376.7 \text{ [Ohm]}.$$

Medium 2 is the conducting medium and consist of wire of the 2MF-100+A185 catenary system. The wire cross-section is equal to $308.82 \cdot 10^{-6}$ [m^2]. At the same time the wire diameter is 19.83 [mm], specific conductivity is $\gamma_2 = \gamma_{\text{eq}} = 5.0 \cdot 10^7$ [Sm/m] and $\mu_{r2} = 1.0$. Feeder current is inverted, 6- or 12-pulsing; and harmonics with the frequency of 300÷2400 [Hz] are dominant in it. Then wave impedance of the material of the equivalent wire, in accordance with (10), is equal to:

$$\underline{Z}_2 = \sqrt{\frac{2\pi \cdot 2000 \cdot 1 \cdot 4\pi \cdot 10^{-7}}{5 \cdot 10^7}} = 177.6 \cdot 10^{-7} \cdot e^{j45^\circ} \text{[Ohm]}.$$

Let's define the reflection coefficient on the border "air – wire surface" according to (2):

$$\underline{\rho}_{\text{rf12}} = \frac{177.6 \cdot 10^{-7} \cdot e^{j45^0} - 376.7}{177.6 \cdot 10^{-7} \cdot e^{j45^0} + 376.7} \approx -1.$$

Then, in accordance with (3) and (4), field intensities of the reflected wave are $\underline{\vec{E}}_{\text{rf1}} \approx -\underline{\vec{E}}_{\text{in1}}$ and $\underline{\vec{H}}_{\text{rf1}} \approx \underline{\vec{H}}_{\text{in1}}$.

Hence, the reflected wave almost completely reflecting, changes the sign at the vector $\vec{E}$. Refraction coefficients according to (5) and (7) are:

$$\underline{\rho}^E_{\text{pn2}} = \frac{2 \cdot 177.6 \cdot 10^{-7} \cdot e^{j45^0}}{177.6 \cdot 10^{-7} \cdot e^{j45^0} + 376.7} \approx 0.94 \cdot 10^{-7} e^{j45^0},$$

$$\underline{\rho}^H_{\text{pn2}} = \frac{2 \cdot 376.7}{177.6 \cdot 10^{-7} \cdot e^{j45^0} + 376.7} \approx 20.$$

Despite the obtained values of ρ_{pn}, a part of the incident wave is refracted. According to (11) it travels into the wire to the depth δ of:
4.11 [mm] for sinusoidal wave with frequency of 300 [Hz];
1.45 [mm] for sinusoidal wave with frequency of 2400 [Hz].

Thus, the equivalent wire of the diameter $d = 19.83\,[\text{mm}]$ of the catenary system is "impenetrable" for a wave; it dies out at the depth of 1.45÷4.11 [mm]. Energy of this wave is spent on the thermal losses in the wire occurring at feeder current flow in it. For comparison, let us calculate these losses in a kilometre of equivalent contact wire using existing (i.e. by the circuit theory) and proposed here "field" approach.

3.1) *The circuit approach.* For resistivity of contact wire $R_{cw} = 42.4 \cdot 10^{-3}$ [Ohm/km] and feeder current $I = 2000\,[\text{A}]$ the losses according to the circuit approach are $\Delta P_{CA} = R_{cw} I^2 = 169.6\,[\text{kW/km}]$.

3.2) *The field approach.* In this case, the losses are determined by the value of Poynting vector $\overline{S}$ at the surface of the contact wire (Fig. 1) as

$$\Delta P_{FA} = S_s = E_{1t} \cdot H_1 \cdot \sin\left(\overline{E}_{1t} \hat{\ } \overline{H}_1\right).$$

The intensity H_1 has to be found according to the total current law as

$$H_1 = \frac{I}{2\pi d/2} = \frac{2000 \cdot 2}{2\pi \cdot 19.83 \cdot 10^{-3}} = 32\ 120\,[\text{A/m}],$$

where d is a diameter of equivalent contact wire. Taking into account calculated above $E_{1t} = 0.112$ [V/m], value of Poynting vector equals

$$S_s = 0.112 \cdot 32120 \cdot \sin 90^\circ = 3597.44\,[\text{W/m}^2].$$

The surface of a kilometre of equivalent contact wire is $62.266\,[\text{m}^2]$, so finally the losses according to the field approach are $\Delta P_{FA} = 225.4\,[\text{kW/km}]$.

As follows from the calculation results, ΔP_{FA} exceeds ΔP_{CA} by 24.7%, which corresponds to so-called "conditional" losses, "leakage" losses, which are still inexplicable and which are not counted by electricity meters, since their principle of operation is based on the circuit theory. Therefore, the electric power analysis of electric transport systems have be carried out on the basis of the electromagnetic field theory.

4) According to (25), the highest power density per time is transmitted in the space of small values x (Fig. 1), that is, in the space around the contact wires and the rail.

It is possible to show that the same condition regarding the direction of electric power is observed also in the space above the catenary system and under the rail base with the ground; but values of the vector $\overline{S}$ will be different, considering non-homogeneity of all traction network.

Considering the above and expression (25), it is possible to make the following strategic proposals regarding devices of electric traction system. According to (25), with the decrease of the value h, i.e. with the decrease of distance between the contact wire and the head of a rail, density of the electric power transmitted from TS to ERS increases. According to the technical safety rules, $h_{max} = 6.8\,[\text{m}]$ and $h_{min} = 5.75\,[\text{m}]$, therefore in the projected electric traction systems it is reasonable to reduce h, and construct low-floor (but probably longish) types of ERS. There is nothing irregular in this proposal if we take into account that nowadays one of the most important indicators of technical progress is the reducing the space occupied by the electromagnetic device [7, 8], as it is for a reason microcomputers, electrical micro machines, different micro devices and the like are created. In the context of energetics for production of the mentioned devices it is reasonable to enter the coefficient of utilization of construction space η_e:

$$\eta_e = S_{min} / S_{max}, \tag{26}$$

where $0 \le \eta_e \le 1$.

4 Electric power flow through the roof bushing insulator and through the roof of electric locomotive

Out of the total electric power, flowing (carried by waves) to ERS out of the air space of the traction network it is possible to identify two flows: the main and the minor. The main flow is the energy flow representing the greatest part of the energy received by ERS. It concentrates around traction network wires and enters to electrical devices of the high-voltage cabinet through the roof bushing insulator, i.e. roof surface (Fig. 2) and through the bottom of the cabinet from the space around the rail. The minor flow is the electric power flow carried by waves which fall on the front part of electric locomotive body, which consists, in its turn, of metal and dielectric (cab windscreen) parts (Fig. 2).

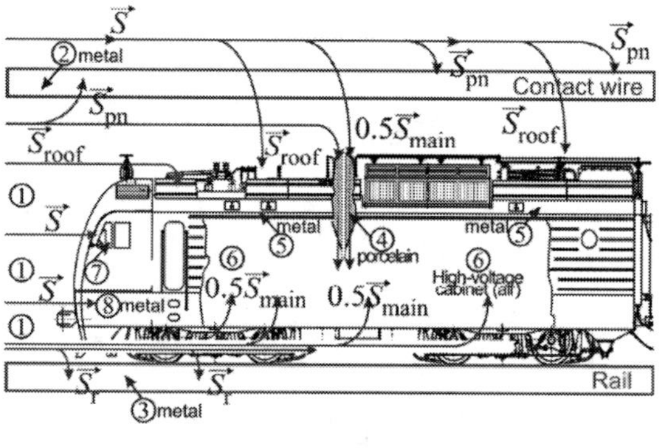

Fig. 2. A qualitative view of the flows of electricity through the roof and the frontal part of ERS: 1 – air around wires and rails of traction network; 2 – contact wire; 3 – rail; 4 – insulator of roof bushing; 5 – metal of the roof; 6 – air in the high-voltage cabinet; 7 – cab windscreen; 8 – metal part.

In the beginning we will analyse the main flow in which it is possible to identify the following components of the Poynting vector carrying it (Fig. 2).

 https://doi.org/10.1051/matecconf/201818001003

1) The main vector $(0.5 \cdot \overline{S}_{\text{main}})$ corresponding to the energy flow density coming into the elements of the traction power circuit of ERS (into the high-voltage cabinet) through the roof bushing porcelain insulator. In order to analyse the electromagnetic waves carrying $\overline{S}_{\text{main}}$ we will define the coefficients of reflection $\underline{\rho}_{\text{rf14}}$ and refraction $\underline{\rho}_{\text{pn14}}$ on the border "air 1 – porcelain 4". For ideal dielectrics like air $(\varepsilon_{r1} = 1.0)$ and porcelain $(\varepsilon_{r4} = 6.0)$, expressions (2) and (5) for $\underline{\rho}_{\text{rf}}$ and $\underline{\rho}_{\text{pn}}$ are simplified as:

$$\rho_{\text{rf14}} = \frac{\sqrt{\varepsilon_{r1}} - \sqrt{\varepsilon_{r4}}}{\sqrt{\varepsilon_{r1}} + \sqrt{\varepsilon_{r4}}} = \frac{\sqrt{1} - \sqrt{6}}{\sqrt{1} + \sqrt{6}} = -0.42,$$

$$\rho^{E}_{\text{pn4}} = \frac{2\sqrt{\varepsilon_{r1}}}{\sqrt{\varepsilon_{r1}} + \sqrt{\varepsilon_{r4}}} = \frac{2 \cdot \sqrt{1}}{\sqrt{1} + \sqrt{6}} = 0.58.$$

Therefore:

$$E_{\text{rf14}} = -0.42 E_{\text{in1}},$$

$$\rho^{H}_{\text{pn4}} = \frac{2\sqrt{\varepsilon_{r4}}}{\sqrt{\varepsilon_{r1}} + \sqrt{\varepsilon_{r4}}} = 1.42,$$

$$H_{\text{rf14}} = -(-0.42 H_{\text{in1}}) = 0.42 H_{\text{in1}},$$

$$E_{\text{pn4}} = 0.58 E_{\text{in1}},$$

$$H_{\text{pn4}} = 1.42 H_{\text{in1}}.$$

As we can see, a part of the energy is reflected from the insulator. To minimise this part, it is necessary to select this insulator with ε_r, close to $\varepsilon_r = 1.0$.

The second half of $\overline{S}_{\text{main}}$ (Fig. 2) enters into the high-voltage cabinet of ERS from the space around the rail (i.e. "from the bottom" of ERS). Wave propagation in this space and power transmission needs more in-depth studying which is not given here yet.

2. The power vector (Poynting) $\overline{S}_{\text{roof}}$ penetrates into the metal roof of ERS through its surface (Fig. 2). The vector $\overline{S}_{\text{roof}}$ is directed normally to the surface. Roof material is a sheet structural steel of 2 [mm] thick with parameters: $\gamma = 7 \cdot 10^6$ [Sm/m]; $\mu_r = 1000$.

Electromagnetic harmonic waves of various frequencies, falling from air 1 onto the border with the surface of metal roof 5 (Fig. 2), are partially reflected, and partially penetrate into the metal of the roof, gradually dying out and at the same time causing losses of electric power in roof metal.

Wave impedance of the roof metal $\underline{Z}_5$, found according to (10) for $f = 2400$ [Hz], equals $164.4 \cdot 10 - 5e^{j45^0}$ [Ohm]. For air it equals to $-\underline{Z}_1 = 376.7$ [Ohm]. Substituting $\underline{Z}_1$ and $\underline{Z}_5$ in (2), the coefficient of wave reflection from the roof surface is $\rho_{\text{rf15}} \approx -1$. Substituting them in (5) the refraction coefficient equals $\rho_{\text{pn15}} \approx 0.87 \cdot 10^{-5} e^{j45^0}$. The wave with the frequency of 2400 [Hz] penetrates on the depth δ_5 of 0.128 [mm], analogically, with frequency of 600 [Hz] – 0.246 [mm], and with frequency of 50 [Hz] – 0.85 [mm].

As far as, the thickness of the steel roof sheet is 2 [mm] and $\delta_{5\max} = 0.85$ [mm], the metal of the roof is "impenetrable" for the waves since the wave dies out completely, without reaching an opposite surface. It means that the metal sheet of the roof can be considered as an endlessly thick sheet or as endless conducting half-space, and the power dissipation in it is equal to the Poynting vector on the surface of the roof $\overline{S}_{\text{roof}}$ multiplied by this surface s_{roof}.

Complex Poynting vector on the surface of the sheet is defined by the expression [8]:

$$\underline{S}_S = (1 + j)\sqrt{\frac{\omega\mu_a}{2\gamma}} \cdot \frac{H^2_{mS}}{2}. \tag{27}$$

Its active component is

$$S_P = \alpha_P \sqrt{\frac{\omega\mu_a}{2\gamma}} \cdot \frac{H^2_{mS}}{2}, \tag{28}$$

and inductive component is

$$S_Q = \alpha_Q \sqrt{\frac{\omega\mu_a}{2\gamma}} \cdot \frac{H^2_{mS}}{2}, \tag{29}$$

where for steel $\alpha_P = 1.4$, $\alpha_Q = 0.85$, H_{mS} is an amplitude of harmonic magnetic field intensity on the given surface.

Let's calculate H_{mS} for a harmonic with frequency of $f = 600$ [Hz] (for a 12-pulse current). The value of H_{mS} for this harmonic is 955.5 [A/m]. Then, applying necessary values and parameters of roof steel sheet in (28) and (29), we get:

– losses of active power from the harmonic per roof surface unit

$$S_P = 371.6 \text{ [W/m}^2\text{]};$$

– consumption of reactive power

$$S_Q = 225.6 \text{ [var/m}^2\text{]}.$$

Total surface of the roof of one ERS section equals ~ 50 [m²], let $s_{\text{roof}} \approx 100$ [m²] for a "medium-active" surface of two-section ERS. Then the power losses are

$$P = 37.2 \text{ [kW]},$$

$$Q = 22.6 \text{ [kvar]},$$

$$S = 43.6 \text{ [kVA]}.$$

Power factor of the roof equals $\lambda = 0.85$ and reactive power factor is $tg\phi = 0.61$.

5 Energy flow falling on the front of the locomotive body

Electromagnetic waves propagating in the middle part of the airspace 1 between the catenary system and the rail, fall onto the front part of the locomotive body which consists of glass 7 and lower metal part 8 (Fig. 2). The thickness of cab windscreen is 15 mm and $\varepsilon_{r7} = 5.5 \div 10$ (let's take 9.0). On the border "air 1 – glass 7" waves are partially reflected with the coefficient of

$$\rho_{\text{rf17}} = \frac{\sqrt{\varepsilon_{\text{r1}}} - \sqrt{\varepsilon_{\text{r7}}}}{\sqrt{\varepsilon_{\text{r1}}} + \sqrt{\varepsilon_{\text{r7}}}} = \frac{\sqrt{1} - \sqrt{9}}{\sqrt{1} + \sqrt{9}} = -0.5,$$

and partially penetrated (refracted) with the coefficient of

$$\rho^{E}_{\text{pn7}} = \frac{2\sqrt{\varepsilon_{\text{r1}}}}{\sqrt{\varepsilon_{\text{r1}}} + \sqrt{\varepsilon_{\text{r7}}}} = \frac{2\sqrt{1}}{\sqrt{1} + \sqrt{9}} = 0.5.$$

Then, $E_{\text{rf1}} = -0.5 E_{\text{in17}}$ and $E_{\text{pn7}} = -0.5 E_{\text{in17}}$.

The metal part of lower front part of ERS body is made from structural sheet steel of 7 mm. And therefore, analogically to the roof metal, it is also "impenetrable" for electromagnetic waves, so a part of the wave, penetrating into metal, creates the power losses [9] determined by (27)÷(29). In order to find them we must know amplitudes of harmonicas of the magnetic intensity H_{mS} on the surface of metal part of the body which are not given here yet.

6 Relationship between reactive-power and standing electromagnetic waves

In addition to the considered above, let's analyse a variable electromagnetic field in the air space of feeder zone. This field is created by the harmonic flat progressing waves of electric and magnetic field intensities that transmit electric power from the TS to the electric locomotive [5]:

$$E(y,t) = E_{m\,\text{in}} \sin(\omega t - \beta y) + E_{m\,\text{rf}} \sin(\omega t + \beta y), \quad (30)$$

$$H(y,t) = H_{m\,\text{in}} \sin(\omega t - \beta y) + H_{m\,\text{rf}} \sin(\omega t + \beta y), \quad (31)$$

where: $\beta = \omega / C_0$ is a phase coefficient; $C_0 \approx 3 \cdot 10^8$ [m/s]. In expressions (30) and (31) the first components of electric and magnetic fields describes the incident (falling) wave, and the second – reflected (reverse) electromagnetic waves.

It is shown above that on the border with the metallic surfaces of the contact wire, roof and front part of the ERS body the wave reflection coefficient is practically equal to 1, i.e. incident wave is reflected practically fully. In this case the amplitudes of incident (falling) and reflected waves are identical and this influences on appearing of *standing waves*.

The propagating harmonic wave with the frequency of 50 [Hz] has length of $\lambda = 6000$ [km] in the air. So, the standing wave on distance between nodes of $\lambda / 4 = 1500$ [km] completely fills a feeder zone. Then, in the running out or idle modes of ERS's traction transformer, or in the regenerative braking mode, when there are no any consumers of recovered power in the section or TS, the active power is not transmitted on a feeder zone. At the same time, on every area which is $\lambda / 4$ long (i.e. on all feeder zone) there is an oscillating process of power exchange of electric and magnetic fields $\left(\frac{\mu_{\text{a}} H^2}{2} \leftrightarrow \frac{\varepsilon_{\text{a}} E^2}{2} \right)$, existing in the zone. An electromagnetic wave at this time in a feeder zone "stands", does not move and does not transfer energy from TS to ERS. Consequently, all time-varying energy of the electromagnetic field, according to (9) and (10), is reactive Q. Thus, in order to compensate Q it is necessary to suppress the standing wave in a feeder zone. One of the ways is to place a medium with a certain electromagnetic properties in front of ERS. The second method is to cover the surfaces (of contact wire and ERS body) with a substance having dielectric capacitance ε_{r2}, close to ε_{r1}, i.e. air, because in this case, according to (2), ρ_{rf} is far from 1, i.e. the reflected wave is practically missing.

7 Conclusions

1) The "field" approach, based on the theory of electromagnetic field, for the proving of transmission and formation of electric power losses in the devices of electric transport systems is more fundamental, and therefore more reliable than the existing "circuit" principles of such analysis based on the theory of electric circuits.

2) Flows of electromagnetic energy go from a traction substation to an electric locomotive through the air between a contact wire and a rail going into a high-voltage cabinet and then into traction engines through the roof porcelain insulator and metal parts of locomotive underbody part.

3) Electromagnetic waves falling from the dielectric (air) onto the surface of contact wire and metal part of the roof are partially reflected, and partially penetrate into the metal, gradually fading and at the same time creating energy losses in them.

4) The appearing and superposition of reflected and incident waves results in appearing of the standing waves in the air of the feeder zone, which create reactive power, as the last characterizes stationary waves. Therefore, in order to compensate reactive power it is necessary to suppress the mentioned waves.

References

1. A. Shadowitz, *The Electromagnetic Field* (Reprint by Dover Publications INC., NY 2010)
2. J. Turowski and M. Turowski, *Engineering Electrodynamics: Electric Machine, Transformer and Power Equipment Design.* (CRC Press Taylor & Francis Group, NY 2014)
3. M. O. Kostin and O. H. Sheikina, *Theory of Electrical Engineering*, Vol. 3, Part 2 (Publ. Dep. of DNURT, Dnipropetrovsk 2011) [in Ukrainian]
4. D. M. Cook, *The Theory of the Electromagnetic Field* (Reprint by Dover Publications Inc., NY 2011)
5. V. Yu. Pimenov, V. I. Volman and A. D. Muravtsov, Technical Electrodynamics (Radio and connection, Moscow 2000) [in Russian]
6. K. G. Markvardt, *Power supply of electrified railways* (Transport, Moscow 1982) [in Russian]

Life cycle assessment: assessing the environmental impact in the railway maintenance

Miro Hegedić[1,*], *Nedeljko* Štefanić[1], and *Mladen* Nikšić[2]

[1]University of Zagreb, Faculty of Mechanical Engineering and Naval Architecture, Ivana Lučića 5, 10002 Zagreb, Croatia
[2]University of Zagreb, Faculty of Transport and Traffic Sciences, Vukelićeva 4, 10000 Zagreb, Croatia

Abstract. The railway sector plays an important role in the European transport sector and its environmental sustainability is a highly important issue today recognized by all the main stakeholders, including the European Commission. EU-28 railway transport network consisted of 220,000 km of railway lines in 2013. Such a big railway transport network requires maintenance. Maintenance of a railway infrastructure is a resource- and cost-demanding activity that has as well a considerable impact on the environment. This paper presents the results of the environmental assessment of an innovative new product which aims to decrease the environmental impact of the railway maintenance processes. Life cycle assessment methodology was used and results show that the biggest environmental impact, in all impact categories, is achieved in the use and maintenance phase. In the end, the normalized data of the environmental impact were presented using the standard functional unit for the freight trains: tonne for kilometre (tkm). Additionally, authors have compared two different functional units that could be used in Life cycle assessment of the self-propelled freight railway vehicles, proposing the use of the new functional unit: tonne for working hour (twh). Use of such customized functional unit is more appropriate because of the specific nature of work that self-propelled bulk carriages have.

1 Introduction

The environmental sustainability of the transport sector in Europe is an important subject today and the railway sector plays an important role, although the majority of the greenhouse gas emission still comes from the road transport [1]. In 2013, EU-28 railway transport network consisted of 220,000 km of railway lines. If the railway lines of China, Russia, Japan, and the USA are added, this number could reach 634,200 km [2] that have to be maintained.

In general, railway infrastructure can be divided into [3]: the track, the power supply system, and the signalling system. Maintenance of the railway infrastructure is resource- and cost-demanding activity with a substantial impact on the environment. This is especially true for the maintenance of the track when diesel locomotives are used for shunting the bulk carriages at the site. More precisely, three locomotives have to be used of which one has to work constantly either shuffling the carriage or in the standby. Although its service may be needed only for a short period of time, the locomotive is taken up all day. The Croatian company, RŽV Čakovec has developed an innovative product called Self-propelled bulk carriage (SPBC) that could make railway maintenance more environmentally friendly. SPBC has the capability to move on rails using its own propulsion system managed via remote control. It uses less energy, compared to the three locomotives, but it still has some environmental impact. To assess the environmental impact of the SPBC, an attributional life cycle assessment methodology was used. LCA is a science-based, comparative analysis and assessment of the environmental impacts of product systems, and it is distinguished from other methods by two basic features: cradle-to-grave analysis (Fig.1) and functional unit [4].

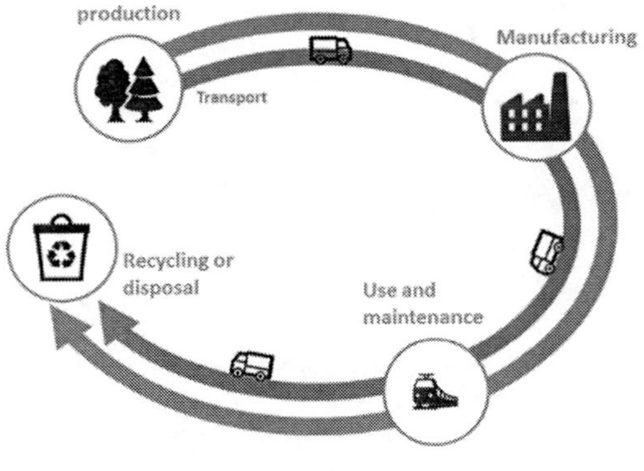

Fig. 1. Cradle-to-grave approach.

LCA methodology has its roots in the late 1960s and early 1970s when the first studies that were focusing on

[*] Corresponding author: mhegedic@fsb.hr

environmental impacts from different types of beverage containers took place in the USA [5]. There were studies analysing the environmental impact of the railway industries in different countries [6], [7], [8], [9], [10], [11], but none of the studies was done for the rail bulk carriage.

LCA study is mostly based on assumptions made by the commissioner, and therefore, it is hard to compare the results of the two different LCAs. For companies to be able to compare the results, it is important that the creation of the LCAs for specific product group follow the same rules. This is why product category rules (PCRs) are created.

PCRs give guidance for the development of Type III environmental declarations and are defined by ISO 14025:2006 [12]. Thus, PCRs serve as guidelines for the creation of the LCA studies for the specific group of products. LCAs built according to specific PCRs can be compared.

General LCA approach used in this study is based on ISO 14040 [13] and ICDL Handbook [14]. In order to analyse environmental impact, CML2001 (baseline) method is used.

According to the PCR, the functional unit for the freight trains is tonne-kilometre (tkm). Although SPBC is a freight carriage, it has different regimes than the normal freight train. Therefore, a different functional unit, such as tonne-working hours (twh) might be more appropriate. The aim of this paper is to briefly present the results of the LCA study and give a comparison between the normalized results given in tkm and twh.

2 Methodology

Life Cycle Assessment methodology (LCA) was used to assess the environmental impact of the SPBC. LCA is a science-based, comparative analysis and assessment of the environmental impacts of product systems [4]. The framework for the LCA method is given by ISO 14040:2006 [13], and includes four main phases (Fig. 2): (1) goal and scope definition, (2) inventory analysis, (3) impact assessment, (4) interpretation.

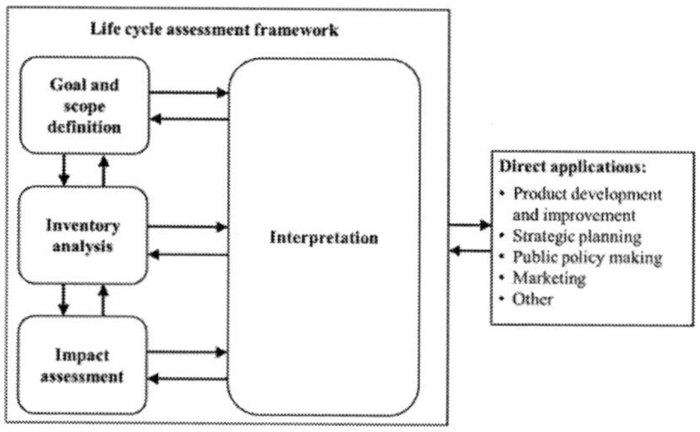

Fig. 2. ISO Life cycle assessment framework.

The first phase of the ISO LCA framework is goal and scope definition. The company wished to gain insights into the environmental impacts of the Self-propelled Bulk Carriage (SPBC). Therefore, the main goal of the conducted study was to calculate environmental impacts using LCA methodology. The studied product system (SPBC) is shown in Fig.3.

Fig.3. Studied product system.

The function of the studied system is to deliver construction materials to the site and discharge the needed amount of materials precisely where they are going to be used. The PCR for rolling stock [15] distinguishes passenger rolling stock and freight rolling stock. The SPBC fits into the category freight rolling stock. Therefore, the basis for the environmental comparison (functional unit) for freight rolling stock used in this study is the transport of one ton of transported and discharged material for one kilometre (1 ton for 1 km).

Product life cycle is divided into three main modules: (1) upstream, (2) core and (3) downstream, and the downstream module is further divided into two phases: (a) use and maintenance (UaM) (b) end of life (EoL) phase. Processes included into the scope of the LCA are shown in Table 1.

Table 1. System boundaries of the studied product system.

Module	Description
Upstream	Raw material extraction. Manufacturing of materials, electricity, and subcomponents.
Core	Rail vehicle manufacturing, energy and auxiliary material use, transportation to final destination.
Downstream	Use and maintenance.
	End of life.

LCA methodology offers to model the environmental impacts in numerous categories. For this study, only five main impact categories were chosen. Categories were chosen based on the recommendation of the PCR for rolling stock [15].

In order to analyse environmental impact, CML2001 (baseline) method is used. According to the PCR [15], five relevant impact categories (IC) were chosen: (1) Global Warming Potential (GWP), (2) Ozone Depletion Potential (ODP), (3) Acidifying Potential (AP), (4) Eutrophication Potential (EP), (5) Photochemical Ozone Creation Potential (POCP).

The most important and most time-consuming phase in performing the LCA is the inventory analysis. Through the life cycle inventory analysis (LCI) phase, the actual data collection and modelling of the system was done following the directions, i.e. goal and scope defined in the previous phase. A brief summary of the information collected through the LCI phase is given further in the paper.

Summarized results of the materials used to produce SPBC are given in Fig.4, from which it can be concluded that more than 97% of the materials are metals.

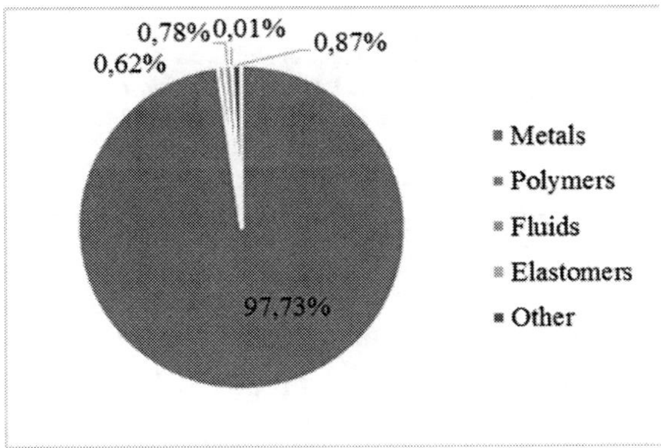

Fig. 4. Materials by type.

Allocation method had to be used to calculate the use of resources, including electricity, gas, and water since the company didn't have information on the amount of resources used per specific product. To allocate the resources, total working hours needed to finish the product were used. First, the total amount of specific resource in a particular time interval was divided by the total working hours in the same period. In this way, a unit amount of specific resource per working hour was obtained. Secondly, this unit amount of specific resource was multiplied by the total amount of working hours invested in the manufacturing of the SPBC. The total amount of working hours for SPBC was 2400 h. Table 2 shows the results of electricity, gas, and water use.

Table 2. Resources included in environmental assessment in core module.

Resource	Amount	Unit
Electricity	5937,30	kWh
Gas	1346,60	m^3
Water	165,60	m^3

Accordingly, the diesel motor technical data obtained shows that it uses 8 litres of fuel per working hour. Taking into account the 35 years long life cycle and approximately 1250 working hours per year of the SPBC, it is estimated that the SPBC will use 350,000 litres of fuel in its whole life cycle.

Based on the data collected and structured during the life cycle inventory analysis phase, life cycle impact assessment (LCIA) was performed using the OpenLCA software version 1.41.

3 Results

Fig.5 shows the results of the LCIA. Environmental impact is shown through five different impact categories and four main phases of the product lifecycle. A detailed description of every impact category can be found in [16].

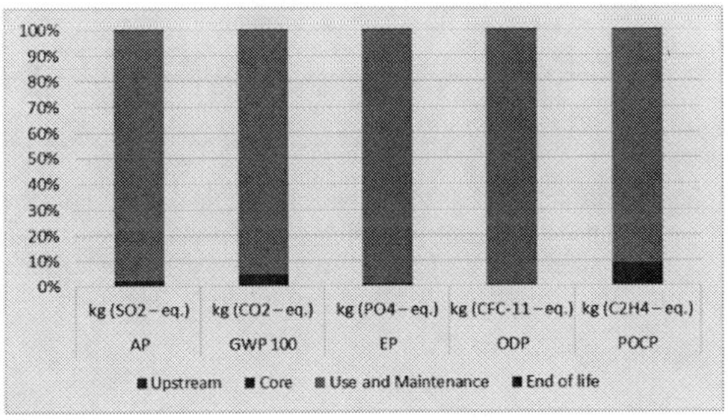

Fig. 5. Environmental impact of the SPBC

As can be seen in Fig.5, the biggest impact on the environment is in the use and maintenance phase in all four categories. The biggest contributing process to the environmental impact in the UaM phase is the "Diesel process". For e.g., for the impact category Climate change – GWP100 through all phases - "Diesel process" contributes 91.84%.

To compare different product systems based on their environmental impact, the values of their environmental impact have to be normalized. Normalization of data was done by calculating the total tkm per SPBC life cycle. The data given in Table 3 were estimated, and based on these data, the functional unit was calculated.

Table 3. Data for calculation of functional unit.

Category	Amount	Unit
Working hours per year	1,250	hours/year
Life span	35	Years
Total working hours	43,750	working hours
Tara weight	37.34	t
Cargo weight (max)	52.66	t
Average weight (80% load)	79.47	t
Average speed (80%)	4.80	km/h
Total kilometres	210,000	km

The functional unit was calculated by multiplying the total kilometres and average weight (80% load) which resulted in 16,688,700 ton per km (tkm). After the functional unit was calculated, the environmental impact results were normalized to the functional unit. Results of normalization are given in Table 3.

Table 4. Normalized environmental impact.

IC	Unit	Upst-ream	Core	Downstream	
				UaM	EoL
AP	kg (SO_2–eq.) / tkm	8,285E-06	3,065E-06	5,264E-04	1,096E-07
GWP-100	kg (CO^2–eq.) / tkm	2,810E-03	7,614E-04	7,021E-02	7,133E-05
EP	kg (PO_4–eq.) / tkm	7,947E-07	3,866E-07	1,198E-04	1,086E-08
ODP	kg (CFC-11 – eq.)/tkm	2,760E-11	6,753E-11	1,262E-08	3,632E-12
POCP	kg (C_2H_4 – eq.) / tkm	1,215E-06	1,688E-07	1,429E-05	6,903E-09

Normalized data obtained from Table 4 represents the environmental impact of the SPBC, measured through different impact categories and achieved by performing the function of transporting one tonne of load over one kilometre.

4 Discussion

Analysing the environmental impact through the given categories, it is obvious that the biggest impact is in the use and maintenance phase of the product lifecycle. As was mentioned earlier, this impact resulted basically because of the use of diesel fuel, i.e., the diesel production and consumption process.

Considering the fact that the very big impact on the amount of the functional unit has an average speed of the SPBC, it will be hard to compare the normalized data of the SPBC with other bulk carriages that are intended for long-distance travel. Therefore, in future calculations, we propose the adoption of the functional unit for the short distance bulk carriage, or more specifically, for the self-propelled bulk carriage such as the SPBC to tonnes for working hour (twh). The functional unit, twh, is more appropriate because the main function of the self-propelled bulk carriage is mainly to discharge the material at the point of use and most of the time it is in a stand-still position. To travel longer distances, it uses a locomotive, but the long-distance travel is not its main function.

If the newly proposed functional unit is used for the normalization, it will result in 3,476,725 twh. The value of the twh functional unit is 4.8 times smaller than the value of the tkm functional unit. Thus, the normalized values of the environmental impact will be increased for the same number.

5 Conclusions

Nowadays, three diesel locomotives are used to manoeuvre bulk carriages during railway track maintenance. Although its service may be needed only for a short period of time, one locomotive is occupied all day. RŽV Čakovec has developed an innovative product called SPBC to tackle this problem and thus decrease the overall impact the railway track maintenance process has on the environment.

SPBC uses 8 litres of diesel fuel per working hour to perform the same activity as would three locomotives. Having in mind that one locomotive uses approximately 300 litres of diesel fuel per hour, it is obvious how big the savings on fuel are. Taking into account the environmental impact of the SPBC, it can be concluded that the biggest percentage of the impact is due to the use diesel fuel. Being able to reduce the amount of fuel used per working hour can directly decrease the impact of railway track maintenance process on the environment. Therefore, the use of the SPBC for certain railway track maintenance activities will reduced the impact on the environment.

SPBC is the prototype of the self-propelled bulk carriage, and it is the only product of this type on the market. But to be able to compare similar products from this product category in the future, experts should consider using functional unit twh (tonnes of transported and discharged material for one working hour) instead of tkm. If self-propelled carriages use twh as the functional unit, a more accurate normalized environmental impact would be obtained. Also it would be possible to compare the environmental impact of the products in the same product category.

The results of the discussed LCA study are used to obtain necessary certificates needed for the commercialization of the SPBC and they are intended for the internal audience. Therefore, there was no comparative analysis done in the study. Also, there are certain data limitations; hence, the study concerns environmental data from Croatia. This means that the availability of this data is limited. Where no relevant sources of data regarding Croatia were available, an EU average data were used.

References

1. European Environment Agency, "Transport in Europe: key facts and trends," Copenhagen, 2016.
2. The European Union; European Commission, EU transport in figures 2016, (Luxembourg: Publications office of the European Union, 2016).
3. N. Jimenez-Redondo, N. Bosso, L. Zeni, A. Minardo, F. Schubert, F. Heinicke and A. Simroth, "Automated and Cost Effective Maintenance for Railway (ACEM–Rail)," *Procedia - Social and Behavioral Sciences,* **vol. 48**, pp. 1058-1067, 2012.
4. W. Klöpffer, Ed., Background and Future Prospects in Life Cycle Assessment, Dordrecht: (Springer Netherlands, 2014.)
5. M. Hauschild, J. Jeswiet and L. Alting, "From Life Cycle Assessment to Sustainable Production: Status and Perspectives," in *CIRP Annal - Manufacturing Technology*, 2005.
6. H. Duan, M. Hu, Y. Zhang, J. Wang, W. Jiang, Q. Huang and J. Li, "Quantification of carbon emissions of the transport service sector in China by using streamlined life cycle assessment," *Journal of Cleaner Production,* **vol. 95**, pp. 109-116, 5 2015.

7. A. Merchan Arribas, S. Belboom and A. Léonard, "Environmental impact assessment of rail freight intermodality," *Current issues in transportation research-Proceedings of the BIVEC/GIBET transport research days 2015,* pp. 77-85, 2015.
8. N. Fries and S. Hellweg, "LCA of land-based freight transportation: facilitating practical application and including accidents in LCIA," *The International Journal of Life Cycle Assessment,* **vol. 19**, pp. 546-557, 3 2014.
9. M. Banar and A. Özdemir, "An evaluation of railway passenger transport in Turkey using life cycle assessment and life cycle cost methods," *Transportation Research Part D: Transport and Environment,* **vol. 41**, pp. 88-105, 12 2015.
10. F. Del Pero, M. Delogu, M. Pierini and D. Bonaffini, "Life Cycle Assessment of a heavy metro train," *Journal of Cleaner Production,* **vol. 87**, pp. 787-799, 1 2015.
11. H. Jones, F. Moura and T. Domingos, "Life cycle assessment of high-speed rail: a case study in Portugal," *The International Journal of Life Cycle Assessment,* **8** 2016.
12. ISO, "ISO 14025:2006. Environmental labels and declarations – Type III environmental declarations – Principles and procedures.," Geneva. Switzerland, 2006.
13. ISO, "ISO 14040:2006. Environmental management – Life cycle assessment – Principles and framework.," Geneva. Switzerland, 2006.
14. European Commission - Joint Research Centre - Institute for Environment and Sustainability, "The international Reference Life Cycle Data System (ILCD) Handbook – General guide for Life Cycle Assessment – Provisions and action steps," (Publications Office of the European Union, Luxemburg, 2010).
15. EPD, "Product Category Rules According to ISO 14025:2006. Product Group: UN CPC 495 – Rolling Stock, 2009:05 – Version 2.11. 2014.," 2009.
16. J. B. Guinée, Handbook on life cycle assessment: operational guide to the ISO standards, Dordrecht; Boston: (Kluwer Academic Publishers, 2002).

Novel analysis methods of dynamic properties for vehicle pantographs

Andrzej Wilk[1,*]

[1]Gdansk University of Technology, Faculty of Electrical and Control Engineering, ul. G. Narutowicza 11/12, 80-233, Gdańsk, Poland

Abstract. Transmission of electrical energy from a catenary system to traction units must be safe and reliable especially for high speed trains. Modern pantographs have to meet these requirements. Pantographs are subjected to several forces acting on their structural elements. These forces come from pantograph drive, inertia forces, aerodynamic effects, vibration of traction units etc. Modern approach to static and dynamic analysis should take into account: mass distribution of particular parts, physical properties of used materials, kinematic joints character at mechanical nodes, nonlinear parameters of kinematic joints, defining different parametric waveforms of forces and torques, and numerical dynamic simulation coupled with FEM calculations. In this work methods for the formulation of the governing equations of motion are presented. Some of these methods are more suitable for automated computer implementation. The novel computer methods recommended for static and dynamic analysis of pantographs are presented. Possibilities of dynamic analysis using CAD and CAE computer software are described. Original results are also presented. Conclusions related to dynamic properties of pantographs are included. Chapter 2 presents the methods used for formulation of the equation of pantograph motion. Chapter 3 is devoted to modelling of forces in multibody systems. In chapter 4 the selected computer tools for dynamic analysis are described. Chapter 5 shows the possibility of FEM analysis coupled with dynamic simulation. In chapter 6 the summary of this work is presented.

1 Introduction

Transfer of the electrical energy from overhead catenary system has to be safe and reliable at different operating conditions. For this purpose a contact force value between the pantograph and the overhead contact line is recommended for different voltage systems of electric railways [1]. A lot of research was realized on design, construction and materials used to make pantographs. In addition such features as: compact profile for saving of space, lower weight, low aerodynamic noise and minimum maintenance are taken into account.

High performance pantograph is usually a single arm pantograph. It consist of a main frame, lower and upper arms, coupling rod, balance rod, pantograph head and drive (Fig.1). The concept of the operation between lower and upper working positions is based on two articulated quadrangles. The first articulated quadrangle includes a part of the main frame, the lower arm, the part of upper arm and the coupling rod. The second one includes a small part of the upper arm, the balance rod, the link element and the upper arm. The moving pantograph is subjected to static and dynamic forces. Hence, the dynamic properties of pantograph are important for its entire working range during both upward and downward movements [2].

The objective of dynamic simulation is to determine the time-related behaviour of all pantograph assembly and analysis of interactions between particular parts (links) in connecting joints [3,4]. The dynamic simulation is useful to determine the contact force acting by the collector strips on the contact wire. Computational methods used in dynamic simulation should be coupled with FEM calculations in order to analyse stress acting on the pantograph parts.

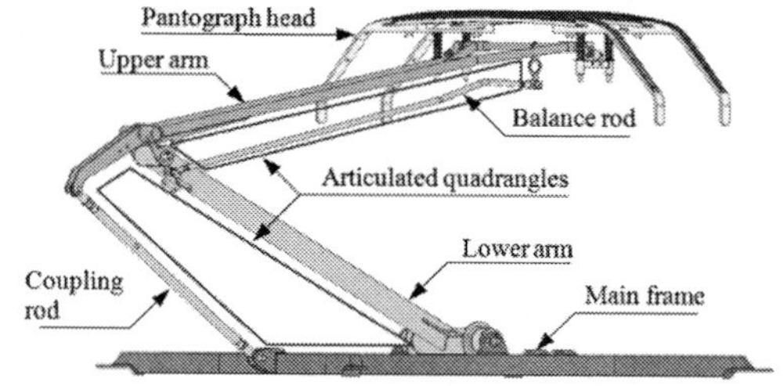

Fig. 1. Railway pantograph as a multibody system with closed loops (two articulated quadrangles).

2 Methods for formulation of equations

For mechanical systems only Newton's second law and D'Alembert principle are usually sufficient. However a pantograph consists of numerous rigid-body links (arms, rods etc.) and different joints, hence the formulation of governing equations from Newtonian approach is very

* Corresponding author: andrzej.wilk@pg.edu.pl

difficult. For complex mechanical systems Lagrange, Hamilton and Gibbs equations of motion are developed based on energy equations. Computer applications need to derive numerical form of governing equations ready for computer implementation. Automated form of the equations of motion can be derived using Kane's equations [5]. In the past two decades the computational methods used in the dynamic simulation of flexible multibody systems were developed based on the Finite Element Methods (FEM) [6] and Boundary Element Methods (BEM) [7].

2.1 Lagrange's equation

Lagrange's equation is derived from energy-state functions. The general form of Lagrange's equation for conservative mechanical system with k degrees of freedom is given by

$$\frac{\mathrm{d}}{\mathrm{d}t}\left[\frac{\partial L(\dot{\xi},\xi,t)}{\partial \dot{\xi}}\right]-\frac{\partial L(\dot{\xi},\xi,t)}{\partial \xi}=f_k, \quad k=1,2,\ldots N \tag{1}$$

where L is the Lagrangian of the system, ξ is the vector of generalized coordinates (including translational and rotational positions), f represents the generalized forces – applied and associated with the constraints.

The diagram of the simplified kinematic system of articulated pantograph with two degrees of freedom is shown in Fig.2. In this kinematic system only angular displacement α_1 of the lower arm and vertical displacement y of the pantograph head are assumed as degrees of freedom.

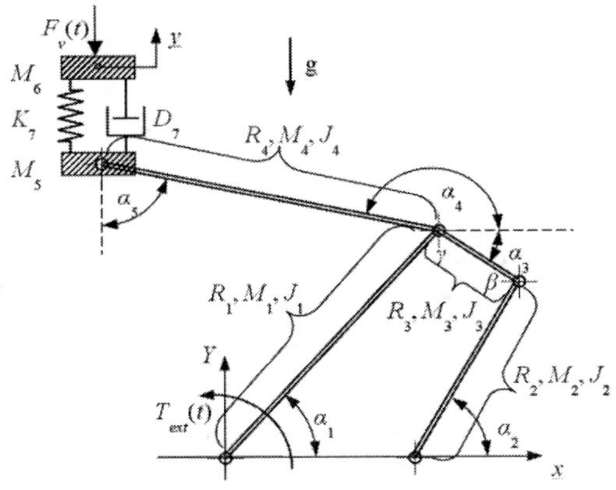

Fig. 2. Kinematic diagram of articulated type pantograph with two degrees of freedom – rotational and translational motion.

For this case the governing equation as system of nonlinear differential equations is [8]

$$\begin{aligned}
&\frac{\mathrm{d}}{\mathrm{d}t}\left(\frac{\partial L(\dot{\alpha}_1,\alpha_1,\dot{y},y)}{\partial \dot{\alpha}_1}\right)-\frac{\partial L(\dot{\alpha}_1,\alpha_1,\dot{y},y)}{\partial \alpha_1}+\frac{\partial P_m(\dot{\alpha}_1,\dot{y})}{\partial \dot{\alpha}_1}=f_\alpha(t),\\
&\frac{\mathrm{d}}{\mathrm{d}t}\left(\frac{\partial L(\dot{\alpha}_1,\alpha_1,\dot{y},y)}{\partial \dot{y}}\right)-\frac{\partial L(\dot{\alpha}_1,\alpha_1,\dot{y},y)}{\partial y}+\frac{\partial P_m(\dot{\alpha}_1,\dot{y})}{\partial \dot{y}}=f_y(t)
\end{aligned} \tag{2}$$

where f_α and f_y are the so-called generalized forces resulting from applied torques and forces.

Generalized force $f_\alpha(t,\alpha_1)$ can be represented as a superposition of three torques acting on the pantograph

$$f_\alpha(t,\alpha_1)=T_{ext}(t,\alpha_1)+T_g(\alpha_1)+T_{fy}(t,\alpha_1) \tag{3}$$

where: $T_{ext}(t,\alpha_1)$ is the torque raising the pantograph, $T_g(\alpha_1)$ is the torque resulting from gravity, $T_{fy}(t,\alpha_1)$ is the torque derived from the contact force $F_y(t)$ as a result of catenary interaction. All this torque components are functions of the rotation angle, even when the forces have a fixed value. The explicit matrix form of equation (2) is given by

$$\begin{bmatrix} J_\alpha(\alpha_1) & k_y(\alpha_1) \\ k_y(\alpha_1) & M_6 \end{bmatrix}\begin{bmatrix} \ddot{\alpha}_1 \\ \ddot{y} \end{bmatrix}=\begin{bmatrix} Q_\alpha(t)-h_\alpha(\alpha_1)\dot{\alpha}_1^2-D_\alpha(\alpha_1)\dot{\alpha}_1 \\ Q_y(t)-h_y(\alpha_1)\dot{\alpha}_1^2-D_7\dot{y}-\dfrac{y}{K_7} \end{bmatrix} \tag{4}$$

where coefficients $J_\alpha(\alpha_1)$, $h_\alpha(\alpha_1)$, $k_y(\alpha_1)$, $h_y(\alpha_1)$ are very complex functions of the α_1 rotation angle. They can be determined from physical properties of materials and dimensions of pantograph parts. The substitute coefficient of viscous damping $D_\alpha(\alpha_1)$ and dry damping are usually determined using an experimental approach.

2.2 Hamilton's equation

The Hamilton principle is related to the Lagrange energy method through transformation called the Legendre transformation. Hamiltonian formalism in the formulation of equations takes into account the total energy of the system. Hamiltonian function H for the system with n degrees of freedom is defined as follows:

$$\mathrm{H}(\xi,\mathbf{p},t)=\sum_{i=1}^{n} p_i\xi_i-L(\dot{\xi},\xi,t) \tag{5}$$

where p is generalized momentum conjugate with ξ_i, that is

$$p_i=\frac{\partial \mathrm{L}(\dot{\xi},\xi,t)}{\partial \xi_i} \qquad i=1,\ldots,n \tag{6}$$

The desired Hamilton canonical equations of motion are then obtained as

$$\begin{aligned}
&\frac{\partial H(\xi,\mathbf{p},t)}{\partial p_i}=\dot{\xi}_i \qquad (i=1,\ldots,n)\\
&\frac{\partial H(\xi,\mathbf{p},t)}{\partial \xi_i}=\dot{p}_i \qquad (i=1,\ldots,n)
\end{aligned} \tag{7}$$

The equation (7) is the set of first-order equations that are particularly symmetrical in form. Hamilton's principle has not provided any new physical theories, but it is sometimes a convenient approach to solve multibody dynamics.

2.3 Gibbs-Apell equations

Gibbs–Appell introduced a function known as Gibbs function for the formulation of the equations of motion. This function deals with acceleration, velocity and position. For a system consisting of N particles with mass m_i ($i=1,\ldots,N$) this function is defined as

$$G = \frac{1}{2}\sum_{i=1}^{N} m_i \ddot{\mathbf{r}}_i^2 \qquad (8)$$

where $\ddot{\mathbf{r}}_i$ is the acceleration of the ith particle.
The equations of motions are then given by

$$\frac{\partial G}{\partial \ddot{\xi}_k} = f_k \qquad (9)$$

where ξ_k is the kth generalized coordinate, f_k represents the kth generalized force.

Gibbs-Appel equation does not introduce any new physics in classical mechanics but may be more convenient in some cases, particularly when nonholonomic constraints are involved.

2.4 Kanes equations

Some methods for the formulation of equations of motion are more efficient than others. Kane's equations [9] enable derivation of the explicit form of equations of motion in a relatively simple and automated way. General form of Kane's equations for a system of rigid-bodies formulated in N generalized coordinates $\boldsymbol{\xi}$ is given by

$$f_k + f_k^* = f_k^c, \quad k = 1,\dots,N \qquad (10)$$

where f_k, f_k*, f_k^c denote the generalized active, inertia, and constraint forces respectively.

Using the kinematics of multibody system in matrix form and the partitioning of generalized coordinates it is possible to obtain the final representation of the equations in an automated way ready for computer implementation

$$\mathbf{A}\ddot{\boldsymbol{\xi}} + \mathbf{B}\dot{\boldsymbol{\xi}} + \mathbf{C}\boldsymbol{\xi} = \mathbf{F}(t) - \mathbf{J}^T\boldsymbol{\lambda} \qquad (11)$$

where matrices **A**, **B**, and **C** contain the contribution from both translational and rotational degrees of freedom. Matrices **B** and **C** also account for all the quadratic velocity terms, Coriolis forces and nonlinear terms associated with the velocity contribution to the equations. **F** is the vector of generalized external forces. **J** is the constraint Jacobi matrix, and **λ** is the vector of the Lagrange multipliers.

3 Modelling of forces in multibody system

3.1 Generalized active forces

Active forces come from pantograph drive system, aerodynamic effects, and contact force as a result of the interaction of collector strips with the overhead contact line. Figure 3 shows a separated link (it can be a part of pantograph) subjected to external forces $\mathbf{F}_1$ and $\mathbf{F}_2$.

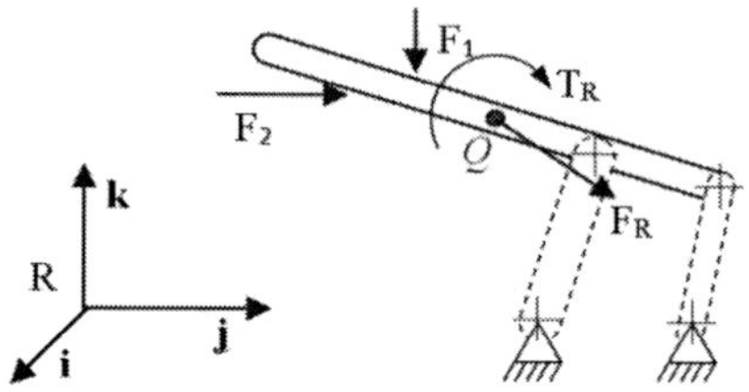

Fig.3. Forces and torques acting on rigid-body.

The equivalent force system can be replaced by the resultant vector force $\mathbf{F}_R$ acting at a point Q (center of gravity) and a couple vector torque $\mathbf{T}_R$. The generalized active force f_k associated with a rigid body is given by

$$f_k = \mathbf{F}_R \cdot \frac{\partial \mathbf{v}_k}{\partial \dot{\xi}_k} + \mathbf{T}_R \cdot \frac{\partial \boldsymbol{\omega}_k}{\partial \dot{\xi}_k} \qquad (12)$$

where $\mathbf{v}_k$ and $\boldsymbol{\omega}_k$ are the vectors of mass center velocity and angular velocity respectively of kth body in the inertial reference frame. The generalized active force involves both the forces and torques applied on bodies of the system.

3.2 Modelling of springs and dampers

Rigid bodies can be connected by springs and dampers in translational motion (Fig.4a) or in rotational motion (Fig.4b).

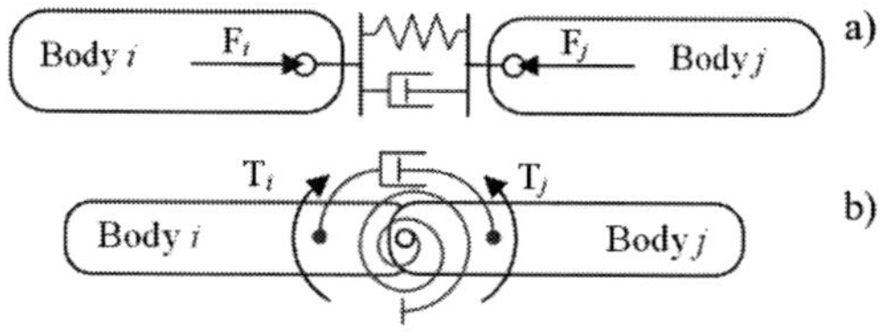

Fig. 4. Spring and damper connecting the rigid bodies in translational motion a) and rotational motion b).

Operation of the linear spring and damper between i^{th} and j^{th} bodies can be replaced by equivalent forces $\mathbf{F}_i$ and $\mathbf{F}_j$ equal in magnitude and opposite in direction.

$$\mathbf{F}_i = F(x,\dot{x})\mathbf{n} = -\mathbf{F}_j \qquad (13)$$

where **n** is the unit vector along the axis of the spring, x is the axial deformation of the spring, and $F(x,\dot{x})$ is the function of physical properties of the linear spring and damper. Operation of the rotational spring and damper between i^{th} and j^{th} bodies can be replaced by equivalent torques $\mathbf{T}_i$ and $\mathbf{T}_j$ equal in magnitude and opposite in direction.

$$\mathbf{T}_i = T(\varphi,\dot{\varphi})\mathbf{n} = -\mathbf{T}_j \qquad (14)$$

where **n** is the unit vector of torque, φ is the relative angle of rotation of the spring, and $T(\varphi,\dot{\varphi})$ is the function of physical properties of the rotational spring and damper.

3.3 Gravitational and contact forces

The forces of gravity acting on a system of N interconnected rigid bodies are usually given by

$$\mathbf{F}_i^g = m_i g \mathbf{n} \quad (i = 1, \ldots, N) \qquad (15)$$

where n is the unit vector of the gravity direction, m_i is the mass of ith body.

The contact forces rise from contact of two rigid surfaces. The contribution of the contact forces to the generalized active forces is zero.

3.4 Inertia forces

The inertia forces $\mathbf{F}^*$ and inertia torques $\mathbf{T}^*$ are generated when the acceleration of the body is not equal to zero. These forces and torques resulting directly from D'Lambert principle

$$\mathbf{F}^* = \sum_{i=1}^{n} m_i \mathbf{a}_i, \quad \mathbf{T}^* = \sum_{i=1}^{n} m_i \mathbf{r}_i \times \mathbf{a}_i\,, \quad (i = 1, \ldots, n) \qquad (16)$$

where the rigid body is assumed to be formed of m_i particles, $\mathbf{r}_i$ defines the position vector from the point mass to the mass center of the body, and $\mathbf{a}_i$ is the acceleration of point mass.

4 Pantograph dynamic analysis in CAD/CAE application

Novel analysis methods of multibody dynamics implemented in CAD/CAE applications give the ability to apply forces to pantograph assembly, which allow thorough evaluation of a pantograph performance. Almost all methods for the formulation of the equations of motion can be implemented in computer CAD/CAE applications. Some methods however are more efficient for numerical implementation. Several computer programs (Solidworks, Solid Edge, Autodesk Inventor, etc.) implement multibody dynamics. In each CAD/CAE environment the process of creating the dynamic simulation study of the pantograph involves following steps [10]:

- creating a pantograph assembly,
- creating joints between links (components/parts),
- creating environmental conditions,
- solving equations and analysis of results.

In this work the dynamic simulation workflow implemented in "Dynamic Simulation Module" (DSM) of Autodesk Inventor will be described.

4.1 Creating an assembly

An assembly of pantograph is needed in order to create kinematic chains (open and closed) and the physical properties of particular bodies. An example 3D computer model of the 160EC type pantograph is shown in Fig. 1. The Dynamic Simulation Module uses the bodies to convey mass, inertia and centers of gravity information to the multibody dynamics solver.

4.2 Modelling joints

There are standard and nonstandard joints in DSM. The selected standard joints are presented in Table 1.

Table 1. Selected standard joints in DSM of Inventor software. DOF means degrees of freedom.

Icon	Name	Description	DOF
	Revolution	No translation Rotation about the axis	1
	Prismatic	Translation along the line. No rotation	1
	Cylindrical	Translation along the axis Rotation about the axis	2
	Spherical	No translation Rotation about 3 axis	3
	Planar	Translation along 2 axes Rotation about the axis	3
	Spatial	Translation along 3 axis Rotation about 3 axis	6

The nonstandard selected joints grouped into the rolling, sliding, and additional joints are presented in Table 2, Table 3, and Table 4 respectively.

Table 2. Selected rolling joints in DSM of Inventor software.

Icon	Name	Description	DOF
	Cylinder on Plane	Motion between a cylinder and plane	NA
	Cylinder on Cylinder	Motion between two cylindrical components	NA
	Cylinder in Cylinder	Motion between a rotating cylinder inside another nonrotational cylinder	NA
	Cylinder-Curve	Motion between a rotating cylinder and rotating CAM	NA
	Belt	Motion of two coupled cylinders	NA
	Cone on Plane	Motion between a conical plane and a planar plane	NA
	Cone on Cone	Motion between two rotating conical parts	NA
	Screw	The same as cylindrical parts with pitch specificat.	NA
	Worm - Gear	Motion between a warm and a helical gear parts	NA

Table 3. Selected sliding joints in DSM of Inventor software.

Icon	Name	Description	DOF
	Cylinder on Plane	Sliding between a nonrotat. cylinder and plane	NA
	Cylinder on Cylinder	Sliding between two nonrotating cylinders	NA
	Cylinder in Cylinder	Sliding between a nonrotat. cylinder inside another norotatational cylinder	NA
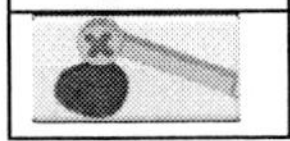	Cylinder-Curve	Motion between nonrotat. cylinder and a rotating cam	NA

Table 4. Additional joints in DSM of Inventor software.

Icon	Name	Description	DOF
	Spring/ Damper/Jack	This enables to create springs, dampers and jacks	NA
	3D Contact	This enables to create contact between two parts	NA
	2D Contact	Motion between two planar curves on two components	NA

When a joint is created there is the possibility to edit:

- initial conditions of starting position and velocity,
- imposed motion to drive the model kinematically by specifying a constant or varying position, velocity and acceleration using the 'Imposed Motion' option as shown in the Fig.5,
- force or torque (in case of a rotation), visco-elastic and dry friction parameters as shown in Fig.6.

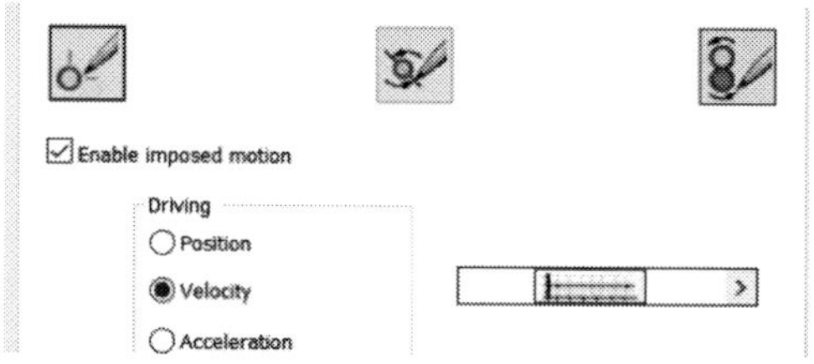

Fig. 5. Defining the imposed motion in a joint.

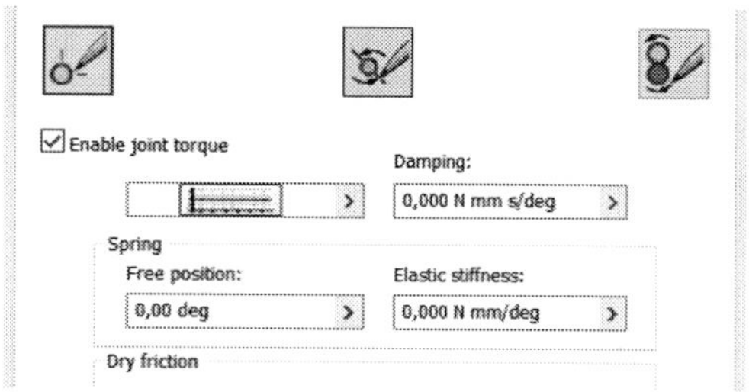

Fig. 6. Editing the imposed force or torque and visco-elastic parameters in a joint.

When assembly constraints are converted to joints it is possible to obtain over constrain the assembly at some joints. The mechanism to determine redundancies in joints is implemented in DSM. All joints are visible in the browser which enables the user to recognize them in the model (Fig.7) and edit their parameters.

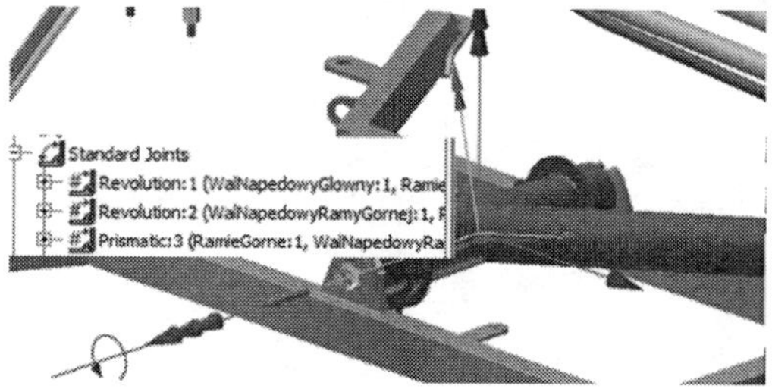

Fig. 7. Set of joints visible in the browser and corresponding local coordinate systems representing joint in the model.

4.3 Creating environmental conditions

Once the joints are created and parameters are defined the next step is to apply external loads. The forces and torques need a location on a body to indicate where they are applied, a direction along which or about which the force and torque is applied and a constant or varying value for the magnitude of the force or torque. Figure 8 shows the definition of torque acting on lower arm of the pantograph as function of angular displacement of the lower arm (1200 Nm at α=0 rad – 250 Nm at α=1,6 rad).

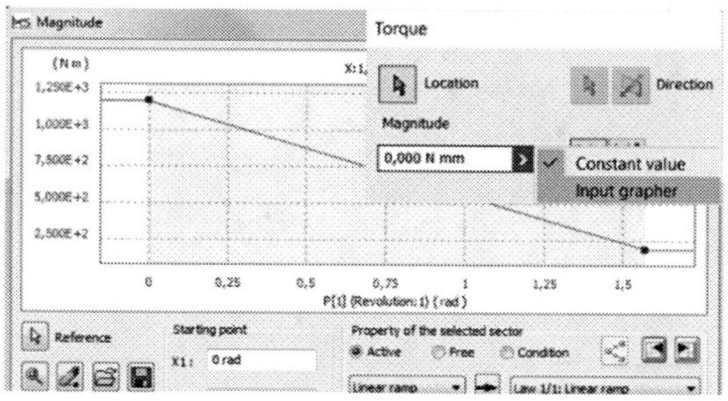

Fig. 8. Characteristic of torque acting on the lower arm as function of angular displacement of this arm.

It is also option to apply the force or torque as a combination of three orthogonal components. The loads can be also defined as functions of position, velocity or acceleration of the body in translational or rotational motion using the 'Input Grapher' editor. The external loads are visible on the pantograph model as vectors indicating the direction and location on the body where they are applied (Fig.9).

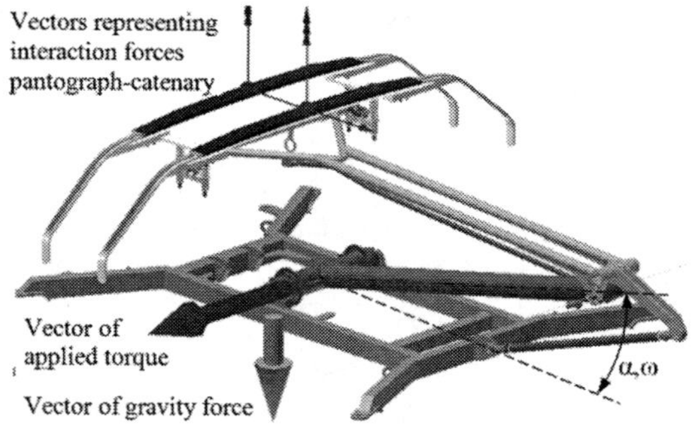

Fig. 9. Visibility of forces and torques as external loads applied to the bodies of the pantograph.

4.4 Solving and analysis

The final step is to run the simulation and do the analysis of the results. To run a simulation the user needs to set the simulation time, the number of time steps to record results during the simulation and activate the solver. The simulation results can be analysed in the 'Output Grapher'. The 'Output Grapher' is an interface for reviewing, plotting and post processing the results of a simulation. Waveforms of the angular displacement α and angular velocity ω of the lower arm are shown in Fig.10.

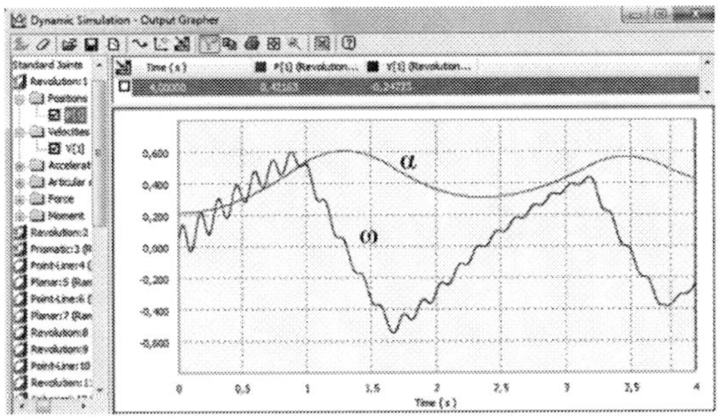

Fig. 10. Results of dynamic simulation presented in the output grapher.

The output grapher gives the possibility to analyse particular joints in terms of: positions, velocities and accelerations; reaction forces and torquies; contact forces; driving forces and torques. The output grapher also enables import and comparison of the results of multiple simulations, export of the results to an excel file, graphs print out.

5 FEM analysis and dynamic simulation

The output grapher has functionality to export the simulation results to FEM analysis in Autodesk Inventor Professional for Simulation [11]. This function passes the resultant forces and torques acting on a part and the boundary conditions (based on the joints) to the part FEM analysis module at a given time step during the simulation. Results of FEM analysis on the coupling rod are presented in Fig.11.

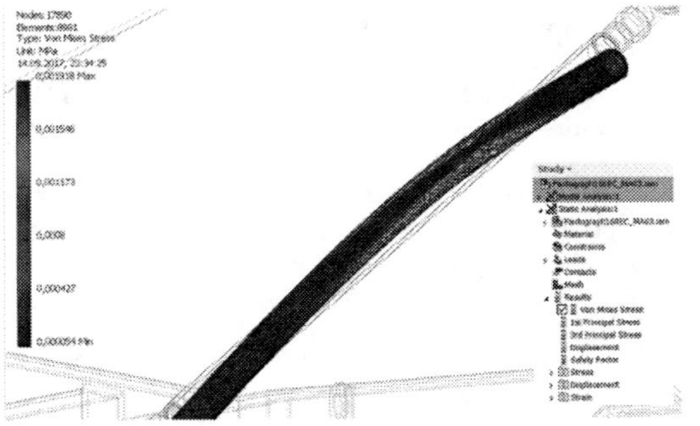

Fig. 11. Von Misses stress results on the coupling rod at some time instant of the dynamic simulation of the pantograph.

6 Conclusions

Novel analysis methods of dynamic properties for railway pantographs are implemented in CAD/CAE computer applications. Computer applications need numerical implementation of the governing equations of motion. There are several methods for formulation of the equations of motion. Some methods are more efficient for numerical implementation than others. Kane's equations enable to derive the explicit form of equations of motion in a relatively simple and automated way.

Novel analysis methods of multibody dynamics implemented in CAD/CAE applications give the ability to apply "real world" external loads to pantograph assembly, which allow thorough evaluation of pantograph performance. Multibody dynamics in CAD/CAE software enable the analysis of: positions, velocities and accelerations; reaction forces and torquies; contact forces; driving forces and torques in joints. Dynamic simulation is also coupled with FEM calculation.

References

1. F. Kießling., R. Puschman, A. Schmeider, *Contact lines for electric railways.* (Publicis Corporate Publishing, Munich Erlanger, 2001).
2. M. Oumri, A. Rachid, *A mathematical model for pantograph-catenary interaction*, Mathematical and Computer Modelling of Dynamical Systems, **22**(5), 463–474 (2016).
3. J. H. Lee, T. W. Park, H. K. Oh, and Y. G. Kim, *Analysis of dynamic interaction between catenary and pantograph with experimental verification and performance evaluation in new high-speed line*, Vehicle System Dynamics, **53**(8), 1117–1134 (2015).
4. S. Bruni, J. Ambrosio, A. Carnicero, Y. H. Cho, L. Finnear, M. Ikeda, S.Y. Kwon, J. P. Massat, S. Stichel, M. Tur, W. Zhang, *The results of the pantograph–catenary interaction benchmark*, Vehicle System Dynamics, **53**(3), 412–435 (2015).
5. F. Amirouche, *Fundamentals of multibody dynamics*, (Birkhäuser, Boston-Basel-Berlin, 2006).
6. W. Sunada, S. Dubowsky, *The Application of Finite-Element Methods to the Dynamic Analysis of Flexible Spatial and Co-planar Linkage Systems*, J. Mech. Des., 103(3), 643–651 (1981).
7. M. Kerdjoudj, F. Amirouche, *Implementation of the Boundary-Element Method in the Dynamics of Flexible Bodies*, International Journal for Numerical Methods in Engineering, **39**, 321–354 (1996).
8. A. Wilk, K. Karwowski, S. Judek, M. Mizan, Numerical implementation of current collector model in rotary motion into simulation in translator motion, *12th International Conference Modern Electrified Transport* 4-7 October 2015, Trogir, Croatia.
9. T. R. Kane, C. F. Wang, *On the Derivation of Equations of Motion*, J. Soc. Ind. Appl. Math., **13**, (1965).
10. D. Banach, *Autodesk Inventor Dynamic Simulation*, (Autodesk University, Autodesk Inc 2012).
11. R. Akella, M. Debono, *An In-Depth Look at Dynamic Simulation and Stress, Analysis Tools*, (Autodesk University, Autodesk Inc 2007).

Rapid services – the Czech high-speed rail project for Central Europe

Jan Ilík[1], and *Agata* Pomykala [2,*]

[1]Ministry of Transport, Department of Railway and Waterborne Transport, Czech Republic
[2]Railway Research Institute, ul. Chłopickiego 50, 04-275, Warsaw, Poland

Abstract. In March 2017, the Czech parliament passed a resolution obliging the government to launch a high-speed line program. Government resolution on launching the program was adopted in May 2017. The High Speed Railway Development Program in Czech Republic is intended to target international connections in several axes: Dresden-Prague-Brno-Vienna/Bratislava/Budapest, Katowice-Ostrava-Brno-Vienna/Bratislava/Budapest, Wroclaw-Prague-Munich. This network is supposed to improve Prague's accessibility to other cities in Czech Republic but also to major agglomerations of neighbouring states. This project is not only railway project but also a project of national development. One-day trip will include, among others, Dresden, Berlin, Vienna, Bratislava, Budapest and Warsaw together with other big cities in Poland. Since 2010 in Polish - Czech cooperation are being prepared studies on lines Wrocław - Praga and Katowice - Ostrava. The high-speed railway program implementation in Czech Republic and in Poland and Baltic Countries will result in an effective connection between Prague and Baltic Countries capitals. Creation of an international fast railway network in Central Europe is a part of the EU's transport policy framework for TEN-T network development and will contribute to the progress of economic links in the region. In addition, it is going to be involved in the European Union social and economic cohesion improvement.

1 Introduction

The construction of a high speed railway system in the Central Europe is very important for the the development of quality and competitiveness of rail transport in Europe. The analysis of the economic potential and transport network in the Czech Republic were the subject of UIC (International Railway Union, *fr. Union International des Chemin de fer*) studies. Already in 1992, they assumed, as the best solution, the construction of high speed lines and their integration into a pan-European network [1]. In 2003, high potential for the development of passenger transport between the Czech Republic and Poland was indicated [2]. In 2006, a study was published regarding the analysis of economic potential and links in the countries of Central Europe indicating the legitimacy of fast construction of high speed lines in the Czech Republic [3].

The aim of the paper is analysing of the work progress in the Czech Republic and presenting them in the aspect of linking with neighbouring countries.

2 The decision on HSR in the Czech Republic

On May 22, 2017, the Government of the Czech Republic approved a key strategic document for the high-speed rail (HSR) development in the Czech Republic, which gives a mandate to many participating stakeholder to work on the preparation of HSR project systematically further on. The name of this document is "Program for the Development of Rapid Services in the Czech Republic" (in Czech "*Program rozvoje rychlých železničních spojení v ČR*"). Its purpose was to decide at the government level on the framework conditions for a further preparation of the HSR project in the Czech Republic and to set out the basic conceptual background for the upcoming decisions and planning.

It should be noted that the HSR ideas have been (in the new socio-economic era) developed for more than twenty years, since mid of 90´s, nevertheless the scope of works has been up to now put only on routing design and its optimization. By adopting this document the Government agreed on further preparation of the HSR in the Czech Republic and confirmed the possibility to launch other preparatory steps and processes necessary to implement this project which envisaged already the EU transport policy TEN-T in the Regulation 1315/2013/EU.

The basis for the decision of the Government was essentially twofold. How to strengthen the role of public transport for the future? In the best case in a sustainable and Fossil Free way. Secondly, how to ensure the further development of rail transport, which has been experiencing a significant increase in performance for almost a decade, or better specifying after the recent

* Corresponding author: a.pomykala@infotransport.pl

https://doi.org/10.1051/matecconf/201818001006

economic crisis? In this respect, the positive trend is facing a new challenge – lack of capacity – as the modernized backbone infrastructure does not allow to accommodate additional trains due to its heavy usage [4]. The lack of capacity on the most important parts of the conventional network is thus representing one of the most critical issues for the railway sector and its development and is becoming critical for its further growth and its further intermodal competitiveness.

In addition, the ambition to construct HSR system in the Czech Republic appears urgent regarding to the need of (i) reorganization of the railway network. This decision takes into account both, the internal requirements of the Czech Republic and the development of the infrastructure abroad and, at the same time, (ii) The objective is to optimize the transport flows in the coming decades, as well as to remedy some of the major deformations, which can be seen on the current network.

The most obvious example is the connection Dresden – Prague – Brno where trains compared to road transport have to travel 100 km longer route, instead of 350 km of "direct" motorway. This fact represents a nonnegligible handicap, which significantly influences competitiveness of the railway transport. Finally. The HRS project should be also seen as the second phase of the Czech railway system/network modernization process following its first part – modernization of the conventional network – which is still ongoing and is expected to be completed by the end of this EU programming period.

On the other hand, it should be mentioned, that the ambition of the Ministry of Transport was not only to discuss different aspects and solutions for the development of railway transport. The approach defining the best possible option for the future is much broader and more complex. The analysis went beyond railway system itself and took into consideration other alternatives, which might have come, at least theoretically, into account e.g. new innovative technologies. Nevertheless, as the only realistic alternative to HSR Rapid Services proposal has been identified further development of conventional railway, covering additional electrification of existing railway lines, their doubletracking etc. with estimated investment of 250-300 billions CZK. Other alternatives (all innovative transport technologies, development of road transport using electric vehicles) have not been found feasible [5, pp 49-55]. For this reason, the development of Rapid Services, including the construction of 700 – 800 km of new high-speed lines was recommended as the most suitable solution.

Finally, an important aspect supporting the governmental decision on the high-speed project in the Czech Republic was influenced by a wider European context (see figure 1). The implementation of the high-speed railway has become a standard not only in EU countries but is getting an important part of railway policy worldwide. Central Europe, including the Czech Republic and Poland, thus might get a new impulse for its development enabling to exploit the geo-economical potential of its location between the West and the East.

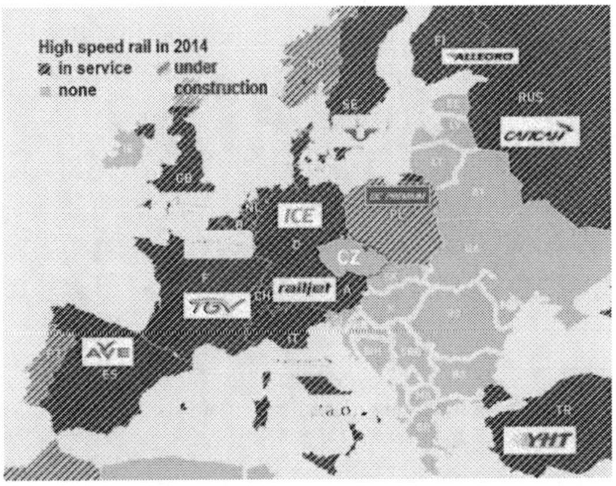

Fig. 1. Present status of implementation of HSR system in Europe. [Source: Ministry of Transport in Czech Republic]

3 Rapid Services

Altogether, the analysis identified the development of HSR system as an optimal solution for the future, which can bring significant socio-economic benefits, which seems to be feasible and include low technological risks [5, pp 75-83]. The solution based on the implementation of new high-speed lines and Rapid Services concept reflects the geographic context of the Czech Republic and Central Europe, their settlement system and density of the railway and other networks, availability of technologies and know-how, costs, spatial planning processes, protection and tradition, legal framework, socio-economic trends and their mutual synergies, dependencies and risks.

Significant factor contributing to the decision in favour of Rapid Services were positive gains for freight transport. Freight transport will get additional capacity on the conventional network and will also be partly able to use some sections (mostly cross-border sections) of the new HS infrastructure, which is going to be designed for mix use. This is not only a case of cross border tunnel sections as e.g. 26 km long tunnel proposed on the line between Prague and Dresden.

Moreover, the concept of Rapid Services does not identify the needs related to infrastructure development only, but also sets up a suitable operational model for the entire future system. Therefore, the whole project is called Rapid Services - reflecting to the real purpose of its building and construction of new HS lines – fast, accessible and high quality connection between major centres within the Czech Republic and abroad. In this regards it must be said that the RS system will not only consist of lines in high-speed parameters (above 250 km/h), which will form its backbone, but we do expect, that many conventional lines modernized for the speed up to 200 km/h will be a part of the system. We do believe this concept will make the whole system very flexible, accessible and primarily attractive because it will allow covering also those regions without direct link to newly constructed high-speed lines [6]

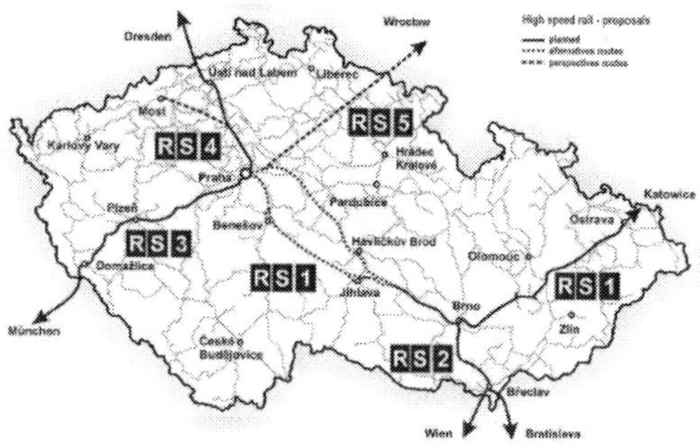

Fig. 2. Proposed routing of the newly build HS lines (line RS1 Prague – Brno in variants). [5, p. 39]

In line with the TEN-T policy in the Czech Republic plans following structure of backbone HSR infrastructure: RS 1: Prague – Brno – Ostrava – Katowice – Warszawa/ - Žilina (- Košice), RS 2: Brno – Břeclav – Wien – Graz / - Bratislava – Budapest, RS 3: Prague – Plzen – Regensburg – Munich / - Nuremberg, RS 4: Prague – Usti nad Labem – Dresden – Berlin – Hamburg / - Leipzig – Frankfurt, RS 5: Prague – Wroclaw – Warszawa (– Rail Baltica).

4 Future position of railway transport

The task of design speeds was not the crucial issue for the document because each of the RS branches will be considered separately in detail. Line speeds are going to be tested and verified for different variants researched, within particular feasibility studies of each RS branches. This means that different directions or even sections might have different speeds. The document only sets the framework, that the newly constructed lines should be designed for speeds of 250, 300 and 350 km/h. Nevertheless, there should be also stated several important facts. Firstly, a homogeneity of speed profile for the whole system must be considered to avoid the situation of excessive variety. Secondly, operational speed is expected to be between 250 – 300 km/h. Nevertheless, the line design should allow including some reserves for the future. The line should accommodate very HS trains as well as slower "IC" trains (e.g. railjet) running at speed 200 – 230 km/h. Finally, the system of clock face timetable between the most important nodes must be also considered.

Another issue discussed in the document was the schedule of works for the preparation and implementation of the whole project. Although the TEN-T policy prescribes the inauguration of several sections of new HS lines by 2030, namely Prague – Litoměrice/Lovosice on Prague – Dresden line, the whole line between Brno and Přerov and the stretch between Brno and Vranovice heading to Austrian and Slovak borders, the document took into consideration the current legal framework and particular procedures given and stated there can be many difficulties caused by this complicated and time consuming procedures given by this legal environment. Therefore, the document concludes that without legislative changes – in particular the acquisition of permitting processes – it will be hardly possible to accelerate the preparation of the whole project (as well as other transport infrastructure development projects). This in other words means only a minimal possibility to start the operation of the new lines before 2030 (except for modernized ones) This reason might place the whole project at risk. The discussion about possible legislative adjustments (e.g. a new act about line constructions), which is now considered to accelerate and make transport infrastructure implementation easier was not taken into consideration due to the fact of uncertainty of its final wording and legal power. Nevertheless, this topic of legal environment has been classified as the top priority to be solved.

5 State of play

As already mentioned, the next steps include elaboration of feasibility studies for individual HS lines in order to assess their economic efficiency depending on proposed routings, speed and territorial throughput. The positive economic evaluation for at least one routing option will be a basic precondition for its further project development.

The key priority represents the most traditional transport axis along the TEN-T Orient/East-Med Corridor, which crosses the Czech Republic from NW to SE and connect (Berlin –) Dresden – Ústí nad Labem – Prague – Brno – Břeclav with Vienna, Bratislava and Budapest. This fact reflects the feasibility studies (FS) ongoing. Both of them – FS for the new HS line Praha – Dresden as well as FS Praha – Brno – Břeclav, which is now going to be contracted, will be finished by 2019. Because of the fact that the new infrastructure can bring significant effects on environment, part of the FSs will be an environmental analysis according to SEA rules and standards. Originally, EIA was considered, but due to the uncertainty of routings, this very detailed assessment has been postponed to a latter preparatory phase. Nevertheless, we do believe that SEA outputs will make the decision-making procedures easier and will contribute to a smoother protection of HSR corridors in spatial planning procedures because many of them need modifications.

Another important sections, which are going to be prepared in a short time, are Brno – Přerov and Přerov – Ostrava, where the latter represents one of the most critical sections on CZ network, especially in terms of capacity. Both stretches are also very important for international interconnection, not only between the Czech Republic and Poland/Slovakia, but also for the relation between Poland and Austria or countries that are even more distant.

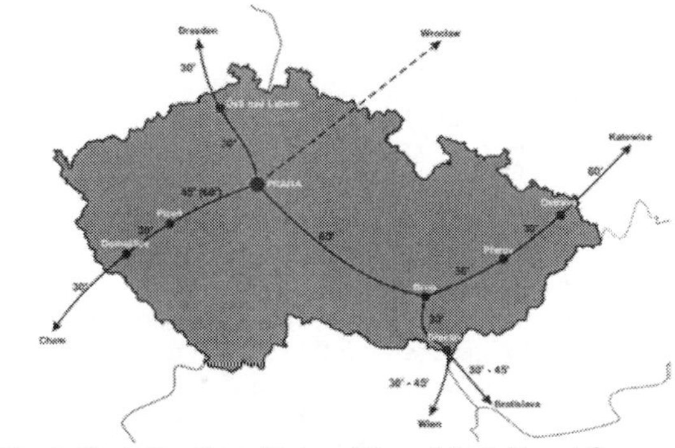

Fig. 3. Clock-face timetable travel times (plan). [5, p. 44]

6 Polish - Czech connections

One of the axis defined in the TEN-T policy is the direction Warsaw – Wroclaw – Prague – Munich/Nuremberg. Whereas the SW section between Prague and Bavaria belongs to the most traditional issues where already many studies have tried to find a proper solution, the northern cross-border part between the Czech Republic and Poland represent a brand new element.

According to the program of development of high-speed rail in Poland the following feasibility studies financed from EU funds for the construction of new lines were done:

- Warsaw - Łódź - Poznań / Wrocław (feasibility study)
- Wrocław - PL / CZ border - (Praha) (preliminary feasibility study)
- Poznań - border PL / DE - (Berlin) (preliminary feasibility study)
- modernization of the CMK line to high speed line parameters and its extension to Katowice and Cracow (feasibility study)
- Katowice - border PL / CZ - (Ostrava) (feasibility study)

In progress is a study for the line Ełk - PL/LT border - (Kaunas and further Riga / Tallinn and Vilnius).

The Polish high-speed line network was initially planned not only for domestic relations, but also for international connections to the South and West.

In view of significant economic relations between Poland and the Czech Republic, as well as large tourist traffic and growing business, current railway connection between these countries do not meet expectations.

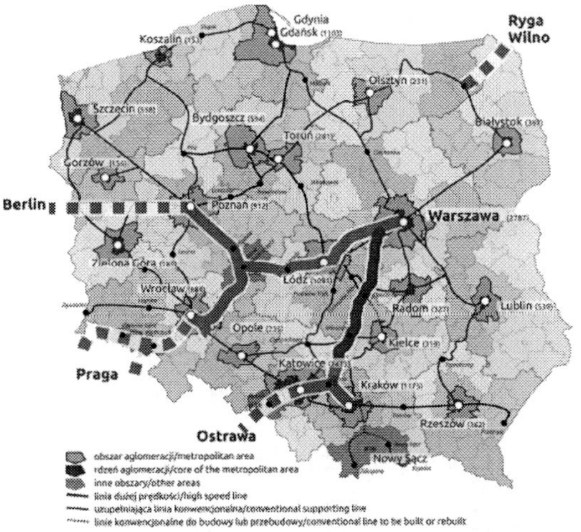

Fig. 4. Agglomerations and cities in direct and indirect influence of high-speed lines in Poland (parentheses of inhabitants of the agglomeration or region). [7 p.18]

The entire railway network on the Polish-Czech border dates back to the 19th century. There are no high-speed rail connections based on lines of high technical specifications. Undoubtedly, there are also difficult topographical conditions; the border crosses the high mountain range of the western and northern border zones, so called Sudetes. A program of building new high-speed lines in a frame of TEN-T is a good opportunity to improve railway connections in the region. TEN-T network defined in 1315/2013 [8] assumes the construction of two new lines: Katowice - Ostrava as part of the corridor Warsaw - Vienna and Wrocław - Prague as an extension of the Polish high speed line Warszawa - Łódź - Wrocław. These lines will allow a significant reduction of the journey time from Warsaw to Prague, to about 3.5 hours and from Katowice and Cracow to Prague to about 3 hours. High speed connections between Poland and neighbouring countries are to be based on the new TEN-T railway lines and the conventional lines that have been upgraded or are planned to be upgraded.

The construction of high-speed lines in Poland and in the Czech Republic will make possible to travel by fast train from the cities of western Poland to Prague. Currently, there are no direct rail connections in this direction due to the lack of railway infrastructure with the appropriate technical parameters. Figure 5 shows the travel times in the perspective of 2050 from Polish cities to Ostrava and Prague and Figure 3 shows travel times between cities in the Republic Czech.

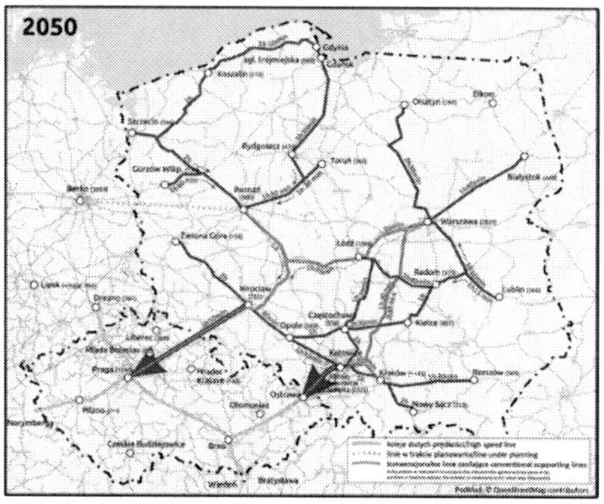

Fig. 5. Times of transit from Polish cities to Prague and Ostrava after implementation of the program of building high speed lines in Poland. [9, p. 29]

Following Poland´s effort to describe all relevant features of such a connection in the pre-feasibility study the Czech Republic launched the routing study recently to identify possible variants on the Czech territory. Generally there are two basic routings possible – via Liberec (with a continuation to Jelenia Gora) and via Hradec Králové and Wałbrzych. This study should be handed over by the end of 2017. This means the discussions about the further steps on Prague – Wroclaw connection can start in 2018.

7 Munich/Nuremberg - Praha - Warsaw - Tallin (Helsinki) Corridor

Currently implemented Rail Baltica project, which assumes connecting Tallinn with Warsaw through Riga and Kaunas by a railway line of 1435 mm track, is a part of the implementation of the European cohesion policy. It is also planned to build a branch from Kaunas Airport (KUN) to Vilnius via Vilnus Airport (VNO) (102 km), also as a new high speed line [10,11].

Thanks to this project, the Baltic countries will be integrated into the European high-speed network. A corridor will be created to connect the countries of Southern and Western Europe with the Baltic States and Scandinavia [12].

It is assumed, that the new line will accommodate both fast passenger traffic and freight traffic. Prospective plans are to lengthen this line from Tallinn to Helsinki by constructing a submarine tunnel. The tunnel project is of great economic uncertainty due to its very high cost and is currently undergoing in-depth economic analysis.
The construction of the high-speed line Rail Baltica creates a good opportunity for fast railway connections from Estonia, Latvia, Lithuania to Poland and Czech Republic.

The transit time between Riga and Tallinn will be reduced to 1 hour 55 minutes, between Riga and Kaunas to 1 hour 25 minutes and Kaunas and the Polish-Lithuanian border to about 25 minutes. However, the low technical parameters of the line on the Polish side will be a barrier to the development of these connections. The maximum speed of 160 km/h on the Warsaw - Bialystok line can only be satisfactory for linked of these cities and not as a national connections, or international connections. In this context, it is necessary to analyse the increase of the maximum speed on the Warsaw - Bialystok line to 200/230 km/h, which is technically and economically feasible while leaving the 3 kV direct current supply system. The section from Białystok to Ełk is planned to upgrade to 160 km/h and from Ełk to the border with Lithuania will be made feasibility study for a new line in the range of up to 230 km/h. With such a real program, it would be time to travel from Warsaw to the border at about 3 hours. For international relations it would be possible to reach the time of transit from Warsaw to Vilnius about 4 hours, to Riga 4 hours 20 minutes and Tallinn 6 hours 15 minutes and if a tunnel was built to Helsinki it would take 7 hours to travel to this city from Warsaw.

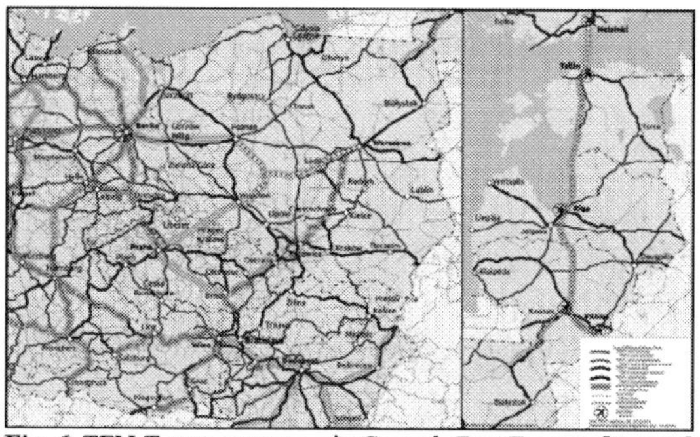

Fig. 6. TEN-T core passenger in Central- East Europe. [6, p.25]

8 Institutional support

It must be clear that such an intensive preparation and huge effort would not be possible without institutional background and funding. In order to ensure the continuation of the work on the HS project a top-level working group in Czech Republic was set up under the direction of the Minister of Transport, Mr. Dan Ťok. The aim of this group is to share information on the status of HS project and to ensure that all the stakeholders have been invited into the process and all of them have information available. This working group consists of representatives of Ministry of Transport, Ministry of Environment, Ministry for Regional Development, Ministry of Industry and Trade, Ministry of Finance, Ministry of Agriculture, State Transport Infrastructure Fund, SŽDC, Association of Regions of the Czech Republic, Economic Committee of the Chamber of the Parliament and Association for Infrastructure Development. Similarly, the Czech infrastructure manager SŽDC established a special HS working group responsible for the further preparation of the project, especially for commission of all the feasibility studies. Moreover, the State Fund for Transport Infrastructure (SFDI) made a special budget chapter available for financing of studies and the whole HS project. This financial guaranty thus represents very important feature, which brings needed stability into all running and prepared processes.

In addition, the Baltic Countries have established RB Rail AS, joint venture company for coordinating the works of beneficients:

- Estonia's Ministry of Economic Affairs and Communications,
- Latvia's Ministry of Transport
- Lithuania's Ministry of Transport and Communication and Implementation Bodies;
- Rail Baltic Estonia OU,
- Estonian Technical Regulatory Authority,
- Eiropas Dzelzcela Līnijas SIA,
- Rail Baltica statyba UAB,
- Lietuvos geležinkeliai JSC

In Poland, in the period 2008-2013 Inter-ministerial Team was appointed to the government Program for the construction and start the operation of high speed railways system. In the 2008 - 2013 the preparatory works was supervised through PKP PLK S.A. by the special department (High Speed Railways Centre). After its dissolution, it only exists at PKP PLK - Infrastructure Manager - team dealing with technical issues of projects.

At present High Speed Project is waiting for further decisions in his realisation and there are no entity nominated for the high-speed rail project.

This investment will be coordinated with construction of the new airport near Warsaw.

9 Conclusions

The dynamics of the HS project preparation has increased dynamically in recent years. The reason to that was not only the revised TEN-T policy, which foresees the development of HSR for the first time also in post-socialist countries but also increasing capacity problems on conventional railway network. In 2017 the focus of studies and related documents left the optics of technical disciplines and was targeted into wider scope, which would offer important answers to politicians.

The team responsible for the HSR project in the Czech Republic believes, that the document approved by the Czech government put a cornerstone for a successful development of the whole project, in a first phase for necessary changes of legal framework, which might significantly help to ease and speed up its realization. For the first time in Czech Republic, the project has been understood as the development project of national importance highly exceeding the transport sector.

The parallel implementation of high-speed line construction projects in Poland and the Baltic States offers the opportunity to achieve a significant improvement in transport accessibility between and within these countries and will support the improvement of the European Union's cohesion. However, to take advantage of these opportunities and achieve synergy effects, it is necessary to coordinate and harmonize these projects across the region in both organizational and technical aspects.

References

1. M. Walrave: The development of high speed rail innovation and tradition. Prospects for the future. *Rivista Internazionale Di Scienze Sociali*, **101**, no. 3, 375–397 (1993) JSTOR, www.jstor.org/stable/41623791 [acess: 28.03.2018]
2. UIC by Intraplan Consult GmbH, IMT Trans and INRETS, Passenger Traffic Study 2020 Poland and Czech Republic (2003)
3. UIC by CENIT Barcelona, Opportunities for high speed rail in Cental and Eastern Europe (2006) [In French]
4. J. Ilík, Railway 21: opportunities-needs-outlook, *ERA 21*, **5**, 52-53, (2015) [In Czech]
5. Ministry of Transport of the Czech Republic,. Program for the Development of Rapid Services in the Czech Republic, (2017) [In Czech]
6. J. Kušnír, J.Ilík, Rapid Service as precondition of competitive railway, in: The Czech Railway in the year 2030, *Bulletin of Czech Railways fair and annex of Železniční magazín*, **5**, 3 (2013) [In Czech]
7. P. Sleszynski, Influence of the structure of the high-speed railway system in Poland to the mutual temporary-spatial availability of the most important urban centres. *Technika Transportu Szynowego, tts* **6**,18-21 (2017) [In Polish]
8. Regulation (EU) No 1315/2013 of the European Parliament and of the Council on Union guidelines for the development of the trans-European transport network and repealing Decision No 661/2010/EU, UE Journal L348, 20.12.2013
9. A. Massel, A. Pomykala, J. Raczynski, Prospects for the development of international passenger rail transport in Central-Eastern Europe in terms of high-speed line construction. *Technika Transportu Szynowego, tts,* **6**, 22-31(2017) [In Polish]
10. AECOM, Rail Baltica Feasibility Study (2011)
11. AECOM, Rail Baltica Feasibility Study. Amendment – Analysis of Vilnius Extension (2014)
12. A. Pomykala, Rail Baltica high speed line in the aspect of the development of railway connection in the Central Europe. *Technika Transportu Szynowego, tts,* **11**, 41-48(2017) [In Polish]

 https://doi.org/10.1051/matecconf/201818001007

Realisation of the mechanical characteristic of a series excited DC machine in the drive with converter – fed induction machine

Grzegorz Skarpetowski[1,*]

[1]Ret. Technical University, Kraków, Poland

Abstract. A control algorithm for converter-fed induction machines, which allows the operation of the drive along of mechanical characteristic similar to the characteristic of series-excited direct current machine especially suitable for adhesion traction drives, is presented.

1 Introduction

Converter-fed induction machines are at present time in the modern energy economical traction drives widely used. In that application, which is still in continuous development are problems in the mechanical part of the drive observed, [1, 2, 3, 4], [8], which are connected with well-known problems of slip and slide dynamic states. The problems are still waiting for final solution. The comparison of properties of old fashioned drives with series-excited DC motor characteristic allows showing of the principal difference between the series and the shunt or separately excited motors. The induction machine belongs to the second group of electrical machines. The rapid change of the developed torque on the mechanical characteristic of induction machine in the region of normal operating points, generates an enormous dynamical stress, not only for electrical, but for the mechanical part of the traction channel, too. Especially difficult, for the control circuits, is to try to avoid, the suddenly change from motoring operation into the braking operation, due to change of wheel speed of the drive. The suddenly sign change of the developed torque, reduce the traction force of the drive and cause overload and fast destruction and devastation of the mechanical parts of the drive.

2 Presentation of the "series excitation" control method of induction machine

The analytical description of the relations used in the control method resulting in creation of series-excited machine properties is based at modified equivalent circuit of induction machine. The classical Kloss equivalent circuit doesn't specially support the main control quantities allowing simple derivation of the relevant relations which are valid for a series-excited mechanical characteristic of induction machines. The modified equivalent circuit of the machine is shown at Fig. 1:

- ω_1 - fundamental harmonic of stator current pulsation;
- ω_2 - fundamental harmonic of rotor current pulsation;
- R_1 - resistance of stator windings;
- L_1 - inductance of stator windings;
- R_2 - resistance of rotor windings;
- L_2 - inductance of rotor windings;
- L_M - main inductance of machine;

$$T_1 = L_1/R_1, \quad T_2 = L_2/R_2, \quad \sigma = 1 - L_M^2 / L_1 \cdot L_2.$$

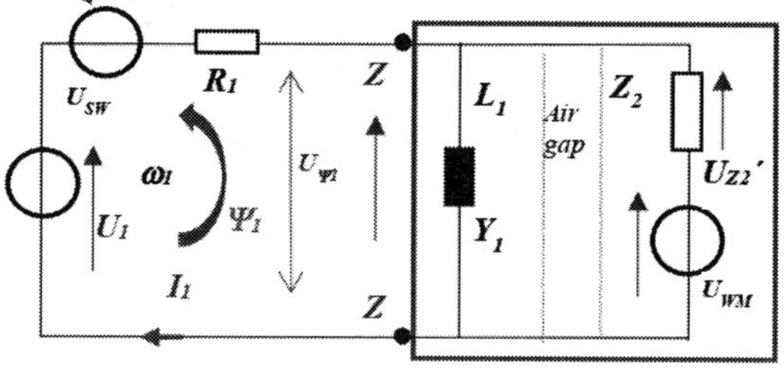

Fig. 1. Modified, valid in frequency domain for the fundamental harmonic of the converter output voltage, equivalent circuit of induction machine. The voltage, current and flux values are presented in the form of the complex amplitudes. In the closed electrical circuit, in frequency domain, the mechanical switch is replaced by equivalent voltage source U_{SW}.

Presented equivalent circuit is used to emphasise the main role of the flux linkage Ψ_1 of the stator winding in the analytical description of induction machines. The value of stator flux linkage $\Psi_1(\omega_1,\omega_2)$ which is a function of converter output voltage fundamental harmonic and pulsation in stator and rotor, allows simple description of voltage drops in rotor equivalent circuit. In the rotor part of modified equivalent circuit, a voltage sink has been introduced. The voltage sink allows explaining the energy flow in the induction machine. The power consumed by the voltage sink is equal to the mechanical power of the machine. The voltage sink can be understood as the equivalent source of the mechanical power, current I_2 in the voltage sink as equivalent of torque and voltage U_{Z2} as equivalent of angle velocity of

* Corresponding author: skarpetowski@hispeed.ch

the machine shaft. The equivalent circuit in Fig. 1 can be analytically described with following U_1/I_1 immittance.

$$\frac{U_1(\omega_1,\omega_2)}{I_1(\omega_1,\omega_2)} = R_1 \cdot \frac{1-\sigma\cdot\omega_1\cdot T_1\cdot\omega_2\cdot T_2 + j\cdot(\omega_1\cdot T_1+\omega_2\cdot T_2)}{1+j\cdot\omega_2\cdot T_2} \quad (1)$$

The stator susceptance is described by equation:

$$Y_1(\omega_1,\omega_2) = \frac{1}{j\cdot\omega_1\cdot L_1} \quad (2)$$

Admittance of the rotor equivalent circuit is described by equation:

$$Y_{Z2}(\omega_1,\omega_2) = \frac{T_2}{L_1}\cdot\frac{(1-\sigma)}{1+j\cdot\sigma\cdot\omega_2\cdot T_2} \quad (3)$$

Three important voltage drops in the equivalent scheme of induction machine depends in very characteristic way of the stator flux linkage $\Psi_1(\omega_1,\omega_2)$. The voltage on magnetic part of equivalent circuit, without voltage drop on stator resistance R_1 depends on flux and additionally of the stator voltage pulsation ω_1:

$$U_{\psi 1}(\omega_1,\omega_2) = j\cdot\omega_1\cdot\psi_1(\omega_1,\omega_2) \quad (4)$$

Voltage on the rotor equivalent voltage sink depends of angle velocity of the machine shaft $\omega_M = \omega_1 - \omega_2$:

$$U_{\psi M}(\omega_1,\omega_2) = j\cdot\omega_M\cdot\psi_1(\omega_1,\omega_2) \quad (5)$$

Voltage drop on rotor passive elements depends on the slip pulsation ω_2:

$$U_{Z2}(\omega_1,\omega_2) = j\cdot\omega_2\cdot\psi_1(\omega_1,\omega_2) \quad (6)$$

Stator current can be described by equation:

$$I_1(\omega_1,\omega_2) = \frac{\psi_1(\omega_1,\omega_2)}{L_1}\cdot\frac{1+j\cdot\omega_2\cdot T_2}{1+j\cdot\sigma\cdot\omega_2\cdot T_2} \quad (7)$$

And rotor current achieves the value:

$$I_2(\omega_1,\omega_2) = j\frac{\psi_1(\omega_1,\omega_2)}{L_1}\cdot\frac{(1-\sigma)\cdot\omega_2\cdot T_2}{1+j\cdot\sigma\cdot\omega_2\cdot T_2} \quad (8)$$

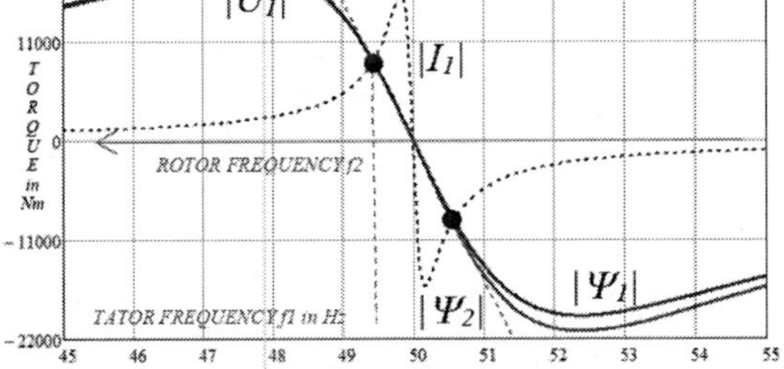

Fig. 2. Selected types of mechanical characteristic of induction machines used at present in the traction system drives. Broad use found the characteristics $|\Psi_1|$ = const. and $|U_1|$ = const. In the practical operating range 3 of the 4 characteristics are very similar. The practical operating range of the machine is located between nominal operating point in motoring and braking. $|\Psi_1|$ =const., $|\Psi_2|$=const. and $|U_1|$=const. are mechanical characteristics of shunt excited machines. The derivate dM/df_1 of these characteristics has very large values. Such behaviour of the traction machine is very unfavourable to the characteristics of the traction drives.

The only chance to improve the behaviour of induction motor drives is to approach of the characteristic of series-excited direct current machine. In that machine, the increase of the angle speed of rotor does not lead to the change of torque sign. Such DC drives do not lose the ability to pull. The solution of the presented problems, observed in converter traction drive systems, seem to lie in the change of the present time used control algorithm of converter-fed induction machines. The mechanical behaviour similar to the series-excited DC machines can be generated by converter-fed induction motor drives too. There is also the important, additional possibility to change the inclination of the mechanical characteristic in any way. The equation [6] describing the change of the torque along the "series" characteristic of a converter-fed induction machine, as a function of stator and rotor frequency, by constant value of stator voltage is as follow:

$$ME(u,\omega_1,\omega_2) = \frac{\frac{3\cdot p}{2\cdot R_1}\cdot|u|^2\cdot(1-\sigma)\cdot T_1\cdot\omega_2\cdot T_2}{(1-\sigma\cdot\omega_1\cdot T_1\cdot\omega_2\cdot T_2)^2+(\omega_1\cdot T_1+\omega_2\cdot T_2)^2} \quad (9)$$

The position of the induction machine operating point is controlled by $|I_1|$=const., ω_2=const. or ω_2 and $|I_1|$ = var:

- $|u|$ - the module of amplitude of stator voltage;
- p - number of magnetic poles of stator windings.

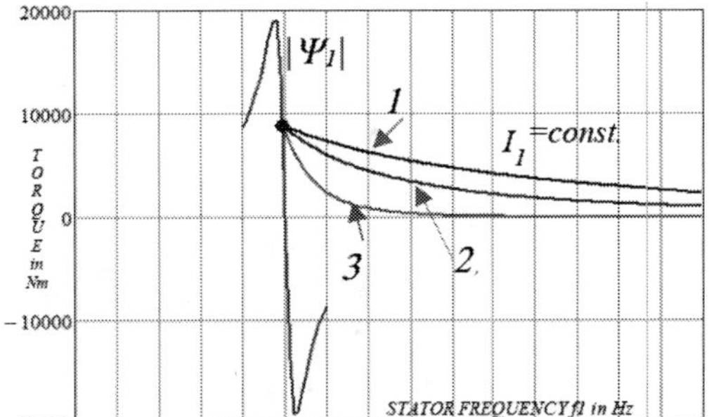

Fig. 3. Three examples of "series" mechanical characteristics realised in drive with a converter-fed induction machine, similar to the characteristic a series-excited DC machine, are shown on the background of the characteristic $|\Psi_1|$=const. The series characteristic marked with number **1** can be generated by control method $|I_1|$=const. The characteristic **2** can be achieved with control method f_2=constant. Characteristic **3** presents the possibility of the additionally change of the characteristic inclination.

Comparison of the characteristics $|\Psi_1|$ = const. and $|I_1|$ = const. carried out on the Figure 4 highlights the suitability of the presented control method for traction

https://doi.org/10.1051/matecconf/201818001007

drives with converter-fed induction machines. The results of the comparison have been presented below.

Properties of the mechanical characteristic of induction machine driven with constant value of flux linkage:

- torque reversal possible even with small change of wheel speed;
- high dynamic loading of the mechanics;
- sudden medium frequency relief and load the elastic elements of the coupling and transmission;
- great steepness of the mechanical characteristic;
- significant reduction of the force transmitted from the motor to the rails in slip and slide condition;
- lower load and minor wear on the wheels of the vehicle;
- large values of derivation $\frac{\Delta Mp}{\Delta fm} = \frac{Mn}{f2n}$

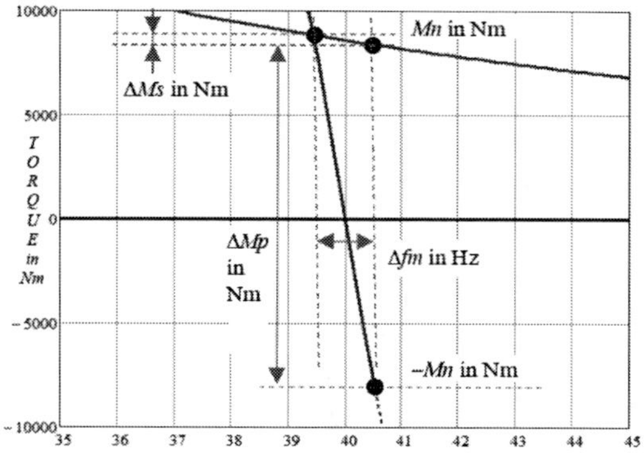

Fig. 4. A difference in the inclination of the mechanical characteristics, caused by using of two different control methods in the same drive with converter-fed induction machine. The very steep characteristic curve is formed by the method $|\Psi_1|$=const. The flat characteristic, corresponds to the characteristic of a series-excited DC machine.

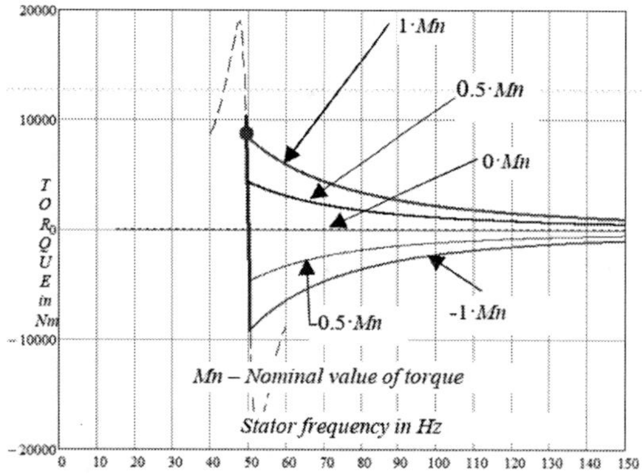

Fig. 5. Controllability of the "series-excited" characteristics of converter-fed induction machines is presented. The characteristics can be varied by change of the stator current or/and of the rotor frequency.

Properties of the mechanical characteristic of the induction machine operated along a series mechanical characteristic of DC motors:

- no change in the sign of the drive torque even with a larger increase of the rotor angular speed;
- even in the slip or slide conditions, the drive, which is in the driving mode remains in the driving and the drive which is in the braking operation remains in the brakes;
- reduction of the dynamic stress of elastic elements in power transmissions channel;
- small steepness of the mechanical characteristic;
- larger wear of the wheels, but the vehicle can pull better;
- small values of derivation $\frac{\Delta Ms}{\Delta fm} = \frac{0{,}1 \cdot Mn}{f2n}$.

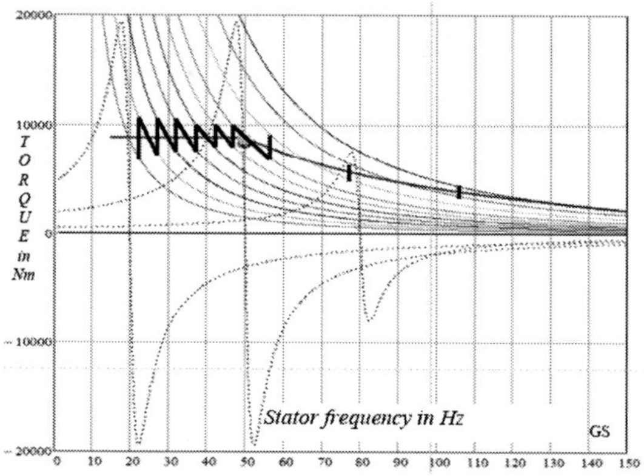

Fig. 6. Application of the "series-excited" characteristics of converter-fed induction machine in control systems comparable with the "tap changer" voltage control used in conventional series-excited DC machine.

The behaviour of series excited, convertor supplied induction machine, in slip conditions, causes the reduction of the torque as result of constant voltage and increasing stator frequency. See Figures 7 and 8.

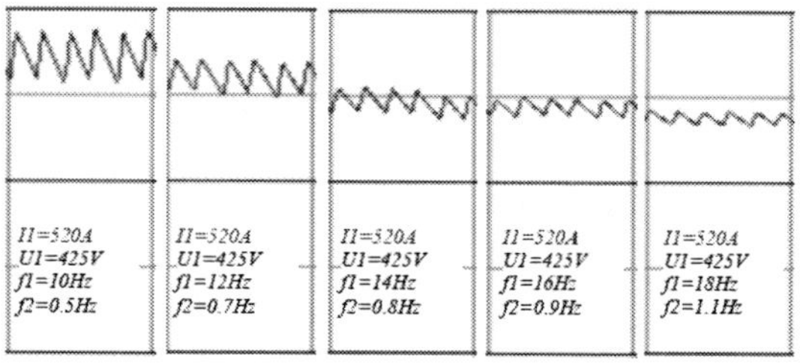

Fig. 7. Slip conditions in motoring state of the traction drive. The reduction of the torque starts as reaction on the rotor angular velocity increase of the induction machine. The change of the torque with control method which keeps previous constant value of the stator current module is presented. Current controller is increasing stator frequency to counteract the sudden current reduction. The stator voltage remains depending on vehicle speed on constant value. The difference between wheel velocity and vehicle velocity has no influence on the value of stator voltage amplitude.

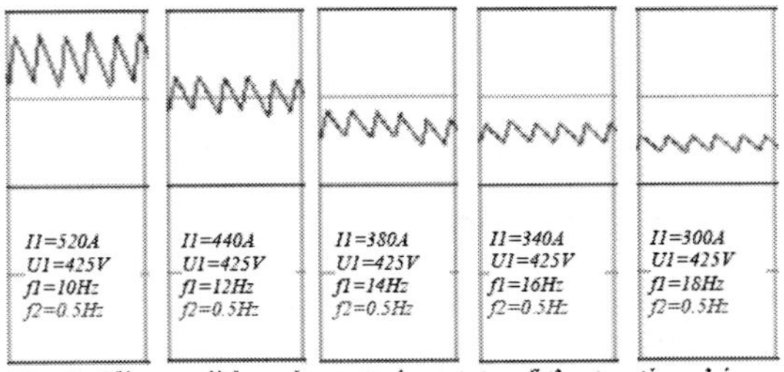

Fig. 8. Slip conditions in motoring state of the traction drives. The reduction of the torque starts as reaction on the rotor angular velocity increase of the induction machine. The change of the torque with control method which is keeping constant value of the rotor frequency is presented. The frequency controller increases the stator frequency to avoid suddenly change of rotor frequency. The controller adds the previous constant value of rotor frequency to the mechanical frequency of the rotor. Due to the simultaneously occurring reduction of stator current in function of stator frequency, the torque decreases faster than by constant current control method.

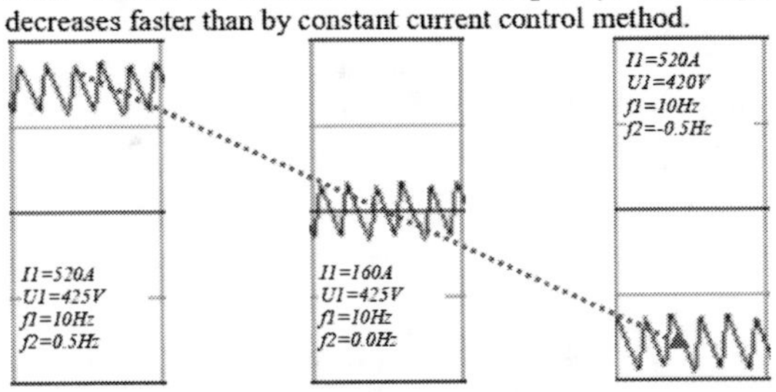

Fig. 9. Figure shows the progress of switching from motoring operation to braking operation which is carried out with constant value of stator flux linkage. $|\Psi_1|$ = const. The harmonics of the torque, which do not depend on average value of torque, are in all operation state nearly constant. The mechanical elements of the power transmission channel are extremely overloaded in all operating states.

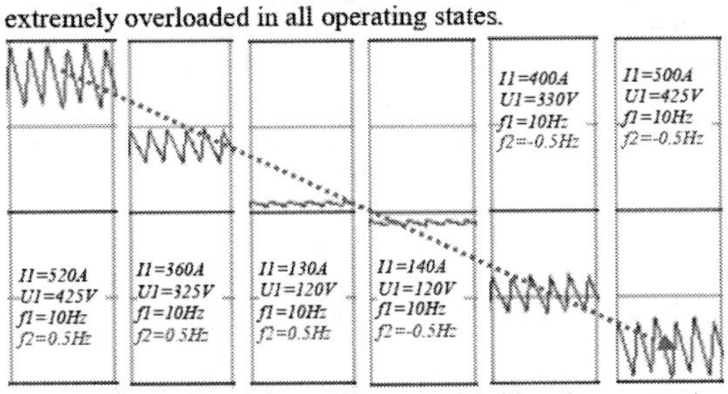

Fig.10. Figure shows the progress of switching from motoring operation to braking which is carried out by switching between mechanical characteristics of a "series-excited" induction machine. The switching between possible mechanical characteristic can be made by change of stator flux linkage or stator voltage amplitude. In zero crossing of torque average value the harmonics of torque are reduced to zero.

3 Conclusion

The described control method can be implemented in vehicles with multi converter-fed drive systems with induction machines. Such drive systems are commonly used in traction system with adhesion power transmission.

The control method allows the realisation of mechanical characteristic of separately-fed induction machine drives similar to the mechanical characteristic of series-excited direct current motor.

During occurring of fast slip or slide condition in the drive, the sign of the torque developed through the machine does not change.

The drive can stationary be operated in the range of macro slip.

The presented method is based on common control of the stator voltage fundamental harmonic which is equal for all motors of the vehicle. The value of the amplitude of the voltage is derived in suitable way from the value of vehicle speed.

The method uses the well-known concept of operating point control trough appropriate control of stator frequency, but the frequency control is made for each motor completely separate.

Each drive is running with own mechanical velocity and with own separately controlled stator frequency.

The stator frequency control circuit can, as far as possible, change the stator frequency with use of one the algorithms listed below:

1) with constant value of stator current module,
2) with constant value of rotor frequency,
3) with custom defined simultaneously change of stator current value or/and rotor frequency.

The described control method is based on fast measurement of 3 phase current module and of the mechanical frequency which is commonly used for description of the velocity of machine shaft.

In multi drive systems the recognition of slip and slid conditions of individual axes is not necessary.

Only in the drive with single motor or in the case of slip of all wheels in multi drive systems it is necessary to recognise on time the hazardous dynamical state.

In the starting state of the drive and in very low stator frequencies where $f_1 < \sim 4 \cdot f_{2N}$, the value of stator current and the value of the torque is controlled directly by stator voltage.

The traction properties of the adhesion drive which are controlled by the presented algorithm will be essentially improved in comparison to the control methods used today.

The state of development in the field of traction drive control is displayed in the specified references.

References

1. R. Frösch, DP. *Verfahren zur Drehmomentregelung einer Drehfeldmaschine*. DE 195 30 622 A1 (1995)
2. P. Good, DP. *Verfahren zur Regelung einer Drehfeldmaschine*. DE 421 59 17 A1 (1992)
3. E. Grossmann, WIPO. *Control device and method for jointly controlled asynchronous machines of a motor vehicle*. WO 2016/119948
4. A. K. Kumar, US. *Method of inhibiting wheel slip in an electric alternating current induction motor powered vehicle*. 5,480,220 (1996)
5. G. Skarpetowski, US. *Method of Controlling Electric Vales of a converter*. Nr. 5,331,537 (1994)
6. G. Skarpetowski, *Przetworniki i przekształtniki energii w napędach trakcyjnych*. Kraków: PIT ISBN 978-83-86219-86-5 (2016)
7. G. Skarpetowski, *Verfahren zum Betrieb einer elektrischen Asynchronmaschine*. Patent pending, EP 17 150 554.8 (2017)
8. N. Tomohito, *Controller for electric vehicle*. EP JPO, 2002095111 (2000)

Track insulation verification and measurement

Jacopo Bongiorno [1], and *Andrea* Mariscotti [2,*]

[1]University of Genova, DITEN, Via Opera Pia 11A, 16145, Genova, Italy
[2]ASTM Sagl, Via Comacini 7, 6830, Chiasso, Switzerland

Abstract. Methods for the measurement of track insulation are reviewed pointing out minimum necessary conditions and practical aspects, as well as consistency and repeatability of results. Track configuration, need of electrical sectioning and use of external electrodes or conductive structure are key parameters to determine suitable methods.

1 Introduction

Track insulation is a requirement for stray current protection and for functional purposes, if track circuits are used. Contractual limits are often specified referring to EN 50122-2 [1], without clarifying margins during commissioning and the acceptable spread of values for variable conditions and different track sections (drying concrete, pollution, moisture, viaduct, tunnel, defects and variability of the fastening system). This is a common problem to all methods that may be addressed by repeated measurements in different environmental conditions and parts of the system (e.g. at different construction times, concrete settlement and drying).

Many measurement methods exist for track to earth conductance, taken from standards and literature. EN 50122-2 (IEC 62128-2) suggests methods suitable for different track arrangements: continuously welded rails, insulating rail joints including verification of their efficiency, presence of civil structures (tunnels and viaducts). In the following these methods are described, including practical considerations on setup issues derived from site experience. Track section length shall always be accurately measured: a metric wheel should be used; combining the number of sleepers and their gauge (distance between sleepers) is also quite accurate. Target accuracy is in the order of 1%.

A reliable and low-resistance connection to the rail should also be ensured, as for some methods the test current may be very large (in the order of 100 A) with the risk of temperature increase and hot spots. Rail clamps and clips are a better solution.

Practical aspects cover necessary logistics and verification of elements extraneous to the measurement itself, influencing results and accuracy: conductive parts touching the rails (e.g. parts of fences and walkways not yet installed), overvoltage protecting devices (OVLDs), substation negative feeder cables, water stagnation, electric noise caused by welding, static converters. etc. For electrically continuous rails all these elements are relevant wherever they occur on the whole line.

Finally, distinction shall always be done between worst cases and average performance (including spread or dispersion), in the perspective of the overall purpose, the evaluation of stray current protection and impact on structures, averaging in realistic exploitation conditions (including to this aim track voltage monitoring).

2 Measurement methods

2.1 EN 50122-2

EN 50122-2 suggests three methods of rail/track-to-earth conductance measurement depending on track section configuration: with and without a civil structure that represents a continuous conductive reference, for sectioned and electrically continuous tracks.

Common to all methods is the prescription that the measured track section shall be without connections to earth, including substation negative feeders, and Overvoltage Limiting Devices (OVLDs) shall be disconnected. In general, repetition of measurements allows evaluating consistency of results and coping with external transients and disturbance.

2.1.1 *Conductance per length for track sections without civil structure*

This method is suggested for electrically continuous track sections shorter than 2 km without metallic reinforcement structures.

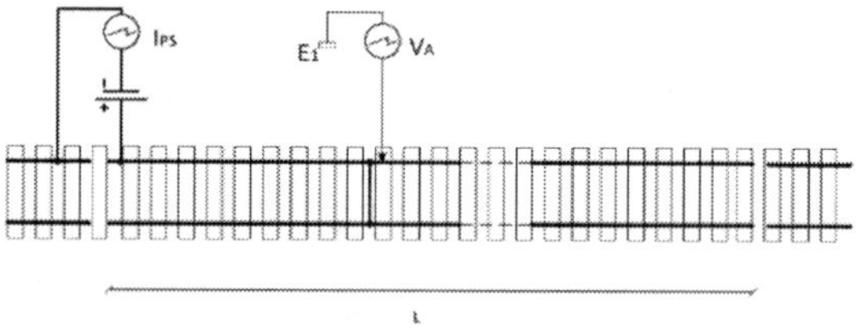

Fig. 1. Track to earth conductance measurement using EN 50122-2 A.3 method.

---
* Corresponding author: andrea.mariscotti@astm-e.ch

https://doi.org/10.1051/matecconf/201818001008

The section is supplied at one end by a power supply (PS) with positive pole connected to the measured track section and negative pole connected to the tracks behind, used as an earth electrode through their conductance to earth. Track voltage is measured at least 50 m away from the PS injection point with remote electrode E_1 as reference, while PS is switched on and off.

Track-to-earth conductance p.u.l. in S/km is:

$$G = \frac{I_{PS}}{L(V_{A,on} - V_{A,off})} \quad (1)$$

where:

G is the track to earth conductance [S/km];

I_{PS} is the current of the power supply [A];

V_A is the rail-to-E_1 voltage [V] for on and off PS;

L is the section length [km].

Section length has impact on measurement accuracy: if close to the maximum length, the measurement underestimates the real conductance with an error of about 1% [2]; optimal results are obtained placing the voltage reading near the opposite end with respect to PS.

Usually the PS consists of batteries; current intensity shall guarantee reliable voltage and current readings. Voltages higher than 50 V are not usually needed and they would turn out in safety issues.

The "behind track section" used for PS negative pole earthing might be not long enough to ensure a low resistance to earth; an electrode may be added to back up PS earthing. Variability of less than 0.1% was found by simulation in [2] for resistance to earth ranging between 0.1 and 20 Ω); experimental variability of few % is reported in [3] for values between 0.5 and 50 Ω.

The location of the remote electrode E_1 is very important. The standard specifies a distance from the track >30 m, but for high resistivity soils it should be increased and fulfilling of "remote condition" shall be checked (the method in sec. 2.5 may be used).

This method is very sensitive to soil polarization: the PS current changes slightly following an exponential decay; simultaneous measurement of voltage and current is thus important and readings should be taken when the current has reached the plateau (also to reduce curve derivative and local sensitivity, improving uncertainty).

Galvanic insulation is necessary for the channels or the use of separate acquisition systems (synchronization of tens of ms is sufficient considering polarization dynamics, see Fig. 2).

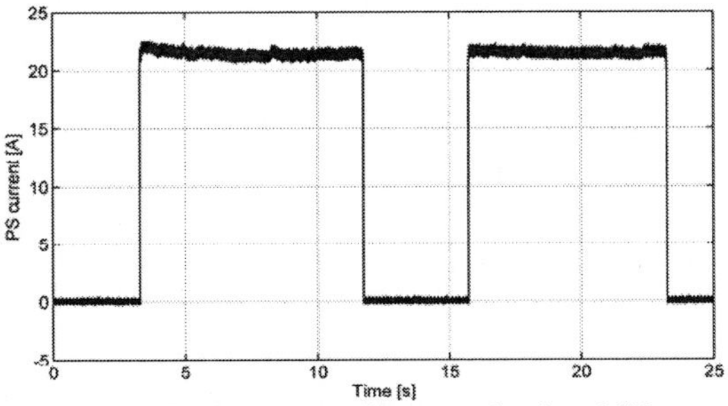

Fig. 2. Example of Power Supply current showing visible polarization in the first pulse.

The method is easy and quick and, although some precautions should be applied for a correct measurement [2], its use, when possible, is suggested.

2.1.2 *Local conductance per length for track sections without civil structures*

In case of continuous track/rail sections longer than 2 km, the standard suggests a "local" conductance measurement. This method is based on field distribution in soil and accurate soil resistivity measurements (see sec. 2.5.3); it needs running vehicles as a source of rail voltage and gives for free an estimate of the average rail voltage to use for EN 50122-2 conductance limit.

The test setup is shown in Fig. 3. The rail voltage (V_{RE}) and the difference of potential due to the electric field in the soil (V_{12}, between two electrodes driven into the soil) are recorded simultaneously and plotted in a graph. The slope of the linear regression of the function $V_{12}(V_{RE})$, named "stray current transfer ratio", is used to estimate track-to-earth conductance.

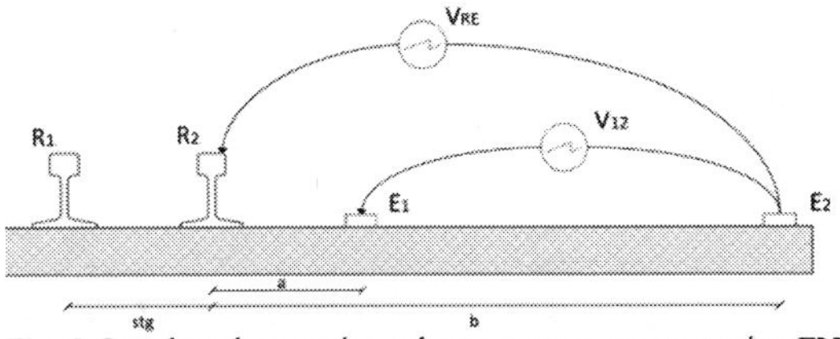

Fig. 3. Local track to earth conductance measurement using EN 50122-2 A.4 method.

The track-to-earth conductance is given by:

$$G_{st} = \frac{m_{sr} 2000\pi}{\rho\left[\dfrac{\ln(b\,(b + s_{tg}))}{\ln(a\,(a + s_{tg}))}\right]} \quad (2)$$

$$G_{dt} = \frac{m_{sr} 1000\pi}{\rho\left[\dfrac{\ln((b + 0.5s_{tg})(b + 0.5s_{tg} + s_{td}))}{\ln((a + 0.5s_{tg})(a + 0.5s_{tg} + s_{td}))}\right]} \quad (3)$$

where:

G_{st} is for single track configurations [S/km];

G_{dt} is the double track configurations [S/km];

m_{sr} is the stray current transfer ratio;

ρ is the soil resistivity between E_1 and E_2 electrodes [Ωm];

a is the distance between external rail R_2 and E_1 [m];

b is the distance between external rail R_2 and E_2 [m];

s_{tg} is the track gauge [m];

s_{td} is the distance between track centres [m].

Running vehicles shall be considered ensuring reliable and stable connections to the rail. Wire fastening is also important when crossing public areas, such as streets with traffic of pedestrians, bicycles, cars, heavy vehicles. Last, measurement of soil resistivity between E_1 and E_2 shall be accurate (see e.g. sec. 2.5.3).

2.1.3 *Conductance per length between running rails and metal reinforced structures*

This method is suitable with a continuous metallic structure along the track (inside tunnels and at viaducts), but may be applied also using soil as reference. The measured section is defined by the distance between the two external measurement points (A and B). Voltages between track and metal reinforcement are measured at three locations: A, B and PS (see Fig. 4).

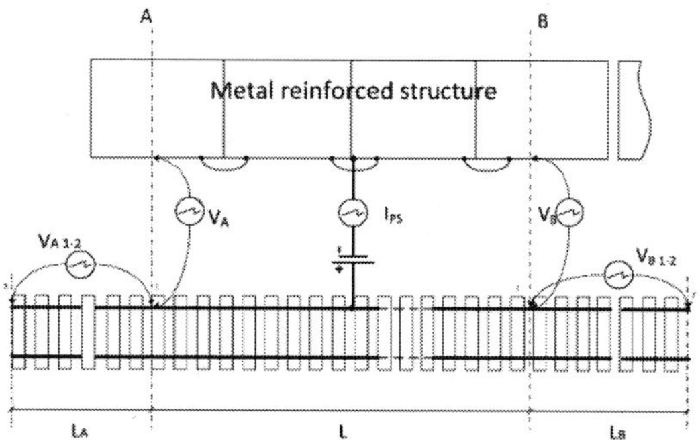

Fig. 4. Track to reinforced metallic structures conductance measurement, using EN 50122-2, A.2 method.

The rail currents flowing away from A and B points are estimated using rail voltage drop measurements and the rail resistance value (measured separately at the same location). All measurements shall be synchronized.

Track to earth conductance is calculated as follows:

$$G = \frac{I_{PS} - \frac{V_{A1-2}}{R_{rail}} - \frac{V_{B1-2}}{R_{rail}}}{L\,\frac{V_A + V_B + V_{PS}}{3}} \quad (4)$$

where:

G is the track to earth conductance [S/km];

I_{PS} is the power supply current [A];

$V_{A1\text{-}2}$ is the potential difference caused by the current leaving the section across measurement point A [A];

$V_{B1\text{-}2}$ is the potential difference caused by the current leaving the section across measurement point B [A];

R_{rail} is the rail resistance measured at the same points where V_A and V_B will be measured [Ω/km]

V_A, V_B and V_{PS} are the voltage between the rail and the structure at points A, B and at power supply PS (voltage $V_X = V_{X\,on} - V_{X\,off}$, difference between on and off state readings) [V];

L is the distance between A and B [km].

The application of this method is not straightforward. Readings shall be synchronized between measurement locations that may be far away. Length L cannot be chosen arbitrarily large (so to reduce the overall number of measurements over the whole line): site experience discourages this choice. The factors to keep under visual control are several, especially when the measurement is performed during construction phases. Equipment and service boogies outside the measured track section are connected to the rail for safety reasons, thus driving into it any leaking ground current, or, worse, connecting the rail to the earth or tunnel structure while operating. Activities affecting the electric status of the track shall be suspended, but coordination between different companies and contractors is always a hard task. Easy communication between operators shall also be ensured. Usually 500 to 800 m section length is adequate.

Mains and high frequency noise is often unavoidable, making post-processing more complicated and impeding accurate interpretation in place (as a check for the need of repeating the measurement). When measurement positions are close to rail cuts or insulating rail joints noise is usually larger. In general, a hardware low-pass filter is beneficial, also as an anti-aliasing measure. Averaging by software may be ineffective and too dependant on averaging window length, if noise is not stationary and features large fluctuations.

For the estimation of I_A and I_B currents rail resistance measurement shall be accurate, including variations due to temperature changes, and the effect of rail welding (see sec. 2.5.1). The best solution would be measuring the two I_A and I_B currents without the need to calculate them as the ratio of voltage drop at A and B and estimated rail resistance, e.g. using bridge methods or dc current probes fit around the rail.

In general, care must be used for long tracks, where PS current may be very large and unbalanced between A and B sides. Reversing PS polarity may be useful to avoid polarization issues and to check repeatability, although less effective than for method in sec. 2.1.1.

2.2 ASTM G165

ASTM standard G165 [4] reports a track-to-earth conductance measurement method, based on empirical considerations. The method applies to rails and turnouts sectioned by simple rail cuts or IRJs. The maximum track length measurable is 2750 m, the minimum 60 m.

A minimum of three measurements per track section is required against operator's errors and noise: the highest and lowest measured values shall not alter the average of the measurement by more than 3%.

The measurement set up is shown in Fig. 5: the power supply positive pole is connected to the track and the negative pole to an earthing electrode with a resistance to earth lower than 1/100 of the expected track-to-earth resistance (for a track of 1 km the required earthing resistance shall be a fraction of Ω). Voltage measurements are performed using ground and the track behind the measured section as voltage reference.

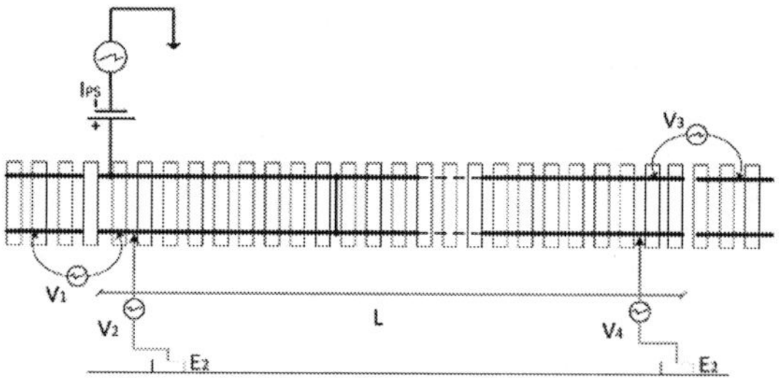

Fig. 5. Track to earth conductance measurement using ASTM G165-99 method.

Track to earth conductance G_n is obtained for each measurement point as the ratio of the current of the power supply and the measured voltage:

$$G_n = \frac{I_{PS}}{V_n} \quad (5)$$

where:
G_n is the conductance of the n-th measurement point [S];
I_{PS} is the power supply current [A];
V_n is the voltage at the n-th measurement point [V].

The method is very similar to sec. 2.1.1, with several voltage readings. G_3 and G_2 are used as a check of the correct insulation of the measured track section, while G_1 and G_4 are the real track-to-earth conductance values (G_4 is the best measuring point). If a large variation of measurement results is observed, too a large PS earthing resistance may be suspected.

This method requires a large number of simultaneous voltage readings and it is thus rarely used. It brings "for free" joint efficiency when measuring G_3 and G_2, that are required to vary less than 3% for a well-insulated track.

2.3 Close Potential survey

This technique is derived from cathodic protection system investigations [5] and, although not commonly used, it may be useful if local "hot points" of track to earth conductance, as caused by defects in the rail insulation, are suspected. The method is an extension of the one in sec. 2.1.1 using many voltage readings along the measured track section. In cathodic protection surveys voltage measurements are relieved from the necessity of synchronization, but for track-to-earth conductance estimation this practice is highly discouraged: voltage readings shall be simultaneous and synchronized with power supply current reading.

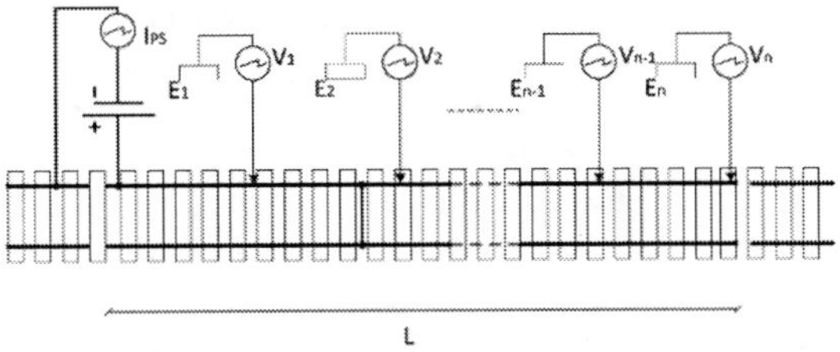

Fig. 6. Track to earth conductance measurement using close potential survey method.

The track-to-earth conductance is calculated as in sec. 2.1.1, taking the ratio of PS current and voltage reading; the same approximations and limitations apply. Significant variations between adjacent voltage readings indicate a local defect with a higher local track-to-earth conductance value; the operator shall be aware that going towards the end of the section voltage amplitude naturally decreases [2], so that attention shall be given to abrupt variations of voltage and not to a degrading voltage when moving toward section ends.

2.4 Resistance to earth method

This method is the application of the commonly used method to measure earth resistance of an earthing mesh in civil buildings: the farthest electrode E_2 applies the test current; a second electrode closer to the rail is used to read the rail potential. This method can be applied only to insulated track sections delimited by rail cuts or IRJs. The length L of the section shall be measured.

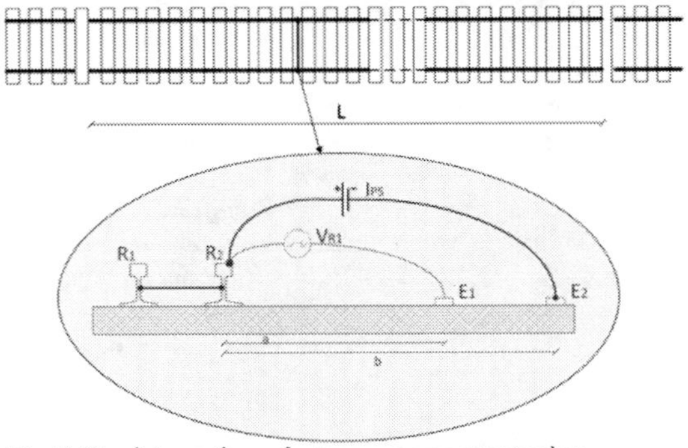

Fig. 7. Track to earth conductance measurement using resistance to earth method.

The voltage between the track and a probe electrode E1 is measured and the track-to-earth conductance is given by:

$$G = \frac{I_{PS}}{L\,V_{R2}} \quad (6)$$

where:
G is the track-to-earth conductance [S/km];
I_{PS} is the power supply current [A];
V_{R2} is the voltage between R_2 (electrically bonded to R_1) and probe electrode E_1 [V];
a is the distance between R_2 and voltage electrode E_1 [m];
b is the distance between R_2 and current electrode E_2 [m];
L is the section length [km].

Care shall be paid in placing the two electrodes as they shall not be influenced by the track electric field. The most widespread rule to place electrodes is the 62% rule (a =0.62b); this choice may be led back to the 62% rule for flatness of potential curve of the Fall-of-Potential method, described in [6], that, however, is strictly valid in homogeneous soil conditions.

The method is very fast and useful as preliminary test to save time later during measurements. It can be performed with a common earth resistivity tester, taking care that the resulting conductance be in the measurement range of the instrument.

2.5 Auxiliary measurements

As mentioned in previous sections, several methods need an estimate of some system parameters, such as rail resistance or soil resistivity. They are reviewed in the following.

2.5.1 *Rail resistance*

The measurement of rail resistance is of fundamental importance for methods described in sec. 2.1.3. The measurement method described here is that suggested by EN 50122-2, Annex A.1 (see Fig. 8).

A track section delimited by rail cut or insulating rail joints is used. The two rails of the track are bonded together at section ends and a current is applied in the center of the 8-shape loop circuit formed by the two bonded rails. In case of electrically continuous track, the two bonding connections alleviate the problem and can define quite effectively the track section under measurement if well bonded to the rails and of low resistance. The voltage drop across two adjacent rail sections on the same rail caused by the flowing current is measured using two voltmetric readings for a given length, L_1 and L_2, that shall be measured accurately.

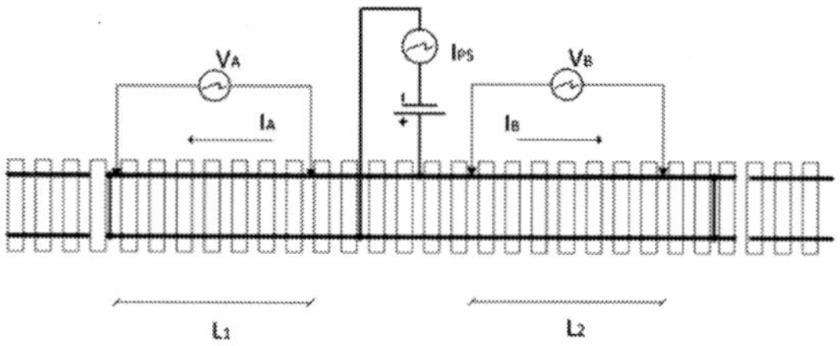

Fig. 8. Rail resistance measurement (EN 50122-2, A.1).

The resistance of the rail is calculated as follows:

$$R_{rail} = \frac{V_A + V_B}{(L_1 + L_2)\, I_{PS}} \tag{7}$$

where:

R_{rail} is the rail resistance [Ω];

I_{PS} is the current of the power supply [A];

L_1 and L_2 are the rail lengths for V_A and V_B voltage drop measurements [km];

V_A is the V_{on}-V_{off} voltage drop in sec. A of length L_1 [V];

V_B is the V_{on}-V_{off} voltage drop in sec. B of length L_2 [V].

Repeated measurements are advisable to evaluate consistency and uncertainty. However, such approach is ineffective for systematic errors: the two lengths shall be measured more than once, also to exclude operator's mistakes; rail temperature influences significantly rail resistivity [7]; rail welds are quite relevant, as they are allowed to increase overall rail resistance up to 5%, so that measuring points shall correspond to those used for the voltage drop measurement of the track insulation test. It is reminded that the so obtained rail resistance value is used to estimate the current flowing from voltage drop measurements; the difference at the numerator of eq. (4) may be quite small for good track insulation and small errors in the estimate of "R_{rail}" are thus amplified.

Current intensity may be very large (e.g. hundred Amps for just 12 V PS) and voltage drops along very short rail and wire lengths already significant: 4 wires techniques shall be adopted and galvanic isolation between channels is a must.

2.5.2 *Remote electrode distance*

This section describes a method derived from cathodic protection measurements [5] to find out the correct distance for placing the electrode to consider it electrically remote from the track.

A track section delimited by insulating rail joints or rail cuts is considered, but the method can be applied also to electrically continuous tracks. PS is connected between track and ground reference at the beginning of the section; at the other end of the section (chosen to avoid influence from PS earthing electrode), the track voltage is measured with respect to a movable electrode driven into the soil at various distances from the track.

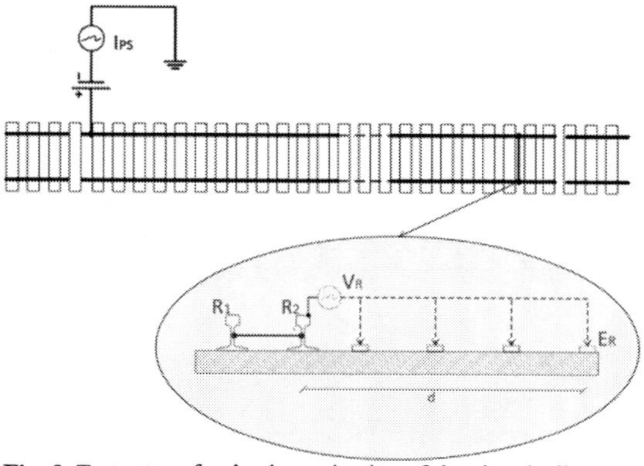

Fig. 9. Test set-up for the determination of the electrically remote electrode position.

The voltage V_R follows a logarithmic decay vs. distance from track and then finally reaches a constant value, corresponding to the minimum distance so to consider the electrode electrically remote. As for other methods where flatness of potential curve is to be evaluated (e.g. Fall-of-Potential method), a suitable criterion for flatness is needed: e.g. the spread of three adjacent points is <5% of the average, where minimum increment of distance is 10-20% at each point. Plotting V_R vs. distance is of great help to interpret results.

2.5.3 *Soil Resistivity*

Soil resistivity may be measured by driving four electrodes into the soil and using a soil resistivity meter implementing Wenner or Schlumberger methods [6][8].

3 Practical tips

Some practical tips may be formulated based on previous considerations and experience.

For measurements inside tunnels, mobility is extremely important, since access to the track during construction may be from a few entry points (e.g. at some stations), requiring long walks to reach the desired location. Measuring equipment should fit into backpacks and trolleys avoiding need of bogies.

If reinforcement is not electrically continuous (e.g. when tunnel sections are gapped or prefabricated tunnel lining segments are used), electrically continuous and earthed cable trays may be used as reference.

Communication inside tunnels may be difficult due to fan noise, on-going activities and poor radio propagation.

When measuring in urban areas, a good earth reference may be hard to find for the soil cannot be reached due to asphalt and constructions: green areas with deep soil shall be considered (flowerbeds are made by landfill and are not a good earth reference). Concrete electrodes of the $CuSO_4$ type may be also considered as reference for voltmetric readings, but the effect of buried pipes and cables shall be verified, e.g. by moving the electrodes to check variability.

Wires connecting to electrodes shall cross the road, and will be damaged and may also cause accidents if not fastened and protected (e.g. by heavy tape glued to the asphalt). High visibility tape should always be used to indicate the presence of wires, including pedestrian and green areas, useful also in case of grass cut.

Connection to rails shall be effective both for mechanical robustness, contact resistance and rapidity to connect and disconnect: there are rail clamps for this purpose, although not all ensure the necessary mechanical stability and that rust is removed when inserted; for rails embedded in concrete or soil the accessible part is the top of rail, for which modified bench clamps or naked conductor pressed against the rail surface may be used.

Standards are not clear if track insulation is to be evaluated also in wet conditions. The correct mix of dry and wet conditions representative of weather conditions for a specific location shall be considered, when comparing to the limits. In all cases, measurements should be done when a suitable time interval has elapsed after track wetting to let surface dry, avoiding the measurement of the conductance of water itself.

When using rail cuts to isolate the track section, rail length may change significantly if temperature varies due to e.g. the position of the sun: isolation may be compromised, also if rubber sheets were interposed, because the applied force is extremely intense.

As an alternative to rail current measurement through voltage drop, current clamps may be used for a direct measurement: clamps with such a large section are all of the Rogowski type and cannot measure dc, so that a suitably low frequency signal shall be used, that can reasonably replace dc (artefacts may be observed when the ac signal changes its polarity at each half cycle due to soil polarization).

After data of a measurement have been recorded, a preliminary in-place calculation of relevant quantities and evaluation of measurement quality is nevertheless important to decide if the measurement shall be repeated and to avoid useless complications in post-processing.

4 Conclusions

Several methods have been reviewed for the measurement of track-to-earth conductance. For all uncertainty is caused not only by instrumental uncertainty, but also variability of system parameters, including soil and rail resistivity. Repeatability is suggested as a means to monitor measurement accuracy. Outliers and aberrations can always occur due to external factors, mostly practical, such as unavoidable working activities and unnoticed earthing of the rails under measurement (e.g. by OVLDs, negative feeder cables, accidental contact with earthed conductive parts).

An electrically insulated track section is always the best configuration for accuracy and to avoid propagation of electric noise from other parts of the line. Examples of continuous conductive structure are several: tunnel structure, walkways, handrails and cable trays, or a longitudinal earthed conductor.

Methods in sec. 2.1.1 and 2.1.3 are quite accurate without and with such conductive structures. Increasing the number of voltmetric readings brings to the method in sec. 2.3, useful to spot out local defects of insulation. An equivalent method based on the difference of measured rail current along the track section is currently under study, using e.g. mentioned Rogowski coils.

For minimum impact on an existing line with running trains method in sec. 2.1.2 is almost a forced choice: thanks to the averaging of several points, the method is quite accurate and copes with sporadic transients; the worst aspect is the lack of a literature reference to determine the extension of the measured section at the left and right ends of the voltmetric terminal.

References

1. EN 50122-2 / IEC 62128-2, *Railway applications - Fixed installations -Electrical safety, earthing and the return circuit - Part 2: Provisions against the effects of stray currents caused by dc traction systems* (CENELEC, Brussels, Belgium, 2012)
2. J. Bongiorno and A. Mariscotti, *Accuracy of railway track conductance and joint efficiency measurement methods*, Acta IMEKO, **4**, 82-87 (2015).
3. J. Bongiorno and A. Mariscotti, *Variability of track to ground conductance measurement*, 22th IMEKO TC4 Int. Symp. proceedings, Iasi, 14-15 Sep. 2018.
4. ASTM G165-99, *Standard Practice for Determining Rail-to-Earth Resistance*, ASTM International, Conshohocken, Pennsylvania, United States (2005)
5. A.W. Peabody, *Control of pipeline corrosion*. 2nd ed., (NACE International, Houston, 2001)
6. IEEE Std. 80, *IEEE Guide for Safety in AC Substation Grounding*, Piscataway, New Jersey, United States (2013)
7. A. Mariscotti and P. Pozzobon, *Resistance and Internal Inductance of traction rails: a survey*, IEEE Trans. on Vehicular Technology, **53**, 1069-1075 (2004).
8. A. Samouelian, I. Cousin, A. Tabbagh, A. Bruand and G. Richard, *Electrical resistivity survey in soil science: a review*, Soil and Tillage Research, **83**, 173-193 (2005)

Traction power consumption as a component of maximum speed choice on High Speed Lines

Andrzej Żurkowski[1,*]

[1] Railway Research Institute, ul. Chłopickiego 50, 04-275, Warsaw, Poland

Abstract. The paper includes briefly the most important methodological considerations for selecting the maximum speed for the newly designed HSL. They provide a context for discussion on energy aspects regarding traction energy costs. The research problem in question includes both the assessment of the absolute values associated with the consumption of traction energy and their reference to other costs and revenues associated with the operation of the High Speed Rail (HSR). The proposed approach is based on an analysis of the experience described in the international literature of the subject and simulation calculations for Polish conditions. Based on the approximate results and dependences in this field obtained by the railways with operational experience in relation to the typical HSR train sets, the records of dependencies which are most likely to be confirmed in Polish conditions are presented. Currently, 3 kV DC is used on the railway network in Poland. This is therefore technically different from the one designed on HSL (25 kV AC), which the calculations are related to. Choosing the maximum speed for HSL must take into account the consumption of electricity because of its significant cost. However, the shortening of travel time by increasing the speed makes the HSR offer more attractive and thus allows for increased revenues. The final decision should therefore be a compromise based on detailed calculations, forecasts and simulations.

1 Introduction

The choice of the maximum speed of trains running on High Speed Lines (LDP) is one of the key elements determining the economic results of such an investment. The set of factors that should be taken into account is very extensive and includes technical, commercial, operational and economic parameters. Among them, the use of traction energy is also an important element. The article presents in a synthetic way the whole problem of choosing the speed on the High Speed Railway starting from the basic concepts, and then focusing on the conditions related to traction and its costs.

The considered batch problem includes both the assessment of absolute values related to the traction energy consumption, as well as their reference to other costs and revenues related to the operation of the High-Speed Railway (KDP).

The proposed approach is based on the analysis of experiences described in the international subject literature and on simulation calculations for Polish conditions. The aim of such a comprehensive approach is to demonstrate that there are a number of rational premises for establishing the maximum speed both on Lines of High Speed and on upgraded lines, which is in contradiction with the often-presented approach. expert, based only on general ideas.

Choosing the maximum speed of trains on HSL (High Speed Lines) is one of the key factors in determining the economic performance of such an investment. The consumption of traction energy is an important determinant in the list of technical, commercial, operational and economic factors.

The paper presents the overall problem of speed selection, focusing on traction and cost aspects.

2 Basic concepts related to railway speeds

Speed classification in rail transport is based on the distinction of two basic types of speed:

- **maximum** (v_{max}) that are possible to achieve depending on the design and technical condition of railway lines, railway signaling and power systems and the type of rolling stock used,
- **average speed** (v_a), which characterizes the operation process on railway.

For evident reasons $v_a < v_{max}$. The exact list of speed types and their definitions are shown in Table 1. The maximum speed of the train for the given operating conditions is equal to the lower of the two speeds: road and rolling stock:

$$v_{tr} = \min(v_r, v_{rs})$$

The average speed of a given train is essentially due to the achievable maximum speed v_{tr} and the method of organization of railway traffic, represented in the form of a timetable and described by means of the technical speed v_t.

From the passenger's traveler's point of view, the most important is the commercial speed v_c, which determines the total travel time on the L route, which consists of n sections. The relationship between the maximum and the average speed can be described by the so-called the maximum train speed utilization rate w_{max} form:

$$w_{\max} = \sum_{i=1}^{n}\left(\frac{v_c^i}{v_{tr}^i}\cdot\frac{l_i}{L}\right)$$

* Corresponding author: azurkowski@ikolej.pl

n – and the number of sections, which divided the entire route,
L – the length of the route [km],
l_i – section i length [km].

The w_{max} value is usually from 0.5-0.6 on classic lines to 0.7-0.9 on high speed lines [2]. It results from longer distances between stations on HS lines, line specialization and homogeneous train traffic, which translates into its incomparably greater liquidity. In practice, the value of w_{max} should be interpreted as a measure of the assessment of the use of the technical capabilities of railway lines in the rail operation process in relation to your rolling stock.

In accordance with the long-standing technical tradition, the maximum speed thresholds on the railways were defined as the total multiple of 20, thus: 80, 100, 120, 140, 160 or 200 km/h. Then, the threshold for High Speed Rail in Europe, previously indicated in Directive 2008/57, is 250 km/h (already without the above rule), and further on 280, 300, 320 and - as of today - 350 (380) km/h.

Table 1. Maximum and average speeds in rail transport.

Speed type	Name of speed	Symbol	Definition
maximum (v_{max})	road	v_r	The speed achievable in a given section of railway line.
	rolling stock	v_{rs}	The speed that can be obtained with a given type of rolling stock.
	train ride	v_{tr}	The speed of the train.
average (v_a)	technical	v_t	The speed achieved by the train on the whole route, with the deduction of commercial stops at stations.
	commercial	v_c	The speed achieved by the train along the whole route, taking into account commercial stops at stations.

3 Overview of the maximum speed key selection factors

The set of criteria for selecting the maximum speed adequate to the conditions of a particular railway line can be systematized in three subsets, the classification proposed by M. Walrave [5, 6] is as follows:

- transport criteria, particularly important from the point of view of the carrier, determining in particular the share of rail in the transport market in a given corridor,
- technical and operational, i.e. related to the technical conditions of both the infrastructure and railway rolling stock understood comprehensively,
- social, including, among others, costs related to ecology (energy acquisition, noise), etc.

The criteria described in the three groups are the basis for the formulation of four functions, which together constitute the "economic balance of speed". All of them are non-linear and can be represented using the formulas listed in Table 2.

The set of parameters $\{A, ..., J\}$ for the presented functions requires a detailed analysis of a given system of connections by High-Speed Rail and taking into account local technical and economic conditions. Then, examining their course allows you to determine the optimum speed, which is shown schematically in fig. 1.

Table 2. Economic balance of speed.

N°	Function's name	Function's formula	Type of function
1	Journey time	$f_c(v) = A \cdot v^B$ A > 0 and B < 0	power
2	Traffic & revenues	$f_p(v) = C \cdot \ln(v) + D$ C > 0 and D > 0	logarithmic
3	Operating costs	$f_k(v) = E \cdot e^{Fv}$ E > 0 and D > 0	exponential
4	Internal benefits	$f_w(v) = G \cdot v^3 + H \cdot v^2 + I \cdot v + J$ G, H, I > 0	polynomial

Own study based on [5, 6]

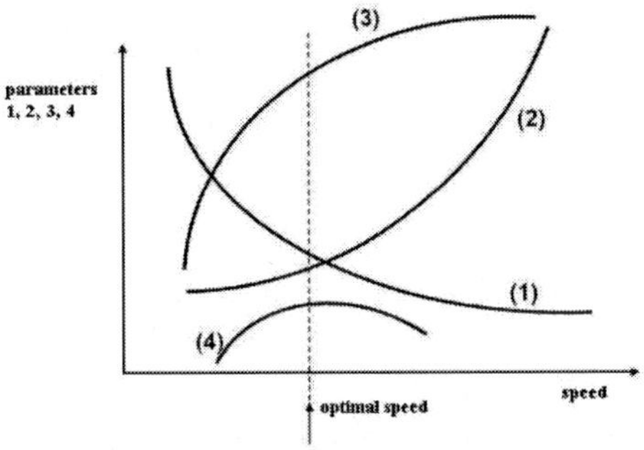

Fig. 1. Economically optimal speed [5].

The transport criteria relate to the issues of transport modeling [1]. The choice by means of a transport means from the so-called preferences, or features of transport services. The full set of preferences includes a dozen or so elements. The logit model [5, 7] describes their influence on the final decision of the traveler in relation to all possible transport means.

The High Speed Railways in relations with the length range from several hundred to about one thousand kilometers compete in principle with air transport. In this case, the market partition describes the so-called price-time model [7]. It is based both on the calculation of travel time and its price, as well as on the individual assessment of the time of the traveler.

The choice of the maximum speed of trains is therefore of major commercial importance, as it decides about participation in the total passenger transport market. Referring to the passenger transport market in Europe, it can be concluded that the share of rail in modal split, amounting to 6.7% for several years, remains steady at this level thanks to the development of the KDP system, which already carries 25.7% of all rail passengers[a].

The choice of the maximum speed of trains on the considered LDP can therefore start from determining the set of permissible solutions and at the same time assume that the maximum speed limit currently is 574,8 km/h. From a theoretical point of view, speed optimization can

[a] *EU Energy & Transport. Statistical pocketbook 2017*, European Commission 2018.

lead to a solution of any value v_{max} from the range of, for example, 200 ... 570 km/h. The maximum speed limits used on the rail limit the actual selection to values divided by the number 10.

The first step in speed selection should therefore be to set several thresholds to be used, no. 250, 280, 300 or 320 km / h. Turning then to the economic plane, it seems reasonable to analyze for each of these speeds the costs and benefits (revenues) associated with the four basic groups of conditions: technical, economic, operational and commercial. Their combination and critical analysis should make it easier to choose the maximum velocity of the most-preferred High Speed Line under certain conditions.

4 Traction power consumption as a factor of maximum speed choice

The amount of energy used for traction purposes depends on many factors related to the investment process of designing and building the High Speed Line, the type of used rolling stock, as well as operational methods.

The impact of the construction of High Speed Line on the amount of energy consumed is related to:

- its geometrical profile, both vertical (elevation) and horizontal (number and radius of arcs,
- engineering structures (tunnels, viaducts) implying possible speed limits or changes in train resistance,
- the number and arrangement of tracks on intermediate stations or outposts and the permissible speeds when crossing them,
- the type of traction power system used (DC - direct current, AC - alternating current, frequency).

The impact of rolling stock construction on energy consumption includes:

- aerodynamic resistance related to the shape of the train bodies,
- the power of traction motors and the manner of their arrangement (concentrated - traction components at the ends of the train or traction deployed on the length of the train - EMU),
- type of the technical construction – mechanical energy consumption.

Both the technology of designing and building the High Speed Line as well as the composition of the sets results from the progress of relevant engineering disciplines. The amount of energy consumption depends on the generation (year of production) of vehicles and the modernity of the innovative solutions used.

The exploitation process has a very large impact on energy consumption. The type of railway traffic control system, cabin signaling, etc. is of particular importance in this regard.

At the same time, the key skills and experience of the driver are key. With high saturation of the lines and the resulting short time of train succession, the energy consumption increases in the case of frequent deceleration and acceleration.

The possibilities of significant progress in this area are related to the autonomy of running trains (ATO – Automatic Train Operations system). Such solutions have been used for many years on subway lines and on aero-rails.

Practical implementation of them on the railway is possible primarily on lines completely separated from road and pedestrian traffic and used for transports realized with a homogeneous type of vehicles. Such conditions exist, for example, on agglomeration railway lines in city centers and on Lines of High Speed.

The existing systematic of train traffic automation (Grade of Automation) covers four levels. The lowest level – GoA 1 refers to the case of a train driven by the driver with the cab signaling, while the highest one – Goa 4 means a completely unmanned train.

The advantages of such a system consist in the possibility of full application of train traffic based on the so-called movable block sections. In this way, the fluidity of train traffic is optimized, which at the same time translates into optimal, minimal traction energy consumption on the whole line.

Prospects for the use of such solutions on the leading European railways concern the next few years. It is worth pointing out that the automatic acceleration and braking of trains (GoA 2, 3 and 4) already allow for the optimization of energy consumption for individual trains to a large extent.

The costs of energy used for traction purposes include:

- charges for energy collected (at the substation's entrance),
- refund for recovered energy (recuperation),
- energy supplier fees for distribution management,
- brokerage costs charged, for example, by the infrastructure manager.

It should be noted that the traction energy consumed also includes transmission losses over the traction network (in the substation-pantograph section), and in the case of recuperation – the possibility of energy recovery recovered at the unit rate of energy supplied. At the same time, distribution costs are usually calculated in proportion to the costs of delivered energy.

Making a comparative assessment of traction energy costs from the point of view of the carrier requires its reference to one inventory place in the train (to the seat-km) or for one train-km [3].

On Polish railways the highest train speeds are obtained on the Central Trunk Line (CMK). For more than 60% of the length of this line, the ED250 Pendolino trains run at a speed of 200 km/h, which in the light of the Railway Interoperability Directive (2016/797, Annex I) means the High Speed Rail.

Using the mathematical formulas [3] elaborated for the HSR [8], the costs of traction energy consumption were referred to the CMK line.

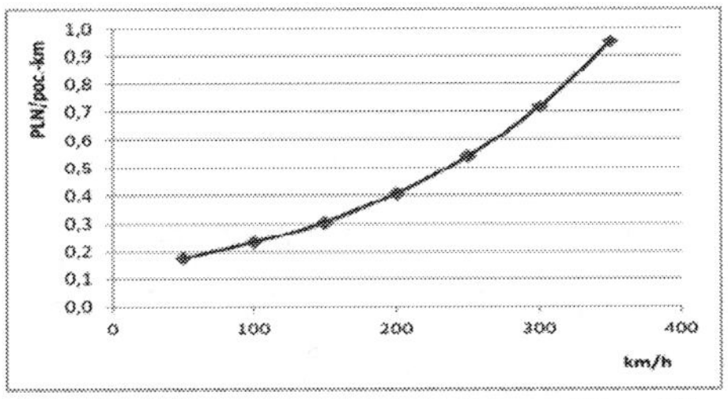

Fig. 2. Traction energy costs in Polish conditions [8].

By denoting C_e as energy consumption, this his relationship can be described by the formula (R^2= 0,9998):

$$C_e(v) = 0{,}1336 \cdot e^{0{,}0056 \cdot v}$$

The above pattern is of course indicative, but confirms the exponential increase in energy costs as a function of the speed of train.

I would like to formulate a comment on this relationship from the position of an expert who has been dealing with the issue of rail passenger transport for nearly forty years: both from the theoretical and practical side – managing the railway operator. The cost of energy is an important item in the budget of the operator, but if such an expense leads to shortening the time of train travel and acquiring additional passengers, then it should not be a limitation in this respect.

This applies both to the situation of direct connections between cities, as well as between interchanges. The speed of trains should be shaped also in relation to the possibility of connecting trains and changing passengers.

Probably greater savings in energy consumption can be obtained through the use of ATO systems, thanks to the optimal acceleration and braking of the train as well as the resulting fluidity of railway traffic on lines and at railway junctions.

5 Conclusions

With the problem of choosing the maximum speed of trains on KDP lines, a set of possible solutions is finite and countable, and the space of acceptable solutions is determined by technical conditions and railway operation rules.

The chosen "optimal" speed belongs to the set of practical "railway" speeds. Energy for traction purposes is an important element of calculation and increases exponentially as the speed of the train increases, In order to determine the optimal speed, the simplest method is to perform an economic calculation and to examine the technical, organizational and commercial consequences successively for all possible optimal speeds.

Currently, the maximum speeds on High Speed Lines are shaped based on the assessment of local conditions and technical possibilities.

Until now, a 3 kV DC voltage is used on the railway network in Poland (including the CMK line). Therefore, it is a technically different situation than the one for the 25 kV high-speed AC line being designed).

The choice of the maximum speed should take into account the consumption of electricity due to its significant cost. Shortening the travel time by increasing the speed increases the attractiveness of HSR connections, and thus allows you to increase revenues. The final decision should therefore be a compromise based on detailed calculations, forecasts and simulations.

References

1. M. Lebœuf, *Grande vitesse Ferroviaire. Cherche Midi*, (Paris, 2013)
2. A. Massel, *Wykorzystanie prędkości maksymalnej pociągów [The use of the maximum speed of trains]*. Problemy Kolejnictwa, **4** (2017)
3. G. Rus (ed.), *Economic Analysis of High Speed Rail in Europe*, (BBVA, Bilbao, 2009)
4. M. Siergiejczyk (ed.), *Koleje Dużych Prędkości w Polsce [High Speed Rail in Poland]*, (Instytut Kolejnictwa, Warszawa, 2015)
5. M. Walrave, *Analyse économique et financière des projets Grande Vitesse, Formation Systèmes Grande Vitesse*, (UIC, Paris, 2007)
6. M. Walrave, 9 main issues about the optimal Speed, *UIC 6th World Congress on High Speed Rail*, (Amsterdam, 2008)
7. A. Żurkowski, *Price-time model Applied in Assessment of Rail Market Share in Passenger Traffic in Transport Corridor*. The Archives of Transport, No **3-4**, Vol. XXI (2009)
8. A. Żurkowski, *Przesłanki wyboru prędkości maksymalnej pociągów na liniach szybkiego ruchu [Premises for choosing the maximum speed of trains on fast lines]*, Logistyka, **4** (2012)
9. A. Żurkowski (ed.), High-Speed Rail in Poland, (Balkema, 2018)

MATEC Web of Conferences **180**, 02001 (2018)
MET'2017
https://doi.org/10.1051/matecconf/201818002001

An influence of a complex modernization of the DC traction power supply on the parameters of an electric power system

Jerzy Wojciechowski[1,*], *Krzysztof* Lorek[2], and *Waldemar* Nowakowski[1]

[1]Kazimierz Pulaski University of Technology and Humanities in Radom, ul. Malczewskiego 29, 26-600, Radom, Poland
[2]Sonel S.A., ul. Wokulskiego 11, 58-100, Świdnica, Poland

Abstract. Urban and suburban public transport constitute basics of functioning of modern, urbanized metropolises. It is a comfortable, economic and ecological means of transportation. Thus, a stable and fast growth of this solution. One of its components is the power supply system. It should allow functioning of the whole transportation system, maintaining the following criteria: energy efficiency and modernity. These criteria have contributed to creation and modernization of power supply systems. Where it comes to DC systems, such an investment includes: replacing 6-pulse rectifiers with 12-pulse rectifiers and raising rated supply voltage. In this article, the authors have presented research results, based on measurements of electrical energy parameters before and after a modernization of a suburban transportation traction system. These parameters have been divided into two groups. The first consists of parameters mentioned in the EN50160 norm. Another group consists of parameters not mentioned in the norm. What has been presented are waveforms and graphs showing these parameters. In the final part of the article a discussion on reasonability of introducing new, modernized traction power supply system for suburban transportation has been performed.

1 Introduction

The issue of energy efficiency in the traction power system constitutes one of the most basic and most important questions of modern means of transportation. The energy efficiency is both the energy consumption of transportation processes, and the influence of the traction power systems on electric power systems. In order to improve the energy efficiency of DC traction systems, among other things, one can modernize traction substations. This includes the following actions: replacing 6-pulse rectifiers with 12-pulse rectifiers (typical procedure in traction power supply systems) and raising rated voltage of the power supply (sporadic procedure). Research results of electrical energy parameters in the municipal traction power system before and after the modernization of the traction substations were shown. The modernization included both changing the rectifiers, and raising the traction power supply voltage from 0,6 kV DC to 3 kV DC. At the same time, traction vehicles used were replaced by newer models. The new vehicles are equipped with the energy recovery function. The measurements carried out concerned primarily the power quality parameters and the parameters related to the electrical energy consumption.

2 The power supply system

One of the most basic issues of electrical transport is its energy consumption and energy safety [1-5]. The issue of real value of the power supply voltage of traction units and electrical energy given back to the units are basic technical problems of DC traction power supply systems [6-14, 18-22]. Additionally, there is an issue of the influence of a DC traction substation on the power supply of an electric power systems. Changing power supply systems, which has already been mentioned at the beginning of the article, serves the purpose of minimizing all negative phenomena related to their functioning.

Replacing 6-pulse rectifiers with 12-pulse rectifiers along with raising rated voltage of the traction network power supply is aimed at achieving the following effects [11-12, 15-22]:

- improving voltage stability,
- reducing losses in the transmission and distribution of electricity,
- reducing electricity deformations retrieved from a commercial power system.

Replacement of rectifiers requires a significant reconstruction of traction substations. Raising rated voltage of the power supply results in the necessity to raise the insulation level of the traction power network. This may involve replacing the traction power network, traction insulators and protection systems.

The mentioned actions constitute modernization costs which should be compensated by achieving planned effect.

The case analyzed by the authors involved a traction substations system modernization, without any changes in the power supply system of the commercial power system. Both before, and after the modernization of the substations, they were supplied from Point of Common Coupling (PCC) - rated voltage of U_n=15kV, work with

* Corresponding author: j.wojciechowski@uthrad.pl

tap-changer 110/15,75 kV/kV. One, representative traction substation has undergone a measurement analysis. Before modernization, two 6-pulse rectifier units with DC U_n=0,6kV worked in the substation. After the modernization there were two 12-pulse rectifier units with DC U_n=3,3kV. Basic blocks of each of the rectifier units are:

- a rectifier transformer type TOTp-6300/15, with nominal power 6300kVA and transformer ratio 15kV/2x1,29kV,
- a diode rectifier type PD-17/3,3, with rated voltage 3,3kV DC and rated current 1700A and class III overload.

A block scheme of the traction power supply of one of the traction substations has been presented in the Figure1.

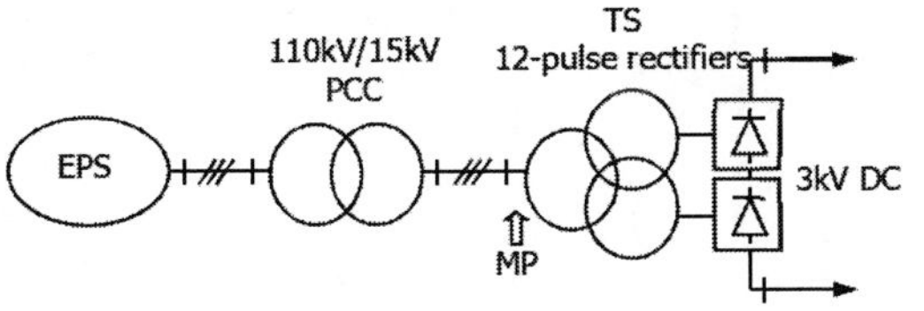

Fig. 1. A block scheme of the DC traction power supply after the modernization (EPS - Electric Power System, PCC - Point of Common Coupling, TS - Traction Substation, MP - Measuring Point).

DC power supply system load consisted of traction units: EN94 (LKc-310 DC motor, 8 pcs. per draft, power 56,5 kW) and EN97 (TME 42-26-4 AC motor, 8 pcs. per draft, power 180 kW).

3 An analysis of electrical energy parameters in the examined traction substation

Quality parameters in electrical energy in the power system are physical quantities describing discrepancies between real waveforms and ideal sinusoidal waveforms of a three-phase circuit. These parameters and their allowed values are defined in the PN 50160 (EN 50160) norm [23]. What was performed in the examined object were 24/7 measurements of electrical energy parameters on the 7th and 8th of April (before the modernization) and on the 7th and 8th of June (after the modernization). Daily measurements were chosen due to the fact of the repetitiveness of the traction load cycle, resulting from invariability of the traction units' time schedule. In the analyses presented in the article, both daily, and hourly periods (daily peak hours – 7:30 – 8:30 a.m.) were taken into account. A high similarity of measurement conditions, necessary for a comparative analysis, was kept. Measurements before and after the modernization were performed on week days, with outside temperature above 20°C (heating in the vehicles was switched off). The measurement equipment were PQM-711SONEL meters, allowing for measuring electrical energy parameters.

A registration was performed with 200ms averaging time, with actual value sampling of 10,24 kHz. Measurements were conducted in indirect systems from current and voltage transformers, on the primary side of the transformer (15kV). For both systems one of the secondary winding was devoted only for metering-billing or billing purposes.

3.1 Power quality parameters in the PN 50160

PN 50160 defines acceptable electrical energy quality parameters with reference to 10-minute periods. In the analysis, the 200ms period measurements were averaged to 10-minute periods. Below, the PN 50160 parameters before and after the modernization have been compared.

- RMS value (PN – 95% 10-minute average RMS values should fit into a ±10% U_n variation band), both "before" (15,406/15,446/15,402kV$_{min}$/15,951/15,981/15,938kV$_{max}$), and "after" (15,469/15,491/15,441kV$_{min}$/15,895/15,918/15,86997kV$_{max}$) PN requirements are fulfilled,
- a negative sequence symmetrical component – (PN – 95% from a 10-minute average RMS value set should fit into a range from 0% to 2% of a positive sequence value component), both "before" (0,08% min/0,22% max), and "after" (0,015% min/0,21% max) PN requirements are fulfilled,
- voltage harmonics – (PN – throughout each week 95% from a 10-minute average RMS set for every harmonics should be lower or equal to the values presented in Table 1 and Table 2), odd harmonics values that are not the multiple of 3, "before" and "after" have been presented in the Table 1, odd harmonics values that are the multiple of 3 "before" and "after", on the other hand, have been presented in the Table 2.

Table 1. Odd harmonics components other than a multiple of 3 according to: PN; "before" and "after" modernization.

Order	PN	Research "before"	Research "after"
5	6%	0,2%	0,9%
7	5%	1,0%	0,8%
11	3,5%	0,2%	0,7%
13	3%	0,2%	0,7%
17	2%	0,1%	0,1%
19	1,5%	0,1%	0,07%
23	1,5%	0,1%	0,2%
25	1,5%	0,03%	0,1%

Table 2. Odd harmonics that are the multiple of 3 according to: PN; "before" and "after" modernization.

Order	PN	Research "before"	Research "after"
3	2%	0,2%	0,3%
9	1%	0,05%	0,07%
15	0,5%	0,02%	0,03%

• THD – (PN – lower or equal to 8 % throughout each week), "before" (0,82/0,82/0,8%min /2/1,98/2,04%max), and "after" (0,99/1,06/0,9%min/1,82/1,92/1,82%max) PN requirements are fulfilled.

Power Quality parameters mentioned above both before, and after the modernization are compliant with the PN 50160 norm. A special emphasis should be put on the two of them, i.e. RMS values and a power supply voltage harmonics set. The consequence of increasing the traction voltage from 0.6 kV to 3 kVDC was the possibility of using rolling stock with the recovery of electricity. Additional reduction of the current consumed from the electric power system was achieved thanks to the work of the PCC transformer with tap-changer 110/15,75 kV/kV. As a result, what happened was a decrease in the value of current taken from the power system and, consequently, a decrease in voltage drops in the power supply system. This has been presented on a graph with the supply voltage "before" and "after" – Figure 2 (a measurement in a daily peak hour with 200ms averaging).

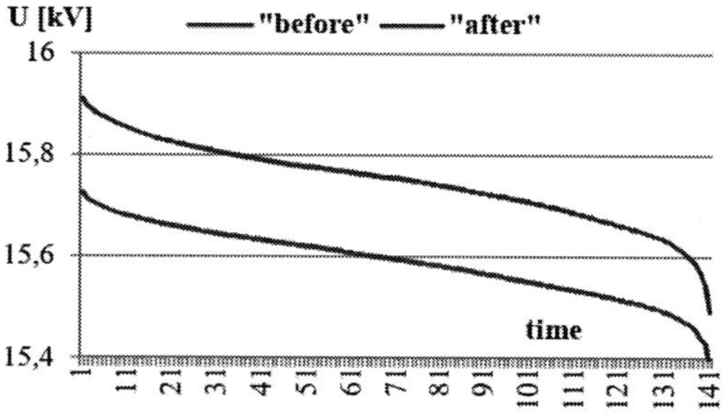

Fig. 2. RMS values of the supply voltage (side 15kV) before and after the modernization.

A comparison of share of harmonics in the supply voltage "before" and "after" is not explicit because there has been a replacement of a 6-pulse rectifier with a 12-pulse rectifier (Table 1 and Table 2). One would assume elimination of harmonics of the 5th, 7th, 17th, 19th order. However, in the measurement results such an effect is not completely visible. The cause of this, most probably, is not the influence of the analyzed rectifier on the power supply system, but a reverse process. In our opinion, in the analyzed power system the 6-pulse rectifier that generated the above-mentioned harmonic components was worked. As a result of the supply voltage including the 5th, 7th, 17th, 19th harmonics, the current flows through a winding of the primary side of the rectifier transformer. Despite of that, share values of these harmonics are low and far from the norm.

Voltage harmonics have low values, however their occurrence should be analysed. There is a possibility that their values will go up (it might be related to a predicted raise in the traction burden), in this case it will have a negative influence on keeping the value of the electrical energy quality parameters. The analysis has been conducted basing on the correlation of the harmonics of current and voltage. Apart from the harmonics characteristic for 6-pulse and 12-pulse rectifiers, harmonics of the 3rd, 35th, 37th, 47th, 49th order were also taken into account. The last four are also characteristic harmonics, which are omitted in the PN norm. However, because of their significant values in the readings taken, the authors have decided to include them in their work. Figure 3 presents spectrum of voltage harmonics (1 hour during maximum burden, without the 1st harmonic) before and after the modernization.

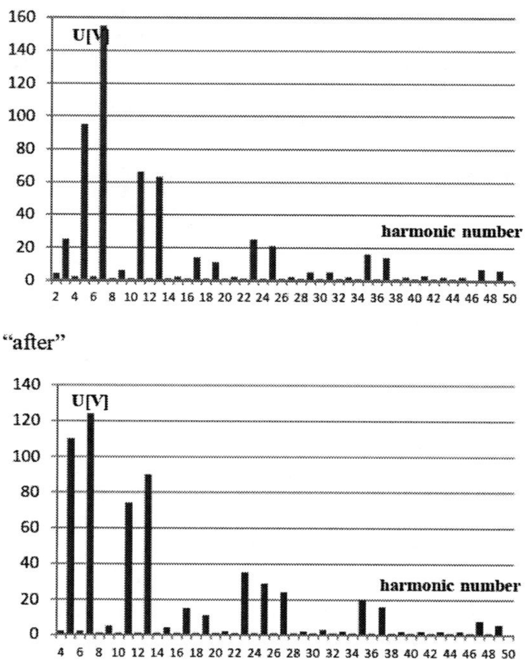

Fig. 3. Spectrum of voltage harmonics before and after the modernization (side 15kV) (picture PQM-711).

The analysis has been conducted for two variants. In the first one, correlation between voltage harmonics and current taken (RMS value) has been calculated. In the second one, correlation between voltage harmonics and corresponding with them current harmonics has been calculated. Graphic results for the chosen harmonics have been shown in the Figure 4, 5 and 6.

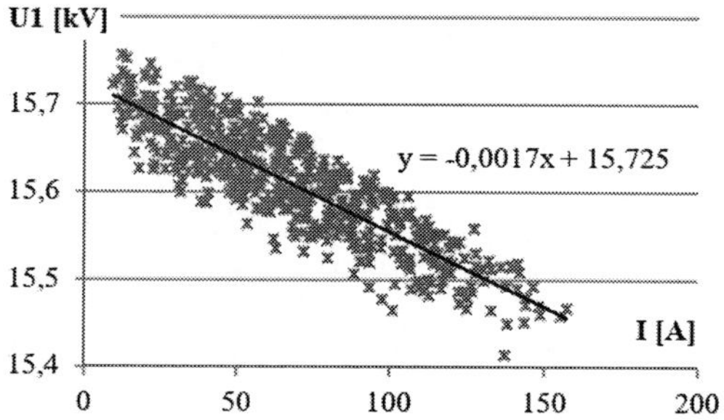

Fig. 4. Correlation between current I (RMS value) and the 1th voltage harmonic – the reading taken in the MP.

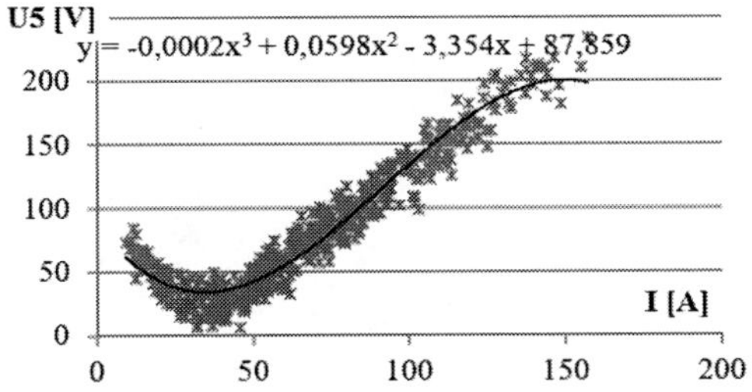

Fig. 5. Correlation between current I (RMS value) and the 5th voltage harmonic – the reading taken in the MP.

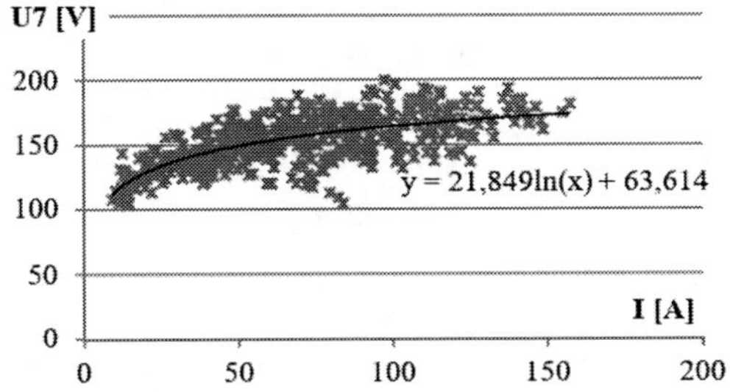

Fig. 6. Correlation between current I (RMS value) and the 7th voltage harmonic – the reading taken in the MP.

Values of the correlation ratio between various readings have been presented in the Table 3 and 4.

Table 3. Values of the correlation ratio between current (RMS) and voltage harmonics – an analysis in the MP.

	U5	U7	U11	U13	U17	U19	U23	U25
"before"	0,91	0,64	0,41	0,33	0,43	0,08	0,12	0,02
"after"	-0,05	-0,26	0,70	0,71	0,53	0,13	0,85	0,78

	U1	U3	U35	U37	U47	U49
"before"	-0,87	0,04	-0,06	-0,01	0,06	0,01
"after"	-0,70	-0,21	0,40	0,49	0,10	0,13

Table 4. Values of the correlation ratio between current harmonics and voltage harmonics – an analysis in the MP.

	U5	U7	U11	U13	U17	U19	U23	U25
"before"	0,90	0,73	0,45	-0,16	0,06	0,34	0,23	0,14
"after"	-0,13	0,36	0,68	0,71	0,49	0,04	0,79	0,71

	U1	U3	U35	U37	U47	U49
"before"	-0,87	-0,15	-0,12	0,01	0,03	0,07
"after"	-0,70	0,02	0,17	0,20	0,18	0,23

A correlation analysis has been conducted in reference to the total current (RMS value) and its harmonics. One can assume that a more detailed description of these phenomena is possible thanks to the correlation using individual current harmonics. The approximation error has been calculated for particular pairs of data. What results from them is that for lower order harmonics (up to 17th) using the RMS value and current harmonics do not vary significantly.

Values of the correlation ratio for the characteristic harmonics of the 6-pulse and 12-pulse rectifiers have high values (over 0,5). Voltage harmonics are strongly dependent on the current harmonics.

Low value correlations between current and voltage of high values are significant. Such connection suggests that a particular harmonics occur in the electrical system voltage power supply. This is a consequence of voltage deformation not resulting from the current taken by a 6-pulse or 12-pulse rectifier. Such cases take place for harmonics of the 35th, 37th, 47th ,49th order.

3.2 Power quality parameters not included in the PN 50160

The parameters mentioned in the 3.1 paragraph mostly result from the value and character of the current taken by the traction power supply system.

- RMS value of the driving current – as a result of the conducted modernization, a decrease in the value of the current taken from the power system has taken place (increased energy recovery level between the units). Figure 8 presents a graph of the current RMS value "before" and "after" in a chosen power supply phase.
- harmonics of the driving current – as a result of the conducted modernization there has been a decrease in the amount of harmonics in the driving current. Figure 7 presents a harmonic spectrum of a single-phase current for the "before" and "after" periods (without 1st harmonic).
- electrical energy value – as a result of the modernization and an increase in energy recovery level by the vehicles, there has been a decrease in value of the taken current and, consequently, power. As the consequence of reducing the power factor tgφ from 0,33 "before" to 0,18 "after" is the reduction of reactive power consumption. The reactive energy consumption was decreased after modernization by 35% in relation to its consumption before modernization. Similarly, active energy consumption decreased by 25% in relation to its consumption before modernization. Decreasing the consumption of active and reactive energy from the electric power system by a modernized rectifier set is the most important argument confirming the rightness of the actions taken.
- increases, dips, blackouts and transients of voltage – because of a high level of randomness of these phenomena, measurements one day "before" and one day "after" do not result in a sufficient amount of data. In the measurement period both "before" and "after" single occurrences of a dip and a transient took place.

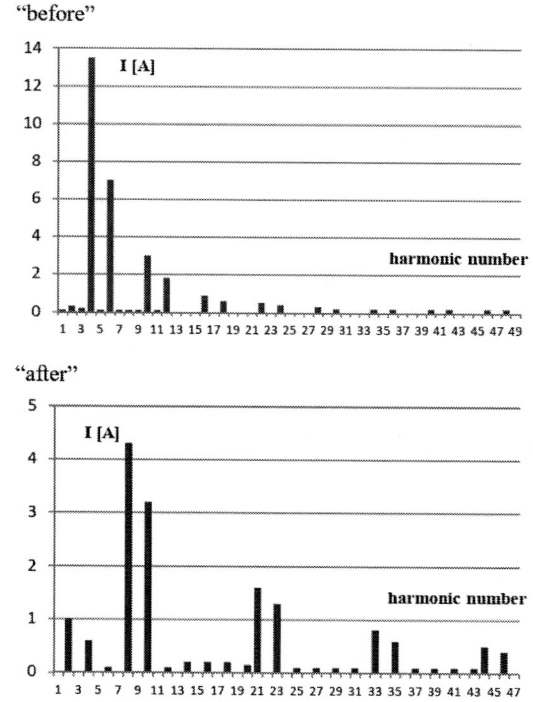

Fig. 7. A spectrum of harmonics of the driving current (side 15kV) before and after the modernization (picture PQM-711).

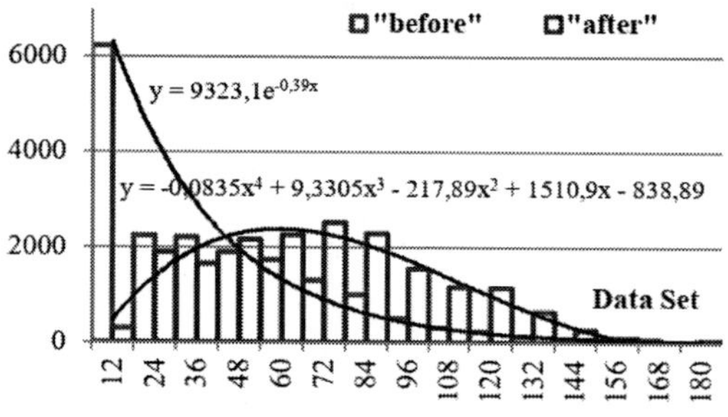

Fig. 8. RMS values of the driving current (side 15kV) before and after the modernization (a single-phase power supply).

4 Conclusions

This article has presented research results of the electrical energy parameters taken from the power system before and after the modernization of the traction substation of public transport. It has shown only the most important parameters that help answering a question about advantages of the conducted modernization. The aim of the modernization was: using in the traction traffic a railway rolling stock with a recovery energy function, minimization of the influence of the DC traction system on the electric power systems, minimization of the voltage drops in the power supply system and increasing the power given back to the traction units. Using energy recovery between the vehicles has given impressive results that has contributed to financial effects. In order to have a complete assessment of this effect one would have to compare expenditures incurred related to the investment and energy savings. Unfortunately, the authors have not been able to access a full financial documentation of the investment. Minimization of the influence of the DC system on the commercial power system is an issue difficult to assess. It is related to the fact that both before, and after the modernization electrical energy quality parameters had values significantly over the PN 50160 norm. One has to assume that the modernization has meaningfully raised the supply of these values. Thus, it is possible to increase a system load without any negative effects where it comes to the electrical energy quality. The situation is similar with regards to the drops of voltage and power given back to the traction units.

In the conclusion of the article, one should wish for further modernizations of the power supply systems in public transport.

References

1. A. Szeląg, *Increasing the Energy Efficiency of Railway Transport*, TTS - Rail Transport Technology, **12**, 12-18, (2008)
2. K. Kawałkowski, J. Młyńczak, Z. Olczykowski, and J. Wojciechowski, *A Case Analysis of Electrical Energy Recovery in Public Transport*, Advances in Intelligent Systems and Computing, **631**, 133-143, (2017)
3. S. Hamacek, M. Bartlomiejczyk, R. Hrbac, S. Misak, and V. Styskala, *Energy recovery effectiveness in trolleybus transport*, Electric Power Systems Research, **112**, 1-11, (2014)
4. L. Łukasik, T. Ciszewski, and J. Wojciechowski, Power supply safety of railway Traffic control systems as a part of international transport safety, *Proceedings of the 16th International Scientific Conference Globalization and Its Socio-Economic Consequences*, Rajecke Teplice, Slovakia, part IV, 1212-1219, (2016)
5. J. Wojciechowski, European Electricity Clearing Systems, *Proceedings of the 17th International Scientific Conference Globalization and Its Socio-Economic Consequences*, Rajecke Teplice, Slovakia, part VI, 3005-3012, (2017)
6. J. Altus, M. Novak, A. Otcenasova, M. Pokorny, and A. Szeląg, *Quality parameters of electricity supplied to electric railways*, Scientific Letters of the University of Žilina-Communications, 2-3, (2001)
7. L. Battistelli, P. Caramia, G. Carpinelli, and D. Proto, *Power quality disturbances due to interaction between AC and DC traction systems*, Power Electronics, Machines and Drives, (2004)
8. L. Mierzejewski, A. Szeląg, and P. Jankowski, *Electrical energy quality studies in DC electric traction systems for different ways of connection traction substation to power system*, Advances in Electrical and Electronic Engineering, **2**, (2005)

9. A. Szeląg and T. Maciołek, *A 3 kV DC electric traction system modernisation for increased speed and trains power demand – problems of analysis and synthesis*, Przegląd Elektrotechniczny, **R.89 3a**, 21-28, (2013)
10. A. Szeląg and M. Patoka, Some aspects of impact analysis of a planned new 25kV AC railway lines system on the existing 3 kV DC railway system in a traction supply transition zone, *SPEEDAM 2014, Int. Symposium*, Ischia, Italy, (2014)
11. A. Horn, R.H. Wilkinson, and J.H.R. Enslin, Evaluation of converter topologies for improved power quality in DC traction substations, *Proceedings of the IEEE International Symposium On Industrial Electronics*, **1 and 2**, 802-807, (1996)
12. M. Bartłomiejczyk and M. Połom, *Napięcie sieci trakcyjnej jako wyznacznik możliwości zwiększenia odzysku energii*, TTS - Rail Transport Technology, **4**, 42-46, (2013)
13. A. Capasso, R. Lamedica, A. Ruvio, M. Ceraolo, and G. Lutzemberger, Modelling and simulation of electric urban transportation systems with energy storage, *16th International Conference on Environment and Electrical Engineering* (EEEIC), IEEE Press, (2016)
14. A. Szeląg, *Railway electric traction in Poland*, TTS - Rail Transport Technology (Special English Edition Innotrans-Berlin/200), (2004)
15. V.G. Synchenko D.O. Bosiy, and E.M. Kosarev, *Improving the quality of voltage in the system of traction power supply of direct current,* Archives of Transport, **35**(3), 63–70, (2015)
16. A. Mariscotti, Statistical evaluation of measured voltage spectra in DC railways, *Proceedings of IMEKO*, (2010)
17. A. Capasso, G. Guidi-Buffarini, V. Morelli, L. Mierzejewski, A. Szeląg, and A. Wach, *Potenziamento del sistema di alimentazione della linea ferroviaria VarsaviaKunowice-Berlin. Mostra Convegno*, La Tecnologia del Transporto su Ferro e L'orientamento al. Mercato, 122-130, (1998)
18. W. Jefimowski, *Simulation research of the influence of the train traffic situations on the rail potential in the power supply system 3 kV DC,* Prace Naukowe Politechniki Warszawskiej, Transport, **111**, 203-213, (2016) [In Polish]
19. V.G. Kuznetsov, O.I. Sablin, A.V. Chornaya, *Improvement of the regenerating energy accounting system on the direct current railways*, Archives of Transport, **36**(4), 35-42, (2015)
20. J. Arrillaga, B. Smith, N. Watson and A. Wood, *Power system harmonics analysis*, John Wiley & Sons, (1997)
21. Y. Baghzouz, R.F. Burch, A. Capasso, A. Cavallini, A.E. Emanuel, M. Halpin, A. Imece, A. Ludbrook, G. Montanari, K.J. Olejniczak, P. Ribeiro, S. RiosMarcuello, L. Tang, R. Thallam, and P. Verde, *Probabilistic Aspects Task Force of the Harmonics*, Time-Varying Harmonics: Part I, IEEE Transactions on Power Delivery, **13/3**, (1998)
22. L. Battistelli, P. Caramia, G. Carpinelli, and D. Proto, *Power quality disturbances due to interaction between AC and DC traction systems*, Power Electronics, Machines and Drives, (2004)
23. PN-EN 50160:2014 Parameters of the supply voltage in the public power networks. Warsaw, Polish Committee of Standardizatiom, (2014) [In Polish]

Application of supercapacitors in electric traction storage systems

Miroslaw Lewandowski[1,*], *Marek* Orzyłowski[2], and *Maciej* Wieczorek[1]

[1] Warsaw University of Technology, Institute of Electrical Power Engineering, Electric Traction Division, ul. Koszykowa 75, 00-662, Warsaw, Poland
[2]University of Social Sciences, Information Technology Institute, ul. Sienkiewicza 9, 90-113, Łódź, Poland

Abstract. Energy storage systems (ESS) are based on electrochemical batteries and supercapacitors (SC). SCs have a lower energy density compared to batteries, but have an advantage over them in power density. The combined use of the two in a hybrid ESS (HESS) gives both high energy density and high power density of ESSs. Fractional-order impedance describes the SC dynamics better than commonly used integer order impedance. Such a description can be based on the dielectric relaxation equations e.g. Cole-Cole equation used in this work. The paper compares results of the impedance approximation of a number of SC's with various fractional and integer models, on the basis of which the time responses are analysed. The methods of capacitance and ESR measurement recommended by the IEC standard are examined. They are compared with the methods used in practice by SC producers. The problem of energy losses in ESS with respect to charging/discharging current pulses is presented. A proper ESS designing requires an accurate approximation of the SC impedance over the frequency band in which the dominant part of the power spectrum of current pulses is located.

1 Introduction

The braking energy of electric vehicles can be to a high degree retrieved and used again. In the case of traction vehicles, it can be transferred through a network to drive other vehicles. However, for full energy retrieval the energy storage systems (ESS) must be used. Numerous publications concern this topic, for example [1] or items listed in the bibliography of the work [2].

ESSs collect energy in electrochemical batteries and supercapacitors (SC), which have a lower energy density compared to that of batteries, but have an advantage over them regarding the power density. The combined use of the two makes it possible to achieve both high energy density and high power density of ESS. This is very important while receiving pulse energy during braking and giving it back during the acceleration of the vehicle.

When using SCs, it is necessary to take into account their specific dynamic properties resulting from the principle of their operation and construction. This article deals with this issue in relation to the ESS. It discusses the practical use of the descriptions of SCs dynamics, presented by different manufacturers and describes the problems of dynamic load of SC also.

2 Energy Storage Systems

Systems using both electrochemical and SC batteries are referred to as hybrid energy storage systems (HESS). An exemplary structure of the HESS system is shown in Fig. 1 [3]. Other HESS structures are described, for example, in [4].

ESS efficiency depends on losses in electrochemical batteries, SCs and DC/DC converters [5]. For example, [6] considers a stationary ESS that was located at the end of the tramline and was based solely on SCs (Fig. 2). It had the task of taking over and delivering energy up to 10 MJ (2.8kWh), necessary to speed up a tram. Losses in the inverters can be divided into the losses in chokes and those in the switching elements. In the system of Fig. 2, the copper and the core losses of the choke L are represented by the equivalent resistance R_L. The losses in transistors T_1 and T_2 and diodes D_1 and D_2 can be divided into conduction losses and switching losses. Analysis showed that the losses in these converter elements and those in the SC are comparable. It has also been estimated that total losses in SCs and DC/DC converters are generally lower than 10% of stored energy.

* Corresponding author: miroslaw.lewandowski@ee.pw.edu.pl

https://doi.org/10.1051/matecconf/201818002002

The mass and dimensions of the ESSs are affected by the used batteries, SCs and chokes. In the stationary ESS located at the end of a tram line [6] the value of stored energy required the use of SC battery made up of e.g. 20 pieces of 125 V Heavy Transportation Modules (Maxwell) with a unit capacitance of 63 F and a weight of 61 kg, i.e. of the total mass of more than 1200 kg.

In this case, the mass of the choke in the DC/DC converter depends on the voltage of used SC battery and in effect depends on the current charging/discharging the battery. It would be 100 kg - 200 kg depending on the parameters mentioned.

In this case, the mass of the choke in the DC/DC converter depends on the voltage of used SC battery and in effect depends on the current charging/discharging the battery. It would be 100 kg - 200 kg depending on the parameters mentioned. In the case of stationary ESS the weight does not much matter, although there is a problem of transmission losses. On the other hand, location of ESS on the vehicle increases the weight to be moved on and associated with it power consumption.

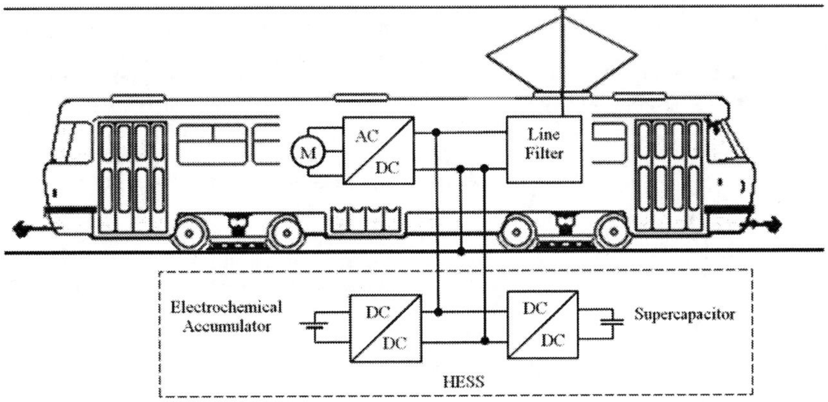

Fig. 1. Tramcar traction drive with hybrid energy storage system (parallel structure).

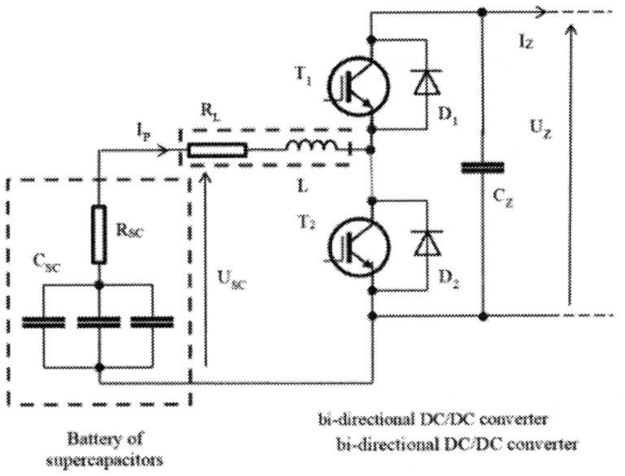

Fig. 2. Simplified scheme of ESS for the tram line, including SC battery and bi-directional DC/DC converter.

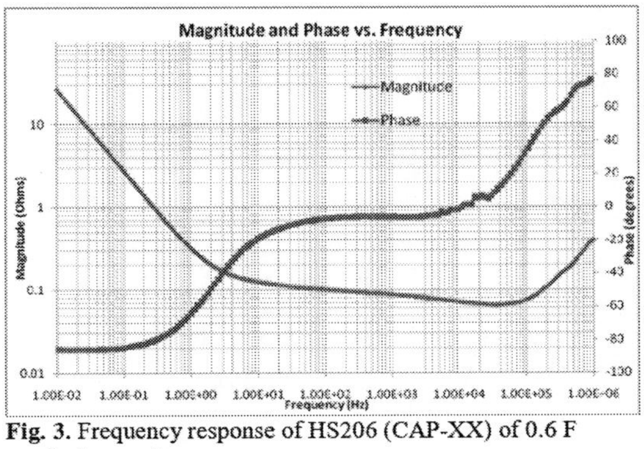

Fig. 3. Frequency response of HS206 (CAP-XX) of 0.6 F nominal capacitance.

3 Supercapacitor models and parameters

The specific properties of SCs result from the principle of their operation, which consists in the accumulation of electrical charge in the double electrical layer. This layer is formed by ions adsorbed on the surface of the porous electrodes. The very large surface of these electrodes allows for large capacities. An exemplary SC impedance graph over a wide frequency range is shown in Fig. 3. It follows from the plot that the capacitive nature of SC impedance exists in the range of lower frequencies. Above this range, the impedance of SC becomes resistive. The highest frequency range, in which the impedance is of inductive character, is of no significance for the matters considered here.

ESS designing, including the use of SC modules and energy flow control, requires the use of an accurate SC impedance model. Commonly used dynamics descriptions use transfer functions with integer exponents. The impedance of SC in this case has the general form

$$Z_I(s) = \frac{U(s)}{I(s)} = \frac{b_m s^m + b_{m-1} s^{m-1} \ldots b_1 s + b_0}{a_n s^n + a_{n-1} s^{n-1} + \cdots a_1 s + a_0} \quad (1)$$

Reducing the number of coefficients of this model while increasing the accuracy can be achieved by replacing this description with a fractional description [7]. The fractional description in this case does not have a fully unambiguous physical interpretation, and the parameter values are determined experimentally. However, it should be taken into account that mathematical models only approximate some important features of systems and phenomena and serve for an effective solution to a given problem.

At the description by fractional differential equations, the SC impedance has the form

$$Z_F(s) = \frac{b_m s^{\beta_m} + b_{m-1}{m-1} + \cdots + b_1 s^{\beta_1} + b_0 s^{\beta_0}}{a_n s^{\alpha_n} + a_{n-1} s^{\alpha_{n-1}} + \cdots + a_1 s^{\beta_1} + a_0 s^{\alpha_0}} \quad (2)$$

where exponents αi and βi are real numbers. Description (2) may be based on the equations of dielectric relaxation in SC, such as the Cole-Cole equation of permittivity [8]

$$\varepsilon_{CC}(j\omega) = \varepsilon_\infty + \frac{\varepsilon_s - \varepsilon_\infty}{1 + (j\omega T)^\delta} \quad 0 \leq \delta < 1. \quad (3)$$

or that of the Cole-Davidson's [9].

$$\varepsilon_{CD}(j\omega) = \varepsilon_\infty + \frac{\varepsilon_s - \varepsilon_\infty}{(1 + j\omega T)^\gamma} \quad 0 \leq \gamma < 1, \quad (4)$$

The impedance description of ZCC(s), using the equation (3), is presented, for example, in [10]. The starting point for this was the equivalent circuit of a real capacitor in the form shown in Fig. 4.

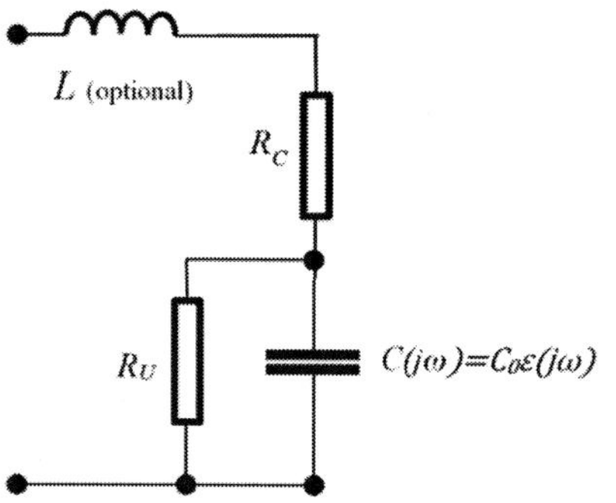

Fig. 4. An equivalent circuit of a supercapacitor.

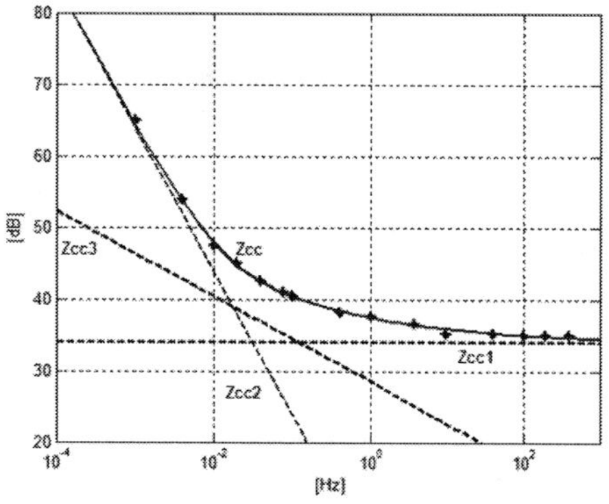

Fig. 5. Component moduli of impedances (6) of 0.1 F supercapacitor.

By omitting the inductance and after some simplification [12] this impedance can be presented as

$$Z_{cc}(s) = \frac{\left(1 + \frac{R_c}{R_u}\right) + s^\delta \left(1 + \frac{R_c}{R_u}\right) T^\delta + s R_c C}{\frac{1}{R_u} + sC} =$$

$$Z_{cc1}(s) + Z_{cc2}(s) + Z_{cc3}(s) \quad (5)$$

where

$$Z_{cc1}(s) = R_c \quad (6a)$$

$$Z_{cc2}(s) = \frac{R_u}{1 + s R_u C} \quad (6b)$$

$$Z_{cc3}(s) = \frac{s^\partial T^\partial R_u}{1 + s R_u C} \quad (6c)$$

To perform the frequency analysis, the Laplace transform should be replaced by its special case, which is the Fourier transform. This is done by substituting s = ω. The graph in Fig. 5 shows the approximation of the real frequency characteristic of the SC impedance by the fractional order model based on Cole-Cole equation. There are shown measured points marked by asterisks and the asymptotes connected with (6a)-(6c) SC impedance components.

The approximation of the real SC frequency response assumed by the authors was based on the minimization of the performance index of the form

$$J_f = \sqrt{\frac{1}{N-1} \sum_{i=1}^{N} \left(\frac{|Z_{CC}(j\omega_i) - Z_P(j\omega_i)|}{|Z_P(j\omega_i)|} \right)^2} \quad (7)$$

where

- $Z_{CC}(j\omega)$ is the impedance approximating the real impedance of SC,
- $Z_P(j\omega)$ is the measured complex impedance of SC,
- ω_i is angular frequency for i^{th} measuring point.

This performance index can be treated as the standard deviation of the approximation error. The values of J_f performance indices for various SC models are collected in Table 1.

The SC impedance can be used to simulate the SC pulse responses, for example the voltage response to a square current wave excitation. The results of such simulations for a 2 Hz frequency, compared to real SC responses in the form of relative approximation errors, are also included in Table 1.

Table 1. Comparison of approximation accuracy of frequency and time responses of supercapacitors.

Capacity	Standard deviation of approximation by model							
	2nd order integer		3rd order integer		Cole-Cole		Cole-Davidson	
Response	freq.	time	freq.	time	freq.	time	freq.	time
0.047 F	9.8%	11.2%	5.3%	6.3%	5.1%	5.1%	3.5%	2.7%
0.1 F	10.5%	16.4%	4.4%	4.9%	4.8%	2.3%	3.8%	2.7%
0.33 F	16.3%	22.6%	7.8%	7.3%	7.8%	2.6%	5.4%	3.4%
0.6 F	13.7%	11.4%	10.0%	8.8%	10.4%	9.9%	14.0%	3.9%
100 F	8.0%	4.4%	5.8%	5.8%	6.2%	6.2%	6.0%	6.0%
2700 F	3.2%	2.4%	1.9%	2.2%	3.4%	5.4%	1.9%	3.1%

The results presented in Table 1 show that the Cole-Cole and Cole-Davidson models provide a similar level of approximation accuracy, with the Cole-Cole model being directly applicable to the standard impedance (2) determination. In addition, these models, with fewer coefficients, are more accurate than the 3rd order integer model.

A certain problem concerns the interpretation of the SC catalogue data of nominal capacity and nominal ESR. The reason is the suggested method of measurement.

IEC 62391-1: 2006 „Fixed electric double-layer capacitors for use in electronic equipment" standard [11] recommends the method that is based on the SC voltage waveform obtained during constant current discharging.

The methods for determining capacitance and internal resistance are illustrated in Fig. 6, as set forth in the description of the standard. In this drawing, the voltage vs. time graph is approximated by straight lines.

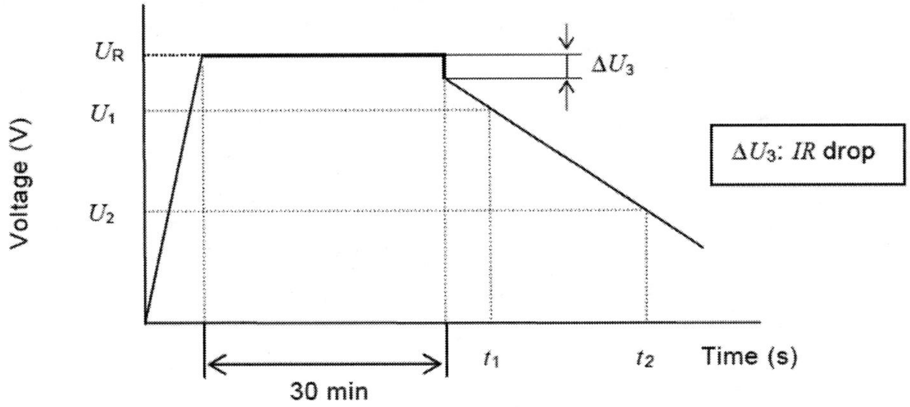

Fig. 6. Voltage on the capacitor terminals during the capacitance and ESR measurements according to IEC 62391-1: 2006 standard [11].

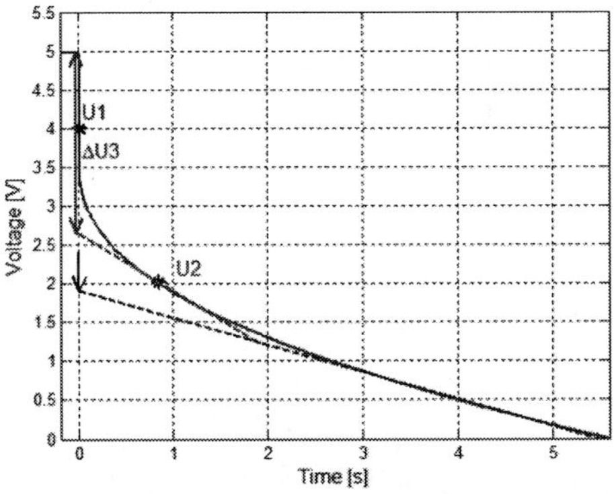

Fig. 7. The discharge of the supercapacitor of 1 F nominal capacitance with 200 mA current according to IEC Standard for Class 4. Measurement results: C=0.08 F, R=14 Ω.

The measurement takes place during constant current discharge I. The capacity is determined by the dependence

$$C = \frac{I \times (t_2 - t_1)}{U_1 - U_2} \tag{8}$$

while the internal resistance by the dependence

$$R = \frac{\Delta U_3}{I} \tag{9}$$

As mentioned, the standard assumes that the SC voltage response to a step current excitation can be well approximated by straight lines. Meanwhile, the component (6c) of impedance (5) can cause a significant nonlinearity of this response. This is illustrated in Fig. 7, showing such a response for one of SC types of 1 F capacitance. The response concerns the discharge current recommended for the class 4 standard described as "instantaneous power". This class corresponds to a pulse mode work of SC in ESS. The voltage levels and the linear approximation are marked on this graph, which are taken into account in (8) and (9). Direct application of this kind of result leads, among other things, to the determination of the capacitance by the order of magnitude lower than nominal. The standard allows to change the value of discharge current, but it makes it very imprecise.

As a result of the occurrence of such cases, producers do not fully apply IEC 62391-1: 2006 standard or do not use it at all. In addition, they supplement the technical data of the measurements according to their own methods. Table 2 shows, for example, the measurement methods of three SC module manufacturers that can be used to build ESS [Lewandowski et al. (2016)]. Because of these differences when designing SC systems, it is necessary to know the individual methods of measuring their parameters.

https://doi.org/10.1051/matecconf/201818002002

Table 2. Methods of determination of capacitance and ESR.

Producer	Method of capacitance measurement	Methods of ESR measurement	
Ioxus	IEC method, discharge current specified for Class 3	DC method: Discharge current specified for IEC Class 4	ΔU_3 drop (IEC method)
			Voltage drop after 10 ms
		AC method: Resistance for 1 kHz	
Skeleton	Method similar to IEC method, current specified for Class 3	DC method: Discharge current specified for IEC Class 3	ΔU_3 drop (IEC method)
			Voltage drop after 10 ms
		AC method: Resistance for 100 Hz	
Maxwell	Own method with sequence of current steps	Own method with sequence of current steps	

4 Energy losses under dynamic conditions

The analysis presented in [6] treats the processes of loading and unloading the SC battery in ESS in a static manner, that is, it does not take into account the duration of these processes and the transient states associated with them. This approach is sufficient to estimate the level of losses in individual ESS blocks. However, for the design of ESS it is desirable to analyze the dynamic conditions typical for the operation of these systems.

The graph in Fig. 3 shows that for lower frequencies the SC impedance is capacitive, while for higher - resistive. The boundary between the two usually corresponds to the time constant of dielectric relaxation and lies generally in the range of 0.01÷10 Hz. The value of those limits are important because of the frequency spectrum of the current pulses during braking and acceleration of the vehicle.

The pulse of unitary amplitude and the duration of $-T_0/2<t<T_0/2$ has the Fourier transform of the form

$$I_p(j\omega) = sinc\left(\frac{\omega T_0}{2}\right) \quad (10)$$

Let us consider a couple of pulses of the current $i_{pp}(t)$ charging and discharging SC that are shown in Fig. 8. These pulses have unitary amplitudes, the duration of T_0, and are shifted from the centre of the coordinate system by $\pm\tau$. The Fourier transform of the signal $i_{pp}(t)$ has the following form

$$I_{pp}(j\omega) = sinc\left(\frac{\omega T_0}{2}\right) e^{j\omega\tau} - sinc\left(\frac{\omega T_0}{2}\right) e^{-j\omega\tau} \quad (11)$$

Using Euler formula it can be written as

$$I_{pp}(j\omega) = sinc\left(\frac{\omega T_0}{2}\right)[e^{j\omega\tau} - e^{-j\omega\tau}] =$$

$$= -2j sinc\left(\frac{\omega T_0}{2}\right) sin(\omega\tau) \quad (12)$$

The energy spectrum of the signal i_p (t) depends on length T_0 of the braking and acceleration current pulses. Significant influence on this spectrum has also the time shift τ of these pulses. Let us consider the cases of different lags, shown in Fig. 9. Exemplary energy spectra normalized to pulse length T_0 are shown in Fig. 10.

It follows from Fig. 10 that the dominant parts of the power spectra of the pulses under consideration are located below the frequency

$$f_0 = \frac{1}{T_0} \quad (13)$$

Considering the energy loss analysis during storage in SC, it is important to take into account the frequency characteristic of SC in the range of the significant part of the current pulse spectrum. The analysis shows that shortening these pulses during braking or acceleration reduces the energy efficiency of ESS. This analysis will be presented in one of the following papers.

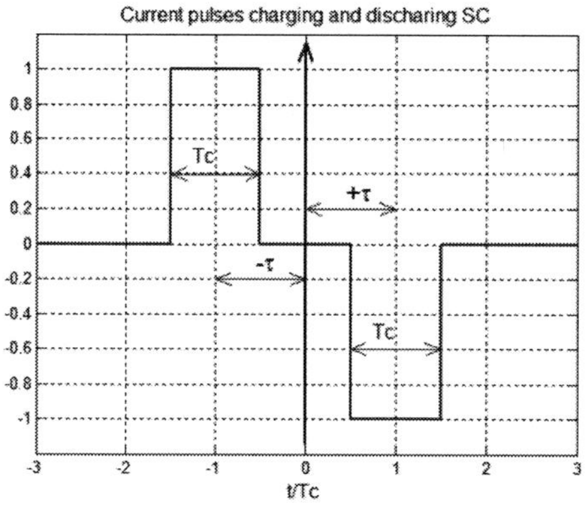

Fig.8. Graph of the normalized current charging and discharging SC.

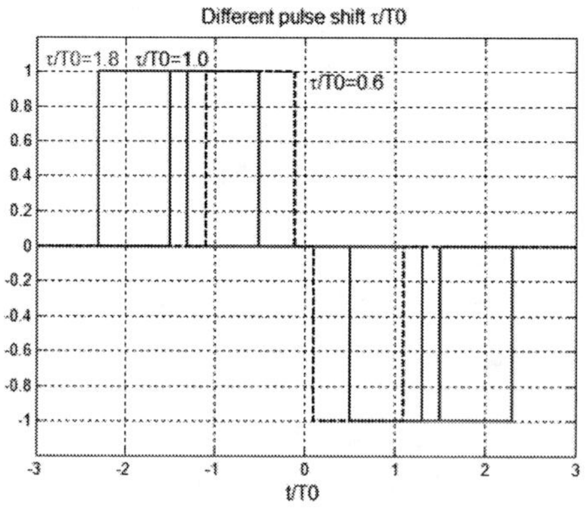

Fig. 9. Pulses of charging and discharging current shifted in time.

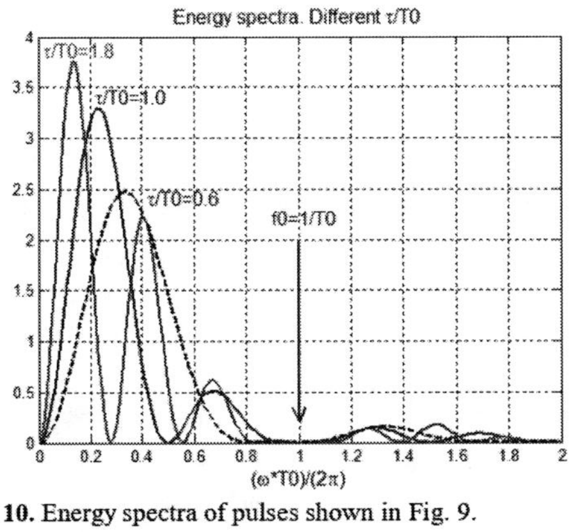

Fig. 10. Energy spectra of pulses shown in Fig. 9.

5 Conclusions

For proper electric vehicle ESS designing, the accurate SC model must be used. Such the model is the fractional order model. Among others this model allows one to analyze the measurement problems of SC parameters. It also allows one to analyze the influence of the dynamics of braking and acceleration of the electric vehicle on the energy efficiency of the ESS.

References

1. A. Szeląg, *The influence of the voltage in the 3kV DC traction network on the energy and traction parameters of supplied vehicles* (INW SPATIUM, 2013) ISBN: 978-83-62805-01-3 [in Polish]
2. M. Wieczorek, and M. Lewandowski, *A mathematical representation of an energy management strategy for hybrid energy storage system in electric vehicle and real time optimization using a genetic algorithm*, Applied Energy, **192**, 222–233 (2017) doi: 10.1016/j.apenergy.2017.02.022
3. G. Zaleskis, V. Brazis, and L. Latkovskis, *Estimation of Traction Drive Test Bench with Energy Storage System Operation in Regenerative Braking Mode*, Electrical, Control and Communication Engineering, **1**, 40-45 (2012) doi: 10.2478/v10314-012-0007-y
4. Ziyou Song, Heath Hofmann, Jianqiu Li, Xuebing Han, Xiaowu Zhang, and Minggao Ouyang, *A comparison study of different semi-active hybrid energy storage system topologies for electric vehicles*, Journal of Power Sources, **274**, 400-411 (2015) doi: 10.1016/j.jpowsour.2014.10.061en
5. J. Leuchter, P. Bauer, P. Bojda, and V. Rerucha, Bi-Directional DC-DC Converters for Supercapacitor Based Energy Buffer for Electrical Gen-Sets, *European Conference on Power Electronics and Applications, Aalborg* (2-5 Sept., 2007) ISBN 9789075815108
6. M. Lewandowski, M. Orzylowski, and A. Buze, *Power losses in energy storage system for* tram line, Przegląd Elektrotechniczny, **R.92**, Nr 12/2016, 289-295 (2016) doi:10.15199/48.2016.12.71 [in Polish]
7. C. A. Monje, Y. Chen, B. M. Vinagre, D. Xue, V. Feliu-Batlle, *Fractional-Order Systems and Controls: Fundamentals and Applications* (Springer, 2010) doi: 10.1007/978-1-84996-335-0
8. J-L. Déjardin, and J. Jadzyn, *Determination of the nonlinear dielectric increment in the Cole-Davidson model*, The Journal of Chemical Physics, **125**, 114503, (2006), doi: 10.1063/1.2346378
9. A. Dzieliński, G. Sarwas, and D. Sierociuk, *Comparison and Validation of Integer and Fractional Order Ultracapacitor Models*, Advances in Difference Equations, (Springer, 2011) doi:10.1186/1687-1847-2011-11
10. M. Lewandowski, and M. Orzyłowski, *Fractional-Order Models: The Case Study of the Supercapacitor Capacitance Measurement*, Bulletin of the Polish Academy of Sciences: Technical Sciences, **65**(4), 449-457 (2017) doi: 10.1515/bpasts-2017-0050
11. International Standard IEC 62391-1:2006, *Fixed electric double-layer capacitors for use in electronic equipment*, Part 1: Generic Application, IEC (2006)
12. M. Orzylowski, and M. Lewandowski, *Computer Modeling of Supercapacitor with Cole-Cole Relaxation Model*, Journal of Applied Computer Science Methods, **2**(4), 105-121 (2013) ISSN 1689-9636

https://doi.org/10.1051/matecconf/201818002003

Bilateral power supply of the traction network as a first stage of Smart Grid technology implementation in electric traction

Mikołaj Bartłomiejczyk[1,*]

[1]Gdansk University of Technology, Faculty of Electrical and Control Engineering, ul. Narutowicza 11/12 G, 80-233 Gdańsk, Poland

Abstract. Since 2001, trolleybus system in Gdynia has been involved in many activities related to the reduction of power consumption, both in terms of implementation and research and development. In PKT, in cooperation with SESTO company, started applications of Smart Grid technologies in supply network: the bilateral supply. The paper presents results of this this novel investment.

1 Introduction

The city of Gdynia, Poland, exploits an extensive, 50-km-long trolleybus network, which is serviced by 85 vehicles. This network is supplied from 10 traction substations with 600 V DC. The trolleybus system is operated by Przedsiębiorstwo Komunikacji Trolejbusowej Sp. z o.o. (PKT) which, for a number of years, has been working towards limiting energy consumption [1]. Vehicles with power electronic drive system are gradually introduced and, consequently, energy consumption has decreased by more than 20% over the past 10 years. In the years 2009-2012 PKT conducted comprehensive modernisation of the supply system, which resulted in the increase of its reliability and in the decrease of transmission losses [2]. PKT also carries out scientific and research activities connected with reducing energy consumption.

2 Bilateral supply system

The possibility to work in reverse mode, as generators of electric energy, is a characteristic feature of electric motors [3-6]. It allows to change kinetic energy of a vehicle into electric energy, which can then be returned to the traction supply system and consumed by other vehicles. However, when there is no possibility to receive recuperated energy we lose the energy which could potentially be recovered.

In a classic supply system each of the contact line sections is supplied solely from one substation, so the supply areas of individual substations are galvanic separated from each other. In the case of low traffic intensity or small substation areas, there may occur a situation where there is no vehicle which could absorb the recuperation energy from a braking vehicle [7]. However, it is possible to increase the change of finding a recipient of the generated energy by joining the supply area with the neighbouring substation, thus allowing for the flow of recuperation energy to the area of that neighbouring substation. Such a situation has been presented schematically in Fig. 1.

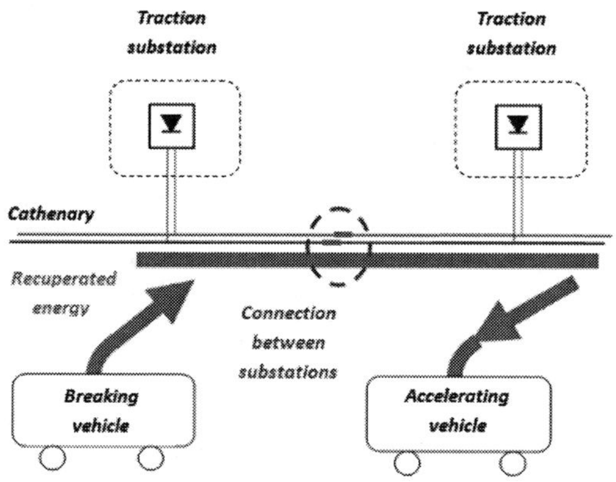

Fig. 1. Diagram of recuperation energy flow between braking vehicle and accelerating vehicle in bilateral contact line supply system.

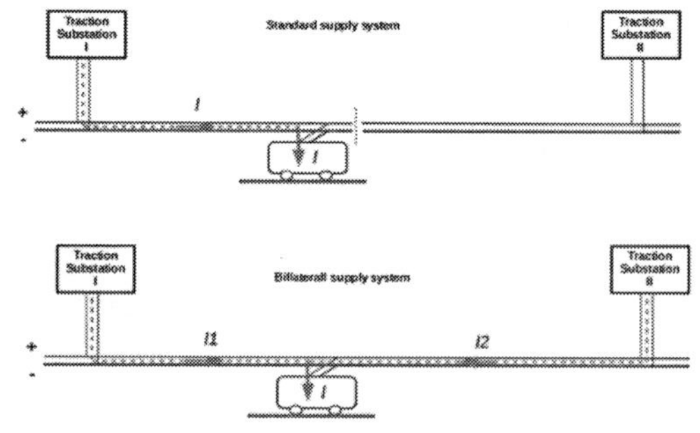

Fig. 2. Comparison of vehicle current distribution between supply traction substations.

* Corresponding author: mikolaj.bartlomiejczyk@pg.edu.pl

Bilateral supply also allows to reduce the voltage losses in overhead catenary, what is presented on Fig. 2. In bilateral supply the vehicle can be supplied in two ways, so the vehicle current I can be distributed between two smaller substation load currents I1 and I2, which reduces the load of catenary and voltage losses.

Bilateral power supply can also be applied for upgrading classic central supply system with large substations. Then it is possible to supply the traction section of the traction network in two places by means of cable feeders coming out of one substation. Such situations shown in Fig. 3.

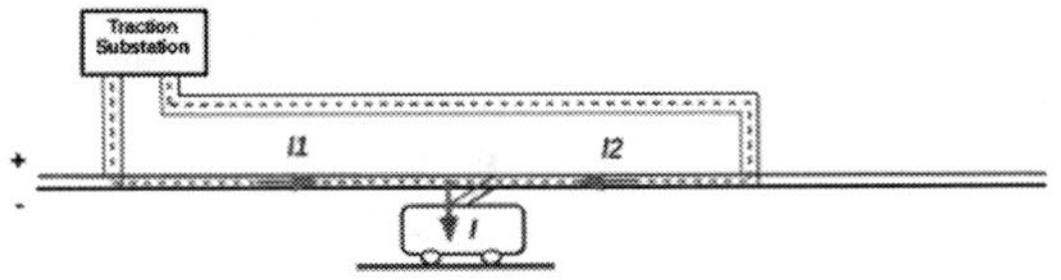

Fig. 3. Bilateral supply system with one substation.

3 Implementation of bilateral supply system in Gdynia's trolleybus network

In June 2016 PKT Gdynia, in cooperation with company SESTO, began practical implementation of Smart Grid technology within the exploited trolleybus system. This task was realised in the year 2016, within the Eliptic "Electrification of public transport in cities" H2020 project. The bilateral supply system was applied in 3 parts of Gdynia's trolleybus network (Fig. 6):

1) between substations Grabówek and Dworzec, on one of the most exploited trolleybus lines,

2) between two sections, which are both supplied from one substation Redłowo,

3) between substations Sopot and Sopot I, on interurban line with low traffic.

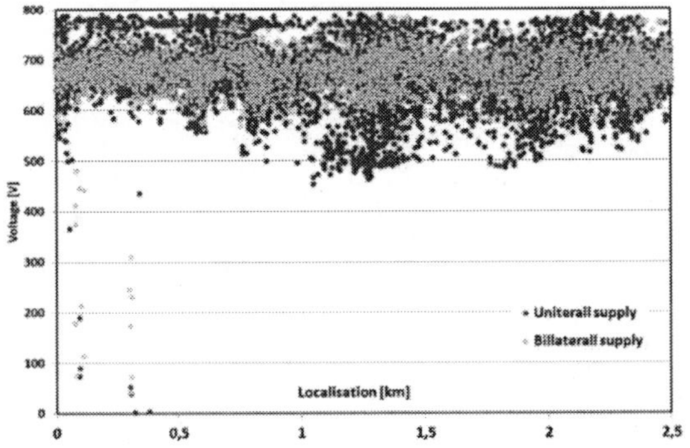

Fig. 4. Spread of the voltage values in area of substations Dworzec – Grabówek in catenary before and after implementation of bilateral supply.

These three places reflect different possible traffic conditions and power supply structures: line with high traffic and high voltage drops (1) which corresponds to situation on Fig. 2, two sections supplied from one substation (2) which corresponds to situation on Fig. 3 and line with low traffic and low usage of breaking energy recovery (3) which corresponds to situation on Fig. 1.

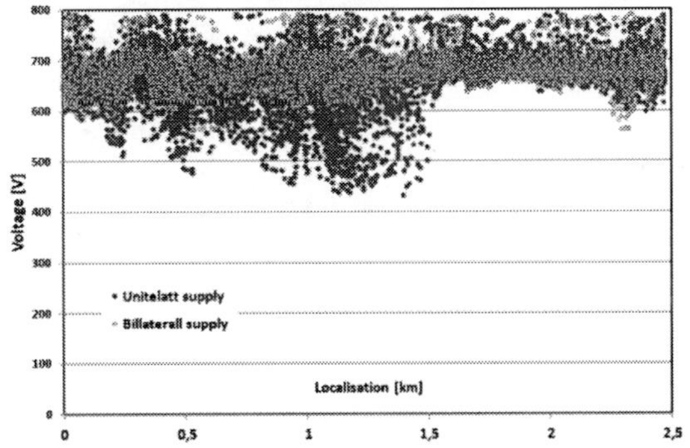

Fig. 5. Spread of the voltages value in area of substation Redłowo in catenary before and after implementation of bilateral supply.

The bilateral supply system was realised as a software integrated with the remote control system of trolleybus substation in Gdynia. The software was developed by SESTO company, which was a supplier of a control system equipment and software of all trolleybus substation in Gdynia. The bilateral supply system and substation protection devices were based only on the internet connection between substation, which was the main challenge of this system.

The introduction of bilateral supply between substation Sopot I and Sopot Reja resulted visible in the increase of recuperation effectiveness and slightly in the reduction of energy transmission losses (Table 1). In case of substations Dworzec – Grabówek and Redłowo, the implementation of bilateral supply significantly reduced the voltage drops (Table 1, Figs. 4 and 5). Voltage stability was improved, what is visible on graph of the spread of voltage values (Figs. 4 and 5). The analysis was made on the basis of the vehicle data loggers registrations. Registrations consist values of electrical parameters of the vehicle and GPS localisation in each second. In case of substations Dworzec and Redłowo after implementation of bilateral supply system the spread of voltage points was narrowed from both side - lower and upper part. Reduction of the lower part means the reduction of voltage drops caused by voltage loses in catenary. The horizontal upper limit of this group means the no-load substation voltage, which corresponds to the situation where there is no receiver for recuperated energy. Consequently, reduction of the upper part means the better usage of recuperated energy.

https://doi.org/10.1051/matecconf/201818002003

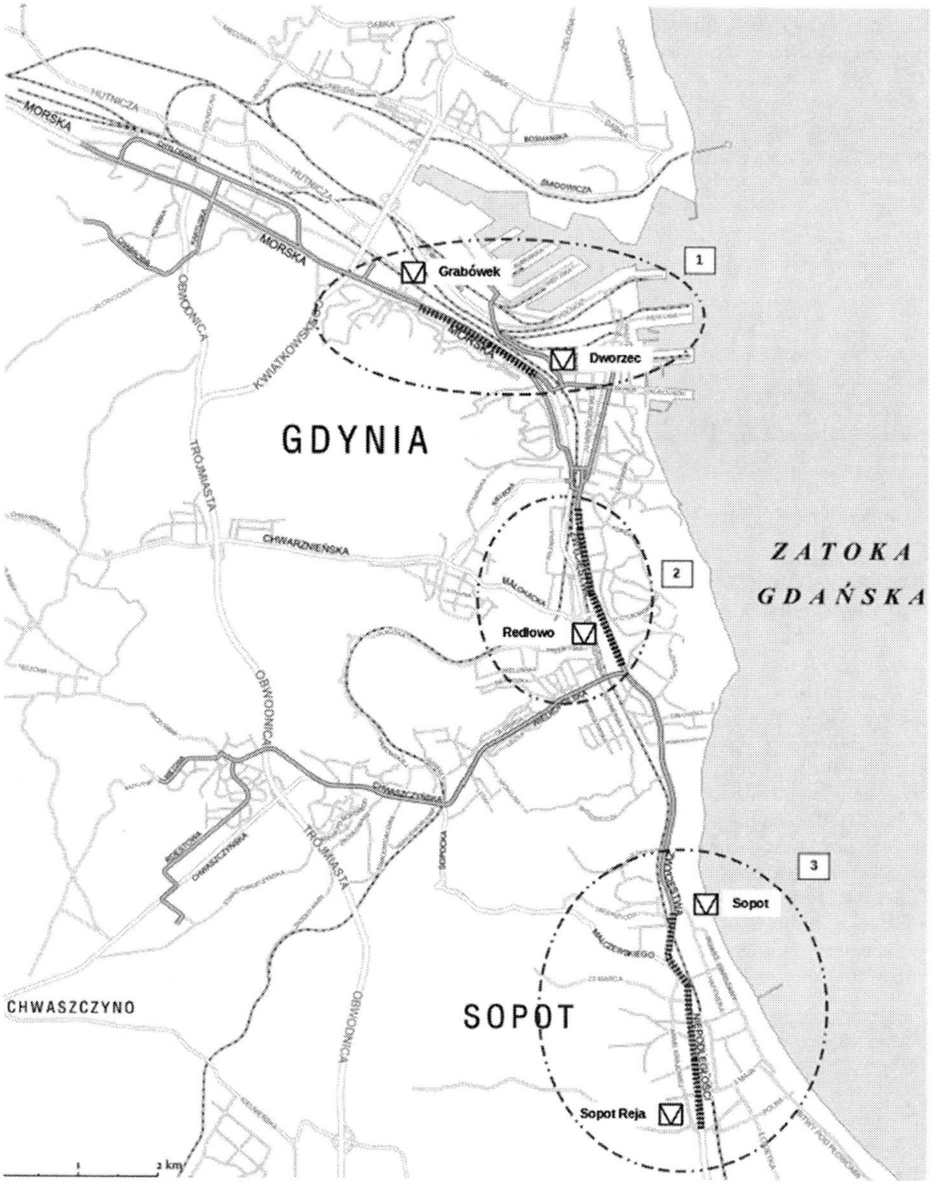

Fig. 6. The scheme of bilateral supply systems, section working in bilateral supply systems are pointed red and blue.

The coordination of several voltage levels plays as important. In order to ensure the energy recuperation between two vehicles the breaking vehicle should generate voltage higher than supply voltage of acceleration voltage. This voltage difference should be enough to overcome the resistance of catenary. In case of too big voltage difference between vehicles the voltage of breaking vehicle should exceed the threshold level of resistance breaking (770 V). Moreover, when recovered energy flows through busbars of traction substation (breaking and accelerating vehicle are localized on two different supply section of one traction substation), the generated voltage should exceed the no-load level of traction substation (720 V). This issue is expressed by voltage dispersion in function of localization.

Fig. 7 presents the voltage values in catenary during acceleration (blue points) and breaking (red points), localization values 0 km (left side of graph) equals to localization of substation Sopot, 2,5 km (right side) means substation Sopot Reja. The two main groups of points are visible: one with the voltage value around 770 V, which equals breaking recovery of vehicle and second with the value under 720 V, mean the acceleration of vehicle. The structure of supply system should enable recovered energy flow which is caused by the voltage differences between these group

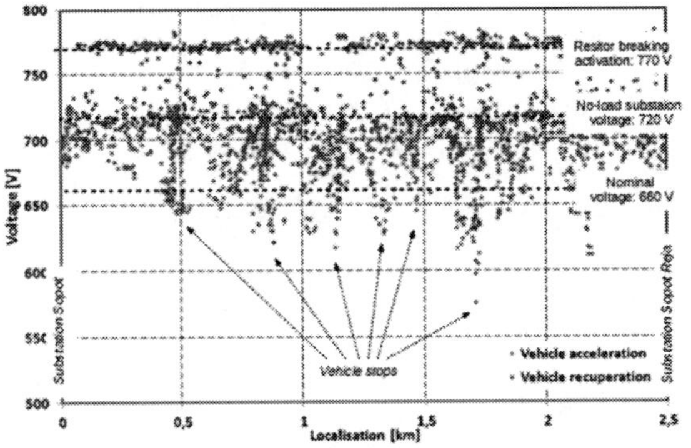

Fig. 7. Spread of the catenary voltage value in area of substations Sopot – Sopot Reja in function of localisation. Red points - vehicle braking (recuperations), blue points - vehicle acceleration. The groups of points, which mean voltage drops during acceleration near stops, are shown.

Figs. 8 and 10 show the time-current characteristics of power supplies before and after the bilateral power supply. Implementation of the bilateral supply resulted in reducing the maximal value of feeder's current up to 40%. Due to asymmetric structure of supply system, the current reduction of feeders working in parallel hasn't the equal value. An important issue relating to the implementation of bilateral power supply in traction supply system is to ensure effective breaking capacity. In the case of long power sections supplied from two substations, there may be a situation where short-circuit current is too low for effective switching of a protection device. However, it is possible to provide adequate protection against short-circuits by means of time-currents. Fig. 10 shows the detection of short circuit current in the two-sided power supply on the example of the section. In Grabówek-Dworzec section, the minimum short-circuit current is 1100 A.

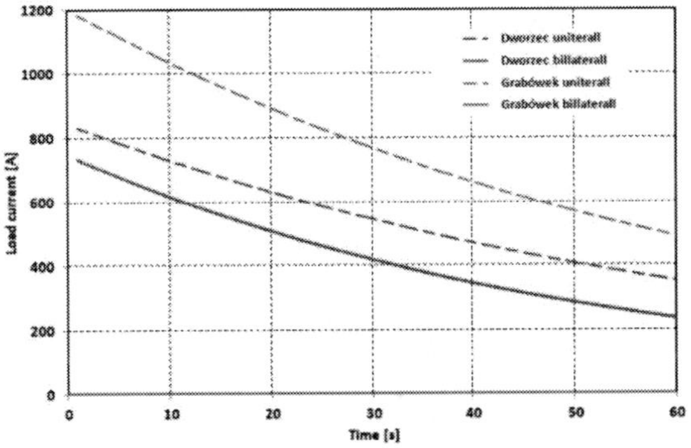

Fig. 8. Time - current characteristic of load feeders from substations Dworzec and Grabówek before and after installation bilateral supply.

Table. 1. Effects of introducing bilateral supply system.

	Unilateral supply	Bilateral supply
	Sopot – Sopot I	
Recovery effectiveness	2,5%	10%
Transmission Losses	4,6%	3,2%
Voltage drops <550 V	0,23%	0%
Voltage drops <500 V	0,01%	0%
	Dworzec – Grabówek	
Recovery effectiveness	19%	22%
Transmission losses	11,2%	7,1%
Voltage drops <550 V	0,93%	0,01%
Voltage drops <500 V	0,38%	0,09%
	Redłowo	
Recovery effectiveness	12,4%	13,3%
Transmission losses	6,8%	2,9%
Voltage drops <550 V	3,6%	0,05%
Voltage drops <500 V	1,1%	0,04%

Under normal conditions in spring months, the maximum load current is 765 A. However, due to the potential for increased energy consumption during winter or complicated traffic conditions, a 50% increase in load should be expected. This corresponds to a maximum current of 1148 A. The current protection may be activated by this load (exclamation mark on Fig. 10). However, based on the 10 seconds average of the protection, it is possible to get a reserve between the maximum load current and the minimum short-circuit current (arrow mark on Fig. 10). Consequently, it is possible to operate the short-circuit protection circuit correctly. It should also be noted that the time-current characteristics of the load (Figs. 8, 9) are well below the thermal characteristics of the power cables and traction networks designated for adiabatic heating.

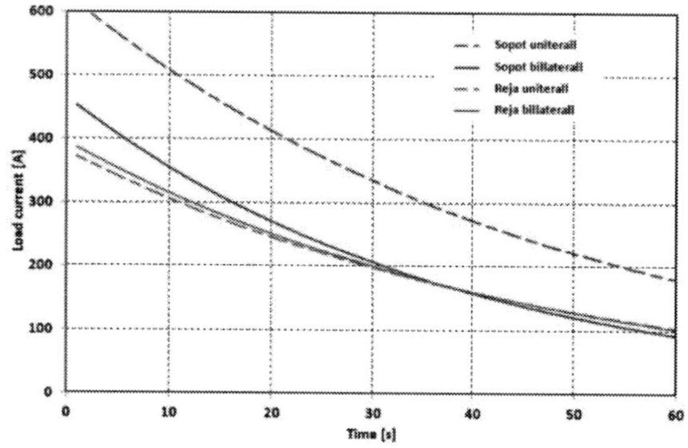

Fig. 9. Time - current characteristic of load feeders from substations Sopot and Reja before and after installation bilateral supply.

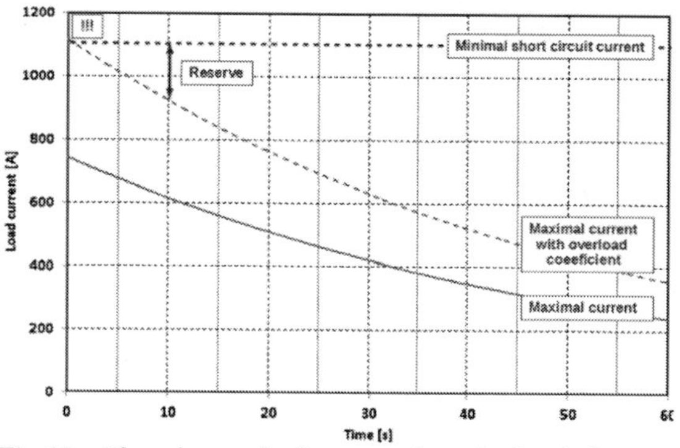

Fig. 10. The scheme of adjustment of overload and short-circuit protection system in bilateral supply.

4 Conclusions

Bilateral supply system allows to reach the energy savings and improvement of supply parameters with small investments. It can be recommended especially for weak supply systems or decentralized systems with many small substations, which makes the flow of recuperated energy difficult. Moreover, the bilateral supply allows to reduce voltage drops in supply network.

To increase the use of energy recovery, at first it is necessary to consider the possibility of reconfiguring the supply system, which will facilitate the flow of braking energy. In many situations, very good results can be achieved at a low cost. Small substation power supply areas galvanically isolated from the rest of the network should be avoided. Supply areas of such substations should be connected to create the largest area of recuperation energy flow. Therefore, it is recommended to introduce bilateral supply of the overhead line to allows for an increase in the utilization rate of recuperation and reduced transmission losses.

This paper has been elaborated in the framework of the project ELIPTIC "Electrification of public transport in cities" co-financed by the European Union's Horizon 2020 research and innovation programme under the grant agreement No 636012

References

1. Hebel, K., Wołek, M., *Perception of modes of public transport compared to travel behaviour of urban inhabitants in light of marketing research*, Scientific Journal of Silesian University of Technology. Series Transport, **92**, 65-75 (2016),
2. Bartłomiejczyk, M., Połom, M., *The impact of the overhead line's power supply system spatial differentiation on the energy consumption of trolleybus transport: planning and economic aspects*, Transport, **32**, 1–12 (2017)
3. Kacor, P., Marek, M., Palecek, J., *Analysis Of Magnetic Field Distribution Inside Tram-Car*, Proceedings Of The IV International Scientific Symposium On Electric Power Engineering - Elektroenergetika 2007, 128-129 (2007),
4. Jarzebowicz, L., Karwowski, K., Kulesza, W. J., *Sensorless algorithm for sustaining controllability of IPMSM drive in electric vehicle after resolver fault*, Control Engineering Practice, **58**, 117-126 (2017)
5. Jarzebowicz, L., Judek, S., *3D Machine Vision System for Inspection of Contact Strips in Railway Vehicle Current Collectors*, International Conference on Applied Electronics (AE), Pilsen, Czech Republic, 139–144 (2014).
6. Hamacek, S., Bartlomiejczyk, M., Hrbáč, R., Misak, S., Styskala, V., *Energy recovery effectiveness in trolleybus transport*, Electric Power Systems Research, **112**, 1–11 (2014)
7. Bartłomiejczyk, M., Połom, M., *Multiaspect measurement analysis of breaking energy recovery*, Energy Conversion and Management, **127**, 35-42 (2016)
8. Hrbáč, R., Kolář, V., Mlčák, T., 2015. *Distributed measurement system with GPS synchronisation and its use in electric traction*, Elektronika ir elektrotechnika, 21, 8-13 (2015)
9. Hrbáč, R., Mlčák, T., Kolář, V., *Improving Power Quality with the Use of a New Method of Serial Active Power Filter (SAPF) Control*, Elektronika ir Elektrotechnika, 23, 15-20 (2017)

 https://doi.org/10.1051/matecconf/201818002004

Case study of effective current estimation in main circuit of tram vehicle

Karol Chyła [1,*], *Marcin* Urbański [1], *Dominik* Sawuła [1], and *Stanisław* Rawicki [2]

[1] Solaris Tram Sp. z o.o, Poland
[2] Poznan University of Technology, Faculty of Electrical Engineering, Poland

Abstract. Currently the trams are experiencing the renaissance and more and more Polish cities decide to renew the fleet. New vehicles are usually characterized by high transport capacity, low floor, climate comfort, crash resistance, low energy consumption and low axle loads. Such requirements can only be achieved through optimization in many areas of the vehicle's construction. One of such tasks is the optimal selection of components of the main circuit installation and traction needs. The paper is oriented on the issue of engineering modelling of the effective values of currents flowing in the main circuit and the tramway traction needs circuits. Due to the non-periodic nature of the vehicle traffic loads in urban traffic, the optimal selection of the performance parameters helps in better selection of conductor cross sections, filters and safeguards. A better selection of these components means a reduction of its weight, which is extremely important for meet basic requirements of modern tram vehicles. The paper is intended to gather the experience of authors in this field of knowledge and compares several modelling strategies that the authors use during designing tram vehicles. The elaboration summarizes the results obtained using different modelling strategies, and shows correlation between effective value and selection of components.

1 Tram main circuit

Effective current (RMS value) which flows in main circuit, is one of basic parameter, which allows to choose the correct components for this circuit. Main circuit in tram vehicle is a part of its electric installation, which main function is to provide power supply from pantograph, so from catenary net, to main loads: Traction converter, Auxiliary converter, HVAC, and others which has supply direct from catenary voltage. This paper focuses mainly over this part of main circuit, which is supplying traction converter. The main circuit of simulated tram is presented in Fig. 1a and in Fig. 1b.

The simulated tram has one pantograph with positive potential, four motorized bogies with two motors each, and four independently supply of traction converters. All other loads are omitted because they are not important according to goal of this paper. Ground potential is reached via brush contacts assembled on axles in each motorized bogie.

2 Components assign

The most visible influence of the effective current could be observed during selection of fuses and wires cross section. That's why this paper focuses mostly on selection those components. Optimized selection of components of main circuit has to assure safety as well as reliable operations during long exploitation. Safety assurance in

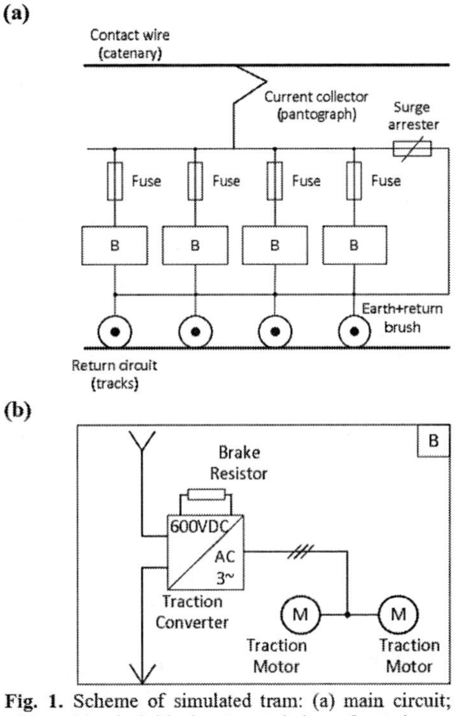

Fig. 1. Scheme of simulated tram: (a) main circuit; (b) one of four identical blocks B consisting of traction converter, braking resistor and two identical traction motors.

* Corresponding author: karol.chyla@solaristram.com

https://doi.org/10.1051/matecconf/201818002004

tram vehicle is mainly defined by standards like EN 60077-1 [3], EN 60077-2 [4], EN 60077-5 [5] EN 50355 [2], EN 50343[1], HD 60364 [6]. however, to ensure the long operation periods is related with the achievement of the best ratio of MTBF and/or the entire circuit MTTF main, and the lowest LCC costs.

In this paper it is present traction converter circuit fuse and traction converter cable selection process. On the first step of components selection to vehicle main circuit is fuse characteristic selection. Generally fuse characteristic "aR" or "gR" is most suitable for traction converter overcurrent protection. In next step we are defined the fuse minimal rated current value by the following equation:

$$I_n \geq I_{rated} \tag{1}$$

$$I_{rated} = \frac{I_B}{A} \tag{2}$$

I_B [A] – traction converter input RMS current;
I_{rated} [A] – fuse minimal rated current;
I_n [A] – fuse nominal current;
A=0,462 – reduction factor depending on: ambient temperature, ventilation air speed, operation mode.

For the most critical route section is necessary to achieve safe distance between traction converter input RMS current and fuse let-through current in specific time. Such verification is made according to the equation:

$$I_{TTC} \geq \frac{I_{Bn}}{B} \tag{3}$$

I_{TTC} [A] – fuse let-through current in specific time is defined according fuse time-current characteristic;
I_{Bn} [A] – traction converter input RMS current in specific time;
B=0,5 – fuse time-current characteristic reduction factor.

Second design phase it is traction converter supply cable selection. During cable selection is necessary to focus of many topics, but in this part we consider only selection in respect to traction converter load current.

When fuse nominal current value was defined we can select traction converter supply cable cross-section (CSA).

Minimal required value of traction converter supply cable cross-section is defined by equation (base on EN50343 [1]).

$$I_n \leq I_z \tag{4}$$

$$I_z = I_{cable} \times K \tag{5}$$

I_n [A] – fuse nominal current;
I_z [A] – maximal permissible current of traction converter supply cable;
I_{cable} [A] - nominal current of traction converter supply cable;
K=0,776 – reduction factor dependent on: ambient temperature, installation type, cable conductor temperature, operation mode.

After selection of traction converter fuse and traction converter power supply cable it is necessary to verify coordination between fuse and cable. Coordination verification is based on EN 50343 [1] standard.

Proper cable protection against overload is described by equation below:

$$1{,}6 \times I_n \leq 1{,}45 \times I_z \tag{6}$$

I_n [A] – fuse nominal current;
I_z [A] – maximal permissible current of traction converter supply cable.

Negative verification results force the changes of main circuit components or other design parameters for example higher class of electrical insulation. Many times it is necessary to use forced ventilation or even liquid cooling.

In some circumstances it is necessary to use additional thermal insulation which eliminates overheating of elements used during maintenance operations i.e. flap handle.

RMS current value also influences on many components not related to components of main circuit such as: hosing dimension, cooling method, thermal insulation solution, vehicle space arrangement.

As we can see the proper calculation of traction RMS current has high influence on safety and value of main circuit components LLC costs.

As we can see traction RMS current influences on many design areas.

The goal of below calculations is to present how traction RMS current has influence on main circuit components selection.

3 Simulation parametrization

In view of the fact that trams movement has non-periodic character, there has to be performed simulations of theoretical substitution movements scenarios, base on which, the reference effective value of current is estimated, and used for further components choosing.

Very interesting is fact that to create correct scenarios and assign right parameters to simulations, designer needs to firstly study infrastructure data like: numbers of his, average stop distance, factor of non-collision tracks, maximal and average speed, and general time schedules, all those data has influence on parameters selected to simulation.

For purpose of this paper there were performed four simulation scenarios:

Scenario 1 – phase of acceleration, phase of constant speed drive, phase of braking without recuperation to catenary (Fig. 2);

Scenario 2 – phase of acceleration, phase of constant speed drive, phase of braking with recuperation to catenary (Fig. 3);

Scenario 3 – phase of acceleration, phase of service braking with recuperation to catenary (Fig. 4);

Scenario 4 – phase of acceleration, phase of service braking with recuperation to catenary (Fig. 5).

For these scenarios 1,2,3 there were defined the same infrastructure and vehicle parameters :

- distance between stops is defined on 500 m;
- slope 1,5%;
- tram loaded on 2/3 of nominal load – 72200 kg.

The time schedule for scenario 1, 2 was defined base on maximal allowed tram speed – 50 km/h, drive time is 50 s and stop time is 20s.

Scenario 3 was defined as case where after maximal acceleration to maximal speed 50 km/h was performed maximal service braking with recuperation.

Scenario 4 was defined by parameters:

- distance between stops is defined on 500m
- slope 6%
- tram loaded on 2/3 of nominal load – 72200kg.

On purpose of choosing fuses and wires it was defined that according to simulated tram, that effective current for one traction converter is ¼ of traction current for whole vehicle. For simulation there was omitted current flows for auxiliaries and HVAC systems.

The assumed values allowed to model the traction current, allowing to complete the timetable on the reference section. The time of each driving phase is shown in Table 1.

Table 1. Time of phases for each scenario.

Scenario / Phase	1 Time [s]	2 Time [s]	3 Time [s]	4 Time [s]
Acceleration to 50km/h	21,23	21,23	21,23	39,8
Constant speed 50km/h	17,58	17,58	0	0
Braking from 50km/h to standstill	12,29*	12,29	12,29	10,64
Stop standstill	20	20	20	20
Travel time	71,1	71,1	53,52	70,44

4 Scenario simulation

4.1 Scenario 1

Fuse nominal current selection (I_n).based on RMS1 current characteristic (I_B)

$$I_n \geq I_{rated}$$

$$I_{rated} = \frac{I_B}{A} = \frac{145{,}05}{0{,}462} = 313{,}31\ [A]$$

$$I_n = 315\ [A]$$

I_B=145,05 [A] – traction converter input RMS current;

I_{rated} [A] – fuse minimal rated current;

I_n [A] – fuse nominal current;

A=0,462 – reduction factor depend on: ambient temperature, ventilation air speed, operation mode.

For each driving phase RMS2 current should be compared with fuse let-through to eliminate fuse melting

$$I_{TTC} \geq \frac{I_{B1}}{B}$$

$$1400\ [A] \geq \frac{304{,}393\ [A]}{0{,}5}$$

$$1400\ [A] \geq 608{,}786\ [A]$$

I_{TTC}=1400 [A] – fuse let-through current in 10,77 [s] is defined according fuse time-current characteristic;

I_{B1}=304,393 [A] - traction converter input RMS current – starting;

B=0,5 – fuse time-current characteristic reduction factor.

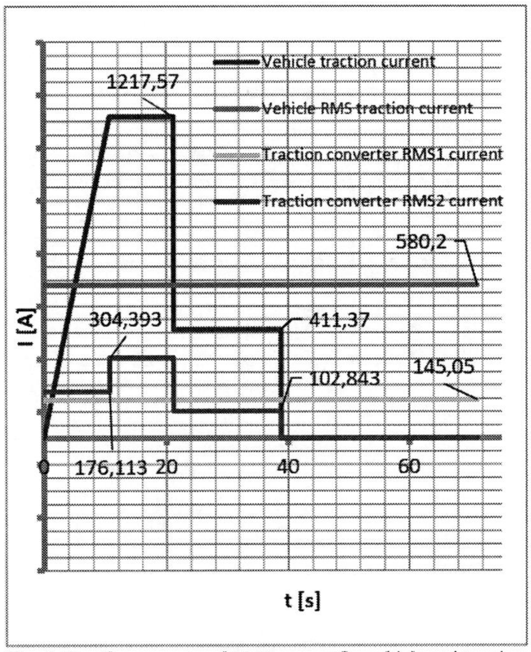

Fig. 2. Traction current and RMS current for vehicle and traction converter.

Cable selection is made in calculations below:

$$I_n \leq I_z$$

$$I_z = I_{cable} \times K$$

$$I_z = 507 \times 0{,}776 = 393{,}43[A]$$

$$315[A] \leq 393{,}43$$

I_n=315 [A] – fuse nominal current;

I_z [A] – Maximal permissible current of traction converter supply cable;
I_{cable}=507 [A] – Nominal current of traction converter supply cable 95 mm^2;
K=0,776 – Reduction factor depend on: ambient temperature, installation type, cable conductor temperature, operation mode.

Proper cable protection against overload is verified by the following calculation:

$$1{,}6 \times I_n \leq 1{,}45 \times I_z$$

$$1{,}6 \times 315 \leq 1{,}45 \times 393{,}43$$

$$504[A] \leq 570{,}47[A]$$

I_n=315 [A] – fuse nominal current;
I_z=393,43 [A] – maximal permissible current of traction converter supply cable.

4.2 Scenario 2

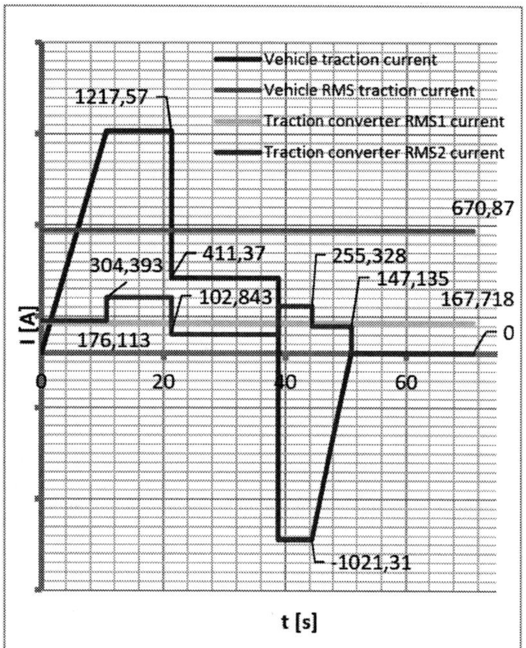

Fig. 3. Traction current and RMS current for vehicle and traction converter.

Fuse nominal current selection (I_n).based on RMS1 current characteristic (I_B).

$$I_n \geq I_{nmin}$$

$$I_{nmin} = \frac{I_B}{A} = \frac{167{,}72}{0{,}462} = 363{,}02\ [A]$$

$$I_n = 400\ [A]$$

I_B=167,02 [A] – traction converter input RMS current;
I_{rated} [A] – fuse minimal rated current;
I_n [A] – fuse nominal current;
A=0,462 – reduction factor depend on: ambient temperature, ventilation air speed, operation mode.

For each driving phase RMS2 current should be compared with fuse let-through to eliminate fuse melting:

$$I_{TTC} \geq \frac{I_{B1}}{B}$$

$$1750\ [A] \geq \frac{304{,}393\ [A]}{0{,}5}$$

$$1750\ [A] \geq 608{,}786\ [A]$$

I_{TTC}=1750 [A] – fuse let-through current in 10,77 [s] is defined according fuse time-current characteristic;
I_{B1}=304,393 [A] – traction converter input RMS current – starting;
B=0,5 – fuse time-current characteristic reduction factor.

Cable selection is made in calculations below:

$$I_n \leq I_z$$

$$I_z = I_{cable} \times K$$

$$I_z = 599 \times 0{,}776 = 464{,}82[A]$$

$$400[A] \leq 464{,}82$$

I_n=400 [A] – fuse nominal current
I_z [A] – maximal permissible current of traction converter supply cable
I_{cable}=599 [A] – nominal current of traction converter supply cable 120 mm^2;
K=0,776 – reduction factor depend on: ambient temperature, installation type, cable conductor temperature, operation mode.

Proper cable protection against overload is verified by calculation below:

$$1{,}6 \times I_n \leq 1{,}45 \times I_z$$

$$1{,}6 \times 400 \leq 1{,}45 \times 464{,}82$$

$$640[A] \leq 674[A]$$

I_n=400 [A] – fuse nominal current;
I_z=464,82 [A] – maximal permissible current of traction converter supply cable.

4.3 Scenario 3

Fuse nominal current selection (I_n).based on RMS1 current characteristic (I_B).

$$I_n \geq I_{nmin}$$

$$I_{nmin} = \frac{I_B}{A} = \frac{185{,}275}{0{,}462} = 401{,}01\ [A]$$

$$I_n = 450\ [A]$$

I_B=185,275 [A] – traction converter input RMS current;
I_{rated} [A] – fuse minimal rated current;
I_n [A] – fuse nominal current;
A=0,462 – feduction factor depend on: ambient temperature, ventilation air speed, operation mode.

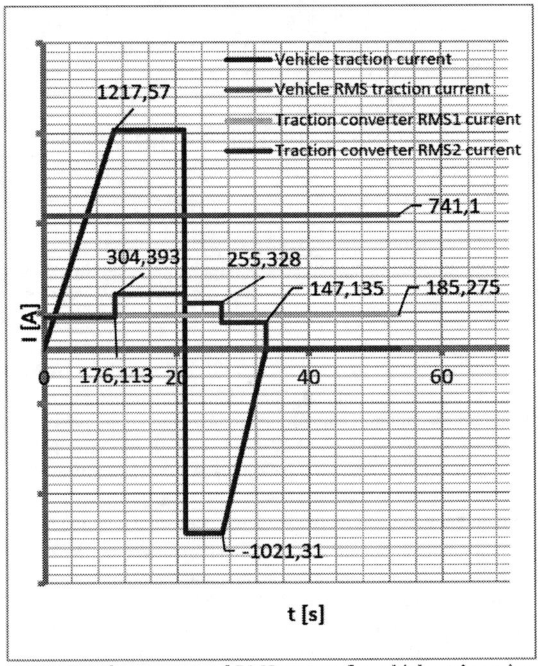

Fig. 4. Traction current and RMS current for vehicle and traction converter.

For each driving phase RMS2 current should be compared with fuse let-through to eliminate fuse melting:

$$I_{TTC} \geq \frac{I_{B1}}{B}$$

$$2800\ [A] \geq \frac{304{,}393\ [A]}{0{,}5}$$

$$2800\ [A] \geq 608{,}786\ [A]$$

I_{TTC}=2800 [A] – fuse let-through current in 10,77 [s] is defined according fuse time-current characteristic;
I_{B1}=304,393 [A] – traction converter input RMS current – starting;
B=0,5 – fuse time-current characteristic reduction factor.

Cable selection is made in calculations below:

$$I_n \leq I_z$$

$$I_z = I_{cable} \times K$$

$$I_z = 691 \times 0{,}776 = 536{,}21[A]$$

$$350[A] \leq 536{,}21[A]$$

I_n=350 [A] – fuse nominal current;
I_z [A] – maximal permissible current of traction converter supply cable;
I_{cable}=691 [A] – nominal current of traction converter supply cable 150mm^2;
K=0,776 – reduction factor depend on: ambient temperature, installation type, cable conductor temperature, operation mode.

Proper cable protection against overload is verified is calculation below

$$1{,}6 \times I_n \leq 1{,}45 \times I_z$$

$$1{,}6 \times 450 \leq 1{,}45 \times 536{,}21$$

$$720[A] \leq 777{,}5[A]$$

I_n=350 [A] – fuse nominal current;
I_z=536,21 [A] – maximal permissible current of traction converter supply cable.

4.4 Scenario 4

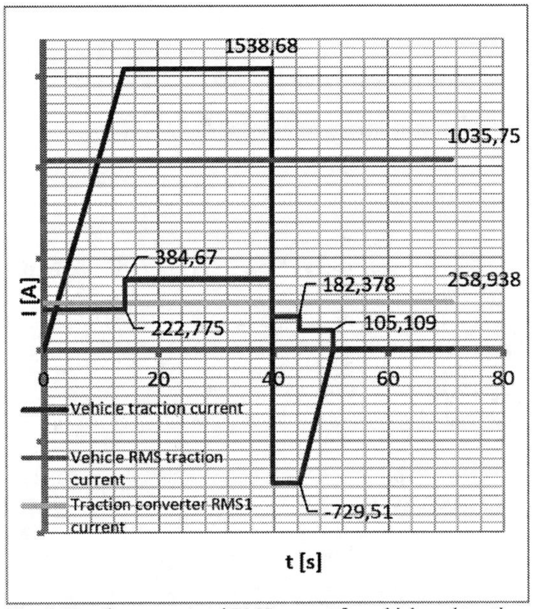

Fig. 5. Traction current and RMS current for vehicle and traction converter.

Fuse nominal current selection (I_n).based on RMS1 current characteristic (I_B)

$$I_n \geq I_{nmin}$$

$$I_{nmin} = \frac{I_B}{A} = \frac{258{,}94}{0{,}462} = 560{,}47\ [A]$$

$$I_n = 630\ [A]$$

I_B=258,94 [A] – traction converter input RMS current;
I_{rated} [A] – fuse minimal rated current;
I_n [A] – fuse nominal current;
A=0,462 – reduction factor depend on: ambient temperature, ventilation air speed, operation mode.

For each driving phase RMS2 current should be compared with fuse let-through to eliminate fuse melting:

$$I_{TTC} \geq \frac{I_{B1}}{B}$$

$$2800\ [A] \geq \frac{384{,}67\ [A]}{0{,}5}$$

$$2800\ [A] \geq 769{,}34\ [A]$$

I_{TTC}=2800 [A] – fuse let-through current in 26 [s] is defined according fuse time-current characteristic;
I_{B1}=384,67 [A] – traction converter input RMS current – starting;
B=0,5 – fuse time-current characteristic reduction factor.

Cable selection is made in calculations below:

$$I_n \leq I_z$$

$$I_z = I_{cable} \times K$$

$$I_z = 950 \times 0{,}776 = 737{,}2[A]$$

$$630[A] \leq 737{,}2[A]$$

I_n=630 [A] – fuse nominal current;
I_z [A] – maximal permissible current of traction converter supply cable;
I_{cable}=950 [A] – nominal current of traction converter supply cable 240mm^2;
K=0,776 – Reduction factor depend on: ambient temperature, installation type, cable conductor temperature, operation mode.

Proper cable protection against overload is verified is calculation below:

$$1{,}6 \times I_n \leq 1{,}45 \times I_z$$

$$1{,}6 \times 630 \leq 1{,}45 \times 737{,}2$$

$$1008[A] \leq 1069{,}94[A]$$

I_n=630 [A] – fuse nominal current;
I_z=737,2 [A] – maximal permissible current of traction converter supply cable.

5 Conclusions

Table 2 presents summary of scenario simulations and differences in selected components provided by differ approaches in simulations of effective current values.

Table 2. Summary of simulations.

Component parameter	Scenario 1	Scenario 2	Scenario 3	Scenario 4
Nominal current [A]	315	400	450	630
Fuse size *WxLxH* [mm]	60x206x60	60x206x60	73x206x73	73x206x73
Fuse weight (1szt) [kg]	1,34	1,34	1,98	1,98
Cable CSA [mm^2]	95	120	150	240
Cable diameter [mm]	17,3	19,6	21,9	29,9
Cable weight (250m) [kg]	242,5	312,5	387,5	612,5
Notes: 250 m is assumed length value of main circuit cable (power supply circuit and return circuit) for simulated tram vehicle.				

Scenario which could be used for further design is scenario 2 which should be the most realistic in terms of real tram operation estimation.

The conclusion is that it is very important at the beginning of project to agree necessary information with tram operator to be sure that simulation parametrization correctly estimate further tram operations. Such deep optimization has also negative effects design is assign to infrastructure of one operator, in future if tram operator would like to sold it to other city it need to be check if some adjustment has to be performed to fit with new scenarios. To omit such situation it would be good to develop common methodology to model effective currents in trams that's all tram vehicles will have common performance and be easy to redistribute after certain ears of operation.

References

1. EN 50343:2014 *Railway applications - Rolling stock - Rules for installation of cabling*
2. EN 50355:2013 *Railway applications - Railway rolling stock cables having special fire performance - Guide to use*
3. EN 60077-1:2003 *Railway applications - Electric equipment for rolling stock – Part 1: General service conditions and general rules*
4. EN 60077-2:2003 *Railway applications - Electric equipment for rolling stock - Part 2: Electrotechnical components - General rules*
5. EN 60077-5:2004 *Railway applications - Electrotechnical equipment for rolling stock – Part 5: Electrotechnical components - Rules for HV fuses*
6. HD 60364-4-43:2010 *Low-voltage electrical installations - Part 4-43: Protection for safety - Protection against overcurrent*
7. Team of Authors, *2007 The Fuse Manual-Ultra-rapid Fuses SIBA-, edited by Siba, Lünen*

MATEC Web of Conferences **180**, 02005 (2018) https://doi.org/10.1051/matecconf/201818002005
MET'2017

Case study of stationary energy storage device in a 3 kV DC traction system

Włodzimierz Jefimowski[1,*], and *Anatolii* Nikitenko[1]

[1]Warsaw University of Technology, Institute of Electric Power Engineering, Electric Traction Division, ul. Koszykowa 75, 00-662, Warsaw, Poland

Abstract. The paper presents the results of economic study of energy storage system (ESS) implemented in 3 kV DC power supply system. Two conceptions of ESS have been investigated: ESS with supercapacitor (SC) and hybrid ESS (HESS) with SC and LFP battery. The investigated locations of energy storage systems are considered among existing traction substations in two railway lines with different density of train operation. The considered aims of energy storage system implementation are decreasing of energy consumption by maximum regenerative energy utilization and reduction of peak 15-min power demand of traction substation. The paper presents a method of regenerative power estimation depending on the location of the considered ESS implementation point. Also the method of optimal location selection of ESS in terms of minimization of Simple Payback Time (SPBT) of investment is presented. Besides the influence of initial cost value as well as energy price on the SPBT value are investigated. The results are compared between two railway lines with different number of trains operating.

1 Introduction

Energy efficiency is increasingly gaining in importance, due to the climate changes and global grow of CO_2 emission. In Poland the issue is highly important due to the structure of electric energy generation sector, where 97% of electric energy is generated by coal power plants. Annual traction energy consumption in Polish railways is 2,4 TWh, which is 1,5% of the whole electric energy consumed in Poland. Meanwhile due to the rolling stock modernization the and replacing programs in perspective of the next years the absolute majority of the rolling stock is going to be equipped with regenerative braking. That condition gives the opportunity of braking power utilization, which in the case of a large number of stops enables to save between 10 % and 30 % of traction energy [1-4].

The regenerative power can be utilized in case if overhead catenary system (OCS) is receptive. In other case in DC systems the additional means of regenerative power utilization need to be introduces. One of them is stationary energy storage device. Apart from energy saving ESS could be used as the mean of 15-min power reduction as well as the pantograph voltage condition improvement [5]. In this paper the conception of ESS implementation in the existing traction substation is considered, which would enable satisfying the first two criteria. The voltage conditions could be improved only in case of substation outage under condition of appropriate ESS parameters and state of charge. The general scheme of the ESS in the 3 kV DC traction substation is presented in Fig. 1.

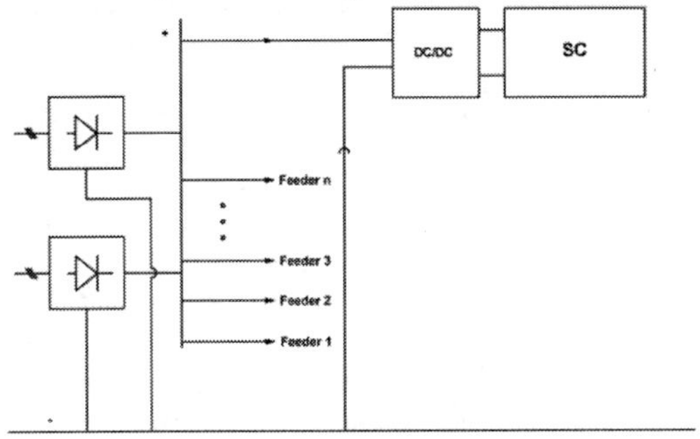

Fig. 1. The scheme of the ESS in the 3 kV DC traction substation with 12 pulse transformer rectifier unit.

Considering the given railway line the significant influence on regenerative energy utilization has the location of the ESS. According to the variable number of train stops and vertical profile along the track the regenerative energy which potentially could be recovered is variable as well. From the point of view of OCS receptivity the number of trains in the specified locations and timetable are important. Thus the factors has also influence on the variation of regenerative energy along the railway track.

In the paper two topologies of the ESS are considered: ESS with supercapacitor (SC) and ESS with SC and LFP battery. In the second topology the SC is used for the utilization of regenerative energy, while the

* Corresponding author: wlodzimierz.jefimowski@ien.pw.edu.pl

LFP battery is used to compensate the 15-min. power demand of the substation in order to reduce power demand charge. The stored energy of the LFP battery is much higher than SC. The LFP battery is charged by transformer rectifier unit. The appropriate energy management strategy has been used for the ESS with SC and LFP battery. For the implemented ESS the rule based deterministic strategy has been used. The energy management strategies have been mainly developed for ESS systems for autonomous electric vehicles, the range of strategies has been developed for the variable purposes [6].

The economic effect of the ESS implementation on traction substation depends on the substation location, the ESS topology, its sizing and the energy management strategy. In the paper the economic analysis has been carried out for the variants with different substation load character. The case study for specific railway line is carried out.

2 Model of the electrified railway line (ERL)

Due to the necessity of taking into account the number of factors important from the point of view of regenerative power availability along the railway track the appropriate simulation model needs to be used. The model of ERL used in the research consists of two parts: the train performance calculation (TPC) algorithm and power flow (PF) algorithm.

The TPC model covers the movement modelling of each train given in the forecasted or current timetable. The train movement parameters could be determined by (1) according to the Newton's second law of motion.

$$\begin{bmatrix} \frac{dx_i}{dt} \\ \frac{dv_i}{dt} \end{bmatrix} = \begin{bmatrix} v_i \\ \frac{F_i(v_i,U_i) - W(v_i,x_i)}{m_i(1+\eta_i)} \end{bmatrix} \tag{1}$$

where:

x_i, v_i, F_i, m_i, η_i, U_i - respectively, i-th train location, speed, tractive effort, mass, rotation mass coefficient, pantograph voltage. The rolling resistance is expressed by (2):

$$W(v_i,x_i) = a_i v_i^2 + b_i v_i + c_i + W_{hor} + W_{ver} \tag{2}$$

Where:

W_{hor} – rolling resistance from hirisontal profile

W_{ver} – rolling resistance from vertical profile

The algorithm determines the acceleration, speed, location, tractive effort and power of each train in each time step of simulation. The power values of train are positive or negative, depending of the mode of train operation and drive cycle phase. The current drawn by each train and pantograph voltage are determined in the power flow algorithm with use of mesh method of circuit analysis and superposition method. Trains are modelled as non-linear current sources, hence the iterative method is used for equivalent circuit analysis. The methodology of power flow calculations has been developed by range of authors [7-9] whose results and descriptions were the source of model development.

3 ESS with supercapacitor

The simulation research of ESS with supercapacitor has been carried out on the SC model based on the equivalent RC model. The mathematical model is based on the capacitor equation. In Table 1 the basic parameters of the used capacitor cells are presented.

Table 1. Supercapacitor cell parameters.

Type	BCAP
Mass	510 g
Nominal voltage	2,7 V
Capacitance	3000 F
Internal resistance	0,29 mΩ
Specific power	5,9 kW/kg
Cost	60 $
Nominal cycle life	1000000

[Source: http://www.maxwell.com]

The price model of the ESS with SC implementation could be obtained based on the formula [10] (3)

$$C_{ESS\_SC} = C_{const} + c\_u_{DCDC} \cdot P_{DCDC\_C} + 1.4 m_C \cdot n_C \cdot c_{sc\_cell}$$
(3)

where:

C_{onst} – constant cost of ESS installation [USD], $c\_u_{DC/DC}$ – unit cost of DC/DC converter [USD/kW], P_{DC/DC_C} – power of DC/DC converter [kW], m_C – number of cells connected in series, n_c – number of cells connected in parallel, c_{sc_cell} – price of the single SC cell.

The profit due to the energy savings is dependent on the tariff group of traction substation. The general active energy cost is calculated according to (4)

$$C_E = \sum_{i=1}^{n} c_i \cdot E_i \tag{4}$$

where c_i is energy price for the specific time zone, and E_i is the energy consumed in the specific time zone, n - number of time zones. The SPBT is calculated according to (5)

$$SPBT = \frac{C_{ESS\_SC}}{365 \cdot C_{E\_24h}} \tag{5}$$

The simulation study has been carried out for two railway lines, with relatively low and high density of train operation. The basic parameters of the investigated railway lines are presented in table 2.

Table 2. The basic parameters of the investigated railway lines.

	Line 1	Line 2
Length [km]	95	50
Number of substations	7	4
Number of trains/24 h/direction	26	115
Nr of train stops	16	15
Average nr of train stops per 10 km	1,68	3
Average distance between substations [km]	13,57	12,5

Fig. 2 shows the graphic timetables of the trains operation in the investigated railway line 1 (a) and 2 (b) for 24 hours cycle. In the figure the locations of train stops as well as the substations and section cabins (red lines) are marked. The substations considered as the possible locations of ESS implementation have been marked in circles.

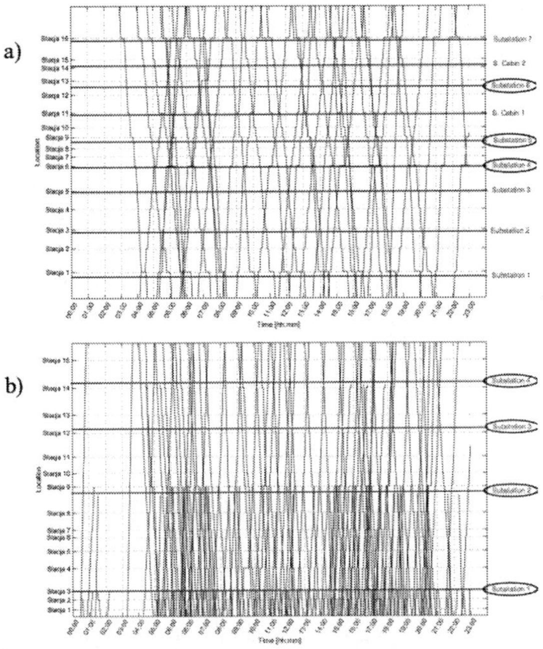

Fig. 2. The graphic representation of timetable of train operation in railway line 1 (a) and 2 (b), investigated in the study.

In railway line nr 1 (Fig. 2 a) two types of rolling stock operates - the 5 cars multiple units and locomotive passenger trains, in railway line nr 2 - 5 cars electric multiple units, locomotive passenger trains, and long distance electric multiple units.

Fig. 3 shows the values of the regenerative energy available in the substations, considered as the possible locations for ESS implementation. The results show that the available regenerative energy is higher in the railway line with the larger number of trains operating, however the difference is not as significant as the difference between number of trains operating.

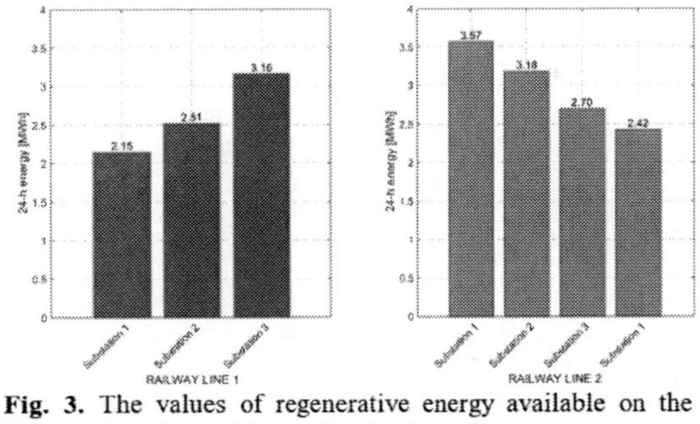

Fig. 3. The values of regenerative energy available on the different locations of the analysed railway lines.

The values of the investment efficiency parameters have been calculated for the ESS implementation, especially simple payback period time (SPBT). The basic assumptions for the calculations are presented in Table 3.

Table 3. The basic assumptions for the cost calculations.

Energy price	80,95 \$/MWh
SC cell cost	60 \$
DC/DC converter unit cost	130 \$/kW
Constant cost C_{const}	100.000 \$
LFP cell cost	21 \$

The assumptions for the cost model presented in Table 3 for Energy price, SC cell cost, LFP cell cost have been taken from the energy tariff and manufacturers catalogues subsequently. The rest of the parameters - DC/DC converter unit cost and constant cost are difficult to evaluate due to a lack of appropriate information. The application of DC/DC converters in 3 kV DC with the energy storage system is prototype and the prices are not established in the market.

The value of C_{const} (3) depends on the cost of all component of equipment necessary for the ESS implementation at traction substation. The most important elements of the facilities are:

- building extension or separate container usage ventilation system adjustment and development,
- DC switchgear extension,
- protection relay and switching equipment installation,
- cable laying.

Cost of each component can vary depending on the substation location and its specific conditions.

Fig. 4 presents the values of the 24-hour energy utilized by the ESS with SC as the function of DC/DC converter power and SC module capacitance for the location of substation 1 in the railway line 2. The energy shown in Fig. 4 is the part of regenerative energy available in the point of substation 1 connection to the overhead catenary. The SC module consists of number of SC cells connected in series, creating branches. The number of parallel branches in ESS module of SC is integer, hence the possible values of SC module capacitance are strictly specified and shown in Fig. 4. In the analysis the assumed number of SC cells connected in series is 500.

Using the formulas (3),(4) and (5) and parameters from table 3, the SPBT was calculated. The SPBT values as the function of DC/DC converter power and SC module capacitance are shown in Fig. 5.

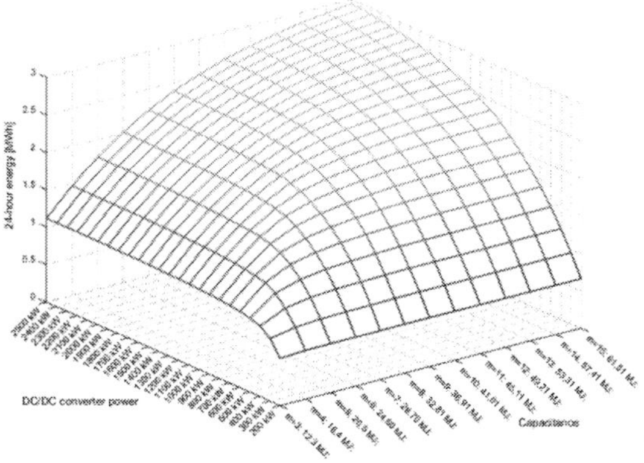

Fig. 4. Energy saving as the function of SC pack capacitance and DC/DC converter for substation 4.

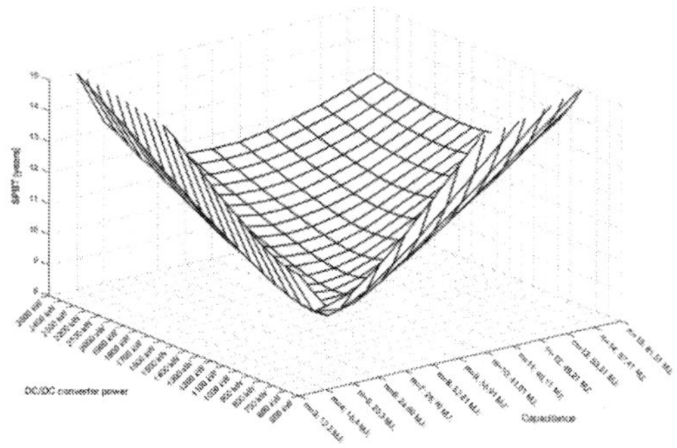

Fig. 5. The value of SPBT for the ESS with SC as the function of ESS capacity and DC/DC converter for substation 4 of railway line 2.

Among the values of the SPBT given as the function of inverter power and SC module for each capacity the minimum value of the SPBT could be found. Fig. 5 shows the minimum values of SPBT for the investigated locations of the railway lines. The results are shown as the minimum values for the specific ESS capacity. Fig. 8 shows that the min. value of SPBT depends on the location of ESS implementation. Comparing the minimum values of SPBT with the values of energy available in the specific points of possible ESS connection, it could be claimed that the SPBT values depend on the energy, however relationship between them is not linear. SPBT depends on the value of 24-hour energy and the shape of the instant regenerative power transient.

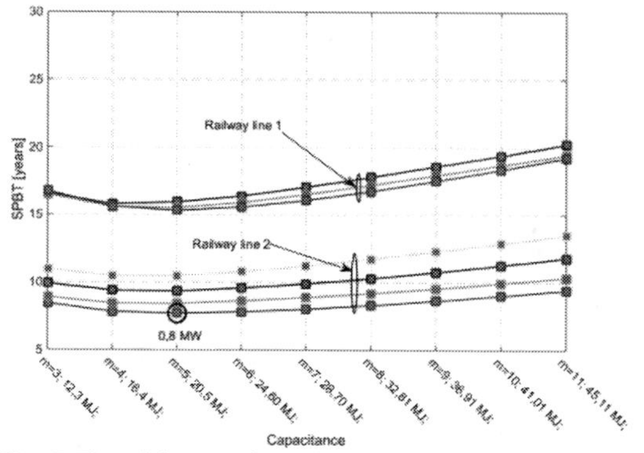

Fig. 6. The minimum values of SPBT for the railway lines and locations.

Fig 7 shows the minimum values of SPBT for the different prices of electric energy, as well as for the current energy price taken from the DNO tariff. The price level of the electric energy is not a constant value, the growth is being observed in the prices. The analysis including the forecast for the energy prices is not included in the scope of the study. The results of the parameter analysis is presented in Fig. 7, where the most reasonable variation of energy price for the perspective of 5-10 years is considered.

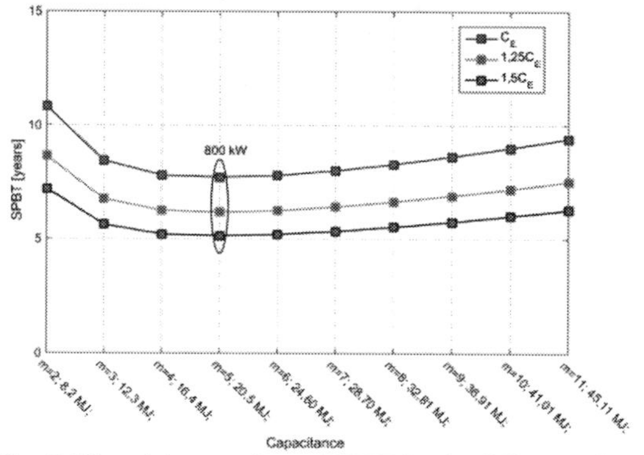

Fig. 7. The minimum values of SPBT for the different values of initial constant cost energy price C_E.

Fig. 8 presents the minimum values of SPBT for the different values of initial constant cost. The initial constant cost could vary depending on the specific substation conditions for implementation of ESS.

Fig. 8 shows that the initial constant cost influence not only the minimum value of SPBT, but as well the parameters of the ESS which are the most beneficial. In the figure values of module capacitance and DC/DC converter power are presented for the different values of initial cost, where the value of SPBT is minimal.

 https://doi.org/10.1051/matecconf/201818002005

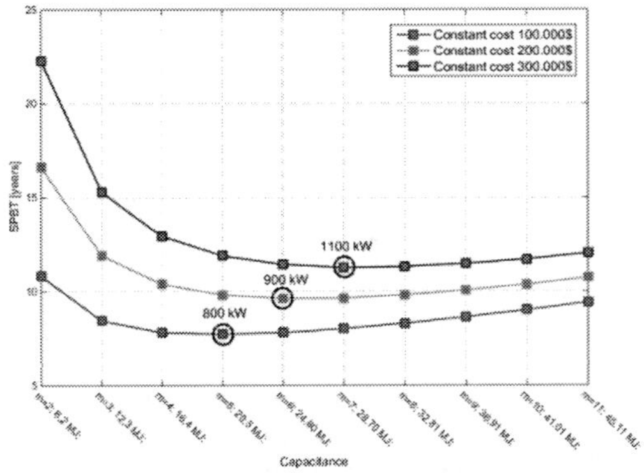

Fig. 8. The minimal values of SPBT for the different values of initial constant cost C_{const}.

4 ESS with supercapacitor and LFP battery

The scheme of the ESS with supercapacitor and LFP battery in traction substation is presented in Fig. 9. The aim of this energy storage system implementation is regenerative energy utilization (mainly by SC part) and 15-min power demand minimization in order to decrease the power demand fee.

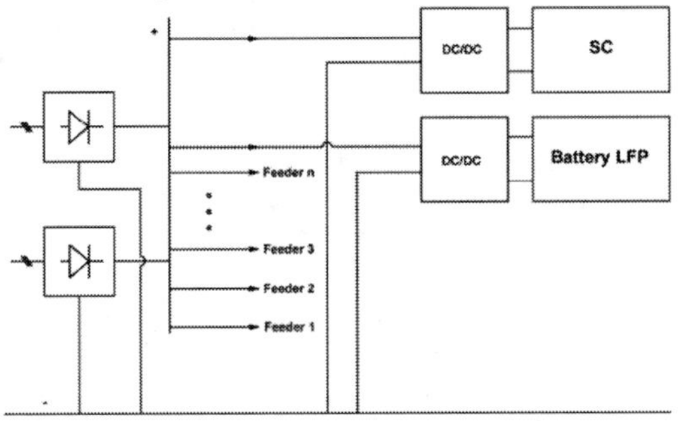

Fig. 9. The scheme of the ESS with SC and LFP battery in the 3 kV DC traction substation with 12 pulse transformer rectifier unit.

The LFP battery remains charged during the most of operation time. The energy of battery is used only during the time when the rolling average power for the last specified period of time excides the specific power. LFP battery could supply the traction load directly and/or charge the supercapacitor. In that way LFP battery compensates 15 min power demand of traction load. The frequency of LFP battery usage, hence the live cycle depends on the difference between the energy demand, which needs to be obtained.

The total cost of energy storage system implementation has been calculated according to next formula [10]:

$$C_{ESS\_SC\_LFP} = C_{const} + c\_u_{DCDC} \cdot (P_{DCDC\_C} + P_{DCDC\_LFP}) + \ldots$$
$$+ \ldots C_{ESS\_SCp} + C_{ESS\_LFPp} \tag{6}$$

where:

$$C_{ESS\_SC\_cell} = 1.4 \cdot m_C \cdot n_C \cdot c_{sc\_cell} \tag{7}$$

$$C_{ESS\_LFP\_cell} = 1.4 \cdot m_{LFP} \cdot n_{LFP} \cdot c_{LFP\_cell} \tag{8}$$

where:

P_{DCDC_LFP} – power of DC/DC converter connected to the LFP battery, C_{ESS_SCp} – price of the SC pack, C_{ESS_LFPp} – price of the LFP battery pack, m_{LFP} – number of LFP cells connected in series, n_{LFP} – number of LFP cells connected in parallel, $c_{LFP\ cell}$ – price of the single SC cell.

Fig. 10 shows the values of fee components for the active energy and the power demand for two variants of railway lines and for two variants of ESS implemented in the traction substation. The results show that connecting of LFP battery to the supercapacitor ESS leads to increasing of the energy fee component and decreasing of demand power fee component. The effect is more significant in the substation of less loaded railway line. The calculations of fees has been carried out based on the data taken from DNO tariff. The assumed energy price is 80,95 $/MWh, and demand power fee 32983,61 $/MW/year.

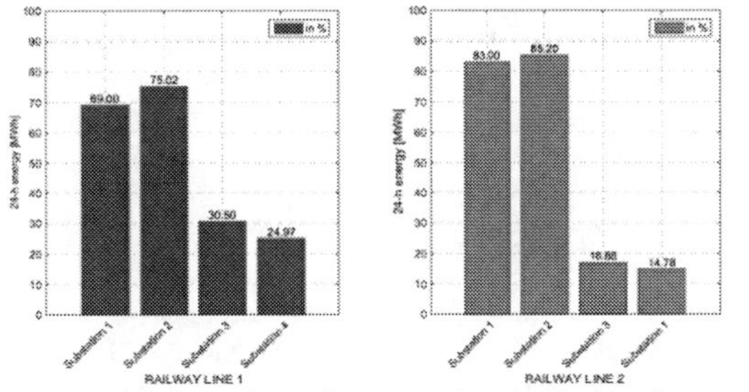

Fig. 10. The comparison of energy and power fee between two differently loaded traction substation and for the different variant of ESS (with SC and LFP battery).

Fig. 11 shows the SPBT value for two investigated railway lines, for two variants of ESS: with SC and with SC and LFP battery. The results show that for both railway lines implementation of LFP battery causes decrease of SPBT values, however the effect is more significant in less loaded railway line, where the power demand fee ratio is higher than in highly loaded railway line.

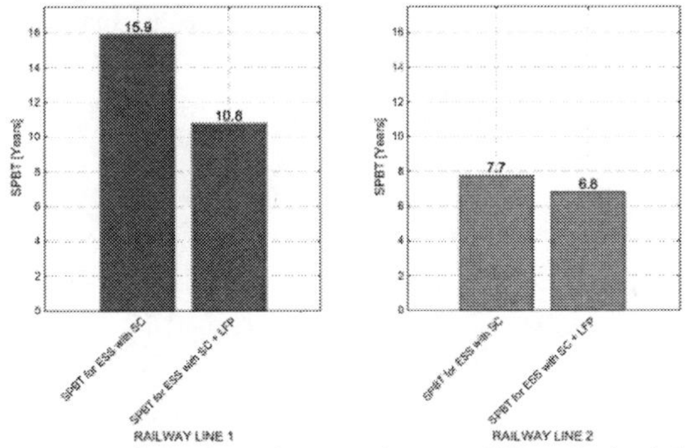

Fig. 11. The value of SPBT for two differently loaded substations and two variants of ESS.

5 Conclusions

The results presented in the paper show that the regenerative energy available in the point of potential ESS system connection depends on its location. The available energy depends on one hand on the number of trains operating and the number of stops , where trains brake with recuperation, but on the other hand also on the overhead receptivity [11-12]. The actual amount of energy which could be utilized could be calculated with use of simulation tools, modelling the electrified railway line. The results show that between two cases of railway line the bigger amount of energy is available in the traction substation of more loaded railway line with larger number of stops, however the difference is not significant.

The SPBT values for the ESS with SC are higher in the railway line with the higher trains operation density, however the relationship between the energy and minimum SPBT is not linear.

The significant influence on the investment effectiveness has the initial constant cost C_{const}. It influence not only the minimum SPBT value, but also the parameters of the ESS with supercapacitors for which the SPBT is minimal.

The conception of HESS with SC and LFP battery is presented in terms of investment effectiveness. The results show that the fee for the power demand could be significantly decreased thanks to the implementation of HESS with SC and LFP battery. Comparing with the ESS with SC, the minimal SPBT values are lower for the HESS with SC and LFP battery. For two investigated examples the implementation of HESS is more reasonable in the railway lines with less number of trains operating.

The research was financed under the Dean's Grant no 504/03411/1041/42.000100 from the statutory activities of the Faculty of Electrical Engineering of Warsaw University of Technology.

References

1. S. Goh, M. Griffith, and K. Larbi, Energy saving by using regenerating braking as normal train operation, *IET Conf. Railw. Tract. Syst. RTS 2010*, 750–756 (2010)
2. M. Bartosik, W. Kamrat, M. Kaźmierkowski, W. Lewandowski, M. Pawlik, T. Peryt, and A. Szeląg, *Magazynowanie energii elektrycznej i gospodarka wodorowa*, Przegląd Elektrotechniczny, **92**(12), 332-340 (2016)
3. S. Lu, P. Weston, S. Hillmansen, H. B. Gooi, and C. Roberts, *Increasing the regenerative braking energy for railway vehicles*, IEEE Trans. Intell. Transp. Syst., **vol. 15**, no. 6, 2506–2515 (2014)
4. S. Song, Q. Wang, and J. Wu, Simulation of multi-train operation and energy consumption calculation considering regenerative braking, *Proc. 33rd Chinese Control Conf. CCC 2014*, 3387–3392 (2014)
5. D. Iannuzzi, F. Ciccarelli, and D. *Lauria, Stationary ultracapacitors storage device for improving energy saving and voltage profile of light transportation networks*, Transp. Res. Part C Emerg. Technol., **vol. 21**, no. 1, 321–337 (2012)
6. M. Wieczorek, and M. Lewandowski, *A mathematical representation of an energy management strategy for hybrid energy storage system in electric vehicle and real time optimization using a genetic algorithm*, Appl. Energy, **192**, 222–233 (2017)
7. A. Szeląg, *Problems of analysis and design of DC electric traction system with application of modelling and simulation technique* (Warsaw University of Technology, 2002)
8. M. R. Irving, and S. H. Case, Iterative techniques for the solution of complex DC-rail-traction systems including regenerative braking, *IEE Proceedings - Generation, Transmission and Distribution*, **143**, 445–452 (1995)
9. R. A. Jabr, and I. Dzafic, Solution of DC Railway Traction Power Flow Systems Including Limited Network Receptivity, *IEEE Trans. Power Syst.*, **33**, 962–969 (2018)
10. The Boston Consulting Group, *Focus Batteries for Electric Cars, Outlook* (2010) http://www.bcg.com/documents/file36615.pdf
11. T. Maciołek, Z. Drążek, and Szeląg A. *Efektywność energetyczna zasobników energii w podstacjach systemu prądu stałego 3 kV DC*, Logistyka, **3**, 2990-3000 (2015)
12. T. Maciołek, *Zastosowanie zasobników energii w trakcji kolejowej 3 kV DC – nieodległa perspektywa?*, TTS Technika Transportu Szynowego, **22**(9), 39-44 (2015)

Comparison of different tram cars in Poland basing on drive type, rated power and energy consumption

Barbara Kulesz[1,*], and *Andrzej* Sikora[1]

[1]Silesian University of Technology, Faculty of Electrical Engineering, ul. Akademicka 2, 44-100, Gliwice, Poland

Abstract. Paper deals with tram rolling stock used in Poland nowadays – overview of car types in operation is given. The detailed calculations of energy consumption by a given tram car running over a given route depend in particular on the work carried out (each route is characterized by, among other factors, its profile, length, number of turns, turn radii, number of stops – scheduled or otherwise, number of transferred passengers, number of brakings with possibility of energy recuperation). Results of energy consumption measurements for different car types in Warsaw are provided and analysed.

1 Tram cars used in Poland

Number of tram cars used in Polish cities is shown in Table 1 [1-4]. In case of popular 105Na trams we differentiated between not modernized and modernized cars whenever possible (modernized in this case means the replacement of standard resistor start-up and speed control circuit with a power electronics device).

Table 1. Tram rolling stock in Polish cities.

City	No. of 105Na cars and their variants* (additional number of modernized cars is given in brackets)	Number of relatively new cars **	Number of other cars***
Warszawa	257 (135)	341	-
GOP	105 (108)	67	40
Gdańsk	34	49	60
Wrocław	92 (168)	103	-
Poznań	61 (20)	109	37
Toruń	55	17	-
Olsztyn	-	15	-
Bydgoszcz	112	14	-
Częstochowa	44	7	-
Elbląg	17 (3)	6	6
Grudziądz	8 (6)	-	10
Szczecin	6 (26)	102	73
Gorzów Wlkp.	-	-	20
Kraków	72	152	172
Łódź	238 (246)	48	29

* Tram cars with dc drives and resistor start-up
**Models such as 122NaL *Swing*, 121N *Tramicus*, *Combino*, S105p *Tramino*, Tatra RT6, 2010NW *Twist*, 116Nd *Citadis*, 16T, 19T, 204WrAs and others, not older than 25 years
***Models such as Duewag M8C, M6S, GT6, GT8N, E1, Pt8, Ptm and others, manufactured e.g. 1962-67, 1972-77, 1978-86, mostly modernized, bought second-hand

2 Energy consumption in a selected tram company

Most electrical energy is used by traction drive system. The remainder goes to heating, air-conditioning when present, lighting, control circuits, auxiliaries. The values set out in Tables 2-4 are calculated from data obtained from biggest tram operator in the country, i.e. Warsaw Tram Company [5]. Total length of tram tracks is c. 280 km. Each tram in Warsaw is equipped with energy measurement device, the new trams with digital recorders. Energy consumed by trams is equal to c. 94.7% of total electrical consumption in this company (2015). The oldest tram cars running in Warsaw date back to 1984, average car is almost 15 years old (*cf.* Table 1). Warsaw tram stock is detailed in Table 2.

Table 2. Tram stock in Warsaw: drive type, energy recuperation possibilities.

Car type	Number of cars	Start-up and control/motor ratings	Energy recuperation/ storage
105Na	257	resistor start-up/4 x 41.5 kW	no/no
105Ni	26	converter/4 x 41.5 kW	yes/no
105N2k	47	converter/4 x 41.5 kW	yes/no
105Nk/ 2000	62	converter/4 x 41.5 kW	yes/no
123N	30	converter/4 x 41.5 kW	yes/no
112N	1	converter/6 x 41.5 kW	yes/no
116Na 116/Na1	29	inverter/4 x 50 kW	yes/no
120N	15	inverter/4 x 105 kW	yes/no
120Na	180	inverter/4 x 105 kW	yes/no
120Na DUO	6	inverter/4 x 105 kW	yes/no
128N	50	inverter/4 x 105 kW	yes/ ultracapacitor 425VDC
134N	30	inverter/8 x 60 kW	yes/ ultracapacitor 400VDC

* Corresponding author: barbara.kulesz@polsl.pl

https://doi.org/10.1051/matecconf/201818002006

Relatively large portion of energy goes to heating and air-conditioning circuits - see Table 3.

In newer trams, in particular 120N series and its variants, 128N and 134 N cars, the ratio of heating/a.c. power to car volume is on the average 2.5-3 times higher than in cars 105Nx, 112N, 123N and 116Nx.

Energy consumed by drive systems varies from c. 67% (123N) to c.83% (120N/Na), while auxiliaries' energy varies from c.19% to c.10%, respectively (see Table 4).

Table 3. Comparison of drive and heating/air-conditioning circuit ratings.

Car type	Total heating/a.c. power in kW	Ratio of heating/a.c. power to motor power in %	Indicator of heating/a.c power per car volume in kW/m³
105Na	12.2/3.5	7%	0.12
105Ni	13/3.5	8%	0.13
105N2k	10.9/3.5	7%	0.11
105N2k/2000	14.2/3.5	9%	0.14
123N	20/3.5	12%	0.18
112N	19.2/3.5	12%	0.12
116N/116Na/116N/1	24.8/3.5	12%	0.13
120N	85.1/49.1	20%	0.33
120Na	75/64	18%	0.31
120Na DUO	80.4/64	19%	0.33
128N	84.4/55	18%	0.35
134N	42.4/40.5	9%	0.27

Table 4. Drive and auxiliary circuits' energy participation in total energy consumed by a tram car.

Car type	Energy consumed by motors	Energy consumed by auxiliaries
105Na	76.2%	16.2%
105N2k, 105N2k/2000	74.2%	17.6%
112N	74.2%	17.6%
116Na, 116Na/1	77.9%	13.2%
123N	67.6%	19.4%
120N/Na	82.8%	10.2%
128N	82.8%	10.2%
134N	82.0%	10.8%

Analysis of energy used for traction purposes was carried out not in terms of total energy consumption but using indicators. e.g. referring energy to the distance per car (unit: car-kms), to vehicle weight (empty car, partial or full load) and finally, to number of passengers transferred. In practice, the indicators related to number (weight) of passengers and not those referring to car-kilometres make it possible to compare energy consumption by different cars in a rational manner. The indicators are set out in Table 5 and shown graphically in Fig.1.

Tram cars 120N and 128N consume more energy by other cars; their consumption is even greater than in case of 105Na – 105Nm cars, which cannot recuperate braking energy back into the network. This phenomenon is due to the fact, that carrying capacity is greater, which results in greater weight of the vehicle. 120N and 128N type cars' weight without passengers (hereafter this will be called weight of empty cars) is also much higher than that of other cars. Therefore, during following comparisons, we took into account number of passengers transferred.

Analysing weight of empty cars and weight of cars with passengers, we determined several intervals for total vehicle weight (car carrying 40%, 70% and 100% of maximum number of passengers, i.e. 40%, 70% and 100% of total carrying capacity). The tram weight per one passenger for these different loads is shown in Fig.2. The smallest weight relates to 100% load, the points marked refer to 70% load, and highest weight per one passenger related to 40% load.

Table 5. The analysed data - energy consumption and car weight.

Car type	Car weight without/ with passengers	Energy consumed (at panto-graph)	Energy consume d by drive system	Recuperated energy	Energy consumed by drive system. recuperation included
	tons	kWh/km		%	kWh/km
105Na.. 105Nf	16.8/25.55	no data	2.50	none	2.50
105Ni	17.5.26.25	1.97	1.91	28.79% *	1.36
105N2k	17.7/26.45	no data	1.91	28.79% *	1.36
105N2k /2000	17.7/26.45	2.17	1.91	10.17% *	1.72
123N	18.2/24.99	no data	1.91	28.00%	1.38
112N	26.0/40.28	no data	3.23	28.00%	2.33
116N	29.0/44.05	no data	3.23	28.00%	2.33
116Na	29.0/44.05	no data	3.23	28.00%	2.33
116N/1	29.0/44.05	no data	3.23	28.00%	2.33
120N	43.4/58.17	no data	4.37	28.00%	3.15
120Na	40.6/55.04	5.7	4.37	29.87%	3.06
120Na DUO	42.7/56.375	5.7	4.37	29.87%	3.06
128N	41.5/56.195	5.45	3.96	29.88%	2.78
134N	30.42/39.45	3.97	2.92	31.45%	2.00

* Ratio of energy recuperated into the network to total energy supplied to the tram car (at pantograph)

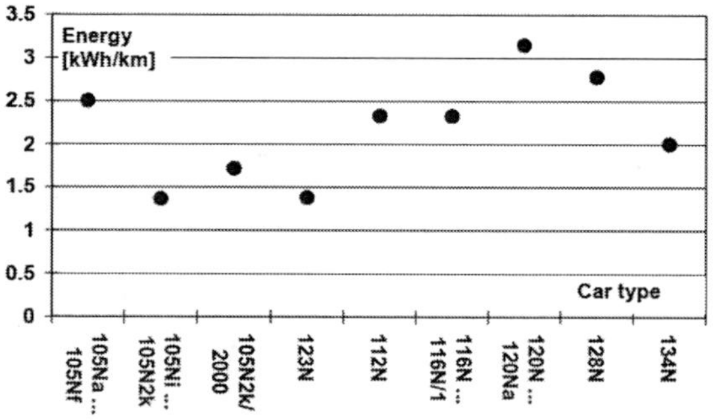

Fig. 1. Energy consumed by drive system – per 1 km of distance, for different cars.

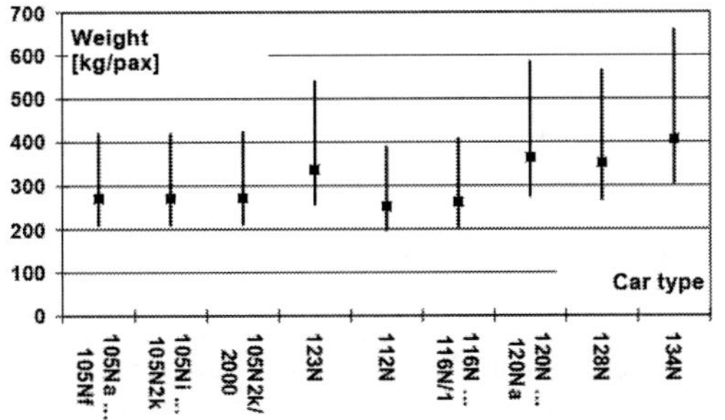

Fig. 2. Weight intervals per one passenger, for different car types. Points refer to 70% load, smallest weight to 100% load, highest weight to 40% load.

120N and 128N tram cars are characterized by higher weight per one passenger than other cars, this is due to the weight of empty cars (i.e. without passengers). Low-floor trams 120N - 134N are equipped with air-conditioning systems for passengers, which improves travel comfort in the summer, but increases vehicle weight at the same time. Cars 128N - 134N are fitted with on-board ultracapacitor storage systems, this also increases vehicle weight. The energy consumed by the drive system has been related to total vehicle weight, taking into account the varying load (see Fig.2). Energy consumption intervals have been determined, for load varying from 40% to 100% of carrying capacity. Energy consumed per 1 km of distance and per 1 ton of car weight is shown in Fig.3. As before, lowest consumption relates to maximum (100%) load, marked points correspond to 70% load and highest consumption occurs, when 40% of tram's carrying capacity is used.

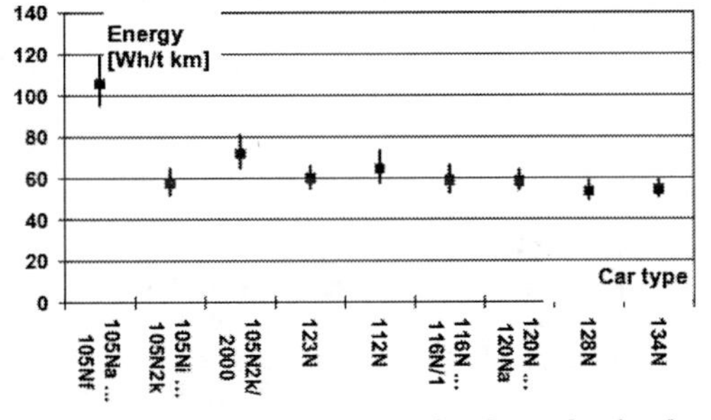

Fig. 3. Drive system energy consumption intervals related to 1 ton of car weight and 1 km of distance, for different trams. Points refer to 70% load, smallest weight to 100% load, highest weight to 40% load.

Tram cars which are able to recuperate energy (105Ni - 134N) are characterized by much lower energy consumption per 1 km of distance and 1 ton of car weight (from *c.* 95 to 130 Wh/tkm) than cars without recuperation. i.e. 105Na – 105Nm (from *c.*50 to *c.*87 Wh/tkm) - see Fig.3.

Next, we calculated energy consumed in relation to 1 passenger, 1 ton of car weight and 1 km of distance. Calculations have been run for load ranging from 40% to 100%. Energy consumed per 1 passenger, 1 ton and 1 km is shown in Fig.4. The lowest consumption corresponds to 100% load, the highest to 40% load and the points refer to 70% tram load.

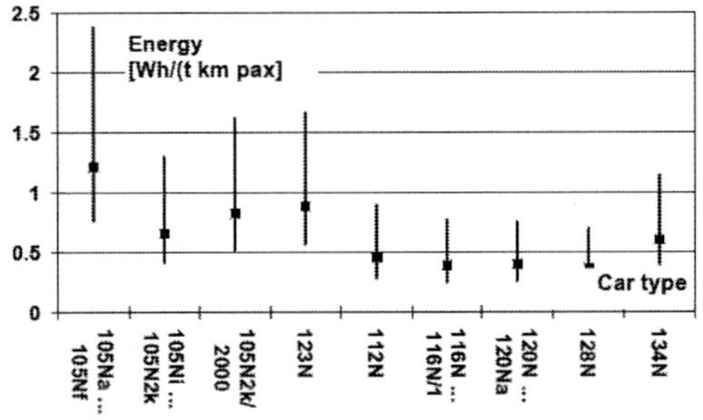

Fig. 4. Drive system energy consumption intervals related to 1 passenger, 1 ton of car weight and 1 km of distance, for different trams. Points refer to 70% load, smallest weight to 100% load, highest weight to 40% load.

Tram cars which are able to recuperate energy (105Ni - 134N) are characterized by much lower energy consumption per 1 passenger, 1 km distance and 1 ton of car weight than cars without recuperation - see Fig.4. The heavier trams (120N - 134N) equipped with air-conditioning of passenger compartments are characterized by low consumption related to 1 passenger, 1 km of distance and 1 ton of car weight.

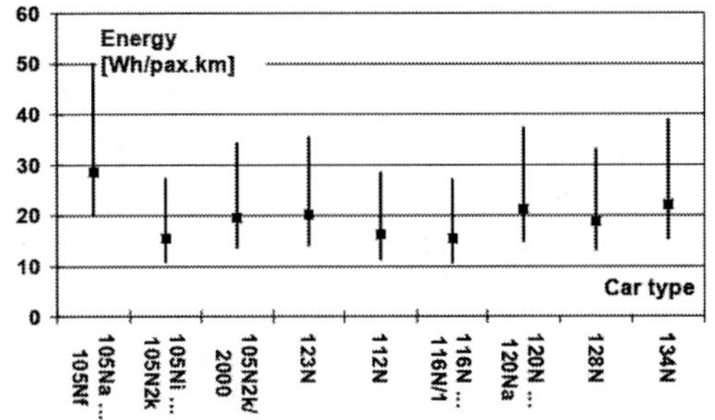

Fig. 5. Drive system energy consumption intervals related to 1 passenger and 1 km of distance, for different trams. Points refer to 70% load, smallest weight to 100% load, highest weight to 40% load.

We have also calculated energy consumption per 1 passenger and 1 km of distance (kWh/pax.km). The calculations were carried out for load varying from 40% to 100% of total carrying capacity. The energy consumption per 1 passenger and distance of 1 km is shown in Fig.5. As before, lowest consumption relates to maximum (100%) load, marked points correspond to 70% load and highest consumption occurs, when 40% of tram's carrying capacity is used.

Tram cars which are able to recuperate energy (105Ni - 134N) are characterized by much lower energy consumption per 1 km of distance and 1 passenger, than cars without recuperation. i.e. 105Na – 105Nm. Low-floor trams 120N - 134N are equipped with air-conditioning systems for passengers and show greater energy consumption per 1 passenger and 1 km than tram cars with energy recuperation possibilities, but without air-conditioning systems. This is due to the value of empty car weight.

 https://doi.org/10.1051/matecconf/201818002006

3 Some observations on tram car energy consumption calculations

Publications on the subject present different procedures for energy consumption. Exemplary data for trams running in Poland are given by Kuminek [6]. He analysed energy consumption for 105Na cars with resistor start-up and speed control, for routes in Silesian agglomeration, for modernized 205WrAs tram cars running in Wrocław as well as manufacturer's data for 120Na tram cars. He differentiated between energy consumed by cars with heating on and off. The appropriate data for 105Na car is 2.45-4.14 kWh/1 car-km, 2.17-3.14 kWh/1 car-km for 205WrAs and 4.1-5.3 kWh/1 car-km for 120Na car. It must be pointed out that these calculations did not account for the actual tram load (i.e. passenger weight) and that measurements were carried out at different routes.

In Cracow the measurements were carried out by experts from Technical University of Cracow [4], energy consumption varied from 3.2 kWh/car-km for N8 tram cars to 6.3 kWh/car-km for 105Na cars. NGT6 tram (without energy recuperation) shows energy consumption lower by 11-28% than 105Na car and taking into account passenger loading capacity this value is c.26% [7,8].

Tackoen et al. [9] have approached the energy saving issue by way of simulation. They adopted a particular route of a given tram in Brussels and analysed different variants of energy storage on-board systems (ultracapacitors with different capacitance and weight). They calculated possible savings taking into account varying load of the tram. The energy consumption indicator was 5 kWh/km, and possible savings equalled 24% (average value).

Chymera et al. [10] analysed possibilities of improving energy consumption by using ultracapacitors. In order to conduct calculations, they measured energy consumed by 22-ton tram car of City Class operating in Blackpool. The total energy consumption indicator was equal to 1.1 kWh/km, and *c.* 0.75 kWh/km was appropriated by the drive system (motors, converter, dynamic braking circuit, gear). The authors did not account for different loading of the tram. However, in another paper [11] they compared the data for City Class tram car equipped with asynchronous motors, with length equal to 29 m, weighing 22 tons, and with loading capacity of 200 passengers, while older type tram (Centenary Tram) with dc motors and comparable weight (17.5t) may carry 74 passengers only. They also considered possible weight (number of passengers transferred) of the compared cars. The respective indicators for 4 drive cycles were: for City Class tram car from 0.0085 to 0.0098 kWh/pax.km, for Centenary Tram from 0.0194 to 0.026 kWh/pax.km. These values result from simulations carried out for a given route.

Erd et al. [12] analysed possibilities of lessening energy consumption in tram by applying flywheels (rotating energy storage devices). They used data for Variobahn tram (6 motors, 95 kW each, carrying capacity 240 passengers) provided by Bombardier company (distance 17.6 km, route in Heidelberg city). The calculated indicators were 4.21 kWh/km for tram without storage and 3.33 kWh/km for tram with flywheel. Again, the varying passenger load was not accounted for.

4 Applications of energy saving strategies

Polish operators and Warsaw Tram Co. in particular at present investigate different ways of possible reduction in energy consumption. This coincides with current world and European trends in public transportation approach.

An overview of different energy recovery strategies in public transport has been given in [13]. Strategies have been classified in accordance with mobility (mobile storage applications such as on-board energy storage systems, stationary storage applications and stationary applications related to grid). Technologies such as batteries, ultracapacitors, flywheels and reversible substations are presented together with examples of prototypes designed and operating in public transport sector. However, the authors point out that apart from the mobile storage systems (ultracapacitors mostly), most of possible solutions have not yet been implemented on a mass scale.

Henning et al. [14] discuss in detail two options for energy storage in light rail vehicles: flywheels and ultracapacitors. Power densities of ultracapacitors and modern flywheels are similar, but flywheels offer higher energy density, which is better if acceleration times are longer (unit cited here offers 4 kWh, 300 kW at 375 kg compared to double that weight back in 1995).

Destraz et al. [15] have studied impact of ultracapacitor storage and high line resistance on energy consumption in urban trams. Again, modelling has been conducted on the basis of existing line data (track in Mannheim, Germany). They have shown that quality of the line has a powerful impact on overall energy consumption, calculating effects of supercapacitor energy storage on a normal network and "weak" network. The results of actual measurements have shown energy saving equal to *c.*25% on vehicle level.

Some trial runs of trams equipped with batteries able to store energy have been carried out in Warsaw. Battery rated at 80 Ah/600 V was used. Possibilities of energy recovery were interesting (over a 12 km run 12-18 kWh could be recovered); however, life cycle of this battery is set by its manufacturer at 13260 cycles, i.e. four years. Simulations have also been conducted on energy saved due to recuperation for city centre and suburban regions, during peak hours and the remainder of the day. Different possibilities have been investigated. Comparison of energy recovered by different tram types has been made (Cityrunner tram, 805Na tram with induction motors, 805Na tram with dc motor and without energy recuperation). Researchers have also compared possibility of installing inverters or flywheels at the substations [16]. Effects of these two solutions are similar. The best effects are shown for regions where the network is most highly loaded; however, in such regions recovered energy may be taken over by other trams and substation devices are not really indispensable. Use of inverters or flywheels in suburban regions may not achieve significant effects, since the network may show high voltage drops and energy recovered during braking cannot be reliably transmitted.

Among the drawbacks of capacitors, we may point out the significant increase of tram car weight (e.g. EDLC ultracapacitors used in Innsbruck trams, rated at 0.85 kWh/288 kW weigh 820 kg). Similar considerations are true for batteries (e.g. NiMH tried in Sapporo Municipal Transport in 2011, rated at 250 kW/120 kWh, weighed 3200 kg).

Recuperation of energy into national grid is not popular as use of energy storage devices mostly on account of high investment costs. Some trials have been done, in Poland notably in Łódź and Olsztyn in 2014-2016, where inverters rated at 0.5-1MW have been used. No solution has been as yet commercially installed.

Another approach to energy consumption is analysis of the greatest tram load, i.e. traction motor.

A thorough overview of motors for light rail traction applications (and other types of vehicles as well) is given by Gieras and Bianchi [17]. Authors compare motors such as standard cage induction motors, standard PM brushless motors with interior type NdFeB magnets, PM brushless motors with short coil spans, hybrid synchronous motors (utilizing permanent magnets and electromagnetic excitation both), permanent magnet transverse flux motor and switched reluctance motor. Different points of motor and drive construction are considered, such as gear, size, volume, motor weight, inverter power. Comparison is valuable as it is provided for 75 kW brushless motors. Efficiency of induction motor is given as 90% and efficiency for remaining motor types varies from 93% to even 97.6%; weight of IM is given as 272kg and corresponding weights of other motors vary from 147 kg down to 73 kg. Apart from technological drawbacks and higher prices, the permanent magnet motor potential is very high.

Demmelmayr et al. [18] have compared energy consumption of standard induction machine with novel prototype permanent magnet synchronous machine (PMSM). Calculations have been made on a basis of typical time graph for electric city tram (total drive type of about 70 seconds, including acceleration phase, maximum speed, deceleration and standstill). Authors concluded that for this given route energy consumed by PMSM was only 49% of induction machine consumption.

Similar issue has been addressed by Mermet-Guyennet [19]. He shows evolution of motors for electric traction used in French railways: from 1981 DC motor rated at 535 kW and weighing 1560 kg (ratio weight/power 2.9 kW/kg), through 1989 synchronous motor (1130 kW, 1525 kg, 1.35 kW/kg), asynchronous motor of 1994(1020 kW, 1260 kg, 1.23 kW/kg) up to synchronous PM motor of 2004 (800 kW, 768 kg, 0.96 kg/kW).

Permanent magnet machines are used almost exclusively in light-duty hybrid vehicles on account of higher power density and efficiency [20]. Trams utilizing PM motors have been designed and built by Czech company SKODA (ForCity Alfa trams, currently operating in Prague, Czech Republic and Riga, Latvia). These trams utilize 16 PMSM motors (surface-mounted magnets) with maximum power 46.6 kW each. Gears have been eliminated. PM motors have also been developed for high-speed rail such as prototype train AGV (2008).

In Poland PMSM motors have not been so far applied for urban rail vehicles (some trials were done by ALSTOM company for X04 tram, but the project was abandoned).

The principal advantages exhibited by PM motors are continuously cited as high power density and efficiency, simple and robust machine construction, low noise, durability, possible elimination of mechanical gearing, possibility of high overload. However, this potential has not been so far utilized fully, the commercial rail vehicles in operations are few in comparison to standard-drive vehicles.

Since substantial part of the energy consumed goes towards heating and air-conditioning, methods for accurate measurement and comparation of different air-conditioning system energy consumption have been developed by Richter [21] in the course of EcoTram research project. Different measures for reducing energy consumption have been researched, such as stepless heat power control, variable supply of air taken from outside (in accordance with CO2 content inside the vehicle), heat pump, elimination of bypass valve from compressor (variable speed control applied), different refrigerator agents used, different quality of window glazing and sun protection films, different colour of outer tram housing surface, predictive-type control for HVAC modules. Development of more energy-targeted HVAC units may lower energy consumption by up to 30%.

Chymera and Goodman [22] give a workable overview of calculating a rail vehicle performance. They enumerate different factors affecting energy consumption and stress the fact that more and more operators will in future require some simulation of rail vehicle performance over a given route while considering offers for stock purchase. These simulations must take into account diverse features such as unscheduled stops and delays.

In Poland and in accordance with peculiarities of Polish operators, the most immediate measures undertaken in order to reduce energy consumption at present are:

- modernization of existing rolling stock, in particular replacement of dc resistor-controlled drive systems with systems providing possibilities of energy recovery, mostly ac induction motors
- installation of on-board storage ultracapacitor systems.

Other possibilities of energy reduction are extensively investigated but have not been implemented commercially.

5 Conclusions

The highest energy consumption in tram cars is due to drive systems, second place is held by air-conditioning and heating systems. In this paper we have analysed energy consumed by drive motors, using indicators expressed in kWh/km (ranging from 2.50 for 105Nx cars with dc motors, resistor start-up and speed control, to even 5.7 kWh/km for 120Na cars and 3.97 kWh/km for 134N cars with asynchronous motors and inverters allowing for braking energy recuperation back into traction network). When recuperation is considered, these indicators range from 2.50 for 105Nx cars to 3.15 for 120 Na cars and 2.00 for 134N cars.

Any car type provided with energy recuperation, irrespective of drive type is characterized by significantly lower energy consumption related to 1 passenger travelling over 1 km distance. Low-floor cars such as 120N, 120Na and subsequent variants, 128N and 134N having air-conditioning for passenger compartments are characterized by higher energy consumption related to 1 passenger travelling over 1 km distance than tram cars able to recuperate braking energy, but not fitted with a/c systems. This heightened specific energy consumption is due to weight of empty trams. To summarize, in our opinion any comparison of energy consumed by trams must be related to the possible passenger load. Such indicators may be nowadays calculated quite easily for different types of trams, since new vehicles commissioned into service are equipped with complex energy recording systems as well as localization systems and passenger count systems.

References

1. MPK Poznań, www.mpk.poznan.pl
2. Tramwaje Gdańskie http://tramwaje.gdanskie.info/
3. Tramwaje Śląskie www.tram-silesia.pl
4. Aspekt ekologii w transporcie szynowym, docplayer.pl/6440201-Aspekt-ekologii-w transporcie-szynowym.html [in Pol]
5. Energy consumption data provided by Warsaw Tram Company - unpublished material.
6. T.Kuminek, *Energy consumption in Tram Transport*, Logistics and Transport, **18 (2)**, 93-100 (2013)
7. S. Jedliński, *Efektywność energetyczna w miejskim transporcie publicznym,* Technika Transportu Szynowego, **3**, 60-64 (2007) [In Pol]
8. W. Czuchra, J.Prusak, and W.Zając, Tramwaje z napędem asynchronicznym – ocena energochłonności. Technika Transportu Szynowego, **1-2**, 66-72 (2006) [In Pol]
9. X.Tackoen, R.Barbero, B.Leduc, and J. Van Mierlo, Energy recuperation and storage on the Brussels tram network, *Proc. of Conference Innovation for Sustainable Production - i-SUP* (2008)
10. M.Chymera, A.C. Renfrew, and M.Barnes, Analyzing the Potential of Energy Storage on Electrified Transit Systems. *Proceedings of the 8th World Congress of Railway Research* (2008)
11. L.Lesley, A.Winstanley, A.Renfrew, M.Barnes, and M.Chymera, Power consumption in a new LRV, *Proc. of 9th International Conference and Exhibition on Railway Engineering* (2007)
12. N.Erd, X.Li and A.Binder, *Power Flow Simulation of Flywheel Energy Storage Systems for Tramways*, Renewable Energy and Power Quality Journal, **1** (15), 149-153 (2017)
13. F.-O. Devaux and X.Tackoen, X., *WP2B Energy Recovery: Overview of braking energy recovery technologies in the public transport field.* Report from Project "Ticket to Kyoto" (2011)
14. U.Henning, F.Thoolen, M.Lamperth, J.Berndt, A.Lohner and N.Jaenig, Ultra low emission traction drive system for hybrid light rail vehicles, *Proc. of International Symposium on Power Electronics, Electrical Drives, Automation and Motion Speedam,* 1068-1072 (2006)
15. B.Destraz, P. Barrade, A.Rufer and M.Klohr, Study and Simulation of the Energy Balance of an Urban Transportation Network, *Proc. of European Conference on Power Electronics and Applications,* 1-10 (2007)
16. T.Solarek, Ocena możliwości odzysku energii hamowania w trakcji tramwajowej, *Proc. of 7th International Conf. Modern Electric Traction in Integrated XXIst Century Europe*, 165-168 (2005) [In Pol]
17. J.F.Gieras and N.Bianchi, *Electric Motors for light Traction*, European Power Electronics and Drives, **14** (1), 12-23 (2004)
18. F. Demmelmayr, M.Troyer, M.Schroedl, Advantages of PM-machines Compared to Induction Machines in Terms of Efficiency and Sensorless Control in Traction Applications, *Proc. of IECON 2011 - 37th Annual Conference on IEEE Industrial Electronics Society*, 2762-2788 (2011)
19. M. Mermet-Guyennet, New Power Technologies for Traction Drives, *Proc. of International Symposium on Power Electronics, Electrical Drives, Automation and Motion Speedam,* 719-723 (2010)
20. A.M.El-Refaie, *Motors/Generators for Traction/Propulsion Applications*, IEEE Vehicular Technology Magazine, 90-99 (2013).
21. R.Richter, *EcoTram – Energy-Efficient Air-conditioning For Trams. New HVAC Units And Additional Measures, Power Measurements, Outlook.* Railvolution, **14** (2/14), 2-4 (2014)
22. M. Chymera, C. J. Goodman, The calculation of train performance, *IET 13th Professional Development Course on Electric Traction Systems*, 1-13 (2014)

Electromagnetic processes of charging of on-board supercapacitor storage system during the regenerative braking mode of DC electric rolling stock with series-wound DC motors

Anatolii Nikitenko [1,*]

[1]Warsaw University of Technology, Institute of Electric Power Engineering, Electric Traction Division, ul. Koszykowa 75, 00-662, Warsaw, Poland

Abstract. The paper is dedicated to the task of improving of regenerative braking of a DC train with the series-wound DC traction motors. It is proposed the schematic solution for the trains with DC motors, which is based on the storing of recovered energy by the on-board supercapacitor storage system and extending the speed range of regenerative braking. The scheme is described by expressions which allowed to build the time dependences of main values. In addition, the experimental researches with the multiple-unit train were performed to find initial data for calculations and verification the results. The transient processes of charging are analyses for different cases of storage capacitances, residual voltages and initial speed of braking. It was proved, that the scheme allows to increase the value of recovered energy up to 40%, but its value strongly depends on the initial conditions and parameters of the system.

1 Introduction

About 36% of electrified railway lines are supplied by 3 or 1.5 kV DC voltage. As much as 47000 DC supplied electric locomotives are currently in use in the world. Many of them, among others in East European countries, still equipped with the series-wound DC motors and rheostatic control due to long life time, comparatively low price, simple design of control system and self-excitation of the field coils [1]. On the other hand, they have numerous disadvantages, designed without regenerative braking or cannot use it due to the frequent faults, which appear in the process of real operation [2]. Therefore, the train drivers frequently have no choice and apply the rheostat braking to reduce the high speed of the vehicle, so the recovered (kinetic) energy is uselessly dissipated in the ballast resistors. Furthermore, the experimental researches of [3] show that the losses of recovered energy in the overhead contact system vary from 6.3 to 20% in case of transmission to the train which operates in traction mode. The losses are equal to 1.5÷43.1% when the recovered energy flows to the traction substation and is inverted to AC grid. Thus, the task of improvement of effectiveness of regenerative braking stays actual for railway vehicles with AC and DC traction. It is a current issue specifically in East and Central European countries with DC supplied railway when thousands of old-type rolling stock with DC motors and rheostatic control are being in use. The problem can be solved by development of the new autonomous from power supply system recuperative braking modes with the application of energy storage systems based on supercapacitor, batteries, flywheel or mixed. Undoubtedly, in recent years the supercapacitors with high capacitances have become very popular due to a low weigh and dimensions, high power density, a large number of charge-discharge cycles, long operating lifetime and comparatively short payback period. This leads to numerous researches of the stationary [4, 5] or on-board storage systems [6-12] in the urban transport. Nevertheless, review of the published researches shows that all of them are dedicated to the electric vehicles with AC traction motors. The only few researches are dedicated to the electric vehicles with DC traction motors.

2 An overview of associated researches

[6] proposes the method of proper sizing of on-board supercapacitor energy storage system for catenary-free operations of a tram. It is based on the measurements made in the process of operation of empty experimental tram vehicle to estimate power and energy losses and find optimal size of the storage. It is equal to 0.75÷3 kWh and allows to overcome the distance of 1 km with the slope of 0÷1%.

The author [7] presents a control strategy for improving the energy recovery of AC traction drives and describes the results of simulation of a railway vehicle equipped with on-board supercapacitor storage system, which is based on *Maxwell* supercapacitor module of *HTM390* type. The control strategy was verified on a physical model and it was established that supercapacitors on the board of train allow to reduce the line current over the 50%. The same result is shown in [8-13]. Furthermore, [8] compares different types of storages and shows the

* Corresponding author: anatolij.nikitenko@ien.pw.edu.pl

results of experimental research with a prototype of light railway vehicle *GT6N* equipped with *Mitrac energy saver*. The system allows to achieve the energy saving of up to 30%. But all these researches are dedicated to electric vehicles with AC motors, accept of [10-13].

[10] investigates different types of electrical vehicles with supercapacitor storages. There are trolleybus *Jelcz M121E* and the *116N* tram with DC motors. The storage allows to reduce the energy consumption on 0.9 kWh/km and nearly 30% of consumed energy. The estimated payback period equals 7÷8 years. At the same time, it should be pointed out, that according experimental investigation in [4] the life expectancy of supercapacitor storages equals 15 years.

The researches of the DC trams of *Tatra T3A* with DC motors are performed in [11, 12]. They describe the prospects of use and modernization of railway vehicles with DC traction motors. The main schematic solutions and results of simulation are shown. The storage allows to store efficiently all the energy recovered during the braking.

In [13] the development of the regenerative braking of series-wound brushed DC electric motors is shown. The main reason is the widely usage of series-wound configuration in electrical vehicles and industry. It compares two schemes, shows semantical solutions with full-bridge modification and results of simulation for the DC motor of *FB1-4001* type, which is used in electric cars.

But all mentioned researches are dedicated to the urban vehicles (metro cars, trams, trolleybuses) and, at the moment, there are no publications which investigate the application of on-board storages in sub-urban trains.

3 Behaviours of regenerative braking of DC ERS with on-board storage

The application of energy storage system on the board of DC electric rolling stock (ERS) may improve the electrical effectiveness of regenerative braking and, simultaneously, stipulates the behaviours of electromagnetic transient processes, which run in the traction power circuits in the charge-discharge modes of the storage. These behaviours are caused by different factors. Firstly, peculiarity of charge-discharge cycles. Their parameters and thus the parameters of the elements of power circuit are changing in wide rage. Secondly, numerous types of the DC ERS apply the regenerative braking mode only till the speed of 51÷49 km/h. The initial speed of the application of regenerative braking mode is equal to 74÷62 km/h, in singular case 91÷88 km/h. Such speed range of regenerative braking is very low and it should be extended till the speed 20 km/h or 10÷5 km/h when the mechanical braking is applicable [2]. This purpose allows to increase the effectiveness of the recovered (kinetic) energy use and may to be achieved with the application of on-board supercapacitor storage system (SSS). Thirdly, in the process of starting up or acceleration the voltage of ERS arises in proportion to the speed, when the voltage of SSS falls down within its recharging. Finally, in the regenerative braking mode (RBM) the speed reduces and thus the total voltage of motors falls down too. But the voltage of SSS increases in proportion to the braking current and the time duration of its flowing. All these behaviours make the analysis of electromagnetic transient processes more complicated and keep finding the alternative approaches for their investigation.

4 Behaviours of regenerative braking of DC ERS with on-board storage

The main purpose of electromagnetic processes investigation is the defining of time dependences of braking current $i(t)$ in the traction electric motors (TEM) and voltage $u_C(t)$ across the SSS. The defining and support of braking current constancy is especially important so long as it is necessary to keep in the constant value of the braking force.

The principal scheme of the regenerative braking with the on-board SSS is shown in Fig. 1 for an DC multiple-unit train (as yet without input and output filters), where Sw1 is a pulse controller-interrupter; Sw2 and Sw3 are the voltage and current pulse controllers (switches).

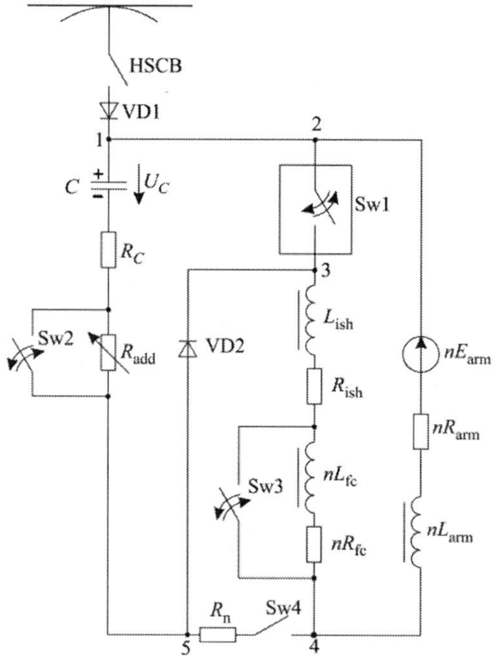

Fig. 1. Principal scheme of the on-board SSS charging.

Let us assume the next conditions for calculation and analysis of the transient processes:

- the inductances of TEMs and inductive shunt are constant values;
- the diodes VD1 and VD2 are an ideal;
- the controllers Sw1, Sw2 and Sw3 operate instantaneously;
- a rotary speed is a constant at the moments of controllers' operation, because the armatures have considerable inertia masses;
- the pulse controllers are changed by switches;
- a self-excitation stage of TEMs has been finished.

4.1 The first stage – zone of high speeds

Let accept that in the high speeds (over 50 km/h) the TEMs' magnet systems operate in the linear part of normal magnetization curve $C_e\Phi(I_{fc})$ within the range from an initial (maximal) speed V_{in} till a final speed V_{fn1}. In the period of speed decreasing the pulse controller Sw3 holds the constancy of armatures' emfs nE_{arm} of n traction motors connected in series. The braking (charging) current $i(t)$ is kept in constancy by the additional resistance $R_{add}(t)$ using pulse controller Sw2. According to calculating scheme in Fig. 2 and taking into account that the braking current $i(t)$ equals the currents in the armature $i_{arm}(t)$ and in the field coil $i_{fc}(t)$, the equation for the first charging stage may to be written as

$$u_C(t)+\left(R_\Sigma+R_{add}(t)\right)\cdot i(t)+L_\Sigma\frac{di}{dt}=nE_{arm}=E_{gen}=const,\quad(1)$$

where $R_\Sigma=n\cdot\left(R_{arm}+R_{fc}+R_{cw}\right)+R_{ish}+R_C=$
$=nR_{TEM}+R_{ish}+R_C$;
$L_\Sigma=n\cdot\left(L_{arm}+L_{fc}+L_{cw}\right)+L_{ish}=nL_{TEM}+L_{ish}$;
E_{arm}, R_{arm}, L_{arm} – emf, resistance and inductance of the armature;
E_{gen} – total generated emf;
R_{fc}, L_{fc} – resistance and inductance of the field coil;
R_{cw}, L_{cw} – resistance and inductance of the commutating winding;
R_{ish}, L_{ish} – resistance and inductance of the inductive shunt;
R_C –internal resistance of supercapacitor storage.

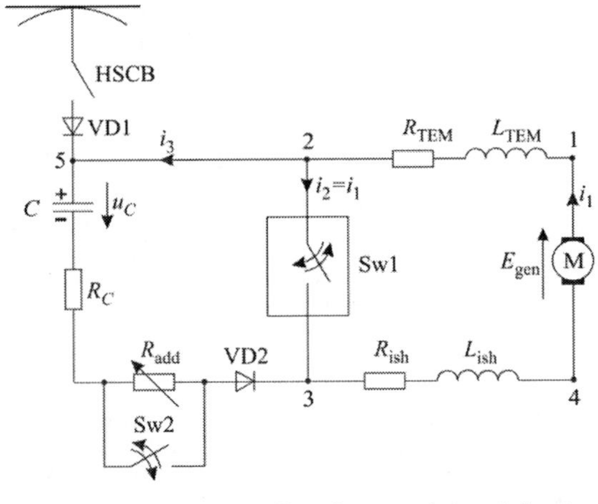

Fig. 2. Calculating scheme of impulse transmission of electric energy to SSS.

As for as $i(t)=C^{du_C}\!/_{dt}$ the equation may to be rewritten as

$$\frac{d^2u_C}{dt}+\frac{R_\Sigma+R_{add}(t)}{L_\Sigma}\cdot\frac{du_C}{dt}+\frac{1}{L_\Sigma\cdot C}=$$
$$=nE_{arm}=E_{gen}=const,\quad(2)$$

which have to be solved for a certain value of initial condition $u_C(0)=U_{C0}$.

Let us calculate (2) for a speed V and find the expression of $u_C(t)$ for the charging mode using Classical approach [14] as

$$u_C(t)=u_{Cst}(t)+u_{Cfr}(t).\quad(3)$$

As far as $E_{gen}(V)$ is an invariable in this stage, thus the voltage $u_{Cst}(t)$ for a steady-state mode is

$$u_{Cst}(t)=E_{gen}=C_e\Phi\cdot V(t)=C_e\Phi\cdot V.\quad(4)$$

According to Fig. 2 the free component $u_{Cfr}(t)$ of the voltage for a general case is

$$u_{Cfr}(t)=A_1e^{p_1t}+A_2e^{p_2t}.\quad(5)$$

As for as $R_\Sigma>R_{critical}=2\sqrt{L_\Sigma/C}$, so the roots p_1 and p_2 are negative and have to be found as

$$p_{1,2}=-\frac{R_\Sigma+R_{add}(t)}{2L_\Sigma}\pm\sqrt{\frac{R_\Sigma+R_{add}(t)}{4L_\Sigma}-\frac{1}{L_\Sigma C}}.\quad(6)$$

Let find the integrating constants A_1 and A_2. Substitute (4) and (5) into (3) and derive it by t as

$$\begin{cases}u_C(t)=E_{gen0}+A_1e^{p_1t}+A_2e^{p_2t},\\ \dfrac{du_C}{dt}=0+A_1p_1e^{p_1t}+A_2p_2e^{p_2t}.\end{cases}\quad(7)$$

After solving of (7), the voltage $u_C(t)$ and the current $i_3(t)=C\dfrac{du_C}{dt}$ in the first stage of charging have to be found according the expressions:

$$u_C(t)=E_{gen0}+\frac{\left(E_{gen0}-U_{C0}\right)p_2}{p_1-p_2}e^{p_1t}-\frac{\left(E_{gen0}-U_{C0}\right)p_1}{p_1-p_2}e^{p_2t},\quad(8)$$

$$i_3(t)=i_1(t)=\frac{E_{gen0}-U_{C0}}{L_\Sigma(p_1-p_2)}e^{p_1t}-\frac{E_{gen0}-U_{C0}}{L_\Sigma\cdot(p_1-p_2)}e^{p_2t}.\quad(9)$$

4.2 The second stage – zone of middle and low speeds

In the previous stage the TEMs' magnet systems reach the full-field mode ($C_e\Phi=const$) till the speed V_{fn1}. As for as the speed continues decreasing till the speed V_{fn2}, the total emf also reduces and it can't be corrected by the current I_{fc} of field coil, therefore the impulse mode of RBM should be applied on the second stage.

According Fig. 2 the regenerative braking operates with short switching on of TEMs by controller Sw1. As a result, the braking current $i_1(t)$ flows in the loop 1-2-3-4 and the equation of this mode may to be written as

$$\left(n\cdot R_{TEM}+R_{ish}\right)\cdot i_1(t)+\left(n\cdot L_{TEM}+L_{ish}\right)\cdot\frac{di_1}{dt}=$$
$$=E_{gen}(t)=n\cdot C_e\Phi\cdot V(t).\quad(10)$$

Using Classical approach [14] with initial data $i_1(0)=i_3(0)=I_{02}$, the equation of current $i_1(t)$ is

$$i_1(t) = \frac{n \cdot C_e \Phi \cdot V(t)}{n \cdot R_{TEM} + R_{ish}} + \left[I_{02} - \frac{n \cdot C_e \Phi \cdot V(t)}{n \cdot R_{TEM} + R_{ish}} \right] e^{-\frac{t}{\tau_1}}, \quad (11)$$

where the time constant equals $\tau_1 = \dfrac{n \cdot L_{TEM} + L_{ish}}{n \cdot R_{TEM} + R_{ish}}$.

While the current $i_1(t)$ is increasing, the TEMs is storing the electromagnetic energy during the period of closed controller Sw1. After some time, the commutator Sw1 is opened so the total emf $E_{gen}(t) = n \cdot E_{arm}$ and the emf of self-induction $E_L(t)$ create the braking current in the loop 1-2-5-C-R_C-Sw2-3-4-1. This process is described by the next equation

$$\left[R_{add}(t) + R_{\Sigma}\right] \cdot C \frac{du_C}{dt} + u_C(t) =$$

$$= n \cdot C_e \Phi \cdot \left(V_{fn1}(t) + \frac{dV}{dt} \cdot t \right) + E_L(t), \quad (12)$$

where

$$E_L(t) = (n \cdot L_{TEM} + L_{ish}) \cdot \left[\frac{n \cdot C_e \Phi}{n \cdot R_{TEM} + R_{ish}} \cdot \frac{dV}{dt} - \frac{I_{02}}{\tau_1} \cdot e^{-\frac{t}{\tau_1}} + \frac{n \cdot C_e \Phi}{n \cdot R_{TEM} + R_{ish}} \cdot \frac{V(t)}{\tau_1} \cdot e^{-\frac{t}{\tau_1}} - \frac{n \cdot C_e \Phi}{n \cdot R_{TEM} + R_{ish}} \cdot \frac{dV}{dt} \cdot e^{-\frac{t}{\tau_1}} \right].$$

Solution of (12) allows to find the charging voltage $u_C(t)$ and charging current $i_3(t) = C \dfrac{du_C}{dt}$.

In the next turning on of the controller Sw1 the TEMs are shorted again and the current $i_1(t)$ increases. After turning off of the controller Sw1 the current charges the SSS and cycle is repeated again. Such braking process reduces the speed V and total TEMs' emf E_{gen} in (10), therefore the duty cycle and the additional resistance R_{add} have to be reduced to keep the barking current $I_1 = const$.

5 Numerical calculations of charging process

The initial data for numerical calculations were found by the experimental researches for the EPL2T multiple-unit train in the process of real operation in the sections of Prydniprovsk railway (Ukraine). All data were recorded using on-board control system and train was driven according standard schedule and driving chart. The motor car of the train has four DC motors of 1DT.003.L8U1 type, which are connected in series and have series excitation. Some parameters of TEM elements are shown in Table 1 [15].

The statistical and probability analyses are done to find the values of voltage, recovered energy, initial and final speeds of the regenerative braking mode and they are shown in Figs. 3-5. The calculations of the voltage across the SSS and the charging current are performed using the abovementioned equations (8)-(9), solution of (12) for a motor car of 3 kV DC multiple-unit train and results of experimental researches. Let us analyse them and describe the selection of the initial data for numerical calculations.

The initial speed V_{in} of regenerative braking varies in wide range with the mathematical expectation equals 72.1 km/h in Fig. 3. Its theoretical curve has normal distribution, therefore this value may to be used in the numerical calculation as the most probably one.

Table 1. Main parameters of the parts of DC traction motor of 1DT.003.L8U1 type [15].

Parts of TEM and their parameters		Values for the exceed temperatures	
		+20°C	+75°C
Armature coil	R_{arm} [Ohm]	0.052	0.068
	L_{arm} [H]	0.0072	0.0072
Commutating winding	R_{cw} [Ohm]	0.017	0.022
	L_{cw} [H]	0.0033	0.0033
Field coil	R_{fc} [Ohm]	0.097	0.127
	L_{fc} [H]	0.014	0.014
Coil of inductive shunt	R_{ish} [Ohm]	0.0306	0.04
	L_{ish} [H]	0.0085	0.0085

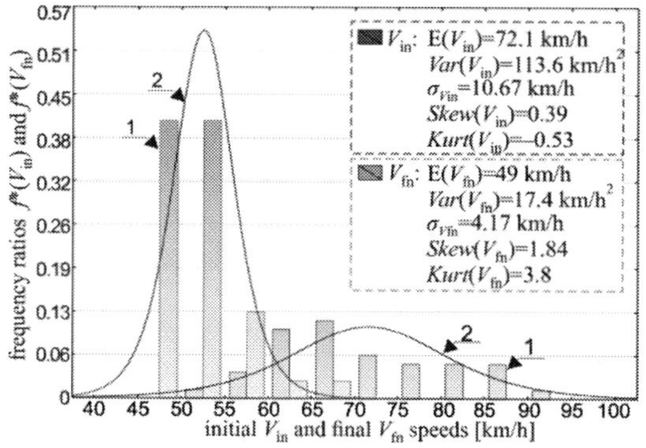

Fig. 3. Statistical (1) and theoretical (2) distributions of the initial and final speeds in the RBM.

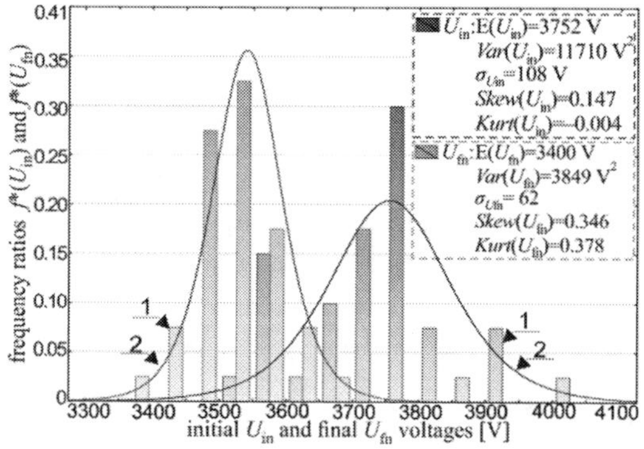

Fig. 4. Statistical (1) and theoretical (2) distributions of the initial and final voltages in the RBM.

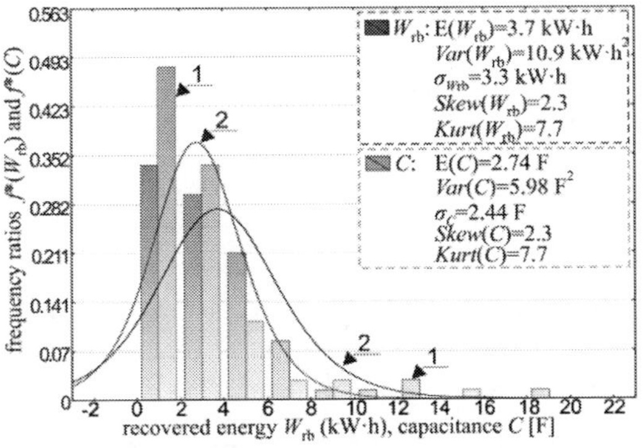

Fig. 5. Statistical (1) and theoretical (2) distributions of the energies recovered in different phases in the RBM and necessary capacities to store it.

The nominal voltage U_{Cnom} of the SSS is one of the most significant quantities, which depends on the volume of recovered energy in a phase of RBM and influences on number of series-parallel cells and, as a result, on total capacitance and internal resistance of the SSS. According different reasons and researches, which are overview above in the paper, it is determined by the nominal voltage of current collector (3 kV), or the nominal voltage of overhead contact system (3.3 kV), or the open circuit voltage of traction substation (3.6 kV), or the total nominal voltage of TEMs (3.75 kV), or the maximal allowable voltage of TEMs (3.95 kV). The statistical analysis presented in Fig. 4 shows that the voltage across the current collector of train in the initial and final moments of the RBM are varying within the range from 3.4 to 4.02 kV with mathematical expectations equal 3.752 kV and 3.4 kV respectively. Both values are not distributed by the normal low, so the numerical calculations were performed for a few nominal voltages (3.6, 3.75 and 3.95 kV) to find the most appropriate of them.

The nominal capacity C of SSS is determined by the energy W_{rb}, which is recovered in a phase of RBM. Fig. 5 shows the statistical and theoretical distributions of these values. The capacities here are calculated for each phase of RBM according the next formula:

$$C = \frac{2 \cdot W_{\mathrm{rb}}}{U_{\mathrm{Cnom}}^2 - (0.5 \cdot U_{\mathrm{Cnom}})^2}. \qquad (13)$$

It is seen, that it has the mathematical expectation of 2.74 F for the voltage of 3.6 kV. As for as its theoretical distribution does not follow to the normal low, therefore it cannot be used as a nominal value. Capacitances for the SSS with nominal voltage of 3.6 kV are 2.25, 3 and 4.5 F. Many companies produce the supercapacitors, for example: MAXWELL, YUNASKO, WIMA, Nesscap, SAMWHA, IOXUS, Cornell Dubilier et al. The YUNASKO supercapacitors were used in calculations and their parameters are given in [16].

The residual voltages of SSS are equal to 25, 50 and 75% of the nominal voltage U_{Cnom}.

The additional resistance is calculated as $R_{\mathrm{add}}(t) = 8.1 - 0.685 \cdot t$ (Ohm) and is necessary to restrict the charging current of SSS, but practically is used only 2÷8 seconds of the first stage of RBM.

Fig. 6. shows of main time dependences of total TEMs' emf $E(t)$, the voltage of SSS $u_C(t)$, the charging current $i_1(t)$ and the speed $V(t)$ only for case when $C = 3$ F and $U_{\mathrm{Cnom}} = 3600$ V.

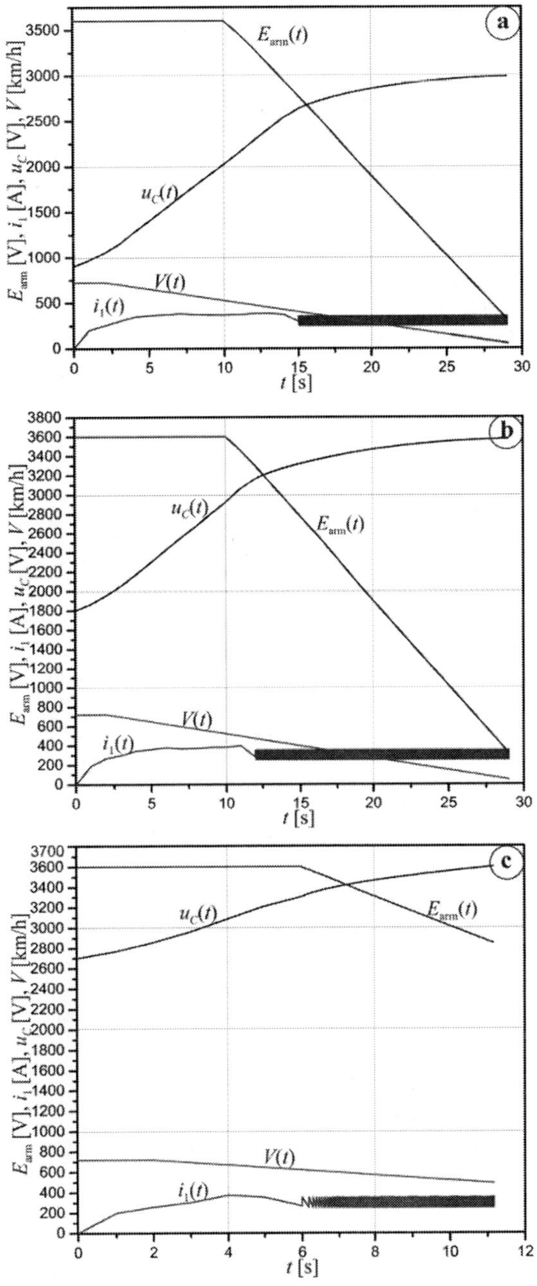

Fig. 6. Results of numerical calculations for C=3 F, U_{Cnom} =3.6 kV and different residual voltages: (a) 25%, (b) 50% and (c) 75% of U_{Cnom}.

6 Conclusions

The regenerative braking mode of ERS with the series-wound DC traction motor may to be performed in two stages with the use of energy storage system. It is seen from the results, that the capacitor of 3 F with residual voltage of 50% is suitable in comparison with all other cases. In this case, the speed changes with the rage of 72÷51 km/h and voltage of SSS increases from 1.8 to 3.15 kV. The second stage (impulse mode) is operable in the all cases and allows to finish charging till the SSS's voltage of 3.57 kV and train's speed of 5 km/h. The calculated time dependences of the total emf, the voltage of SSS and the charging current look as authentic thus the mathematical models are applicable and scheme is operable for practical application.

The research was financed under the Dean's Grant no 504/03412/1041/42.000100 from the statutory activities of the Faculty of Electrical Engineering of Warsaw University of Technology.

References

1. A. Szeląg, L. Mierzejewski, J. Altus and J. Drabek, Railway electric traction in new EU members of Central and East European countries, *Proceedings of 7th International Scientific Conference "Modern Electric Traction in Integrated XXIst Century Europe" MET'2005, Warsaw, Poland*, 192-205, (2005)
2. M. Kostin and A. Nikitenko, *Recuperative braking autonomy is a basis of reliable and efficient energy recuperation in DC electric rolling stock*, Railway Transport of Ukraine, **3**(106), 15-23, (2014) [in Russian]
3. A. Szelag, M. Kostin and A. Nikitenko, Losses of recovered electric energy in the elements of a DC electric transport system, *Proceeding of Power Electronics and Motion Control Conference (PEMC), 2016 IEEE International, Varna, Bulgaria*, 412-418 (2016)
4. J. Morand, D. Bergogne, P. Venet, A. Sari and P. Bevilacqua, An energy saver for tramway networks using double active bridge and supercapacitors, *Proceeding of 15th European Conference on Power Electronics and Applications, Lille, France*, 1-9 (2013)
5. M. Bartołomiejczyk and M. Połom, *Possibilities of reducing of energy consumption in urban transport on example of Gdynia Trolleybus Transportation System [Możliwości ograniczenia zużycia energii w transporcie miejskim na przykładzie gdyńskiej komunikacji trolejbusowej]*, Railway Vehicles *[Pojazdy Szynowe]*, **3**, 86-88 (2011) [in Polish]
6. A. Del Pizzo, L. Fratelli and D. Perna, A method for "design to range" energy storage systems in catenary free operations of light railway vehicles, *Proceeding of 2015 International Conference on Clean Electrical Power (ICCEP), Taormina, Italy*, 761-766, (2015)
7. D. Iannuzzi, Improvement of the energy recovery of traction electric drives using supercapacitors, *Proceeding of 13th International Power Electronics and Motion Control Conference EPE-PEMC, Poznań, Poland*, 1469-1474 (2008)
8. M. Steiner and J. Scholten, Energy storage on board of DC fed railway vehicles, *Proceeding of 35th IEEE Power Electronics Specialists Conference PESC, Aachen, Germany*, 666-671 (2004)
9. L. Streit and P. Drabek, Simulation and emulation of tram with on-board supercapacitors on Pilsen tram line, *Proceeding of International Conference on Clean Electric Power, Alghero, Czech Republic*, 703-706 (2013)
10. Z. Giziński et al., *Power supply circuits or traction vehicles with energy storages [Zasobnikowe układy zasilania w pojazdach trakcyjnych]*, Railway Vehicles [Pojazdy szynowy], **3**, 125-133 (2011) [in Polish]
11. V. Bražis and J. Greivulis, The application of supercapacitor energy storage devices in DC drives, *Proceedings of 7th International Scientific Conference "Engineering for Rural Development", Jelgava, Latvia*, 76-80 (2008)
12. L. Latkovskis and V. Bražis, Simulation of the regenerative energy storage with supercapacitors in Tatra T3A type trams, *Proceeding of 10th International Conference on Computer Modelling and Simulation, Cambridge, United Kingdom*, 398-403, (2008)
13 Y. Xiao et al., Regenerative braking of series-wound brushed DC electric motors for electric vehicles, *Proceeding of 7th IEEE Conference on Industrial Electronics and Applications ICIEA, Singapore*, 1658-1662 (2012)
14. M. Kostin and O. Sheikina, *Theory of Electrical Engineering*, Part 2, (Pub. Dep. of Dnipropetrovsk National University of Railway transport, Dnipropetrovsk 2007) [in Ukrainian]
15. H. Basov and S. Yatsko, *Progress of Electric Multiple-Unit Rolling Stock*, Part 2, (Apeks+, Kharkiv 2005) [in Ukrainian]
16. YUNASKO Ltd. *EDLC Product: 3000 F E-cell*, (2016) http://yunasko.com/images/data/YUNASKO_Datasheet_cell_3000F_E.pdf

Energy storage system in traction vehicle

Maciej Wieczorek[1,*], and *Mirosław* Lewandowski[1]

[1]Warsaw University of Technology, Institute of Electric Power Engineering, Electric Traction Division, ul. Koszykowa 75, 00-662, Warsaw, Poland

Abstract. The paper compares three different types of energy storage system (ESS) in a tramway. It was assumed that the tram has to travel without catenary for 5 km. Two homogeneous energy storage systems were designed to provide energy for the ride: the first made of lithium-ion batteries and the second made of supercapacitors. The third solution is a hybrid ESS (HESS) composed of both of the aforementioned technologies. The comparison was made in terms of weight, price and power consumption when driving with and without a catenary. Using a well-fitted energy management strategy, 50% lower HESS mass was achieved compared to the homogeneous ESSes. The estimated price of the HESS is 29% lower than the battery ESS and 59% from the supercapacitor ESS. A comparison of the energy consumption of the tram with and without the ESS is also presented, depending on the amount of energy absorbed by other vehicles on the same power section. It follows that the use of the ESS allows for significant energy savings for driving with the catenary.

1 Introduction

Energy storage systems (ESS) are increasingly being used in electric traction as a means of more effectively utilizing regenerative braking energy which, in case of rail vehicles, is a significant part of energy taken from power system because of their large mass, or to maintain proper voltage [1]. For this purpose, supercapacitors are most often used because of their high specific strength.

In some applications a tram is required to be able to travel part of the route without power supply. This situation can occur in e.g. historic parts of towns where it is impossible to set up a traction network. A supercapacitor ESS would have to be much bigger to provide the energy. That would translate into a high price of the ESS. The use of a lithium-ion battery will allow to achieve the energy needed for the travel. However, due to low specific power of the lithium cells it is necessary to oversize the ESS in terms of stored energy. This will also translate into a high price of the device.

The solution to the optimum ESS selection is the use of a hybrid ESS (HESS) consisting of lithium batteries and supercapacitors. HESS is a matter widely discussed in literature. Especially when electric vehicles are considered [2], [3]. HESS effectiveness is, however, strongly dependent on the energy management strategy EMS used.

HESS sizing and EMS optimization for an electric car was presented in [4], [5]. Similar topic was also addressed to urban vehicles such as buses in [6-8] and tramway in [9]. HESS's profitability analysis was performed [10].

This article focuses on the selection of ESS to tramway, which must move without catenary for a short distance and compare different solutions of the ESS in terms of mass, price and energy consumption.

2 Trams ESS configurations

It was assumed that the ESS in tram should enable it to drive for 5 km without catenary. To size the ESS the method presented in [11] was adopted. For calculating the initial cost of ESSes the prices of cells were taken from one distributor. The total price of energy storage pack was estimated on the basis of modules and packs assembly cost presented in BCG report [12]. The mass of the components of the packs was assumed at 24% of the cell mass based on the data presented in [13].

In the case of a homogeneous battery ESS, the maximum power is the decisive factor for the final size of the battery pack. Therefore, the stored energy is much higher than the demand. Less than 5% of capacity is used. On the other hand, this has a positive effect on the battery life since it is strictly dependent on the depth of discharge [14]. Nevertheless, such a large oversizing of battery inflates the price and weight (Table 1).

Sizing a supercapacitor ESS should be based primarily on the energy capacity of the package. Due to the significantly higher power density of supercapacitors than lithium batteries, the power of the supercapacitor will be much higher than necessary. In this case, it also results in a high price and the weight of the ESS.

The process of sizing a HESS requires simultaneous design of EMS. The Gamma Based Strategy described in [4] was used in the analysis because of its simplicity and the fast process of selecting the coefficients that shape

* Corresponding author: maciej.wieczorek@ee.pw.edu.pl

https://doi.org/10.1051/matecconf/201818002008

the power of individual devices. It also allows for effective use of supercapacitor capacity.

Power demand of a tram is largely predictable, due to fixed route and minor interaction with traffic. This makes it possible to avoid oversizing ESS. The analysis carried out in this paper has included the tram ride in the UITP sort2 (easy urban) cycle. An example of the power of the devices in HESS is shown in Figure 1. The detailed parameters of the trays are given in Table 1. Red colour indicates the parameters that have the decisive influence on the final size of the pack.

3 ESS models and simulation

The energy ESS installed on board of a vehicle, besides enabling drive without catenary, should accumulate regenerative braking energy. In situations where other trams on the same power section are unable to receive the braking energy, it is dissipated as heat. Therefore, the use of an ESS in a tram should reduce energy consumption. The EMS in HESS differs for driving with the catenary (Fig. 2) and without it (Fig. 1). During no-catenary operation, the supercapacitor provides high power during acceleration due to low maximum power of the battery. It is important not to discharge the supercapacitor prematurely as this would result in a power shortage. Therefore, during regenerative braking and at standstill, the supercapacitor is recharged from the battery.

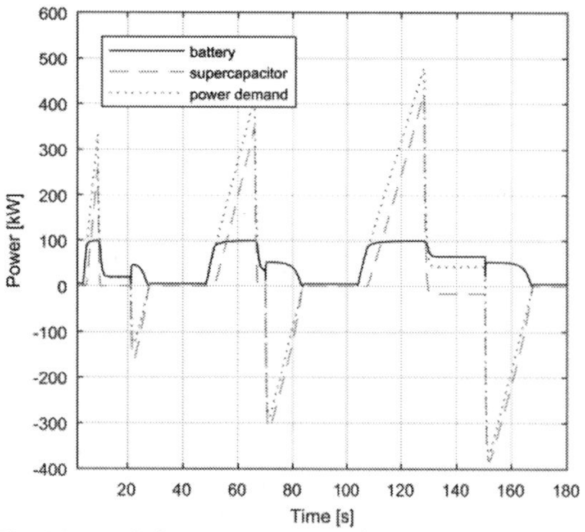

Fig. 1. Power during no-catenary operation.

While driving with catenary, the tram does not use the battery pack of HESS. Only supercapacitors work. The analysis assumes that all the regenerative braking energy is accumulated by the ESS and then used for acceleration. The strategy of using ESS when traveling with catenary is identical for trams with homogeneous ESS.

Table 1. ESSes parameters.

	BATT TRAM	HESS TRAM	SC TRAM
VEHICLE			
total vehicle mass (without passengers) [kg]	45983	44768	45946
motor power [kW]	200	200	200
reference autonomy in UITP sort 2 (easy urban) [km]	5	5	5
calculated autonomy in UITP sort 2 (easy urban) [km]	120	23	5
DC voltage [V]	600	600	600
auxiliary load [kW]	5	5	5
BATTERY MODULE			
manufacturer	HEADWAY		-
chemistry	LFP	LFP	-
number of cells connected in series	186	186	0
number of cells connected in parallel	30	6	0
stored energy [kWh]	178.5	35.71	0
cells mass [kg]	1841	368.2	0
module mass [kg]	2283	456	0
nominal voltage [V]	595.2	595.2	0
cells cost [$]	117180	23436	0
module cost [$]	164052	32810	0
nominal power for 2000 cycles (I=0.5C) [W]	89280	17856	0
Pmax_in	357120	71424	0
Pmax_out	535680	107136	0
SUPERCAPACITOR MODULE			
manufacturer	-	MAXWELL	
number of cells connected in series	0	222	222
number of cells connected in parallel	0	4	16
nominal voltage [V]	0	599.4	599.4
capacitance [F]	0	54	216
cells mass [kg]	0	453	1811
module mass [kg]	0	561	2246
cells cost [$]	0	53280	213120
module cost [$]	0	74592	298368
stored energy [Wh]	0	2699	10798
usable energy [Wh]	0	2024	8098
nominal power [kW]	0	503.5	2014.0
maximum power [kW]	0	2672	10688
DC/DC CONVERTERS			
number of converters	1	2	1
cost of converters	15000	20000	15000
mass of converters	100	150	100
ENERGY STORAGE SUMMARY			
Total energy storage system mass [kg]	2383	1168	2346
Initial energy storage system cost [$]	179052	127402	313368

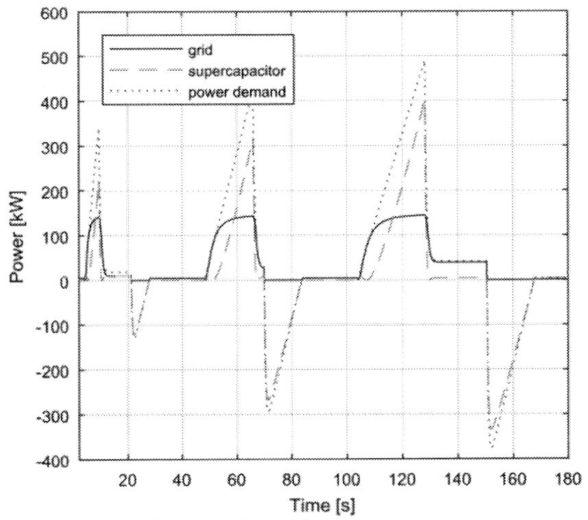

Fig. 2. Power during operation with catenary.

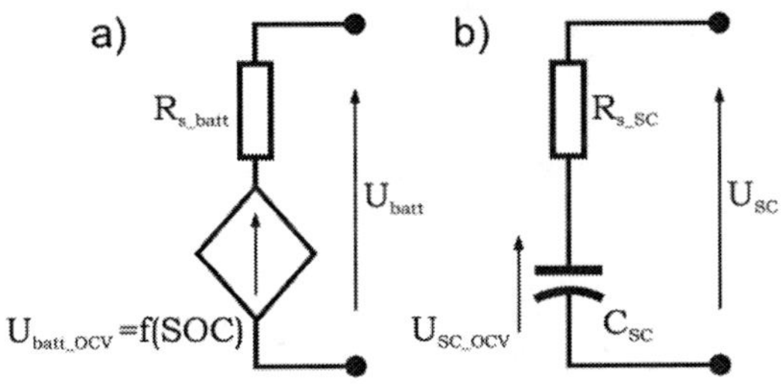

Fig. 3. Equivalent circuit of a) lithium battery and b) supercapacitor.

In order to estimate the energy consumption, simulations of each tramway were performed. The energy storage device models used here affect the energy consumption estimate. Thus, the most important parameters for the calculation are the source voltage and the internal resistance. Therefore, a series models have been used, i.e. of the voltage source and resistance for the battery (Fig. 3a) and of the RC model for the supercapacitor (Fig. 3b).

The internal resistance and capacity of lithium batteries and supercapacitors varies depending on a number of factors, such as state of charge (SOC), value and direction of load current, temperature, and state of health (SOH) [14-18]. These changes, varying from 70% to 200% of rated values given by manufacturers [14], [19-22] have minor influence on trams energy consumption comparison. Therefore in the analysis they were neglected.

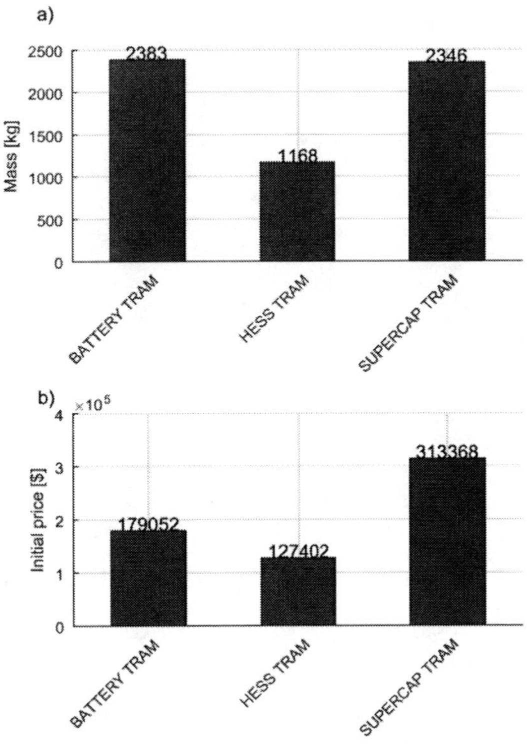

Fig. 4. Comparison of a) mass and b) prices of each ESS.

4 Results and discussion

Fig. 4a shows masses of ESSes in described vehicles. Masses of the homogeneous ESSes are similar. Comparing with them the HESS has 50% lower mass. Compared to the total weight of the tram, this difference seems small, but it has an impact on energy consumption. More interesting result from the point of view of both the producer and buyer is the price of the energy storage. This is highest in the case of the supercapacitor (Fig. 4b). The cheapest solution is the HESS. It is 29% cheaper than the battery ESS and 59% cheaper than the supercapacitor ESS.

The analysis compares the energy consumption of the proposed trams. Fig. 5a shows the energy consumption per kilometre when driving without catenary. Differences are small. The smallest energy consumption was achieved by a vehicle with a supercapacitor. This is because this ESS has the lowest internal resistance.

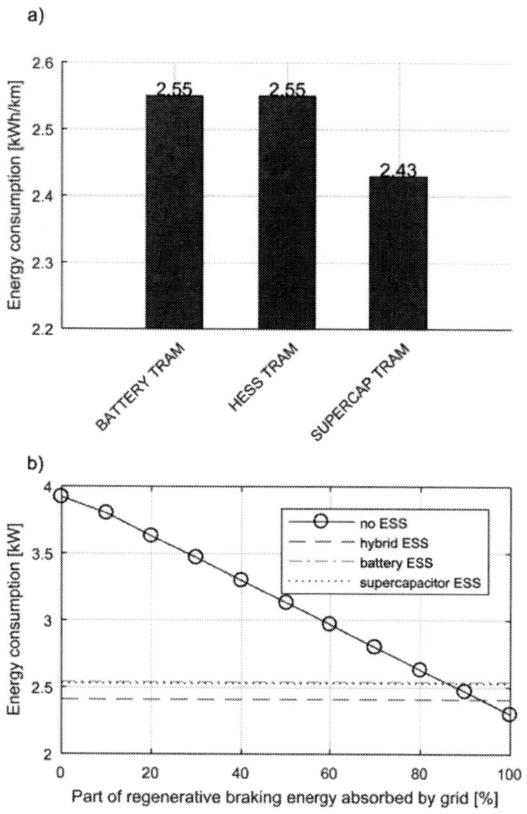

Fig. 5. Energy consumption comparison for a) operation without catenary and b) operation with catenary.

Fig. 5b shows the energy consumption when traveling with the catenary. The results for trams with different ESS are compared to the energy consumption of a tram without ESS, whose braking energy is received by other vehicles at the power section. It can be seen that even in the case of driving with a grid power supply, the energy storage can significantly reduce energy consumption. The energy consumption of the tram without ESS was lower when other vehicles received more than 90% of its braking energy.

5 Conclusions

Basing on the results presented in the paper it can be stated that the use of a hybrid energy storage system with a well-fitted energy management strategy allows to achieve much better parameters than the uniform tanks. This applies both to energy and economic parameters.

References

1. A. Szelag, *Wpływ napięcia w sieci trakcyjnej 3 kV DC na parametry energetyczno-trakcyjne zasilanych pojazdów* (Instytut Naukowo-Wydawniczy SPATIUM, 2013) [In Polish]

2. X. Luo, J. Wang, M. Dooner, and J. Clarke, *Overview of current development in electrical energy storage technologies and the application potential in power system operation*, Appl. Energy, **137**, 511–536 (2015)

3. R. Hemmati and H. Saboori, *Emergence of hybrid energy storage systems in renewable energy and transport applications* - A review, Renew. Sustain. Energy Rev., **65**, 11–23 (2016)

4. M. Wieczorek and M. Lewandowski, *A mathematical representation of an energy management strategy for hybrid energy storage system in electric vehicle and real time optimization using a genetic algorithm*, Appl. Energy, **192**, 222-233 (2017)

5. A. Castaings, W. Lhomme, R. Trigui, and A. Bouscayrol, *Comparison of energy management strategies of a battery/supercapacitors system for electric vehicle under real-time constraints*, Appl. Energy, **163**, 190–200 (2016)

6. V. I. Herrera, A. Milo, H. Gaztañaga, and H. Camblong, Multi-objective Optimization of Energy Management and Sizing for a Hybrid Bus with dual Energy Storage System, *2016 IEEE Vehicle Power and Propulsion Conference (VPPC)* (2016).

7. Z. Song, H. Hofmann, J. Li, X. Han, and M. Ouyang, *Optimization for a hybrid energy storage system in electric vehicles using dynamic programing approach*, Appl. Energy, **139**, 151–162 (2015)

8. Z. Song, J. Li, X. Han, L. Xu, L. Lu, M. Ouyang, and H. Hofmann, *Multi-objective optimization of a semi-active battery/supercapacitor energy storage system for electric vehicles*, Appl. Energy, **135**, 212–224 (2014)

9. V. Herrera, A. Milo, H. Gaztañaga, I. Etxeberria-Otadui, I. Villarreal, and H. Camblong, *Adaptive energy management strategy and optimal sizing applied on a battery-supercapacitor based tramway*, Appl. Energy, **169**, 831–845 (2016)

10. D. Zhu, S. Yue, S. Park, Y. Wang, N. Chang, and M. Pedram, Cost-effective design of a hybrid electrical energy storage system for electric vehicles, *Proc. 2014 Int. Conf. Hardware/Software Codesign Syst. Synth. - CODES '14*, 1–8 (2014)

11. R. Sadoun, N. Rizoug, P. Bartholomeus, and P. Le Moigne, *Optimal sizing of hybrid supply for electric vehicle using Li-ion battery and supercapacitor*, Int. Rev. Electr. Eng., **9**(2), 332–340 (2014)

12. The Boston Consulting Group, *Focus Batteries for Electric Cars*, Outlook, **1**, 1–18 (2010)

13. P. Miller, *Automotive Lithium-ion Batteries*, Johnson Matthey Technol. Rev., **59**(1), 4–13 (2015)

14. N. Omar, M. Abdel, Y. Firouz, J. Salminen, J. Smekens, O. Hegazy, H. Gaulous, G. Mulder, P. Van Den Bossche, and T. Coosemans, *Lithium iron phosphate based battery – Assessment of the aging parameters and development of cycle life model*, Appl. Energy, **113**, 1575–1585 (2014)

15. M. A. Danzer, V. Liebau, and F. Maglia, *Aging of lithium-ion batteries for electric vehicles* (Elsevier Ltd., 2015)

16. K. Jalkanen, J. Karppinen, L. Skogström, T. Laurila, M. Nisula, and K. Vuorilehto, *Cycle aging of commercial NMC/graphite pouch cells at different temperatures*, Appl. Energy, **154**, 160–172 (2015)

17. N. Omar, Y. Firouz, H. Gualous, J. Salminen, T. Kallio, J. M. Timmermans, T. Coosemans, P. Van den Bossche, and J. Van Mierlo, *9 – Aging and degradation of lithium-ion batteries* (Elsevier Ltd., 2015)

18. M. Lewandowski, M. Orzyłowski, *Fractional-order models: The case study of the supercapacitor capacitance measurement*, Bull. Polish Acad. Sci. -Technical Sci., **65**(4), 449-457 (2017)

19. D. Anseán, M. González, M. Ieee, J. C. Viera, M. Ieee, J. C. Antón, and C. Blanco, *Evaluation of LiFePO 4 batteries for Electric Vehicle applications*, IEEE Trans. Ind. Appl., **51**(2), 1855–1863 (2015)

20. H. Gualous, R. Gallay, M. Al Sakka, A. Oukaour, B. Tala-Ighil, and B. Boudart, *Calendar and cycling ageing of activated carbon supercapacitor for automotive application*, Microelectron. Reliab., **52**(9–10), 2477–2481 (2012)

21. P. Kreczanik, P. Venet, A. Hijazi, and G. Clerc, *Study of supercapacitor aging and lifetime estimation according to voltage, temperature, and RMS current*, IEEE Trans. Ind. Electron., **61**(9), 4895–4902 (2014)

22. D. Yan, L. Lu, Z. Li, X. Feng, M. Ouyang, and F. Jiang, *Durability comparison of four different types of high-power batteries in HEV and their degradation mechanism analysis*, Appl. Energy, **179**, 1123–1130 (2016)

Experimental study of the use of electric car powered with stationary solar and electrochemical batteries in Northern Poland

Dariusz Karkosiński[1,*], *Michał* Pacholczyk[1], and *Łukasz* Sienkiewicz[1]

[1] Gdańsk University of Technology, Faculty of Electrical and Control Engineering, Laboratory of Innovative Power Technologies and Integration of Innovative Energy Sources: LINTE^2, ul. Sobieskiego 5, 80-216, Gdańsk, Poland

Abstract. The subject of this paper is an experimental analysis of a Nissan LEAF electric car equipped with 24 kWh battery, powered from the standalone photo-voltaic (PV) charging station in Gdańsk, Poland. The calculations of charging process efficiency and range of test drives were conducted in two extreme situations (winter and summer) of sunlight and road conditions. Experiments were performed in the Laboratory of Innovative Power Technologies and Integration of Innovative Energy Sources (LINTE^2) at Gdańsk University of Technology. The car was night-charged with the stationary electrochemical battery, which, in turn, was powered with PV panels during the day. The impact of sunlight and ambient temperature changes on daily urban and highway drive ranges was analysed. The efficiency of energy conversion in multiple current converters was determined.

1 Properties of electric vehicles: charging process and usage

Electric vehicles (EV) were one of the first car for carrying people. In the years 1832 - 1839 the first carriages and cars with the voltaic pile were announced and tested. Between 1859 and 1871, vehicles with a lead-acid battery and a DC motor were built. EVs were produced along with the steam engines vehicles. Internal combustion engines have not been in use yet. At the beginning of the twentieth century, they EVs were pushed out of the market by vehicles with internal combustion engines, including the Ford T, mass produced from 1908 to 1912. Developing oil lobby has later lead to the almost complete collapse of the EV concept.

The present decade should be considered as a breakthrough in the worldwide development of technology and sales of electric cars. Further advancements in power electronics [1] and the evolution of the EV propulsion battery (from lead-acid to almost exclusively higher efficiency, lighter and characterized by low self-discharge rate lithium battery [2, 3]) can be considered as factors that brought the industry's attention back to electric vehicles. However, the environmental concerns in terms of greenhouse gases (GHG) emissions can also be identified as a major accelerator of the EV market [2]. As reported in [4] the transportation sector in 2012 consumed almost 27% of the global energy production and was responsible for 33.7% GHG emissions. Replacement of the conventional and inefficient internal combustion engine vehicles (ICEV) with hybrid electric vehicles (HEV) and all electric vehicles including battery electric vehicle (BEV) and fuel cell electric vehicle (FCEV) can be a way to not only limit emissions but also to provide cleaner and quieter ambiance as well as to drastically reduce the operational cost compared to gas-powered cars [5]. Since HEV use both electric motor (EM) and ICE to move a vehicle and FCEV face problems with hydrogen storage, BEV are the most interesting regarding environmental protection and in this paper the definition of EV refers to that kind of vehicles.

Although promising, EV capabilities are still limited, additional and continuous development is required to confront market's needs. The specific characteristics of modern electric vehicles include:

a. Replacement of combustion engine vehicle with an EV shifts the emissions from the car to the power plant. Moreover, charging car batteries from a central power station is more efficient than burning fuel in separate engines [6]. Unless produced in a "green" way, GHG are still emitted;
b. Electric vehicles do not emit pollutants while in use but due to not yet fully mature technology and high utilization of rare earth metals, the production process generates much more emissions compared with the ICE cars production process;
c. Limited real drive range: under 250 km for cheaper models and above 400 km - only for the most expensive ones; with electric heating or air-conditioning decreasing the range drastically. Regenerative breaking is often used

* Corresponding author: dariusz.karkosinski@pg.edu.pl

to increase the range. Battery can be also supplemented with additional energy sources e.g. vehicle-integrated PV [7];

d. Long charging time of the battery pack: 10-12 hours from home grid and not shorter than 20 minutes from fast charging stations;
e. The battery capacity drops to approximately 50% under -20 °C;
f. The asking price of an electric vehicle is much higher than the combustion engine equivalent. The huge part of the EV cost lies in battery pack which can exceed 45% of total vehicle cost [8]. On the other hand, electric cars are much simpler and have fewer parts compared to existing vehicles, [6];
g. Limited durability of the battery pack – approx. 200,000 km or 10 years with high replacement cost. The driving range is decreasing due to the wear of the battery pack;
h. The drive range of EVs also depends on the rolling resistance caused by deflection - the flattening of the tires. In every kind of car, the reduced pressure, e.g. by 0.5 bar, increases the rolling resistance by about 15%, which in electric cars translates into a proportional reduction of range.

There are a few models of EVs on the polish market. One of the bestselling models is Nissan LEAF (Leading, Environmentally friendly, Affordable, Family car - produced since 2010). Nissan LEAF (2015 model tested by authors: fig. 1) can be charged via 3 kW, slow charger (charging duration 8 h), 44 kW, 480 V DC fast charger (up to 80% of capacity in 30 min) in SchAdeMO standard or via 16 A, 230 V AC home grid (12 h). The producer gives the rated power of the permanent magnet motor as 80 kW, while the maximal power given by the manufacturer is 90 kW. The capacity of car's battery pack is 24 kWh.

Fig. 1. Nissan Leaf tested by authors.

2 Use of solar power plants for charging electric vehicles

One of the main issues that EVs have to face is that related to charging. Beside cost of the battery pack and charging duration, common availability and accessibility of the charging facilities is another challenge [1]. Especially in Poland, the development of electro-mobility requires a huge investment in the construction of an automobile charging station infrastructure. Clearly, along with EV's market growth more charging stations will be installed. That however, along with increasing number of EVs, will impose extra burden on electrical utility due to the large current drawn from grid while charging [7]. To cope with rising demand, smart grids with renewable energy sources are seen as one of the main sources for charging stations [9].

Among them, solar power can be recognized as competitive since the price of PV modules is still falling [10]. Furthermore, solar technology is found reliable and almost maintenance free. Numerous charging methods using PV have been proposed. However, approaches called PV-grid and PV-standalone are currently most wide-spread.

The first one refers to the combination of PV and the grid, which during insufficient irradiance ensures continuity of the charging process [2]. Produced power can also feed electrical utility while no EV is present at the charging station. PV-grid charging system can operate in the following modes:

a. Mode 1 - EV charged by PV alone;
b. Mode 2 - EV charged by grid alone, inverter in rectification mode;
c. Mode 3 - EV charged by both PV and grid: inverter in rectification;
d. Mode 4 - no charging: inverter in inversion – generated power is transferred to grid;
e. Mode 5 - Vehicle to grid: inverter in inversion. Although attractive, this process can shorten the EV's battery life.

In the second approach it is proposed to use the PV solely with no grid connection. This approach is known as the PV-standalone charger [5, 11]. It is particularly convenient in remote areas where utility supply is too expensive or not available at all [2]. In a typical set-up, the charger is not directly powered by PV but via an intermediate storage battery bank. This is due to the fact that mass produced electric cars communicate with a charger using the Master-Slave model, where the charger is the slave unit. It means, that the car "demands" specific charging power and the charger supplies it. In the case of power being fed directly from the solar power plant, even a temporal overcast limits the supplied power and usually interrupts the charging process, which then has to be re-activated. Additionally, a battery bank enables to utilize the charger when there is no sun e.g. at night [12]. It smooths out the abrupt changes in the PV output power as well [13]. Recently, hybrid PV standalone systems have been introduced [10, 13-16]. The hybrid system consists of a PV generator and secondary power source e.g. fuel cell.

In the context of charging methods that utilize PV, the northern areas of Poland, including Gdańsk, are a big challenge, due to their location at 54° north latitude. They are characterized by low solar radiation compared to the southern parts of Poland and Europe, where solar power plants are much more efficient. Moreover, the large annual variation of solar irradiance is also characteristic for Gdańsk (fig. 2).

https://doi.org/10.1051/matecconf/201818002009

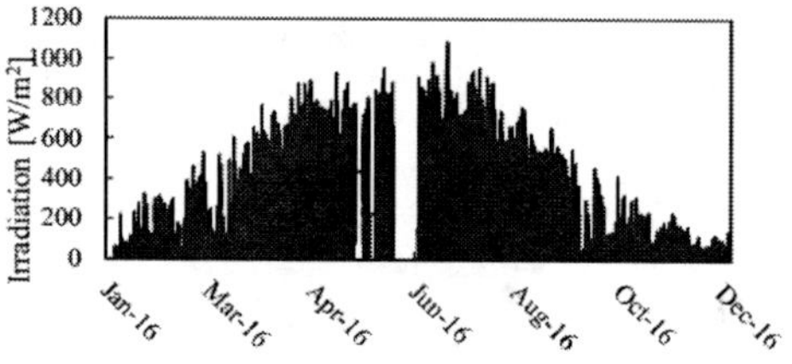

Fig. 2. Annual irradiation in Gdańsk (Breaks due to lack of service of weather stations).

Therefore, the authors conducted an experimental analysis of charging EV from the PV-standalone system with a battery pack for two extremely different weather conditions. Trials with charging an EV from the battery pack (charged by PV) through a slow charger (8h) in winter and through a fast charger (30 minutes to 80% capacity) in summer were performed. The experimental analysis was conducted in the Laboratory of Innovative Power Technologies and Integration of Innovative Energy Sources (LINTE^2). Extensive laboratory equipment within LINTE^2 allows the researchers to freely configure various types of sources and energy storages, as well as to create physical models of transmission lines. The study of modern power systems for EVs composed of 3 solar power plants with a rated powers of 5 kW, 8 kW and 20 kW and an electrochemical battery with capacity of 25 kWh can be performed. LINTE^2 is also equipped with Nissan LEAF EV, which can be powered via the fast DC and slow AC chargers.

A scheme presenting a part of the laboratory's infrastructure is shown in fig. 3. The solar power plant installed on the laboratory's roof consists of three photovoltaic packs of panels, each with different rated power. The PV inverters were connected through appropriate contactors with the bidirectional converters of the stationary battery pack during the day. Winter and summer tests of charging and driving cycles were conducted.

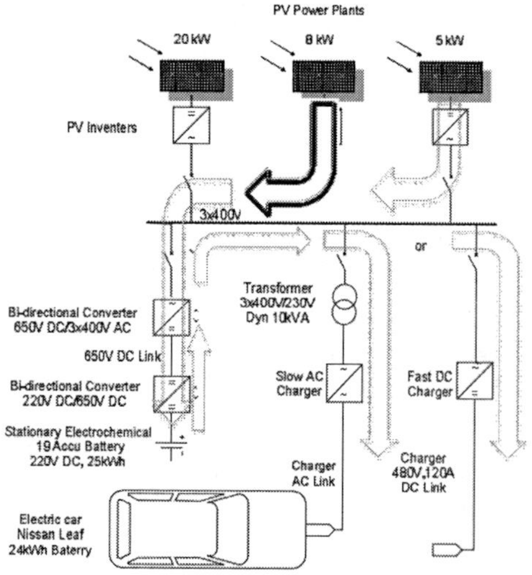

Fig. 3. Experimental setup using LINTE^2 infrastructure.

As shown in fig. 4, the highest irradiance of the sun on a clear day in January did not exceed 250 W/m^2 and on cloudy day 100 W/m^2. The effective working time of the solar power plant did not exceed 5 hours per day. Figures 6 and 7 show fragments of the battery charging cycle (from the total discharge state) fed from the solar power plant on a sunny day (irradiation profile marked by a solid line in fig. 4). Plots marked as PV (fig. 6 and 7) correspond to the AC power generated by the three inverters of the solar power plant. Plots marked as stationary electrochemical battery pack (BA) correspond to the charging power (DC) of the electrochemical battery pack. The difference between PV and BA shows the losses of conversion from alternating current (AC) to direct current (DC). As can be seen, the charging power ranged from 2 to 7 kW with interruptions caused by transient overcasts. The power of the losses is practically constant at about 1.5 kW. The energy supplied to stationary the battery pack during a sunny day in January was 11.4 kWh.

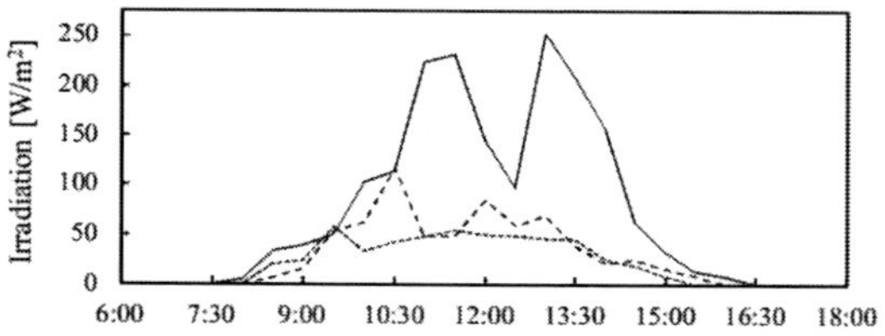

Fig. 4. Examples of daily irradiation profiles in Gdańsk for 3 January days with different overcast.

3 EV charging and test drives

Driving tests were performed for each of the two seasons (winter, summer), in two extreme driving conditions. The first one was the "eco" mode for urban driving and the second one was the highway dynamic mode - with "eco" mode off. In addition, the pressure was lowered in the summer tires.

During the winter tests, the Nissan LEAF was charged from a stationary battery pack with a slow charger. The process was started at dusk. A part of the charging plot is presented in fig. 5. The power supplied was 3.5 kW, at the battery pack power level of 5 kW. In this case conversion losses were also about 1.5 kW. Charging ended automatically after the discharge of the battery pack, i.e. after 3 hours and 48 minutes. At that time, the energy accumulated in the car's battery increased by 10.2 kWh.

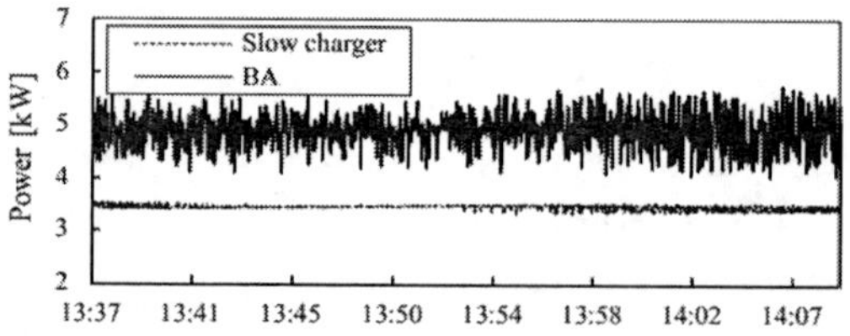

Fig. 5. EV Charging with stationary battery pack (BA) via slow charger – fragment.

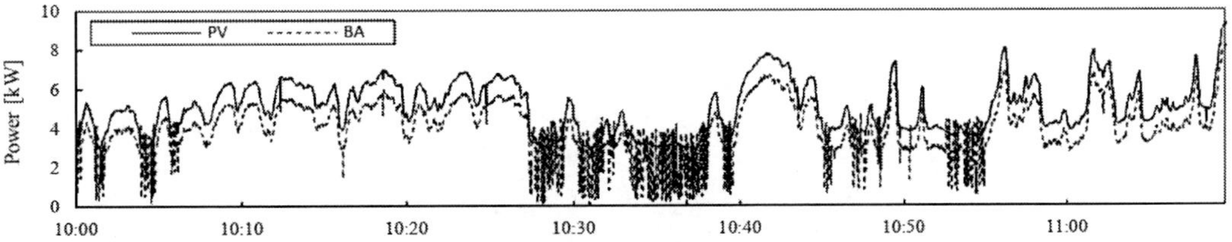

Fig. 6. Stationary battery pack (BA) charging from PV plant (20+8+5 kW) on January morning (fragment).

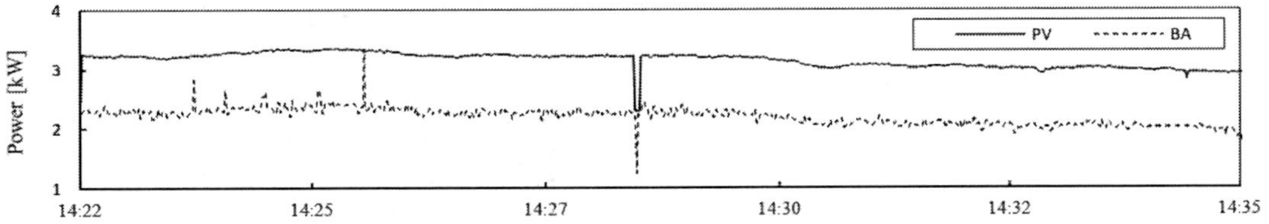

Fig. 7. Stationary battery pack (BA) charging from PV plant (20+8+5 kW) on January afternoon (fragment).

During the summer tests Nissan LEAF was charged from the stationary battery pack with a quick charger. Fragments of the power plots for the following stages of the test configuration are shown in fig.8. The quick charger began charging with a power of about 36 kW, which resulted in loading of the battery pack with 44 kW. Although the charging power was rapidly decreasing, the losses of multiple current conversions were constant and amounted to approximately 8 kW. The biggest losses were measured at 650 V DC Link to 3x400 V AC Busbar conversion stage. The charging process was interrupted after 30 minutes, i.e. after reaching 80% of EV's capacity, which is approximately 19 kWh.

The current and voltage plots in the individual charging stages (fig. 9) indicate that the charging of the car caused the discharge of the battery pack with 175 A DC. This resulted in a fairly rapid decrease of voltage level at battery terminals. At 650 V DC stage, the current and voltage were almost constant. Despite this, the voltages at 3x400 V AC Busbar and charger's DC Link stages increased as the charging current decreased.

Based on the analysis above, the current conversion efficiency from a solar power plant to a car battery through a stationary battery pack using a slow charger can be estimated at a 54%. The efficiency increased to 78-81%, when a fast charger was utilised.

During the test drives, the drive route, the mechanical and electrical parameters of the EV were recorded. Parameter registration and vehicle position data was performed using a mobile device running Leaf Spy Pro application under the Android environment. The application communicates wirelessly with the adapter connected to the vehicle's internal CAN network via the OBDII connector, from which measurement data is collected and stored in the internal memory of the mobile device.

Additionally, the position and accurate speed of the vehicle is recorded using the GPS receiver of the mobile device. A Windows application "LeafLog" has been developed to allow reading, presentation and analysis of the collected measurement data.

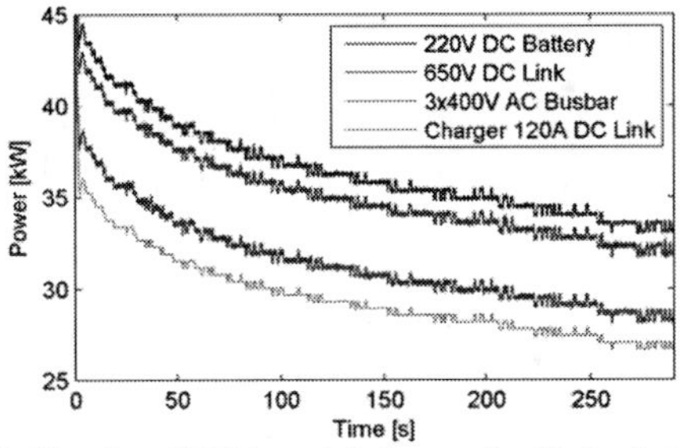

Fig. 8. Charging of EV through battery pack with the fast charger. Power of the following stages (fragment).

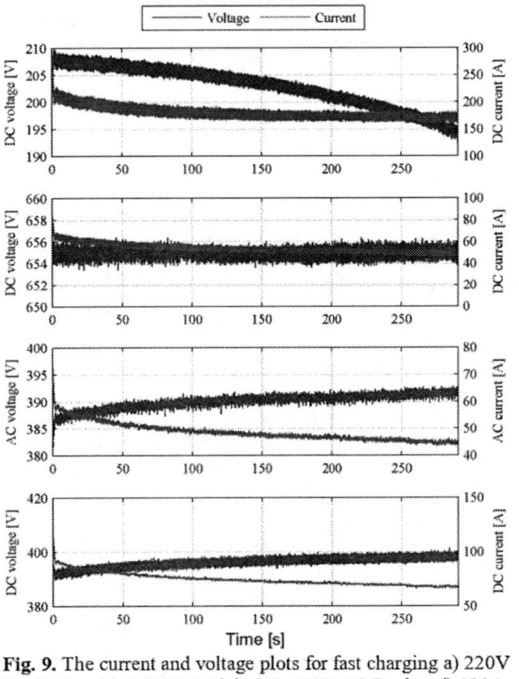

Fig. 9. The current and voltage plots for fast charging a) 220V DC Battery b) 650V DC Link c) 3x400V AC Busbar d) 120A DC Link of the Charger.

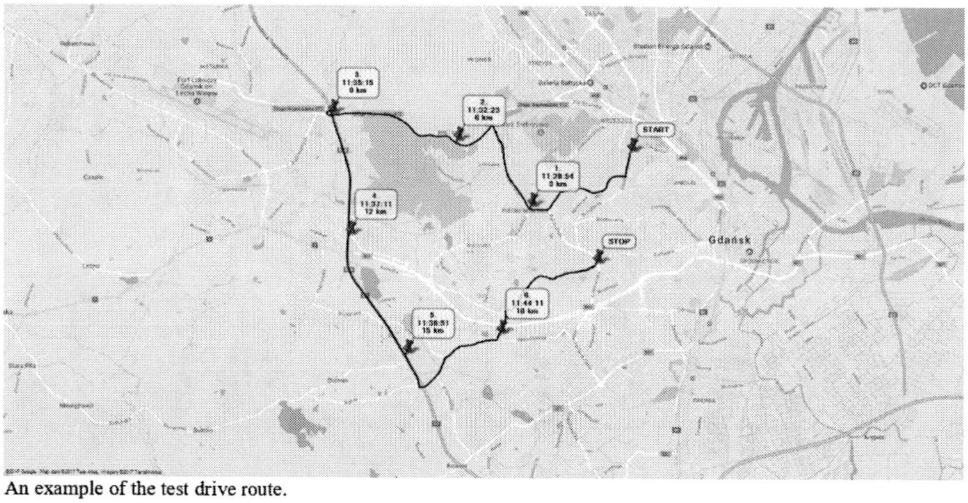

Fig. 10. An example of the test drive route.

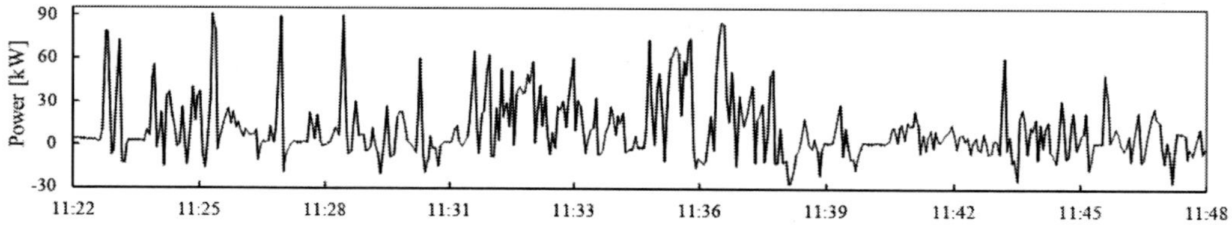

Fig. 11. Example of eco mode test drive; EV battery power.

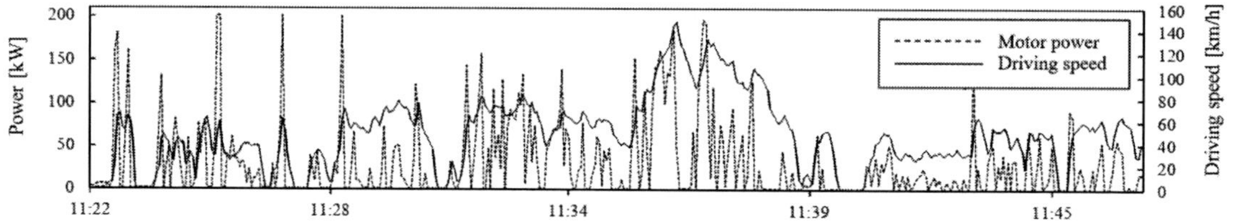

Fig. 12. Example of dynamic mode test drive; engine power [kW] and speed [km/h].

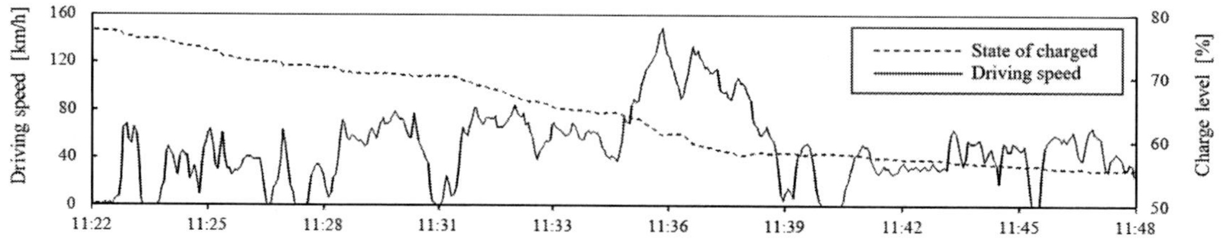

Fig. 13. Example of dynamic mode test drive; speed [km/h] and EV's battery level [%].

An example of a test drive route including selected streets in Gdańsk is presented in fig. 10. Figures 11, 12 and 13 present examples of registered EV parameters. As shown in the fig. 11, the maximal motor power in eco mode reaches 90 kW and during regenerative braking 30 kW. The parameters shown in Table 1 were estimated on the basis of 20 winter Nissan Leaf test drives made in January and 10 summer drives. The test drives were conducted around noon (before the rush hours) in typical urban traffic and on the highway.

This means that even in the short but sunny January days in the Gdańsk area, authors were able to use the EV completely free of charge (due to the solar energy utilization) through 29 km on the Tricity (Gdańsk - Sopot - Gdynia) Highway and 49 km in urban traffic. However, this required the use of a solar power plant with a rated output of 33 kW. This range could be significantly increased after optimization of current conversion losses by the elimination of multiple current conversions.

The summer tests revealed that after 30 minutes of charging the accumulated energy from the solar battery it is possible to use the EV through 100 km on the Tricity Highway and 146 km in urban traffic. This range would increase slightly after optimization of current conversion losses.

Table 1. Results of test driving - Nissan LEAF (2015 model).

	January - 20 tests		**July/August - 10 tests**	
Average ambient temperature	+1^{0}C		+24^{0}C	
Other activated devices	heating, wipers, heated window, daily lights		air conditioning, daily lights	
People on board	Driver & 3 passengers		Driver & 1 passenger	
Tires, pressure	Barum Winter 2.49 Bar		Michelin EnergySaver 2.57 Bar	
Type of drive	Urban eco mode	Highway dynamic mode	Urban eco mode	Highway dynamic mode
Average velocity	31 km/h	87 km/h	35 km/h	113 km/h
Energy [kWh/km]	0.21	0.35	0.13	0.19
Driving range for one charge (10.2 kWh)	49 km	29 km	-	-
Driving range for one charge 80% (19 kWh)	-	-	146 km	100 km
Predicted range for 100% capacity	114 km	69 km	185 km	126 km

4 Conclusions

In order to carry out research on the efficiency of charging an electric car from a solar power plant, part of the infrastructure of the LINTE^2 Laboratory of the Gdańsk University of Technology was used. It included: a solar power plant with a total rated power of 33 kW, a stationary battery pack of 25 kWh and a Nissan Leaf EV with a battery capacity of 24 kWh. As a result of the research, it was determined that the solar energy collected at worst winter conditions, but in a sunny day in northern Poland area, allows for about 30 km of a highway dynamic driving and about 50 km drive in urban eco mode without any costs of energy. For 100% of the EV battery capacity, the extended range would be adequate for 70 km and 115 km drives. In the best summer conditions one quick charge to the 80% capacity allowed the authors to drive in urban eco mode 146 km and dynamic highway on 100 km on average. The predicted summer range for a fully charged battery (100%) at of the EV is adequate for 185 km and 126 km.

The result of power registers during dynamic driving - Fig. 12 - is very interesting. It latches on the car drive maximum power of 200 kW. The manufacturer does not specify this value and gives a maximum power of 90 kW. Despite long-lasting and very dynamic driving, the authors failed to activate the thermal limitation of this power.

References

1. F. S. Tie, C. W. Tan, *A review of energy sources and energy management system in electric vehicles*, Renewable and Sustainable Energy Review, 20, 82-102, (2013)
2. R. Bhatti, et al., *Electric vehicles charging using photovoltaic: Status and Technological Review*, Renewable and Sustainable Energy Review, 54, 34-47, (2016)
3. Xiaopeng, et all., *An overview of lithium-ion batteries for electric vehicles*, Proceedings of IPEC Conference on Power & Energy, 230–235, (2012)
4. R. Schmidt, *Information technology energy usage and our planet.* Thermal and Thermomechanical Phenomena in Electronic Systems, ITHERM, (2008)
5. Abu-Jasser, *A stand-alone photovoltaic system, case study: a residence in Gaza*, Journal of Applied Sciences in Environmental Sanitation, 5:81–91, (2010)
6. "Roadkill", 7-8, (The Economist, 12.08.2017)
7. Kelman, *Supporting increasing renewable energy penetration in Australia - the potential contribution of electric vehicles*, Proceedings of 20th Australasian Universities Power Engineering Conference, 1–6, (2010)
8. S. E. Letendre, *Vehicle integrated photovoltaic: exploring the potential*, Proceedings of the 23rd International Electric Vehicle Symposium, (2007)
9. Z. Li, et al., *Optimal charging control for electric vehicles in smart microgrids with renewable energy sources*, 2012 IEEE 75th Vehicular Technology Conference, 1-5, (VTC Spring, Yokohama, 2012)
10. P. P. Barker, J. M. Bing, *Advances in solar photovoltaic technology: an applications perspective*, Proceedings of Power Engineering Society General Meeting, Vol. 2, 1955–1960, (2005)
11. S. J. Chiang, S. Hsin-jang, C. Ming-chieh, *Modelling and Control of PV Charger System with SEPIC Converter*, IEEE Transaction of Industrial Electronics, 56:43–53, (2009)
12. S. Mesentean, et al., *Smart charging of electric scooters for home to work and home to education transports from grid connected photovoltaic system*, Proceedings of IEEE International Energy Conference and Exhibition, 73–78, (2010)
13. J. MOSSOBA, et al., *Analysis of solar irradiance intermittency mitigation using constant DC voltage PV and EV battery storage*, Proceedings of IEEE Transportation Electrification Conference and Expo, 1–6, (ITEC, 2012)
14. M. Robalino, et all., *Design of a docking station for solar charged electric and fuel cell vehicles*, Proceedings of the International Conference on Clean Electrical Power, 655–660, (2009)
15. Z. Mokrani, D. Rekioua, T. Rekioua, *Modelling, control and power management of hybrid photovoltaic fuel cells with battery bank supplying electric vehicle*, International Journal of Hydrogen Energy, 39: 15178–15187, (2014)
16. R. Parsons, et al., *Willingness to pay for vehicle to grid (V2G) electric vehicles and their contact terms*, Energy Economics, Vol. 42, 313-324, (2014)

Life cycle cost as a criterion in purchase of rolling stock

Jan Raczyński[1,*]
[1] Railway Research Institute, ul. Chłopickiego 50, 04-275, Warsaw, Poland

Abstract. The key parameter in assessing the economic viability of purchasing and operating rolling stock is the total cost incurred by the ordering party, starting from the preparation of the rolling stock purchase plan to the end of its operation, including its disposal. The methodology for calculating these life cycle costs (LCCs) has been developed especially in recent years based on the experience of the next-generation vehicle deployment. In Poland first vehicles purchase using the LCC method was carried out in few recent years - by PKP Intercity and by Lodz Agglomeration Railway. Despite the use of simplified criteria for the economic efficiency of vehicles, optimal offers were selected. This article describes examples of rolling stock purchase based on the low cost of life of the vehicle and the economic and operational results that have been achieved after several years of operation. Further possibilities for optimizing these costs were also indicated. Examples showing the proportion between the cost of vehicle purchase, its technical maintenance and its operation, where the electricity consumption is the main factor are given.

1 Introduction

Investment decisions, in it concerning the purchase of rolling stock, require costs and the benefit evaluation of applying different, alternative solutions [2].

They can concern among others technical parameters of rolling stock, expected maintenance costs and operational cost, the reliability or sources of financing among others.

Technical parameters of rolling stock are determining not only investment costs. They have also a significant influence on the level and the deployment of maintenance costs in the life of vehicle [8].

They are also a factor of creating the competitiveness of the public transport, in it railway, towards the individual motorization [6].

In consequence they are determining the competitiveness of states and regions [5], contributing to fulfil the idea of the sustainable development.

Studies and empirical examinations are confirming that correlation between maintenance costs exists and of use of rolling stock, and with level of their technical wear and tear [4]. Along with increasing the accumulated mileage of rolling stock is being increased labour intensity of repairs, resulting in an increase in employment and the fall in the technical available of vehicles. Also a risk of the defects grows and a reliability of providing services is lowering. It next is limiting benefits of the railway operator [3].

The total cost for the contracting authority, from rolling stock purchase project preparation till the end of its operation, including its scrapping (utilization) it the most reliable parameter in assessing the investment decision of rolling stock purchase.

is. The methodology of calculating these costs, defined as life cycle cost (LCC), has been improved in recent years based on experience from the exploitation of new generation vehicles. The presentation of methodological principles of projecting these costs is an essential purpose of this article.

In Poland in recent years took place first purchases of vehicles using the LCC methodology in PKP Intercity and Łódź Agglomeration Railway. Several years of operation of these vehicles allowed verification and assessment of suitability of the applied methodology. These experiences are positive. Even the use of simplified criteria for assessing the economic efficiency of vehicles allowed for the selection of optimal offers

2 Vehicle life cycle costs

The investment costs of purchase and operation of rolling stock are significant and amount to approximately 1/3 of the railway carrier's costs. For this reason, when purchasing rolling stock, it is necessary to apply a systemic approach going beyond the simple assessment of the vehicle purchase cost. The construction of the vehicle, its susceptibility to maintenance, accepted reliability and availability indicators, the way the service is organized, determine the total cost of vehicle operation. It is estimated that in the period of 25 - 30 years, the costs of rolling stock maintenance may be equal to the vehicle investment cost. The costs of electricity consumption can be even higher. While preparing specification for the purchase as well as organization management should take into account all aspects necessary to ensure maximum economic efficiency.

* Corresponding author: j.raczynski@infotransport.pl

Lifecycle costs of a vehicle (LCC) are commonly used for rolling stock investment efficiency analysis in vehicle operation theory.

However, it should be noted that the definition of cost essentially means the purchase price or service during its operation. The actual expenses for purchase of rolling stock is higher than the cost of raising capital. Methodology is simplified because methods and related costs of raising capital for the purchase and operation of the vehicle are different. In the further part of the study will be carried out an analysis of the actual costs of purchase and operation of the vehicle.

Life cycle vehicle cost (K) can be defined as follows [7]:

$$K = K_i + K_m + K_u$$

where:
K_i – investment costs of vehicle purchase
K_m – technical costs of vehicle maintenance
K_u – vehicle operation costs

The investment costs of K_i include the vehicle purchase cost, spare parts (the first batch for current maintenance), workshop and diagnostic tools, technical documentation, instructions for use and maintenance, modernization and additional investments in the workshop facilities. This group also includes withdrawal and utilization costs.

The costs of the K_m maintenance process include implementation costs, vehicle preparation, maintenance costs (planned and scheduled repairs as well as ongoing repairs).

K_u use costs include, in particular, power supply from network costs – catenary and other purposes as well as costs incurred as a result of the vehicle's impact on the track (this costs are usually omitted).

K_m maintenance costs and K_u operating costs are the costs of vehicle operation.

The cost group K_m and K_u should refer in principle to the entire vehicle life. Calculation periods are usually defined for 20 to 30 years, with a practical indication for30 years period.

3 Cost costs of vehicle maintenance

The total cost of vehicle maintenance during its 30-year lifetime is generally similar to the cost (price) of a vehicle purchase. This could mean that the average annual maintenance cost amount at 3.3% of the vehicle purchase cost. This should be considered as a starting point for further analysis. The cost of vehicle maintenance can be adjusted up or down depending on the following factors:

- the size of the fleet park (the smaller it is, the higher the unit maintenance costs);
- the technical complexity of vehicles;
- the cost of maintaining technical facilities (different depending on the region, land costs and other charges);
- labour cost (depending on the region).

These costs are not evenly distributed over the entire lifetime of the vehicle. The costs of back office preparation and the first period costs related to spare parts, materials and IT systems purchase as well as staff training are high.

In terms of rolling stock technical maintenance management there are two possible solutions:

- maintenance own workshops by operators, with possibility of shorter periods contracts with the rolling stock provider for the maintenance services provision;
- maintenance contract together with the contract for the rolling stock purchase for the first major repair period (about 15 years).

The second solution is the more advantageous for carriers without own technical depot developed and lack of an experienced staff. It transfers the risk of design and other errors to the rolling stock manufacturer. Such contract should require from the contractor to provide for peak hours a given quantity of rolling stock. In case of less than 20 vehicles two of them should be in reserve. In case of failing to meet the requirements, the supplier must provide replacement rolling stock with similar characteristics at their own cost along with a possible contractual penalty.

Rolling stock maintenance takes place in a manufacturer depot or the one lent to him by the carrier but with equipment provided at the expense of the supplier.

Usually are concluded 15-year maintenance contracts as such period is necessary for cost stabilisation due to their long-term predictability. In recent years appeared maintenance contracts for 30 years which means for the vehicle lifetime.

There are also in Poland contracts for periods shorter than 15 years, in example until first inspection repair (maintenance level P4). Such contract doesn't solve fundamental problems related to the maintenance costs for the ordering party. The first inspection repair means the warranty period for the vehicle and subassemblies. Higher maintenance costs may appear later, especially in 10 years from the date of purchase of the vehicle.

4 Rolling stock operational costs

Power supply costs are the key element of operational costs. They can be significantly different in individual countries, due to the differentiation of energy unit prices.

While choosing technical solutions it is possible to reduce power supply by changing such factors as:

- reduction of the train's weight;
- using drive devices with high energy efficiency;
- an effective recuperation system and the use of energy lost in braking for auxiliary purposes;
- optimization of energy consumption for auxiliary purposes (heating, air conditioning, lighting).

Energy consumption information may be required in the declaration provided by the rolling stock manufacturer.

In modern vehicles are being introduced more and more advanced methods of energy saving. Power supply from network is used for:

- traction purposes, i.e. for vehicle drive;
- auxiliary purposes and to ensure comfort for passengers and drivers.

The energy consumption for auxiliary purposes and ensuring comfort is relatively high in modern vehicles and may amount to 20 - 30% of the total energy consumption of the vehicle. For this reason, the effective use of energy for purposes (i. e. air conditioning control) is becoming more and more important. It is also important to ensure the highest energy recuperation. This energy can be:

- distributed directly to the traction supply network;
- stored in the vehicle (in batteries or capacitors) and re-used for traction or auxiliary purposes, when needed;
- directly used for auxiliary purposes.

The use of energy management systems can provide savings at standstill amounting to 3 - 5% of total energy used (15 - 20% for devices that provide comfort to travellers).

The energy consumption assessment method may be direct or indirect. The direct method consists in estimating the energy consumption under certain conditions (theoretical transfers), and the value declared by the manufacturers then verified in practice at rolling stock delivery. This methodology is defined in the document CLC-TSI 50591, Technical Specification for verification of energy consumption of railway rolling stock [1]. This methodology is still not widely disseminated because of difficulties in its application. More simplified is an indirect evaluation methodology using specific mass index. It is the subject of the UIC 345 leaflet. The specific mass is defined in it as follows [9]:

Table 1. Definition of specific mass.

Type of rolling stock	Indicator
Locomotives	Vehicle weight
Multiple units and passenger cars	Vehicle weight per one seat
Freight cars	Vehicle weight per load

For passengers vehicles is used a following formula:

$$P = S + A \times B$$

where:

P – passengers number in vehicle

A – defined standing places per 1 m^2 of available vehicle space

B – available vehicle space for standing passengers [m^2]

The A indicator of standing places per 1 m^2 depends on the type of traffic. In long distance service it is equal to zero, and for agglomeration railways even the value of 4 is used.

The method recommended in UIC 345 leaflet in a modified version was applied in several tenders in Poland: in which the mass per meter of vehicle length in a part available for passengers instead of mass per seat was used. This was justified by defining by ordering parties a minimum value of the seat pitch.

However, this indicator does not take into account all elements affecting the energy consumption of the vehicle. Vehicles with the same specific mass may also have other technical solutions that affect their energy consumption. These are, for example:

– vehicle resistance movement related to its construction, e.g. aerodynamics;
– way of controlling drive train;
– so called intelligent energy consumption management for auxiliary purposes in the vehicle (air conditioning, heating, lighting, auxiliary equipment).

In order to take these factors into account in an accurate and objective manner it is necessary to perform in comparable conditions vehicle tests on selected routes.

5 Taking into account life cycle costs in tenders

Selection of the best offer for passenger rolling stock should be the result of a multi-criteria analysis taking into account all or at least the most important factors affecting the cost of the entire life cycle of the rolling stock.

The main factors to be taken into account and their possible weight in offers evaluation based on the example of selected tenders in Poland are as follows:

– rolling stock cost – 40 – 50% of value;
– cost of rolling stock maintenance in a cycle of at least 15 years (that is till the first major repair but without its cost) – 40 – 50% of value;
– energy consumption (can be calculated in few ways, i.e. by vehicle mass per one passenger indicator) – 15 – 20% depending from the character of the route which are to be operated by the new rolling stock.

Adopting specific percentage depends on the specificity of the rolling stock ordered, the conditions of its use and the expected economic results to be achieved. These values should be defined on the basis of technical and economic analyses in a feasibility study for each rolling stock purchase.

6 Summary

Using vehicle life cycle costs method is the most reliable for rolling stock purchase or modernization tender's offers assessment. This method up to date implementation in Poland have allowed gathering positive experience which allows its further improvement. There are still some problems to be solved particularly whose concerning methodology development in case of rolling stock maintenance by the operator or when it is partly subcontracted by the rolling stock manufacturers. Up to date cases concerns only rolling stock purchase with long term maintenance contracts. Also it should be considered taking into account vehicle energy efficiency based on the CLC-TSI 50591 specification [1]. This should particularly apply to agglomeration and regional traffic vehicles.

References

[1] CLC-TSI 50591, *Specyfikacja techniczna służąca potwierdzaniu zużycia energii przez pojazdy szynowe*

[2] T. Dyr, and P.R. Kozubek, 2013. *Ocena transportowych inwestycji infrastrukturalnych współfinansowanych z funduszy Unii Europejskiej*, (Spatium 2013)

[3] T. Dyr, and P. Misiurski, *Forecasting of costs of maintenace and use of a bus fleet*. Transp. Logist. Probl. **35**, 19–28 (2016) [in Polish] doi:10.18276/ptl.2016.35-02

[4] I. Dziaduch, *Analiza kosztów okresu istnienia (LCC) obiektu technicznego w aspekcie jego niezawodności*, Logistyka, **2** (2011)

[5] V. Maráková, T. Dyr, and A. Wolak-Tuzimek, *Factors of tourism's competitiveness in the European union countries*, E a M: Ekonomie a Management, **19**(3) (2016) doi: 10.15240/tul/001/2016-3-007.

[6] P. Misiurski, *Koszty eksploatacji taboru autobusowego w przedsiębiorstwach komunikacji samochodowej*. Autobusy – Technika, Eksploatacja, Systemy Transportowe, **11** (2012).

[7] M. Sitarz, K. Chruzik, K. Banaszek, and J. Raczyński *Uwarunkowania planowania rozwoju pasażerskich zasobów taborowych operatorów kolejowych. Cześć 2 – zagadnienia techniczno-eksploatacyjne*. Technika Transportu szynowego, 7-8 (2016)

[8] M. Szkoda, *Ocena efektywności ekonomicznej środków transportu szynowego z zastosowaniem analizy LCC*. Technika Transportu Szynowego, 11-12 (2012)

[9] UIC 345, *Leaflet. Environmental specification for the new rolling stock. UIC* (2006)

Modelling of operation of a stationary energy storage device in metro rail transport for selection of its parameters

Petru Valentin Radu [1,*], and *Adam* Szeląg [1]

[1] Warsaw University of Technology, Institute of Electric Power Engineering, Electric Traction Division, ul. Koszykowa 75, 00-662, Warsaw, Poland

Abstract. The paper presents a Simulink model of a DC metro traction supply system with a stationary energy storage device (SESD). The simulation model consists of traction substations, a train model, and an energy storage device (ESD) with supercapacitors (SC). A new energy management strategy considering the line voltage and current, SC state of charge (SOC) and SC charging and discharging current is proposed. This method can improve the energy savings and manage the remaining energy. Simulation results provided in this paper justify the control method. The proposed model can be used with different ESD, such as batteries.

1 Introduction

In 2014, energy needed for transportation was around 24 % in the United States of the overall energy consumption [1]. Nowadays most of the electrical vehicles have the capability to return the braking energy to the traction system [2]. Reutilizing the vehicle braking energy can reduce the transportation costs, power needed for the traction system and the carbon footprint. The key to saving more energy is the optimal design for the energy storage devices including: selection of an appropriate type, location, size and energy management strategy.

During the last few years, a large number of energy management strategies were proposed in different papers for example: adaptive fuzzy logic strategy [3], real time optimization [4], global optimization [5], rule based strategy [6], genetic algorithm strategy [7] and load power strategy [8, 13]. The energy management strategies can theoretically lower the energy consumption up to 20 % and reduce the size of the energy storage devices [9, 10]. Experimental tests with ESD have justified their application in urban transport [14]. On the other hand, these methods have several drawbacks, e.g. the complexity of the solution and the input data needed obtained from additional costly equipment in the vehicle for its location, velocity etc. In this paper a new management strategy is presented and tested on a simulation model build in Matlab / Simulink for the SESD on metro transport.

2 Matlab / Simulink model of a traction supply system with SESD

A simulation model was built in Matlab / Simulink, for a light rail power supply system which includes 15 kV AC power supply source, 750 V DC traction substations with transformer - rectifier, trains and SESD with SC.

2.1 Traction substation model

In Poland the power to the metro traction supply is provided from District Point Supply (RPZ - Rejonowych Punktow Zasilania). The transformer rectifier substations convert the 15 kV, 50 Hz to 750 V DC which is supplying the metro cars by third rail. The metro supply is "dual feed" on AC and DC side to provide backup power in the event of a fault.

Figure 1 presents the polish metro supply configuration.

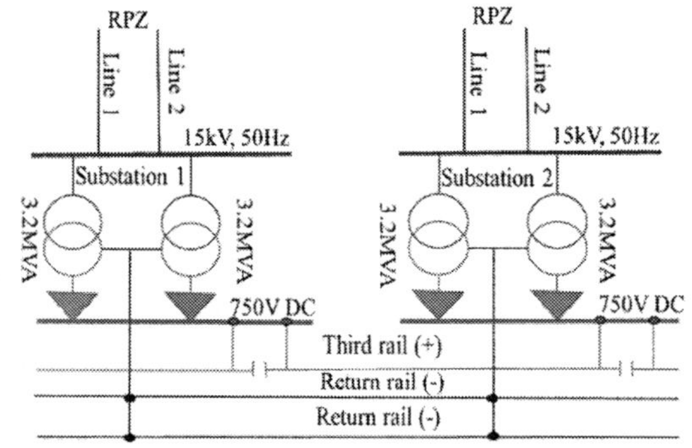

Fig. 1. Metro 750V DC supply configuration.

The metro supply configuration from figure 1 was built in Simulink. The model consists of a three phase power source (PS), three phase transformer - rectifier Y/D connection (TR).

In table 1 is given the assumed specification for the power supply model.

* Corresponding author: petru.radu@ee.pw.edu.pl

Table 1. Traction supply simulation parameters.

Parameter	Unit	Quantity
PS Nominal Power	MVA	20
PS Voltage	kV	15 (50Hz)
TR Nominal Power	MVA	3.2
TR Primary Voltage	kV	15 (50Hz)
TR Secondary DC Voltage	V	750

2.2 Line model

The running rail and the third rail are simulated by linear resistance and considered pure resistive. The running rail has an assumed resistance of 35.7 mΩ/km and the third rail has an assumed resistance of 11.7 mΩ/km.

2.3 Train model

A simplified model of a train was built using a controlled current source and the input signal for the source is provided with a Signal Builder. In the Signal Builder the values of the assumed train current with respect to time is provided. Real values of the train current can be provided to the model.

Figure 2 shows the assumed train currents used for the simulation purposes. For the train model all 3 stages were considered: acceleration, coasting, braking with regeneration. A simulation time of 200 s was proposed to verify the Simulink model.

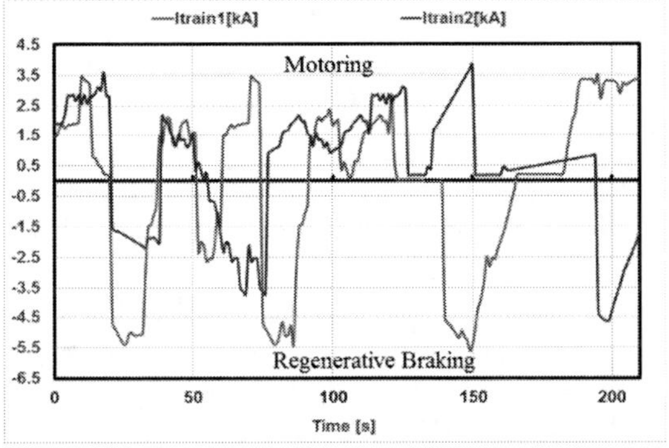

Fig. 2. Train current.

If the train is braking with regeneration and there is no other train or SESD to absorb the energy the voltage in the third rail will increase and exceed the limits. European Standard (EN) 50163 [11] regulates the maximum and minimum allowable voltage for the 750 V DC system as below:

- Un = 750 V DC - nominal voltage;
- Umin1 = 500 V DC - lowest permanent voltage;
- Umax1 = 900 V DC - highest permanent voltage;
- Umax2 = 1000 V DC - highest non-permanent voltage;

To prevent the overvoltage, a braking system with braking resistor was implemented for the train. A control system using Matlab Function and compliant to the EN 50163 standard was designed. The input values are the train voltage values measured by a voltage meter. The output consists of control signals to ideal switches which in turn switch OFF or ON different braking resistance (Rb) according with third rail voltage as per table 2. The 1 means that the braking resistor is switched ON and the 0 means that is switched OFF.

Table 2. Braking logical table.

Train Voltage [V]	Rb_1 [0.5Ω]	Rb_2 [0.4Ω]	Rb_3 [0.3Ω]	Rb_4 [0.2Ω]	Rb_5 [0.1Ω]
≥ 950	1	0	0	0	0
≥ 1050	1	1	0	0	0
≥ 1100	1	1	1	0	0
≥ 1150	1	1	1	1	0
≥ 1270	1	1	1	1	1

2.4 SESD model

The SESD model consist, of supercapacitors, DC-DC charging converter and DC-DC load converter. The Maxwell (125 V module) SC parameters are presented in table 3.

Table 3. Maxwell SC parameters [12].

Parameter	Value
Rated Voltage	750 V
Modules in series	6
Modules in paralel	3
Capacitance	189 F
Energy	40 MJ
Maximum Charge / Discharge Current	6 kA

The SC voltage values need to be between 40 % (300 V) and 100 % (750 V) of the nominal value in order to maintain converters operation. The SC state of charge was set between 40 % and 100 % for the simulation and the maximum SC charging / discharging current has to be kept below 6 kA in order to protect it.

An energy management control system considering the third rail voltage and current, regenerated power, load power, SC SOC, SC charging current and SC discharging current was designed and presented in figure 3.

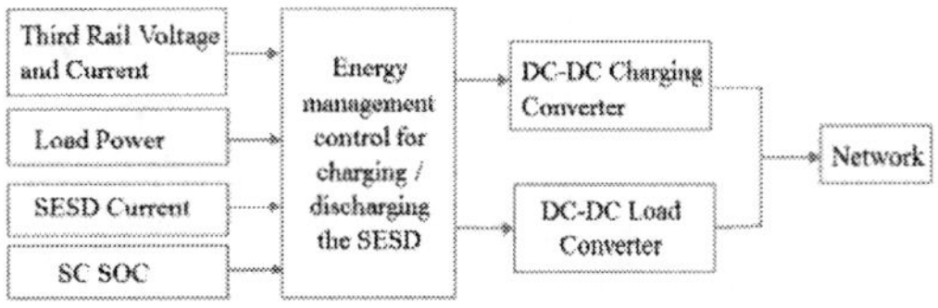

Fig. 3. SESD energy management system.

The power control of SESD charging and discharging cycles is based on the line voltage as shown in figure 4.

2.4.1 DC-DC charging converter

The DC-DC charging converter was built with ideal switches. A diode was used to make the current to flow from the braking train to the SESD. An inductor was designed for keeping the charging current stabile.

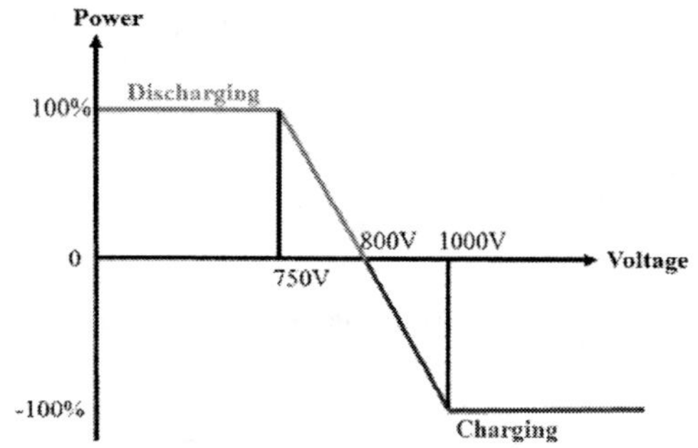

Fig. 4. Power control diagram for SESD.

The block marked "X" in figure 5 and figure 7 is used for multiplication of the logical values of "ESD charging / discharging" for all decision blocks.

The SESD charging control have been implemented as below:

- The third rail voltage (UL) and current (IL) is measured. The regenerated power (PL) is calculated. The train braking with regenerations is detected when the third rail voltage is higher than 800 V and the power is lower than 100 W. The SESD is charging if the network demand power is negative (<0W) which means that the train is braking with regeneration. To avoid fluctuations in the control block the control power for SESD charging was selected arbitrary to be 100W.
- SESD current (ISESD) is measured and compared with the sum current of the running trains. This function allows the SESD to charge only from the regenerative energy and not from the substations.
- SESD current (ISESD) is measured and compared with the maximum allowable charging current 6 kA. This restrictions keeps the discharge current below the maximum value in order to protect the SC.
- SC state of charge (SOC) is measured and when reaches 100 % the SESD stops from charging.

If all conditions are met then SESD will charge from the third rail (SESD charging = ON).

The energy management strategy flowchart for the charging control is shown in figure 5.

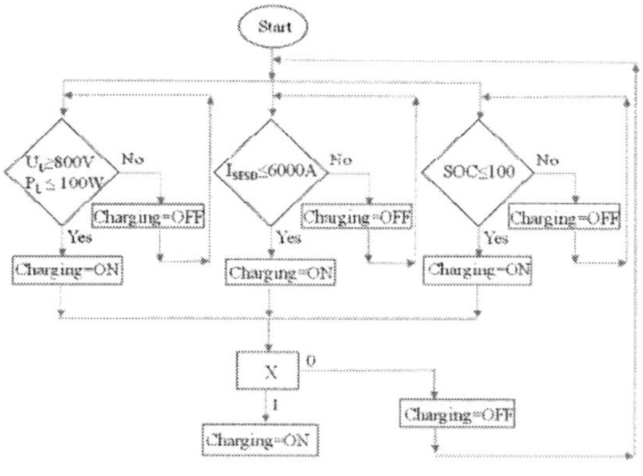

Fig. 5. Energy management strategy for charging the SESD.

2.4.2 DC-DC load converter

A Cuk converter is used for the DC-DC load converter described and tested in [13]. The converter uses a closed loop cascade control with two PI regulators for load current and output voltage.

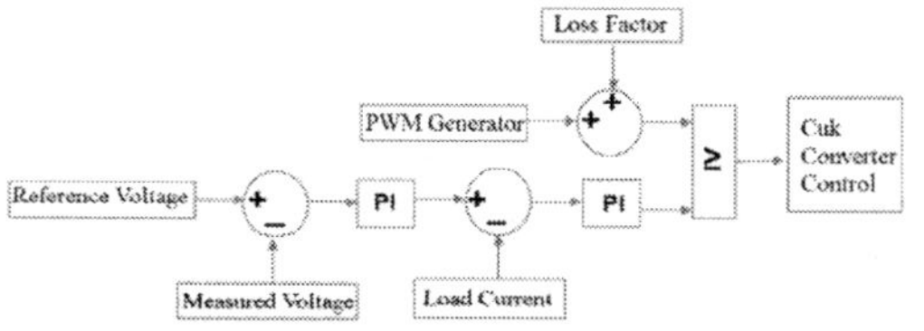

Fig. 6. Control architecture for the DC-DC load converter [13].

The controller objective is to stabilize the current supplied by the SC during sudden changes in load current and keep the output voltage constant to 750 V.

The DC third rail voltage is measured and compared with the voltage threshold value 750 V. If the third rail voltage is lower than 750 V the SESD provides with the rest of the power. To prevent PI regulators saturation during the SESD charging, they are enabled by the energy management load control.

The SESD load control has been implemented as below:

- The third rail voltage (UL) and current (IL) is measured. The load power (PL) is calculated and compared against the minimum value of 1 MW to protect the SESD from simultaneous charging and discharging. The minimum power value allow the SESD to provide with energy only on heavy demand load conditions.
- SESD current (ISESD) is measured and compared with the maximum allowable load current 6 kA. This condition keeps the load current below the maximum value to protect the SC.
- SC state of charge (SOC) is measured. The minimum SOC is set to 40 %.

If all conditions are achieved then the third rail will be fed by the SESD (SESD discharging = ON).

The energy management strategy flowchart for the load control is presented in figure 7.

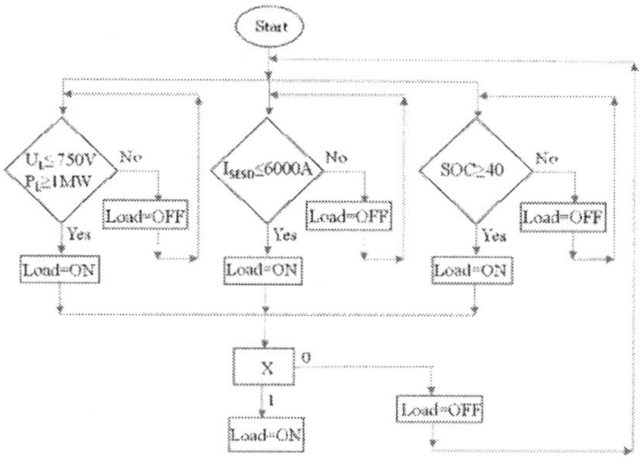

Fig. 7. Energy management strategy for discharging the SESD.

3 Simulation results

Figure 8 presents the Matlab / Simulink model used to simulate the metro 750 V DC supply system from figure 1 with SESD installed within Substation 1. The simulated model is a single metro line with 2 trains running between Substation 1 and Substation 2. To simulate different load for each substation a resistance was designed for third rail and running rail. The total length of the line is considered 2.1 km and the line resistance was designed according with the train distance from the substations. The location of the train is given in table 4.

Table 4. Train location.

Train No.	Distance from Substation 1 [km]	Distance from Substation 2 [km]
1	0.2	1.9
2	1	1.1

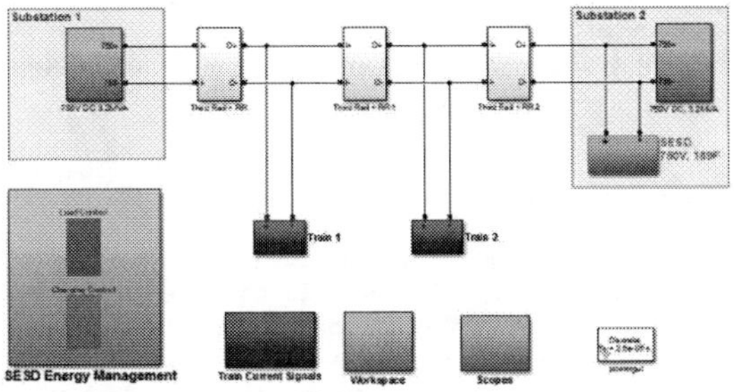

Fig. 8. Metro 750 V DC Matlab / Simulink model.

In figure 9 is presented the simulation results for the SESD installed within Substation 2 for a simulation time of 210 seconds and a step size of 0.5 seconds.

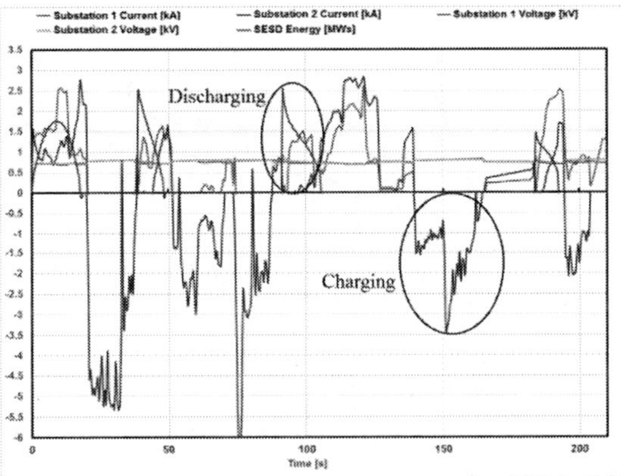

Fig. 9. SESD located within the Substation 2.

During heavy demand of traction energy the SESD successfully provides the auxiliary energy for keeping the third rail voltage stable and under the limits. The maximum discharge current is 3 kA. During braking process the SESD stores the energy with maximum charging current up to 6 kA.

In figure 11 it can be observed that the substations voltage did not exceed the 1 kV voltage limit according with EN 50163 standard.

On figure 13 the SC SOC is charging / discharging according with the train powering or braking.

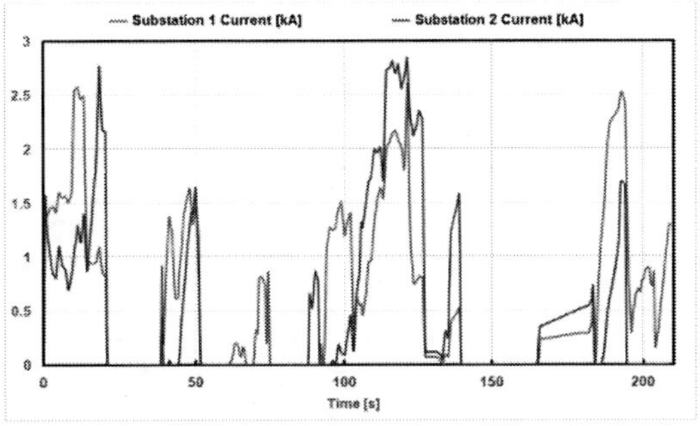

Fig. 10. Substation 1 and 2 current.

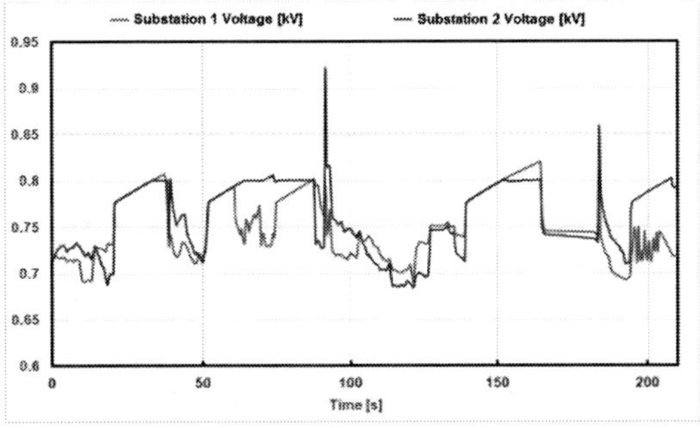

Fig. 11. Substation 1 and 2 voltage.

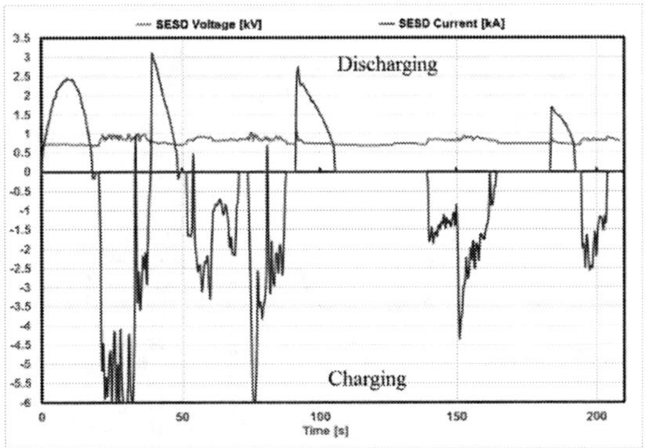

Fig. 12. SESD current and voltage.

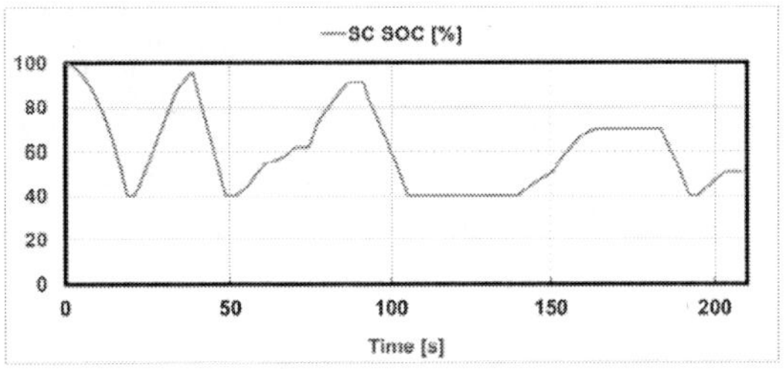

Fig. 13. SC SOC.

From figure 14 and figure 15 it can be observed that the control system for the braking system has a good time response and is able to suppress the overvoltage during train braking. The third rail voltage is below the 1kV maximum allowed non-permanent voltage limit.

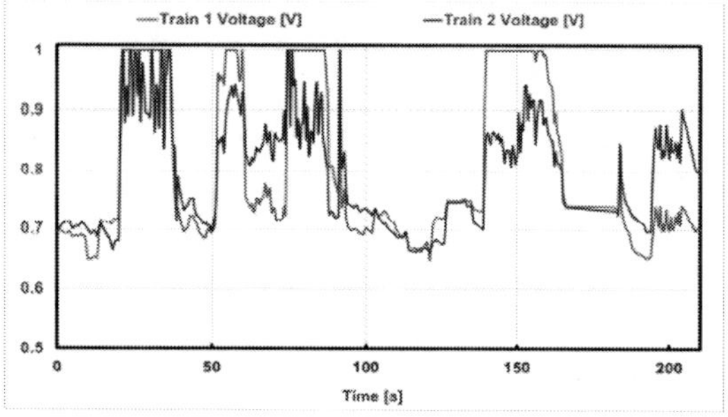

Fig. 14. Train 1 and 2 contact shoe voltage.

During braking process excessive part of the regenerated energy is lost on the braking resistors. This is due to lack of receptivity of power supply.

4 Study case

A study case was performed for a 8 km section of a metro line with a SESD installed in the mid-point between 2 traction substations (TS) distanced 3 km one from another. Traffic of metro trains was organized regularly with density dt = 90, 150, 180 and 240 s between trains on both directions. There were assumed 3 energy capacity of SESD with maximum charge / discharge current rated: 2.5 kA (minimum), 4 kA and 6 kA (maximum). For dt = 90 s 1 hour obtained energy savings were 0.32 MWh, 0.28 MWh and 0.18 MWh respectively for 6 kA, 4 kA and 2.5 kA SESD. The results of simulation of operation of SESD for traffic density dt = 90 s and dt = 240 s are presented in figure 15 - energy Es stored and current Is of SESD for 6 kA SESD having 42 MJ available energy capacity.

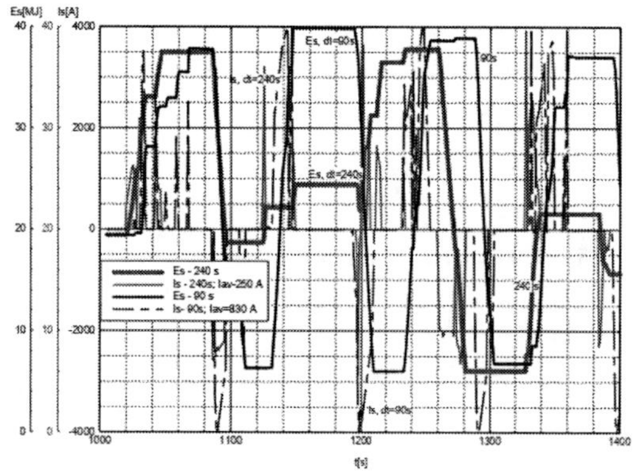

Fig. 15. Energy Es and current Is of a 6 kA SESD for dt = 90 s and dt = 240s.

Values of energy transferred by SESD during 6.5 min of simulation for dt = 180 s was 77 MJ, while for dt = 240 s energy was 44 MJ. It gives, assuming 5.5 h of rush hours with traffic density dt = 180 s and 11 hours with density dt = 240 s about 2.4 MWh of energy saved per working day.

Measurements taken from supercapacitors SESD installed in II Warsaw Metro line shown week-day savings around 2 MWh. Similar results were reported in [14] for high power capacity battery type SESD installed in a TS in metro line in Washington USA, with practically double value of energy saved when this SESD was installed between 2 TS.

5 Conclusion

In this paper a simulation of a metro rail power supply system with SESD was presented. A new dynamic energy management strategy was proposed. The strategy uses the measurements of the third rail currents and voltage and the SC SOC. The simulation results in a study case performed shown influence of density of traffic and SESD capacity on amount of braking energy savings. The simulation tools could be used for designing of the SESD parameters and location to obtain high effectiveness of its installation.

References

1. D. Vine, Transportation emissions roll over power sector emissions (Center for Climate and Energy Solutions 2016), https://www.c2es.org/2016/06/transportation-emissions-roll-over-power-sector-emissions/
2. A. Szeląg, *Efficiency of the regenerative braking in the electrificated rail transport,* Institute of Railway Transport TABOR [Instytut Pojazdów Szynowych TABOR], **4**, 9-16 (2009).
3. J. Yang, J. Zhang and P. Song, Fuzzy Energy Management for a Catenary-Battery-Ultracapacitor based Hybrid Tramway, *2nd Asia Conference on Power and Electrical Engineering (ACPEE), Shanghai, China,* 1-7 (2017)
4. X. Hu, N. Murgovski, L. M. Johannesson and B. Egardt, *Comparison of three electrochemical energy buffers applied to a hybrid bus powertrain with simultaneous optimal sizing and energy management.* IEEE Trans Intell Transp Syst, **15**, 1193–1205 (2014)
5. Z. Song, H. Hofmann, J. Li, X. Han and M. *Ouyang, Optimization for a hybrid energy storage system in electric vehicles using dynamic programing approach,* Applied Energy, **139**, 151-162 (2015)
6. A. Chauvin, A. Hijazi, E. Bideaux and A. Sari, Combinatorial approach for sizing and optimal energy management of HEV including durability constraints, *Proceeding of IEEE 24th International Symposium on Industrial Electronics (ISIE), Buzios, Brazil,* 1236-1241 (2015)
7. M. Wieczorek and M. Lewandowski, *A mathematical representation of an energy management strategy for hybrid energy storage system in electric vehicle and real time optimization using a genetic algorithm,* Applied Energy, **192**, 222–233 (2017)
8. J. Shen and A. Khaligh, A *Supervisory Energy Management Control Strategy in a Battery/Ultracapacitor Hybrid Energy Storage System,*

IEEE Transactions on Transportation Electrification, **3(1)**, 223-232 (2015)

9. M. Bartłomiejczyk and L. Jarzębowicz, *Energy savings by application of supercapacitor storage in trolleybus supplying station – analysis of experimental results*, Technical Transactions electrical engineering, **1-E(2)**, 17-32 (2016)
10. T. Weitzel and H. C. Glock, *Energy management for stationary electric energy storage systems: A systematic literature review,* European Journal of Operational Research, 1–25 (2017)
11. EN 50163, *Railway Applications. Supply Voltages of Traction Systems* (2007)
12. Maxwell Technologies Inc., Datasheet 125V Heavy Transportation Module, http://www.maxwell.com/images/documents/125vmodule_ds_1014696-7.pdf
13. P. Radu and A. Szelag, A Cuk converter integrated with lead-acid battery and supercapacitor for stationary applications, Proceeding of IEEE 18th International Scientific Conference on Electric Power Engineering (EPE), Kouty nad Desnou, Czech Republic, 1-6 (2017)
14. M. Ouattara and Y. J. Gordon, FTA *Report No. 0086, WMATA Energy Storage Demonstration Project, Final Report* (June 2015)

The latest generation drive for electric buses powered by SiC technology for high energy efficiency

Janusz Biliński[1,*]

[1]MEDCOM Sp. z o.o., ul. Jutrzenki 78A, 02-230, Warsaw, Poland

Abstract. High energy efficiency and increasing the working frequency of the converter will make it possible to minimize the size of the cooling system and reduce energy consumption. The use of the SiC technology in propulsion drives increases the efficiency of the converter by 1÷1.5%. Simultaneously, higher working frequency reduces losses in the traction motor, significantly lowering the cost of the energy consumed by the vehicle. Auxiliary converters using the SiC technology are a new quality. The reduction of weight and size is very significant (ca. 40÷50%). Higher switching frequency reduces the size of magnetic components (ca. 80%), and higher converter efficiency minimizes the size of the cooling system. The overall efficiency of the converter is extremely high (94÷96%). This paper presents comparison of Si and SiC parameters which are important for modern EV solutions. Paper presents also parameters of SiC traction inverter and auxiliary converter, designed and manufactured as state-of-the-art product for modern electric bus.

1 Technology of silicon carbide components (SiC)

Modern electric buses, which are operated in cities around the world, use traction inverters and static converters to power on board equipment of the e-bus. In the case of modern e-buses, they provide higher power than traditional rotating inverters. This device, commonly used in electrical traction units, is currently designed and manufactured using IGBT (Insulated Gate Bipolar Transistor) transistor technology. By achieving high reliability and electrical efficiency, technology widely used since the mid-1990s is now recognised as a market standard and used by most manufacturers of electronic components [1-13].

According to the predictions of research institutes, vehicle manufacturers and component manufacturers, in the near future IGBT technology will be replaced by SiC technology [14-16].

Power elements based on the new SiC technology are built on silicon carbide semiconductors (Silicon Carbide, SiC), while the current technology is based on silicon semiconductors. Intensive R & D work carried out by component manufacturers has enabled the availability of SiC components with a reliability similar to that of conventional components and has led to a gradual reduction in production costs, which now allow these components to be commercially used in power electronics.

In traction applications where low weight, small size and high efficiency are the most important parameters of power electronics, the new silicon carbide (SiC) semiconductor technology is changing the game rules. The new semiconductor material has significantly better physical properties than silicon (Si), which allows to build on its base the best power instruments: diodes and transistors. It can be noted that compared to commonly used silicon elements, SiC power elements:

- have shorter switching times,
- have lower voltage drops during conductivity,
- can operate at higher joint temperatures.

As a result, power losses in transformer systems designed with the new technology are significantly lower, which allows to reduce energy consumption, but also to reduce the weight and size of cooling systems. You can also increase the switching frequency of semiconductor components and reduce the weight and size of passive components [17-19].

Application of the new SiC technology in drive inverters allows to increase their energy efficiency by 1÷1.5%. At the same time, the higher switching frequency reduces power losses in the propulsion engine, reducing the power consumed by the vehicle [20, 21].

Static converters using silicon carbide components are a new quality – dimensions and weight can be reduced by up to 50%. The higher switching frequency allows to minimize the size of magnetic elements up to 80%, while lower losses reduce the dimensions of cooling systems. The overall energy efficiency of the inverters is very high, ranging from 94÷96%.

* Corresponding author: janusz.bilinski@medcom.com.pl

Three types of semiconductor components are now available for traction applications using the new silicon carbide technology (Fig. 1):

- Silicon carbide Schottky diodes
- Hybrid modules (silicon IGBTs with SiC Schottky diodes)
- "Full-SiC" modules (SiC MOSFET + SiC diodes).

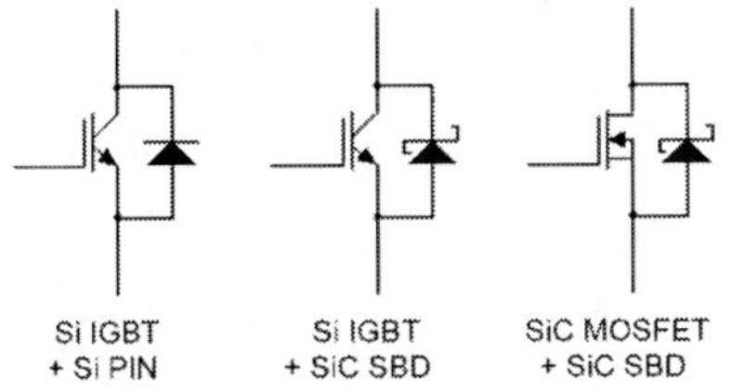

Fig. 1. SiC semiconductor component types – IGBT, hybrid IGBT (transistor with SiC diode), full SiC (SiC MOSFET with SiC diode).

In Fig. 2, a comparison of parameters of IGBT Si and SiC elements for electrical traction power equipment applications is presented.

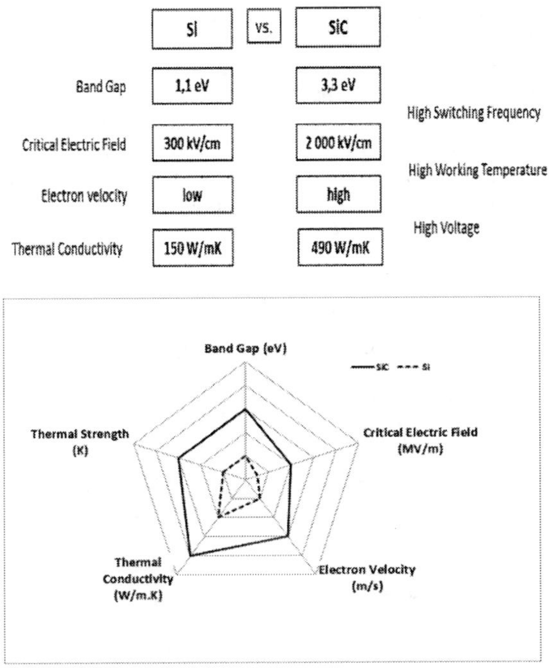

Fig. 2. Comparison of basic parameters of IGBT Si and SiC elements.

Utilizing the advantages of the latest generation SiC components also requires a new approach to device design, as shown in Fig. 3.

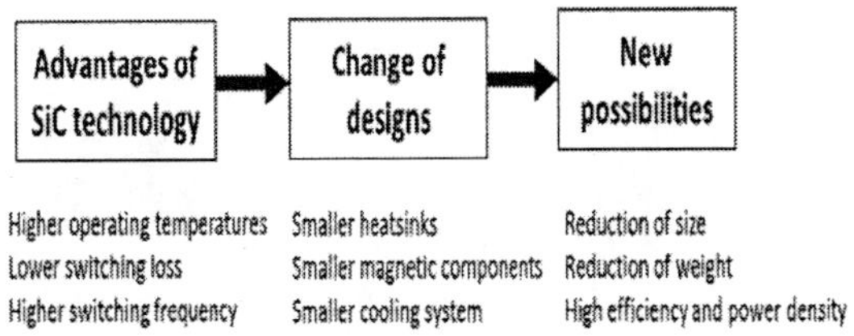

Fig. 3. A new approach to the design of SiCs.

2 Traction inverter integrated with the auxiliary converter with SiC power elements

The FT-160-600 traction inverter is designed for driving electric buses. The inverter is made in full-SiC technology. The drive inverter is controlled by a DSP (Digital Signal Processor) controller using the FOC SVPWM (Field Oriented Control Space Vector Pulse Width Modulation) algorithm. The controller guarantees optimum control of the asynchronous motor and ensures very good traction and high level of driving comfort. The design of the traction inverter allows it to be used for supplying the "classic" traction motor as well as motors integrated with the bus drive axis. The block diagram of the e-bus traction system with SIC inverter and converter is shown in Fig. 4.

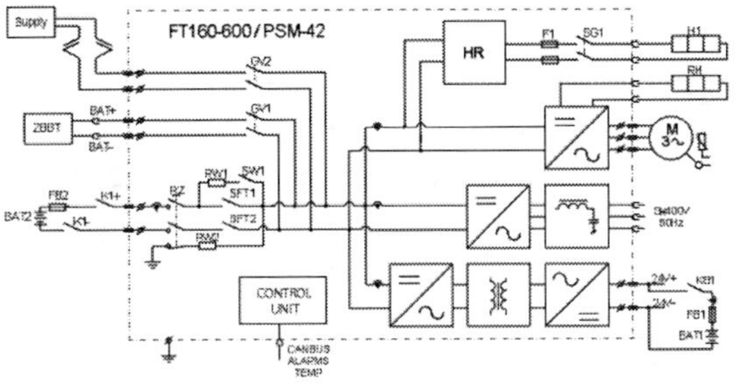

Fig. 4. Block diagram of traction system with inverter and converter FT-160-600 / PSM-42 SiC.

The inverter is equipped with natural air cooling. The system meets all European standards and the requirements of Regulation No. 100 (R100) on safety and radio interference. The system is installed on the roof of the vehicle. It is equipped with an integrated high voltage switchgear connecting the traction battery, buffer power supplies and precharging system in one box. The latest generation of applied elements and advanced diagnostic system guarantee a high level of reliability and low operating costs. The PSM-42 static converter with a power of 30kVA/3x400V AC and 12kW/24V DC is placed in the same box. Fig. 5 shows the device view. The technical parameters of the SiC inverter and SiC converter are shown in Table 1.

 https://doi.org/10.1051/matecconf/201818002012

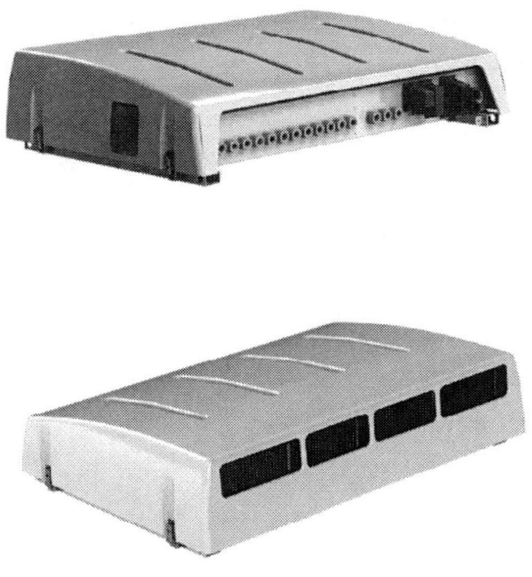

Fig. 5. View of the FT-160-600 inverter.

Table 1. Parameters of the FT-160-600 SIC inverter and the PSM-42 SiC converter.

Specification of FT-160-600 inverter	
Input voltage	520 - 750 VDC
Auxiliary voltage	24 VDC + 30%... – 40%
Rated current	300 Arms
Maximum current	500 Arms
Rated power	160 kW
Frequency	0 ÷ 350 Hz
PWM frequency	2 - 6 kHz
Insulation strength	2,5 kV
Cooling	natural, air
Dimensions	1430×990×330 mm

Specification of the PSM-42 converter	
Input voltage	520 - 750 VDC
Auxiliary voltage	24 VDC + 30% – 40%
AC output nominal power	30 kVA
AC output voltage	3x400 VAC +/- 10% (sinus)
AC output frequency	50 Hz
AC output current	43 A
Overcurrent	1.5In/30 sec.
Electronic protection	overcurrent/ overvoltage/ overheating
DC output nominal power	12 kW
DC output voltage	24 VDC
DC output current	500 A
Insulation strength	2,5 kV
Cooling	natural, air
Weight of the set	165 kg

Summary

Electric vehicles are an important part of life around the world, and this is why it is important to save energy. Power elements based on SiC technology allow for significant reduction of energy losses, increasing the efficiency of the inverters. The expected reduction of dynamic losses is approx. 50% compared to IGBT technology. In addition, power elements based on SiC technology have a significantly higher operating speed, which allows higher frequencies to be used, allowing the use of smaller and lighter magnetic components, which can reduce the weight and size of devices by more than 40% compared to traditional designs. Increased switching frequency reduces traction engine losses by lowering the energy cost of the vehicle.

Improvements in efficiency combined with a decrease in the weight and size of equipment is now one of the key improvements sought by electric vehicle manufacturers. Components based on SiC technology reduce the weight of the whole vehicle, reducing energy consumption. The overall efficiency of SiC converter is ca. 94÷96%, SiC technology of inverter gives 1÷1.5% better value compared to IGBT.

An additional advantage, which is particularly important for vehicle passengers, is the reduction of noise generated by devices based on SiC components, allowing for greater comfort of travel.

The use of semiconductor components in SiC technology in the propulsion system and power supply will reduce energy consumption by about 5÷15% compared to traditional silicon technology. The use of SiC technology is a true technological revolution in the design of innovative power electronic devices for public transport.

References

1. P. Baginski, *Power Modules for Charger Applications* (2016). Viewed 20 September 2017 https://www.vincotech.com/support-and-documents/technical-library.html
2. D. Aggeler, *Bidirectional Galvanically Isolated 25 kW 50 kHz 5 kV/700V Si-SiC Super Cascode/Si-IGBT DC-DC Converter* (PhD thesis, ETH Zurich, Switzerland, 2010)
3. M. Frisch, *Advantages of SiC Schottky Diodes in Fast Switching Power Electronics Solutions* (2010). Viewed 20 September 2017 https://www.vincotech.com/support-and-documents/technical-library.html
4. M. Frisch, and E. Temesi, *Topologies Benefits of NPC Inverter* (2010). Viewed 20 September 2017 https://www.vincotech.com/support-and-documents/technical-library.html
5. J. Liu, K.L. Wong, A. Scott, and J. Mokken, *Performance Evaluations of Hard-Switching Interleaved DC/DC Boost Converter with New*

Generation Silicon Carbide MOSFETs (2016). Viewed 20 September 2017 https://www.mouser.jp/pdfDocs/Cree-Power_Article_4.pdf
6. Mitsubishi Electric, *Mitsubishi Electric to Begin Shipment of Silicon Carbide Power Module Samples* (2012). Viewed 20 September 2017 http://www.mitsubishielectric.com/news/2012/0709.html
7. Mitsubishi Electric, *Mitsubishi Electric Delivers World's First SiC Auxiliary Power Supply Systems for Railcars* (2013). Viewed 20 September 2017 http://www.mitsubishielectric.com/news/2013/0326-a.html
8. K. Mrinal, et al., 10 kV, 120 A SiC Half H-Bridge Power MOSFET Modules Suitable for High Frequency, Medium Voltage Applications, *2011 IEEE Energy Conversion Congress and Exposition, Phoenix, AZ, USA*, **4**, 2689-2692 (2011)
9. P.G. Neudeck, *Silicon Carbide Technology. The VLSI Handbook* (NASA Glenn Research Center, CRC Press LLC 2006)
10. R. Singh, *Silicon Carbide Thyristors usher in the Smart Grid Revolution* (2012). Viewed 20 Sept. 2017 https://www.genesicsemi.com/images/references/trade_journals/bodo_02-12.pdf
11. R. Singh, et al., 1200 V-class 4H-SiC "Super" Junction Transistors with Current Gains of 88 and Ultra-fast Switching capability, *Materials Science Forum*, **717-720**, 1127-1130 (2012)
12. S.G. Sundaresan, et al., 12.9 kV SiC PiN diodes with low on-state drops and high carrier lifetimes, *Materials Science Forum*, **717-720**, 949-952 (2012)
13. S.G.Sundaresan, et al., Integrated SiC Anode Switched Thyristor Modules for Smart-Grid Applications, *Materials Science Forum*, **717-720**, 1159-1162 (2012)
14. S. Taranowich, *Si vs. GaN vs. SiC: Which process and supplier are best for my power design?* (EDN Network, Power Management Design Center, 2013). Viewed 20 Sept. 2017 https://www.edn.com/design/power-management/4409627/Si-vs--GaN-vs--SiC--Which-process-and-supplier-are-best-for-my-power-design
15. How2Power Today, *Dual SiC MOSFET Modules Promise High Efficiency* (2009). Viewed 20 Sept. 2017 https://www.how2power.com/newsletters/0911/products/PowerexSiC.pdf
16. Yole Development, *Power SiC 2017: Materials, Devices, Modules, And Applications report* (2017). Viewed 20 Sept. 2017 http://www.yole.fr/STATUS_SIC_MARKET_TECHNOLOGY_TRENDS.aspx
17. ROHM Semiconductor, *SiC Power DeviceTechnology* (2013). Viewed 20 Sept. 2017 https://www.rohm.com/documents/11308/2420552/SiC_power_device-catalog.pdf
18. ROHM Semiconductor, *SiC Power Devices and Modules. Application Note* (2014). Viewed 20 Sept. 2017 https://d1d2qsbl8m0m72.cloudfront.net/en/products/databook/applinote/discrete/sic/common/sic_appli-e.pdf
19. ROHM Semiconductor, 2016. *Silicon Carbide Schottky Barrier Diodes. White Paper*, vieved 20 Sept. 2017 https://www.rohm.com/documents/11308/12928/ROHM_SiC+Diodes_wp.pdf
20. E. Hagesawa, et al., *Hybrid SiC Power Module With Low Power Loss*, *Power* Electronics Europe, **6**, 20-23 (2012)
21. Circuits Today, *JFET and MOSFET Comparison* (2017). Viewed 20 Sept. 2017 http://www.circuitstoday.com/jfet-and-mosfet-comparison

Stationary energy storage system in a 3 kV DC – the conception comparison

Włodzimierz Jefimowski[1,*]

[1]Warsaw University of Technology, Institute of Electric Power Engineering, Electric Traction Division, ul. Koszykowa 75, 00-662, Warsaw, Poland

Abstract. The paper presents the research results of a few different conception of stationary energy storage system in a 3 kV DC system. The most attention is focused on the comparison between two topologies of the ESS: energy storage system with supercapacitor and with supercapacitor and LFP battery. The variants are compared in terms of energy saving and peak power demand reduction. The implementation of ESS with SC results the decrease of active energy drawn from traction substation. Meanwhile the implementation of ESS with SC and LFP battery leads to achieving of two aims - decreasing of active energy consumption by maximization of regenerative energy utilization and reduction of 15 - min. peak power demand of traction substation.

1 Introduction

The significant part of the railway industry operation are the costs of traction energy. The traction energy costs in railway mainly consist of the active energy component and the peak power demand component costs.

Besides due to climate changes the energy efficiency has become an important problem for the last two decades. The issue takes an important part in EU regulations, which demand the energy efficiency audits to be carried out in a large companies. The environmental aspect of the energy efficiency is of utmost importance in Poland, where 97% of the electric energy is generated in coal power plants.

At the same time the growing number of the rolling stock in Polish railway is equipped with the regenerative braking, which gives the new opportunities of energy saving. In this case the use of regenerative energy depends in the main degree on the receptivity of overhead catenary system (OCS). The overhead catenary is receptive if during regenerative braking of the train if there is the load on the same supply section able to drawn and utilise the regenerative power. If there are no any trains in start up or constant speed mode the energy could be saved in the trackside energy storage system (ESS). The track-side ESS apart from energy savings could contribute to reduction of peak power demand as well as improvement of pantograph voltage conditions in specific situations [1, 2]. Depending on the purpose of the ESS implementation it could be fixed in traction substation or in other place along the track. The general scheme of the ESS connected with 12-pulse transformer rectifier unit in 3 kV DC traction substation is presented in Fig.1.

The problem of stationary energy storage system sizing and location optimization has been undertaken by the range of authors [3-6]. The literature analyse the ESS with supercapacitors and different types of batteries [7].

Not much attention of researchers is dedicated to the problem of energy management strategy in trackside ESS. In [8] the optimization method of energy management based on the genetic algorithm is presented.

The paper presents the analysis of two variants of trackside ESS: with supercapacitor and hybrid ESS with supercapacitor and LFP battery. The analysis covers the comparison among different energy management strategies in both ESS configurations. The deterministic rule based strategies are developed for the specific ESS applications. In order to estimate the effects of energy storage system operation the indication parameters are defined and proposed.

2 ESS operation indicators

As mentioned many researchers focus their attention on the conception of improvement of pantograph voltage conditions by implementation of ESS trackside. The aim could be achieved in railway lines with the big headway times of train operation, where the traction substation could be replaced by stationary ESS. However the paper is focused on the option of ESS

* Corresponding author: wlodzimierz.jefimowski@ien.pw.edu.pl

implementation in the existing substation in 3 kV DC system in order to maximise the regenerative energy utilization and minimise 15-min power demand. In order to estimate the value of ESS operation effects the below parameters are proposed:

- 24 -hour efficiency of energy recovery (1)

$$\gamma_E = \frac{\int_{t=0}^{24h} P_{Dis}(t)dt}{\int_{t=0}^{24h} P_{\mathrm{Re}\,c}(t)dt} \cdot 100\% \quad (1)$$

where:
$P_{Dis}(t)$ - power of ESS discharge at the point of connection of ESS to the catenary,
$P_{\mathrm{Re}\,c}(t)$ - regenerative power possible to utilize due to the deficiency of OCS receptivity;

- maximum 15-min power drawn by traction substation.

The indicators of ESS presented above are used to compare the effects of ESS operation for the different option of ESS structure, energy management strategies, locations and size parameters. The first indicator specifies the ratio between the energy recovered by the energy storage system and the energy available to recover due to the deficiency of overhead catenary receptivity. The second indicator specifies the maximum average 15 minute power, among 15 minutes period during 24 hours. The parameter is important due to the power demand fee which refers to the 15 min. average power.

3 ESS with supercapacitor

The general scheme of ESS with supercapacitor is presented in Fig 1. The most important parameters of ESS with supercapacitor are:

- number of cells connected in module in parallel and serious,
- power of DC/DC converter,
- minimum state of charge (depth of discharge).

In the investigation the equivalent RC model of supercapacitor is used . The equivalent scheme is shown in Fig. 2. The assumed model of supercapacitor is widely approved and used [9].

The supercapacitor current could be expressed by (2)

$$i_2(t) = \frac{u_1(t)}{u_2(t)} \cdot i_1(t) \cdot \eta_{DC/DC} \quad (2)$$

where:

$$u_2(t) = u_{ESR}(t) + u_{SC}(t) \quad (3)$$

and

$$u_{SC}(t) = \frac{1}{C}\int_{t_0}^{t_1} i_2(t)dt + u_{SC}(t_0) \quad (4)$$

where:
$u_{ESR}(t)$ - voltage drop on the internal resistance of the SC pack,
$u_{SC}(t)$- state of charge of SC module,
$u_{SC}(t_0)$- initial state of charge of SC module
C - SC capacitance.

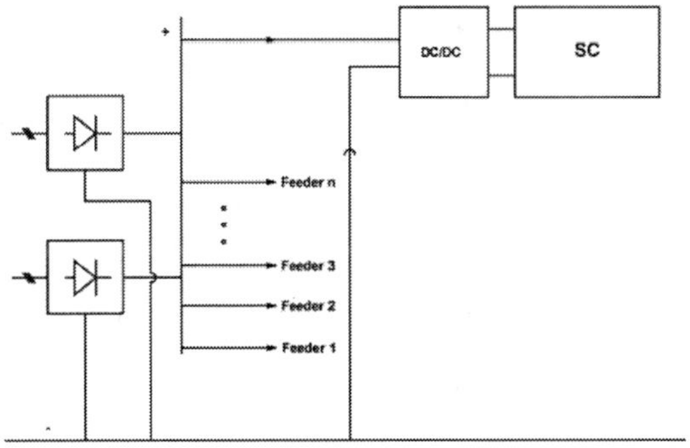

Fig. 1. The scheme of the ESS with supercapacitor connected in to 12-pulse transformer rectifier unit in 3 kV DC traction substation.

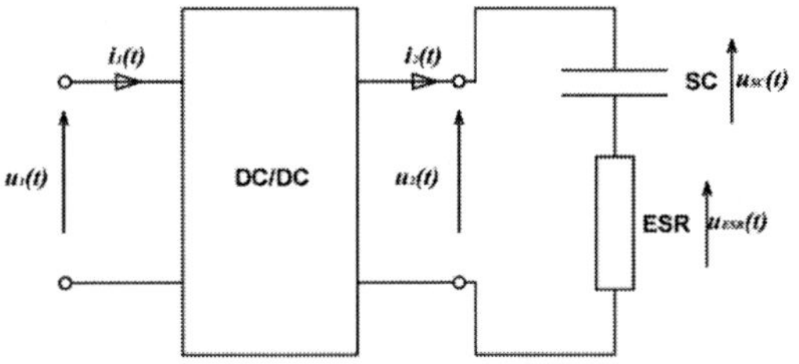

Fig. 2. The equivalent circuit of the supercapacitor ESS.

The module of the supercapacitor cells in proposed solution consists of 500 cells connected in series. The number of branches connected in parallel depends on the capacity of the converter needed in the specific location. The number of cells connected in serious and parallel branches determines the nominal voltage, the resultant capacitance as well as the resultant internal resistance of the SC module.

In Table 1 the parameters of supercapacitor cells used in the module are presented.

Table 1. Supercapacitor cell parameters.

Type	BCAP
Mass	510 g
Nominal voltage	2,7 V
Capacitance	3000 F
Internal resistance	0,29 mΩ
Specific power	5,9 kW/kg
Cost	60 $
Nominal cycle life	1000000

The investigation has been made based on the power demand of substation on the specified location. The power demand of substation including the regenerative power $P_{\mathrm{Re}\,c}(t)$ available is obtained from the simulation model of electric railway line. In the model the real timetable and infrastructure parameters has been implemented. The simulation has been carried out for the substation location on the railway line with the relatively large number of trains operating (115 trains/24h/direction). Three train types operates in the modelled railway line (suburban electric multiple units, long distance multiple units and locomotive passenger trains). For the suburban electric multiple units the average number of stops is 3 per 10 km. The 24-hours energy consumption on the analysed substation is 24,85 MWh, the regenerative energy available during 24-hourse period is 3,56 MWh.

For the given power demand the investigation of the influence of main ESS parameters (usage energy and power of DC/DC converter) under different energy management strategies has been carried out. For the ESS three deterministic strategies was taken into account. The algorithm of the strategies are presented in Fig 3. The purpose of the strategies is minimisation of the energy consumption and minimisation of 15-min. power demand.

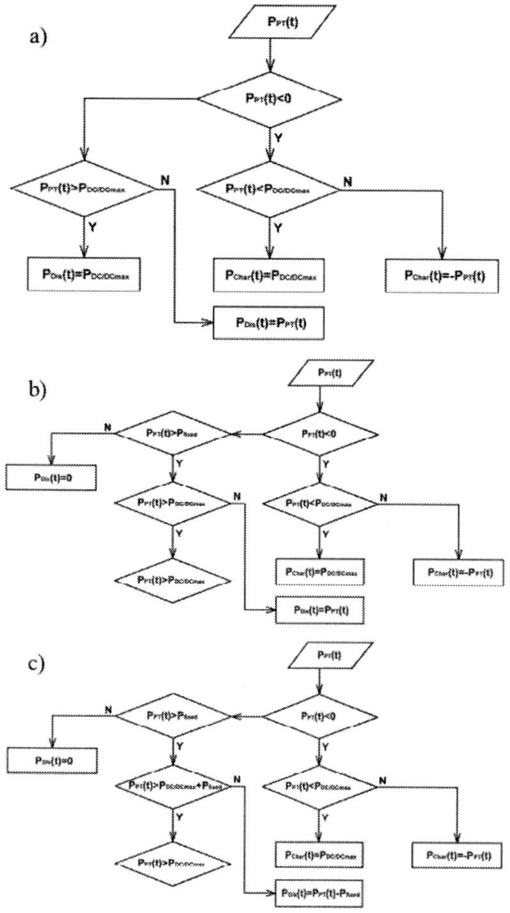

Fig. 3. The algorithms of the investigated energy management strategies for ESS with ultracapacitor.

Strategy a is the most effective from the point of view of energy consumption. The part of regenerative energy recovered γ_E is limited by the ESS and DC/DC converter features. Fig. 4 presents the instant power against time of ESS charge and discharge at the point of its connection to DC busbar of traction substation and regenerative power available at the same point. The figure shows, that part of the available regenerative power is beyond the power of ESS DC/DC converter (blue).

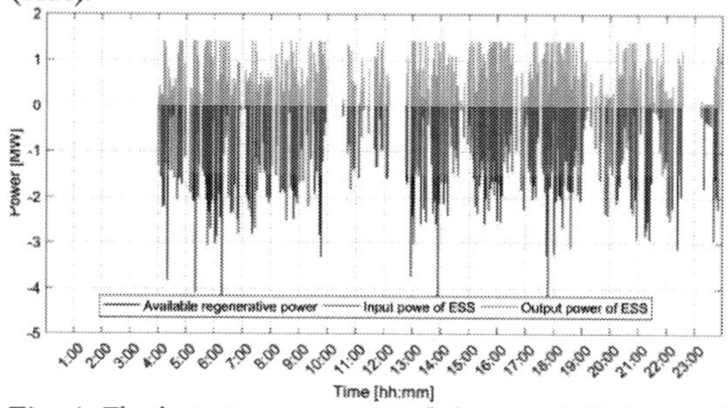

Fig. 4. The instantaneous power of charge and discharge of SC ESS at the point of connection to the DC busbar of traction substation and regenerative power available at the same point in 24-hourse time period.

The further investigation present the influence of the ESS parameters on its operation features. Fig. 5 presents the value of the 24 hour efficiency of energy recovery γ_E (1) for strategy a as the function of ESS capacitance and DC/DC converter.

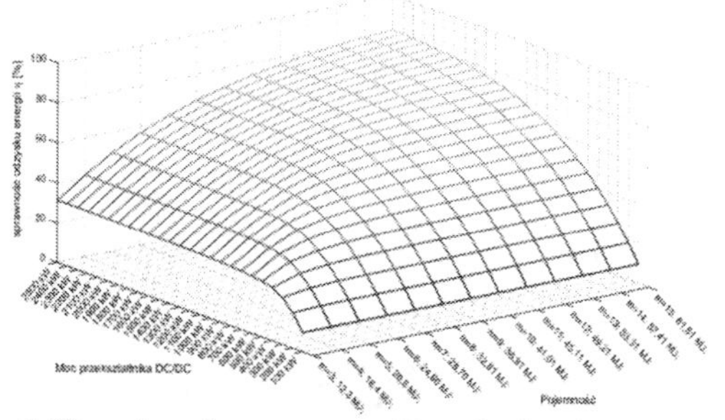

Fig. 5. The value of γ_E parameter (1) as the function of ESS capacitance and DC/DC converter.

The results show that the strategy a is effective from the point of view of energy efficiency. Fig. 5 presents the comparison between the histogram of 15-min power occurrence for strategy a. The result show that in terms of the maximum 15-min power drawn from traction substation the ESS system with SC and the investigate strategy doesn't give any improvement.

The strategy b and c are the modifications of strategy a. In strategies b and c the discharge takes place in case if the instant power of traction substation exceeds the assumed power P_{fixed}. The results of the proposed strategies implementation, presented in Table 2 show that the strategies b and c don't reduce the 15 min power demand.

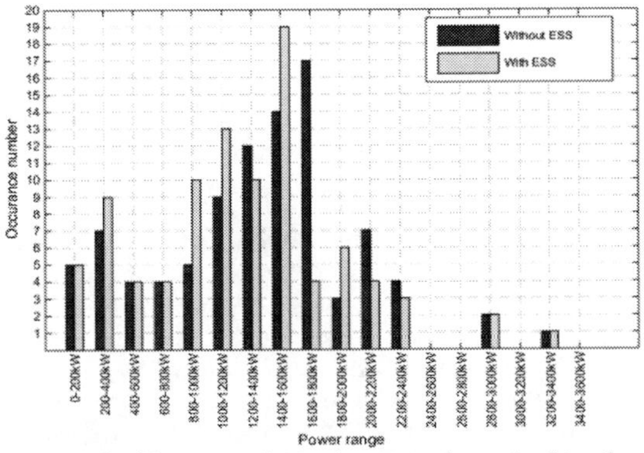

Fig. 6. The histogram of 15-min power demand of traction substation with ESS with SC and without it.

Table 2. The comparison of the ESS operation parameters for investigated strategies.

Parameter	Strategy a	Strategy b	Strategy c
γ_E	75,48 %	71,03 %	67,56 %
P_{15max}	3,335 MW	3,32 MW	3,32 MW
E_{24_PT}	22,16 MWh	22,32 MWh	22,44 MWh

In Table 2 the last row shows the 24 hour energy E_{24_PT} consumed by traction substation. The 24 hour energy consumption on the analysed substation without ESS is 24,85 MWh.

4 ESS with supercapacitor and LFP battery

The scheme of the hybrid energy storage system connected to the 12-pulse transformer rectifier unit in a 3 kV DC system is presented in Fig 7. Among the range of batteries types the LFP batteries was chosen according to its most appropriate parameters in terms of the specific demands.

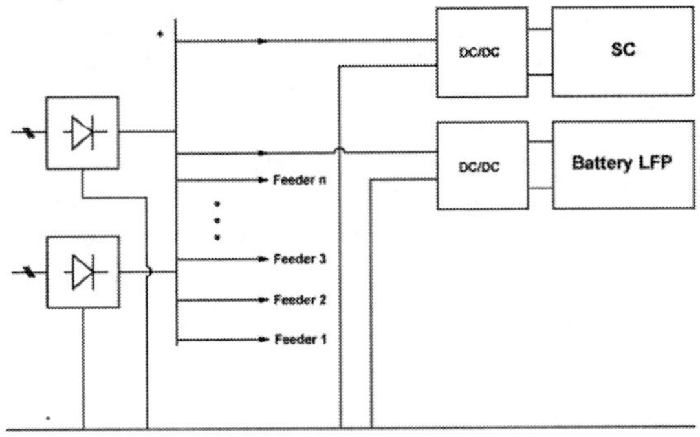

Fig. 7. The scheme of the ESS with supercapacitor and LFP battery connected in to 12-pulse transformer rectifier unit in 3 kV DC traction substation.

The operation effects of that type of energy storage have been analysed. The parameters of its operation have been compared with the corresponding parameters of SC ESS operation.

Fig. 8 presents the histogram of 15-min power demand of traction substation for variant with ESS and without it. The results show that the maximum 15-min current could be limited significantly comparing to the SC ESS. The 24 hour energy consumed by the assumed traction substation with the proposed ESS is 22,59 MWh.

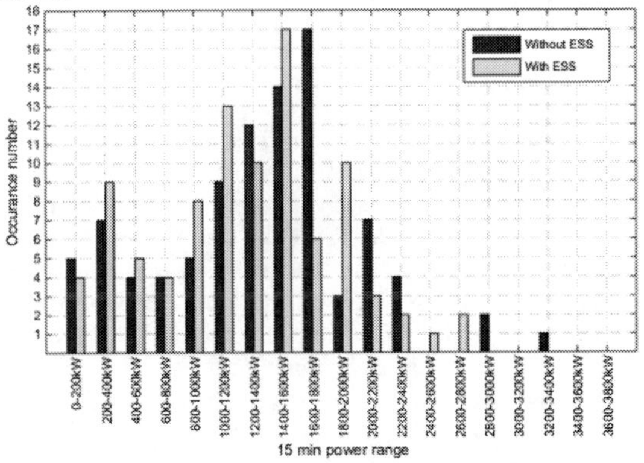

Fig. 8. The histogram of 15-min power demand of traction substation with ESS with SC and LFP battery and without it.

In Fig. 9 the rolling 15-min power drawn from the traction substation for variants with and without ESS. On the figure the operation way of the ESS is shown on the example of 15-min power. The battery discharge and compensation of the 15-min power is recognisable during the peak of the power drawn by the traction substation. After which the charge battery occurs, which lead to the increase to the rolling 15-min power demand.

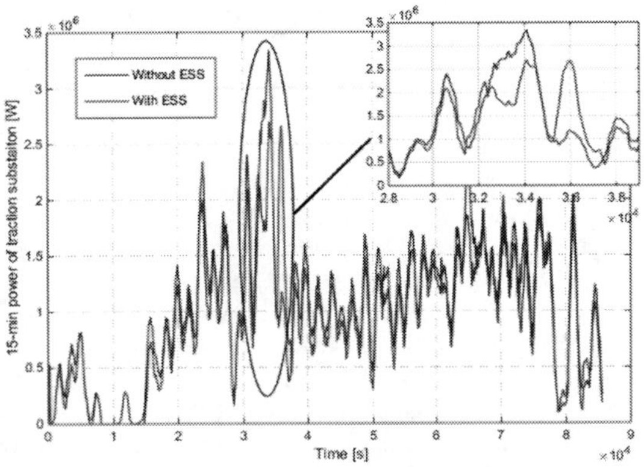

Fig. 9. Rolling 15-min power of traction substation with and without ESS with SC and LFP battery during 24 h period.

In Table 3 the values of the operation parameters are compared for the variant with and without hybrid ESS on the traction substation. The result show the significant improvement of all of the investigated parameters of the ESS operation effect. The value γ_E for this type of strategy is not to be found.

Table 3. The comparison of the ESS operation parameters between the case without ESS and the hybrid ESS.

Parameter	Without ESS	ESS with SC and LFP battery
γ_E	0	-
P_{15max}	3,34 MW	2,7 MW
E_{24_PT}	24,85 MWh	22,59 MWh

5 Conclusions

The paper presents the comparison analysis between two conceptions of ESS in the 3 kV DC electrification system. Both conceptions covers the solution of implementation of the ESS in the existing traction substation. The first is ESS with supercapacitor, the second is ESS with supercapacitor and LFP battery. The aim of the study is finding the solution of ESS which allows to achieve the following aims:

- maximise the utilization of the regenerative energy
- minimise the 15-min power demand of the traction substation

The solution with the SC is effective in terms of the first criteria. It allows to utilise the most significant part of the regenerative energy available in the point of ESS connection, as it is shown in Table 2. The part of the regenerative energy which could be utilized by using of this type of ESS if limited practically only by DC/DC and charge/discharge cycle efficiencies.

However the ESS with SC is inefficient in terms of minimising of 15-min power demand due to the insufficient energy capacitance. The energy capacitance of the ESS with SC, which is sufficient from the point of view of regenerative energy utilization is one order of magnitude smaller than needed for decreasing the 15-min power.

The proposed solution of HESS with SC and LFP battery meets both above mentioned criteria. The value of 15-min. power reduction influence the number of charge/discharge cycles and the discharge current, hence the live cycle of the LFP battery.

There is demanded to insure the high level of reliability for HESS in order to avoid 15-min demand power exceedance. In case of demand power exceedance, the penalty fees are to be paid according to the agreement between DNOs and the energy regulator office guidelines.

The research was financed under the Dean's Grant number 504/03411/1041/42.000100 from the statutory activities of the Faculty of Electrical Engineering of Warsaw University of Technology.

References

1. T. Maciołek, Z. Drążek, and A. Szeląg, *Efektywność energetyczna zasobników energii w podstacjach systemu prądu stałego 3 kV DC*, Logistyka, **3**, 2990-3000 (2015)
2. T. Maciołek, *Zastosowanie zasobników energii w trakcji kolejowej 3 kV DC – nieodległa perspektywa?*, TTS Technika Transportu Szynowego, *22*(9), 39-44 (2015)
3. D. Iannuzzi, Improvement of the energy recovery of traction electrical drives using supercapacitors, *Proceeding 13th Int. Power Electron. Motion Control Conf. EPE-PEMC*, 1469–1474 (2008)
4. S. De La Torre, A. J. Sanchez-Racero, J. A. Aguado, M. Reyes, and O. Martinez, *Optimal Sizing of Energy Storage for Regenerative Braking in Electric Railway Systems*, IEEE Trans. Power Syst., **vol. 30**, no. 3, 1492–1500 (2015)
5. R. Barrero, X. Tackoen, J. Van Mierlo, V. U. Brussel, and B. Elsene, Improving energy efficiency in public transport: stationary supercapacitor based energy storage systems for a metro network, *IEEE Vehicle Power and Propulsion Conference* (2008)
6. S. J. Kashani, and E. Farjah, *Applying Neural Network and Genetic Algorithm for Optimal Placement of Ultra-Capacitors in Metro Systems*, 35–40 (2011)
7. P. Radcliffe, J. S. Wallace, and L. H. Shu, Stationary applications of energy storage technologies for transit systems, *Proc. IEEE Electr. Power Energy Conf.*, 1–7 (2010)
8. H. Xia, H. Chen, Z. Yang, F. Lin, and B. Wang, *Optimal Energy Management, Location and Size for Stationary Energy Storage System in a Metro Line Based on Genetic Algorithm*, 11618–11640 (2015)
9. N. Devillers, S. Jemei, M. Péra, D. Bienaimé, and F. Gustin, *Review of characterization methods for supercapacitor modelling*, J. Power Sources, **246**, 596–608 (2014)

* Corresponding author: wlodzimierz.jefimowski@ien.pw.edu.pl

MATEC Web of Conferences **180**, 03001 (2018) https://doi.org/10.1051/matecconf/201818003001
MET'2017

Batteries in a vehicle – conditions, capabilities and limitations

Piotr Julian Biczel[1,*] and *Maciej* Kwiatkowski[1,2]

[1]Warsaw University of Tehcnology, IETiSIP
[2]Impact Clean Power Technology S.A.

Abstract. Electric buses are now seen as the primary means of public transport in cities. However, their exploitation is associated with a number of limitations. The article presents the problem of battery use in buses. The design differences between the vehicle with the internal combustion engine and the electric motor are discussed. In particular, authors compared the number of passengers that both types of buses can take, as well as the difference in vehicle mass and range, and the reasons for these differences. The types of electrochemical cells that can be used to power vehicles and their basic features are presented. The article focuses on the NMC, LFP, LTO and LIC type cells. Next, the battery pack structure and its components were described. The specific energies of cells and batteries constructed from these cells were compared. Next, the most important features of the vehicle equipped with batteries with various types of cells are discussed. The mass, range, charging time and cycle lifetime of the batteries were compared. The basic dependence was noted: the greater the specific energy, the smaller the number of battery cycles. So if the bus has to take more passengers, it must have light batteries, which should be frequently charged, which requires high cyclic lifetimes.

1 Introduction

The increasing number of population as well as migration to cities, causes more and more difficulties in moving the population. Individual transport is inefficient and must be replaced by collective transport in the shorter term. In addition to lengthening travel time and cost increasing, environmental impact of transport means is an important factor. At present, mainly contamination with dust, nitrogen oxides and noise is mentioned here. These problems can be largely resolved by introducing common electric transport, including road transport. It can be said that the development of electric transport is not at risk, because electric vehicles are, for the time being, the best remedy for pollution and noise.

It should be expected that also in public transport, electric traction, including road traction, will develop. However, more and more often we will drive without traction network. This is evident from the dynamic development of electric bus constructions and the increasing number of cities interested in introducing electric vehicle fleets [1]. The figure 1 presents the forecast of the bus market development in Europe [2].

Electricity can be easily generated, transferred, regulated, and processed, but cannot be saved for later or transported without connection to the source. Completely different from liquid fuels, both fossil and biofuels. At present, electric energy is exclusively stored only in electrochemical batteries. Unfortunately, they have many drawbacks compared to liquid fuels. Above all, small specific energies and limited cycle lifetime.

*e-mail: biczel@ee.pw.edu.pl

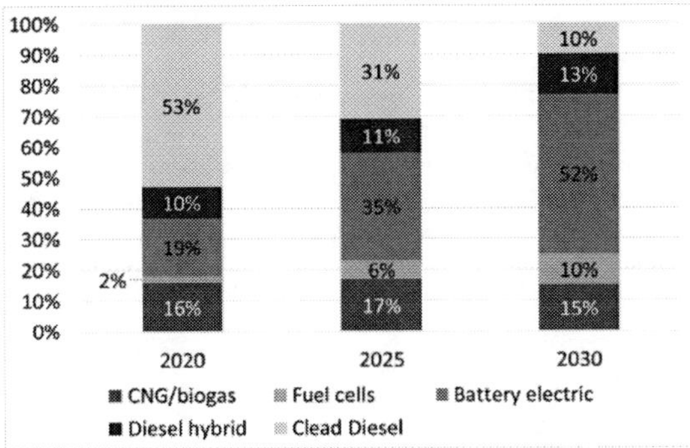

Figure 1. Forecast of buses technology development in Europe [2]

It is also worth mentioning that electric vehicles can be powered by low-emission energy. It is enough to ensure that the power used in them will be produced in zero-emission and renewable sources.

There are therefore many reasons for switching from oil-based transport to electric one. One should wonder how the above-mentioned problems affect the bus's ability and how it can be dealt with.

2 C rate

In the further part of the article there will often be a coefficient written as $\frac{1}{x}C$ or C/x used, where x is the time the battery is discharged and C its nominal capacity in [Ah]

Table 1. Composition of lithium ion cell electrodes of various types

Cell type	Cathode	Anode
NMC	$LiNi_xMn_yCo_zO_2$	graphite
LFP	$LiFePO_4$	graphite
LTO	$LiCoO_2$	$Li_4Ti_5O_{12}$

(charge). We say that the battery is discharged with C/x current, so called x-hour discharging current, if a nominal charge of C is taken from it within x hours. This coefficient has the dimension in [A] and is used to compare the discharge currents of different types and capacities.

The form $\frac{1}{x}C$ is used mainly for discharge currents shorter than one hour and C/x for onger one. I.e. 5-hours current, C/5 or 0,2C, for 20 Ah cell is 4 A:

$$C/5 = \frac{20\,Ah}{5\,h} = 4\,A \quad (1)$$

and 6-minutes current (1/10 of hour), C/0,1 or 10C, for the same cell:

$$10C = \frac{20\,Ah}{0.1\,h} = 200\,A \quad (2)$$

Lithium-ion batteries are mainly intended for short discharges, from 3 hours to several minutes (C/3 – 10C), therefore the form of $\frac{1}{x}C$ will be used here more often in following sections.

This coefficient is formally defined, for example, in [3].

In practice, the current given in the form C/x (coefficient C/x) allows to compare cells of different types and capacities. It makes the charging current independent of the battery capacity.

3 Recharchable cell for vehicles

The basic parameters that determine the use of a given cell in a vehicle are: discharge current, specific energy, number of cycles and operating temperature.

Among the many types of lithium-ion cells used in vehicles [4], buses currently use the following three types of lithium-ion cells: NMC, LFP [5], LTO [6] (table 1). In the near future LIC [7] and new NMC battery types - NMC811 [8] will join the group. The cells differ in the structure and chemical composition of electrodes and electrolyte, and hence the capacity, charging and discharging power, environmental conditions of work and - above all - energy density [Wh/dm^3] and specific energy [Wh/kg].

The LIC cells behave almost like capacitors. And they have similar power parameters and number of cycles. However, they have a very small capacity and specific energy. The bus, supplied with those cell, would have to be loaded every 2-3 stops.

Currently used NCM batteries have an equal amount of nickel, cobalt and manganese, what is marked as NMC111. Soon, cells with a different ratio of these elements will be introduced to the market, the so-called NMC811 cells [8]. They are very light and cheap, but they do not have enough cycles. Pending replacement of batteries every 3 – 4 years is considered, Total Cost of Ownership would be larger, but the bus would have better performance – range or payload.

Table 2. Parameters of selected types of lithium-ion cells

Cell type	Disch. curr.	Char. curr.	Temp. of oper.	No. of cycles
NMC	2C	2C	-20°C–+60°C	300 – 7000
LFP	4C	5C	-30°C–+50°C	2000 – 7000
LTO	20C	10C	-30°C–+40°C	approx. 50000
LIC	200C	200C	-30°C–+70°C	approx. 500000

Typical parameters of the above mentioned cells are presented in the table 2. It can be seen that cells with a higher number of cycles can be discharged and charged with higher powers (high C-rate) in relation to capacity than cells with lower number of cycles.

However, the price of a large number of cycles and large currents is low specific energy. This is due to the fact that the cell has a higher C-rate current and number of cycles, the more they behave like capacitors, not electrochemical cells. Capacitors, however, have small specific energies.

Figure 2 presents a comparison of the specific energy and energy densities of the cells in question. Figure 3 presents the change in the energy density of the individual cell types as a function of the number of cycles.

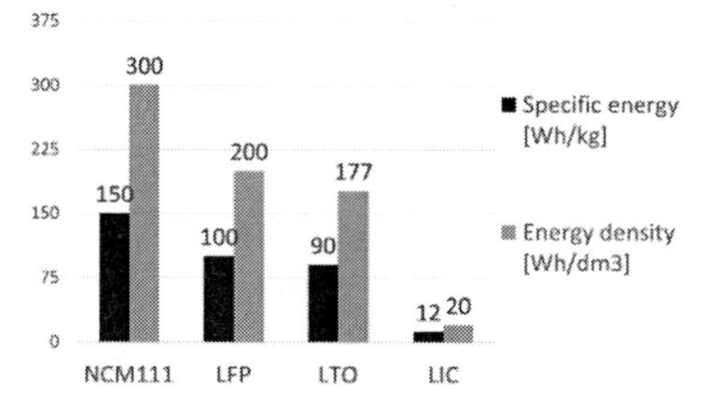

Figure 2. Energy density and specific energy of selected cells

4 A bus

A typical urban bus, most frequently used in Europe, is 12 m long (figure 4) [1]. A bus, produced by one of the leading manufacturers, in the diesel engine version can take 105 people on board. Its gross vehicle mass (GVM) is 18 tons, and its deadweight load is in the order of 10.4 – 13.2

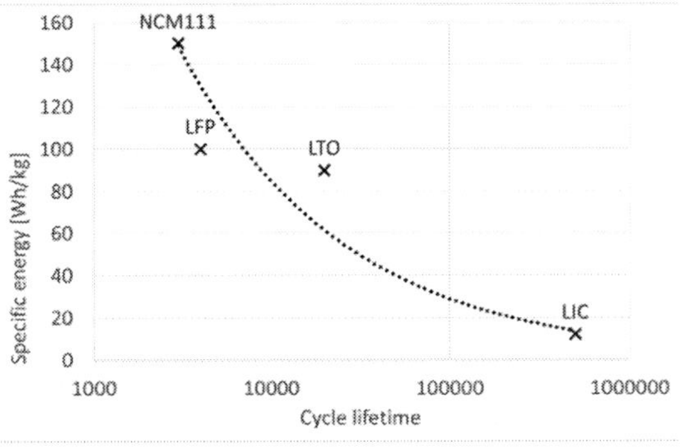

Figure 3. Energy density vs. cycle lifetime of selected cells

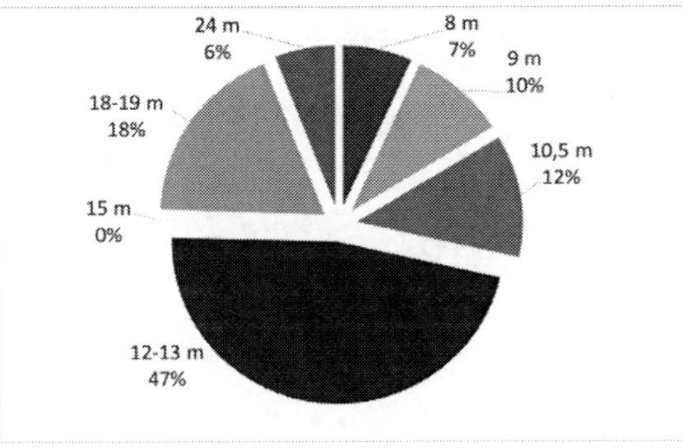

Figure 4. The share of electric buses of different lengths in the European market [1]

tons. It is driven by a 200 kW engine. An electric bus built on the same chassis and with the same body can take only 83 passengers according to the manufacturer. His GVM is also 18 t and it is driven by a 160 kW engine.

So, what difference does the number of passengers have if the GVMs are the same and the other parameters are similar?

To answer this question, compare the weight of both vehicles. The difference in the number of passengers taken is 22 people. An average mass of these people is about 1700 kg. Changing the electric bus into an electric one turns the part of the internal combustion drivetrain:

- comustion engine along with the equipment,
- gear box,
- fuel tank

into electric one:

- electric motor,
- drive powerelectronic converter,
- 24 V auxiliary powerelectronic converter,
- baterries along with the equipment.

The comparison of the mass of these elements is presented in the table 3.

Table 3. Comparison of the mass of the combustion and electric bus components

Combustion bus		Electric bus	
Component	Mass [kg]	Component	Mass [kg]
Engine	500	Motor	350
Gear box	270	Drive converter	100
Fuel tank	350	24 V aux. converter	150
		Battery	?
Total	1120	Total	600 + ?

It is easy to notice that the combined mass of the listed components from the combustion engine bus is 1120 kg. Components of the electric bus, in addition to the batteries have a total mass of 600 kg. Thus, these electric vehicle components are 620 kg lighter. The secret is hidden in the mass of the battery, which must give a mass increase of approx. 1100 kg - as much as the mass of missing passengers reduced by the mass of electrical components.

How is it constructed and what is the battery of the bus composed of?

5 Battery in a bus

The battery in the bus is made of several hundred cells connected in series and in parallel. In addition to the cells, it consists of many other elements. Following can mention here:

- security housing,
- electrical protection,
- contactors,
- BMS battery management system,
- cooling and heating systems,
- wiring.

All this causes that the battery in the bus has the resulting specific energy lower than the cells it is made (figure 5). This difference is particularly high in the case of high C-rate current batteries, which require particularly more intensive cooling than low C-rate current ones.

Batteries, produced by reputable manufacturers, are made in such a way that different cells can be used in the same housing. Thanks to this, one can put more or less energy in the same volume, more often or less frequently charge, with higher and lower currents, depending on the needs of the operator.

In the mentioned bus producer can install batteries with cells of different types. The bus holds from 2 to 6 batteries. This allows to take on board from 60 to 300 kWh, depending on the number of batteries and cell type. The energy can be from 60 kWh (2 LTO batteries) to 300 kWh (6 NMC batteries). This gives a range of 50 to 250 km on a single charge. Bus parameters for 6 batteries of different

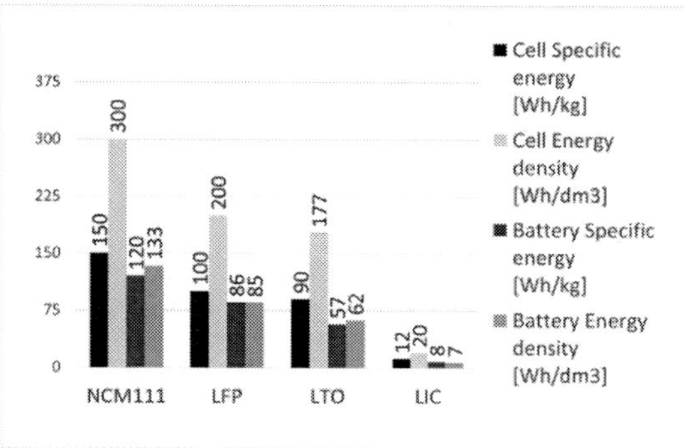

Figure 5. Energy density and specific energy of selected cells and batteries made of them

Table 4. Comparison of bus performance equipped with batteries of various types

Cell type	Max. batt. capacity	Range	Min. char. time to 80% DoD	Max. char. curr.
	[kWh]	[km]	[min]	[A]
NMC	300	250	45	900
LFP	240	200	15	1500
LTO	120	90	5	1800
LIC	10	5	0,33	3000

types are presented in the table 4 and figure 6 comparison of range.

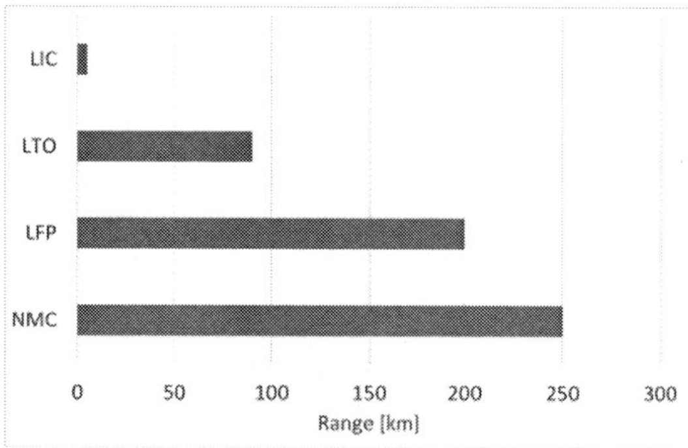

Figure 6. Comparison of bus ranges equipped with batteries made using different types of cells

How does it translate into mass? This is represented by the graph 7 and table 5. One can see that the batteries with LFP cells add the most mass to the vehicle, and do not give the largest range. This is why they are being replaced with NMC cells batteries, although they are safer and can work at lower temperatures. LTO and LIC cells, on the other

Table 5. Comparison of the parameters of buses equipped with batteries of various types

Cell type	Max. batt. capac-ity	Range	Max. batt. mass	Max. mass of elec-tric drive	Mass differ-ence of drives
	[kWh]	[km]	[t]	[t]	[t]
NMC	300	250	2.5	3.1	2.0
LFP	240	200	2.8	3.4	2.3
LTO	120	90	2.1	2.7	1.6
LIC	10	5	1.9	2.5	1.4

hand, give the smallest gains, but require frequent and fast charging.

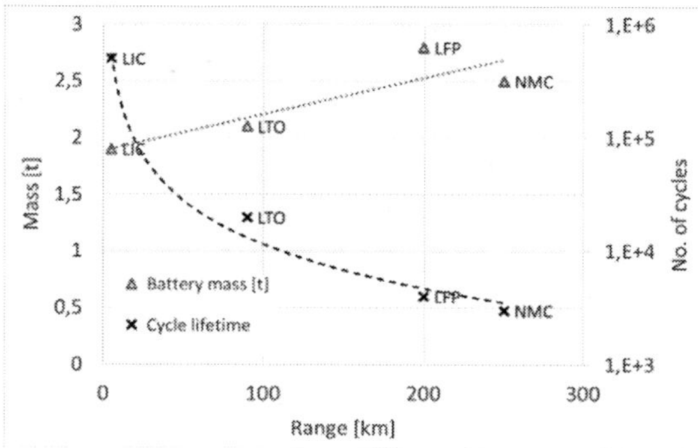

Figure 7. Mass and number of battery cycles as a function of the bus range for various cell types

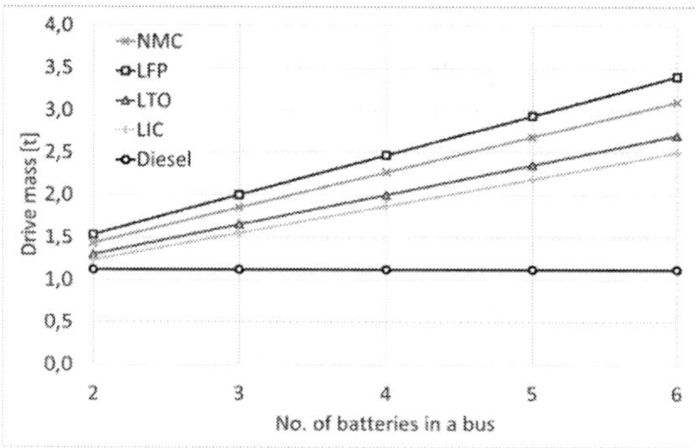

Figure 8. The mass of the drivetrain as a function of the number of batteries and the mass of the diesel engine drivetrain

The figure 8 shows the mass of the electric drive depending on the cell type of the number of batteries used in a bus. For comparison, the mass of the diesel dravietrain bus was added. It is clearly visible that in every variant an electric bus has higher mass than a bus with diesel drive-train. Hence, fewer number of passengers.

The figure 9 shows the range of an electric bus as a function of the number of batteries for different types of cells. These values are several times smaller than the range of the vehicle driven by a diesel engine.

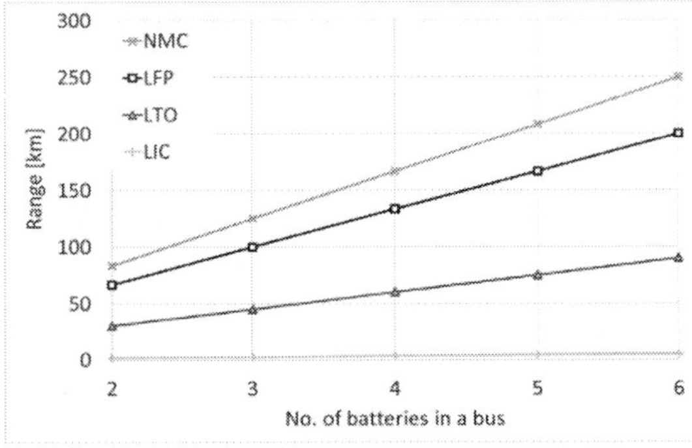

Figure 9. Electric bus range versus no. of installed batteries

The investor should choose which bus will be better for it, what type of cells should be used and the number of batteries installed on board. For long routes with little load, buses with NMC batteries will be better. And for shorter routes, with a large number of passengers better are ones with LTO cells. However, they will require frequent fast charging at the ends of the line.

6 Charging strategies

Depending on the type of cells used and the resulting range of the bus, it is necessary to adapt the charging strategy and to build the appropriate infrastructure.

In general, one can distinguish the following charging strategies:

- daily low current night charging,
- fast recharging at the end of the route and slow night equalizing charging,
- fast charging at the ends of the route,
- fast recharging at selected stops and fast charging at ends of the line.

The basic method is to charge the battery overnight. Large capacity batteries can be charged with small C-rate currents. This allows one to use low power chargers, but it requires a lot of time due to the high battery capacity. Thus it is made at night in the depots. This charging strategy is used for batteries with NMC and LFP cells.

In order to avoid the installation of many charging systems in bus depots and to increase the availability of buses, the second strategy is applied: night charging takes place once or twice a week to equalize the battery and restore the capacity, and is topped up with recharging at the ends of the route. Thanks to this, the battery performs more shallow cycles, which can increase its durability and allow one

Table 6. Charging stretegies and cells types used in the bus

Charging strategy	Cell type
daily, low current full charging, @night	NMC, LFP
recharging @route end, equalizing charge @night	NMC, LFP
fast charging @route end	LTO
fast charging @route end and recharging @bus stops	LTO, LIC

to extend the daily distance traveled by the bus. This is the strategy used for NMC, LFP as well as LTO.

Small capacity batteries can, however, do so many cycles that one can abandon slow night charging. The bus is only charged at the ends of the route, and charging in the depot is incidental, serving only battery maintenance. This charging strategy can be used for LTO cells.

The last charging strategy is necessary when using LIC cells. These cells have so small capacity that the vehicle must be recharged every few kilometers. These strategies can also be used for LTO cells to reduce the depth of discharge cycles and increase lifetime.

7 Conclusions

In public transport, the number of transported passengers is one of the most important coeficients as well as the distance to which they can be transported. The individual route lines differ in the terrain profile, the number and distance between the stops and the number of streetlights. All these parameters influence on power consumption per kilometer and passenger and the selection of batteries for the bus.

An electric bus of a fixed size has a larger mass and a smaller range than an internal combustion engine bus. As shown above, currently the mass of an electric bus is always greater than that of a comparable diesel one. This means that an electric bus can carry fewer people without increasing the total weight.

The problem may be solved by installing of a smaller number of lighter batteries. This will allow one to take a larger number of passengers. But the bus will be able to drive a smaller distance on one charge. The greater the number of passengers, the battery is smaller and the number of charging and discharging cycles is higher.

Therefore, it is necessary to associate the bus charging strategy with the number of passengers to be transported, required range and route profile. Large capacity and small C-rate current batteries (NMC, LFP) must be charged longer, but less frequently. This is done at night in depots. To increase the service life, these batteries can be recharged at selected points in the route to reduce the depth of discharge of the cycles. Small capacity batteries and high C-rate charging currents (LTO, LIC) must be charged more often, even every few kilometers. Charging must take place at selected points of the route. In the case of LIC cell, it must even take place every few stops.

At present, buses are equipped mainly with NMC and LTO batteries, moreover LFP. The first are used for vehicles that are to travel long distances on a single charge. The second onea are used where one need to carry more passengers. This is payed by a larger number of short loads. In both cases, the number of cell cycles, performed by battery, means that the battery life, calculated in years, is similar.

The trend for giving up batteries made of LFP cells is surging. These cells, despite better electrical performance and safety, are gradually replaced by batteries with NMC cells, because they have a significantly larger mass with less capacity.

In the discussion, the authors focused on technical parameters. The financial aspect was omitted. In practice, the dominant indicator is the cost of the bus, vehicle-kilometer and accompanying infrastructure. In the case of lithium-ion cells, the price of cells is the smaller, the more energy they have.

All these problems cause that choosing a battery type for an electric bus is a difficult matter and one general and always true solution can not be given. Thus, the diversity of solutions offered even by one manufacturer.

References

[1] zeeus, *ZeEUS eBus Report. An overview of electric buses in Europe*, Tech. rep., ZeEUS project, UITP, the International Association of Public Transport (2016), [access: 2018-01-10 13:00CET, http://zeeus.eu/uploads/publications/documents/zeeus-ebus-report-internet.pdf]

[2] zeeus2, *ZeEUS eBus Report 2. An updated overview of electric buses in Europe*, Tech. rep., ZeEUS project, UITP, the International Association of Public Transport (2018), [access: 2018-02-20 13:00CET, http://zeeus.eu/uploads/publications/documents/zeeus-janvier2018-final.pdf]

[3] D. Linden, T.B. Reddy, *Handbook of batteires*, 3rd edn. (McGraw-Hill Professional, 2001), ISBN 0-07-135978-8

[4] K. Young, C. Wang, L.Y. Wang, K. Strunz, *Electric Vehicle Battery Technologies* (Springer, 2013), chap. 2, pp. 15–56

[5] D. Anseán, M. González, J.C. Viera, J.C. Álvarez, C. Blanco, V.M. García, *Evaluation of LiFePO4 batteries for Electric Vehicle applications*, in *2013 International Conference on New Concepts in Smart Cities: Fostering Public and Private Alliances (SmartMILE)* (2013), pp. 1–8

[6] J. Mei, E.K.W. Cheng, Y.C. Fong, *Lithium-titanate battery (LTO): A better choice for high current equipment*, in *2016 International Symposium on Electrical Engineering (ISEE)* (2016), pp. 1–4

[7] J. Ronsmans, B. Lalande, *Combining energy with power: Lithium-Ion Capacitors*, in *2016 International Symposium on Power Electronics, Electrical Drives, Automation and Motion (SPEEDAM)* (2016), pp. 261–264

[8] P. Lima, *SK innovation to start producing NCM 811 battery cells soon*, [online] (2017), [access: 2018-01-10 12:00CET, https://pushevs.com/2017/09/02/sk-innovation-start-producing-ncm-811-battery-cells-soon/]

Conditions for using of trams on railway tracks sections in agglomeration communication in Poland

Zbigniew Durzyński [1,*], *Marta* Pachołek [1], and *Rafał* Cichy [1]

[1] Rail Vehicle Institute "TABOR", ul. Warszawska 181, 61-055, Poznań, Poland

Abstract. At the beginning the justification for searching for an effective, energy-efficient mass communication system in agglomerations in Poland are described. One way of relieving the density of the means of communication on the streets of Polish cities is to extend the offer of tram transport by the partial use of railway networks. The literature is reviewed for already implemented two-system tram-train systems and presented solutions for a such tram system in Europe. It is described the technical and organizational problems related to implement the tram-train vehicle in Poland. How-ever, it is important to overcome the formal barriers related to the authorization of vehicle for placing in tram and railway service (new technical standards for this type vehicle). In order to indicate the scope of technical problems, in the further part of this paper it is made the comparisons by analysing the current regulations and standards of basic requirements for tram and passenger railway vehicle. At the end of paper, alternative measures are proposed to introduce tram-train vehicles. That will influence the attractiveness of public transport in urban agglomerations.

1 Introduction

For example, Fig. 1 shows levels of air pollution in Poznan in early 2017. Maintaining such a trend, which is pre-sent not only in Polish cities, will cause catastrophic climate change presented in Fig. 2, which may result in, for example, in the flooding of Venice by the year 2100.

In the described situations it is necessary to seek technical and organizational measures to limit urban car traffic as a result of developing more efficient public transport systems based on energy-efficient means of transport.

One such way is to increase the share of trams in transport in the urban agglomerations, also as a result of the partial use of railway infrastructure in agglomeration and metropolitan areas for tram traffic.

Such systems, called tram-train systems, have already been implemented in several European cities. This subject has already been discussed by specialists in various technical journals [1÷7, 10, 14, 17, 20], but some aspects of this problem were discussed.

2 Review of existing solutions for the use of dual-mode vehicles in public transport

2.1 Karlsruhe

The most popular tram system is the tram in Karlsruhe, Germany [15]. Since September 25, 1992, the first time a Siemens dual-mode tram course took place in the city for the first time in the world. Over time, the dual-track tram route began to expand and in the end there were 12 dual-mode tram lines with-in a radius of up to 70 km from the city centre, serving 190 stations. The Karlsruhe model is mainly based on the vehicle's ability to travel on lines of two different supply voltages: 15 kV 162/3 Hz for rail and 750 V for tram. On urban roads, vehicles follow traditional tram traffic rules and on railways according to the DB rules. This solution uses already existing sections of railway lines. The introduction of the twin-track tramway in Karlsruhe has been successful, among other things, by the fact that the train station is about 2 km from the city centre. This change allowed people to reach the exact centre with-out having to switch to a tram [9].

2.2 Zwickau

The German city of Zwickau has adopted a completely different model of integration of the tram and rail system than Karlsruhe. In order to ensure smooth and uninterrupted commuter traffic to the city centre, it was decided to allow light rail vehicles on tramways. To make it possible, it was necessary to rebuild about one kilometre length of the track in order to allow vehicles wider than the trams to travel through. In addition, it was necessary to add a third rail to the track - the width of the tram tracks was 1000 mm and the rail was 1435 mm. Additionally, the construction of tram tracks has been strengthened to with-stand higher loads. From the side of the rolling stock it was necessary to adapt it to the requirements of tram wagons, among others to equip the

* Corresponding author: z.durzynski@tabor.com.pl

https://doi.org/10.1051/matecconf/201818003002

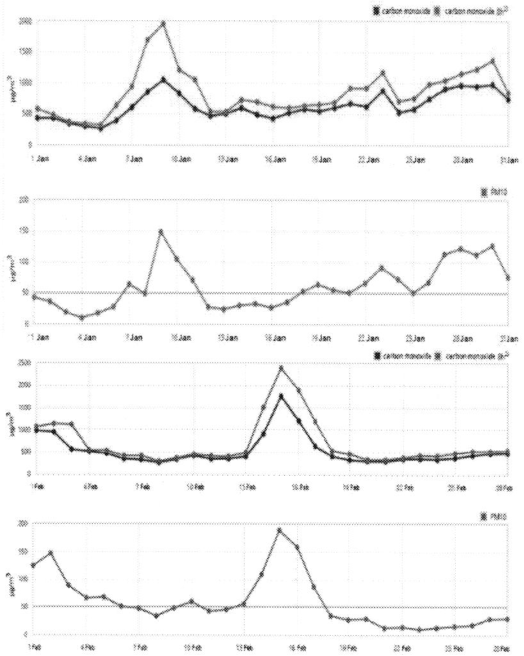

Fig. 1. Poznan air pollution data between 01-02.2017 (measurement: Polanka station) [12].

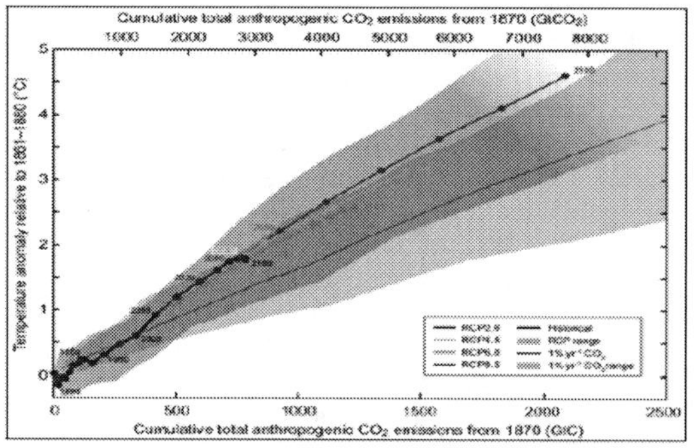

Fig. 2. Forecast of temperature increase due to global CO_2 emissions [17].

direction indicators. But the most important difference in the solution introduced in Zwickau compared to Karlsruhe is the use of rolling stock in urban traffic. There was no need to adapt the vehicle to feed the various traction network voltages [16].

2.3 Kassel

Another city also in Germany, which effectively implemented two-system tram is Kassel. Two types of tram-way connections are used here. In the first case, it consisted of the modernization of the single-track approach from Kassel to Lichtenau used in the past to transport freight. This line was electrified with a tram voltage of 600V. The stops are equipped with so-called. Additional track weights to make it possible to continue to operate the railway in this area. The second way to introduce trams to the railway was the introduction of two-system trams adapted to move on the tram and railway network. The supplier of such vehicles was Alstom, offering a RegioCitadis vehicle. Up to now, two-system trams in Kassel serve four lines on a total length of 145 km, but on the RT4 section, which has not been electrified, only hybrid-electric vehicles operate. Other lines support two-system vehicle solution with a 600V power supply system for trams along a 15 kV – 162/3 Hz [18].

2.4 Nordhausen

The concept of a dual-mode tram was used in the Ger-man city of Nordhausen in a slightly different way, compared to Karlsruhe. In this case, the choice was a hybrid fleet equipped with an additional diesel engine used for travel-ling on the off-street traction. In addition, in order to make this connection between tram and rail infrastructure possible, it was necessary to build a link between the tram loop and the narrow-gauge railway station in 2002. Two more years have passed before the dual-mode tram line No. 10 to Ilfeld was launched. This line is 11 km long, and the time it takes to travel using a the two-system tram ranges be-tween 27 and 35 minutes. Three Siemens Combino Duo vehicles with a total capacity of 34 tonnes were purchased for this line. The vehicle has 27 seats and a capacity for additional 68 standing passengers. It moves at a maximum speed of 70 km/h [11].

2.5 France

Not only German cities observe the trend in the use of rail-way lines for tram vehicles. France has followed the trend of its neighbors. In 2006 Ile de France launched the first dual track-tram system in France. A fragment of a closed railway linking Aulnay sous Bois and Bondy was used. Stops along the lines have been adapted to handle trams by mounting a passenger information system and ticket machines. Two-system trams were provided by Siemens. These trams are powered by 750 V for tram traction and 25 kV 50 Hz for rail traction. Apart from adapting vehicles to these changes the city has decided to introduce a number of infrastructure improvements. One of them was to prioritize tramways on pedestrian crossings by using ap-propriate traffic lights and GPS devices. For pedestrians, a red light signalling on the STOP sign board was added and marking the nearest crossing with black and white chickened pattern [8].

2.6 Others

Apart from France and Germany, two-mode trams can be found in other European countries such as Italy, Spain, the Netherlands as well as North America. These systems are the most well-known models. Other solutions in other cities rely directly on these systems to match the infrastructure to their requirements [10, 11].

https://doi.org/10.1051/matecconf/201818003002

The concept of building a two-system tram using the rail-road tracks to move trams also appeared in Poland. Such solutions have been considered in Wroclaw[10], Cracow [2, 3, 19], Gdańsk among others [13].

3 Barriers to putting a tram on the railroad tracks

3.1 Technical difficulties

3.1.1 Different wheel and track profiles

Different shape of the wheel rim and tilt of the rail. On the rail track, the tilt is 1:20 and 1:40 inward, while in the tram network the inclination is contained in the rail profile by the appropriate configuration of the rail head (the most commonly used rails are Ri60N, S49, LK1). In addition, the issue of track crossing is different. For trams, this takes place along the perimeter of the gently rising ramp inside the crossing groove – see Fig. 3. The crosses are done by passing the rim over the groove in the crossroads of rail-road tracks – see Fig 4. Possible solution is the selection and use of an intermediate wheel profile.

Dual-track trams can be extended with 135 mm wide rims, so it is necessary to raise the rail above the track base by 10-15 mm to avoid overriding the rail head and avoid contact of the extended wheel rim with the track. It would also be advisable to use rails with a groove depth of 45-47 mm and it would be impossible to run wheels on the rim in the event of significant vertical wear of the rails.

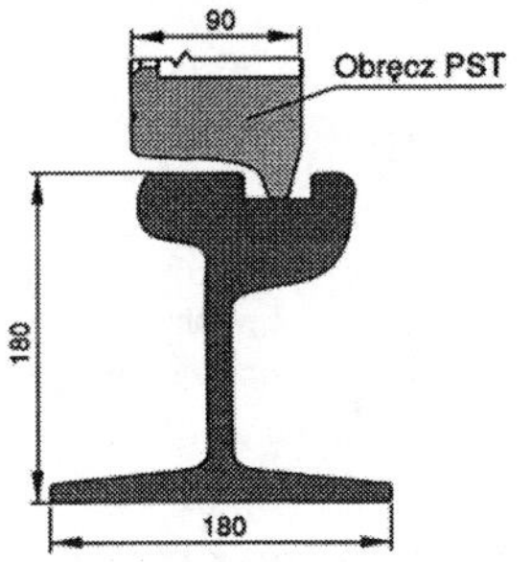

Fig. 3. Image of a tram wheel when crossing a switch or a tram intersection [4].

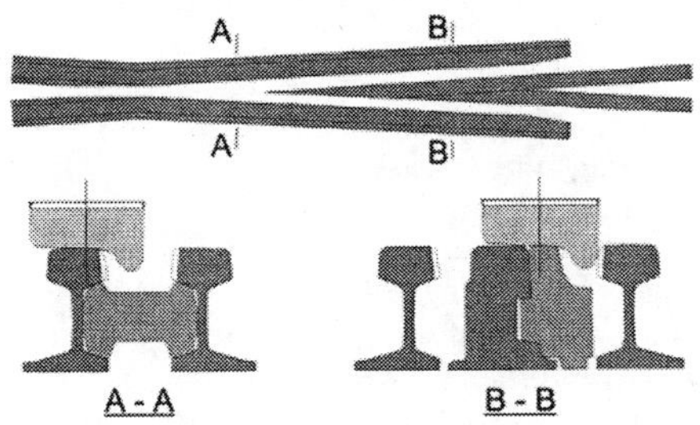

Fig. 4. Image of a rail wheel when crossing a switch [4].

3.1.2 Expansion tolerances in arches

The rail uses wooden sleepers to widen the rail track to ensure smooth passage through arches. The use of reinforced concrete reinforcements is not possible due to the unequivocal point of fixing the anchorage. For example, for an arc of less than 160 m, the width can be as high as 25 mm, giving a 1460 mm track spacing [11]. In contrast, in tram tracks there is no such widening, and the wheelbase tolerance is ± 2 mm on the straight and -0 and +4 on the bends (according to PN-K-92011:1998).

3.1.3 Different drive system

The electric railway traction voltage in Poland is 3 kV and the tram system is 600 V. It will be necessary to adjust the vehicle to operate using two voltages. Inverters and circuit switches must be used. This solution also requires adapting the infrastructure, which must have an electric traction section of the overhead contact line. An alternative to this solution is to equip the vehicle with an internal combustion engine. Then the vehicle could move on tram tracks using traction and train tracks using an internal combustion engine and vice versa.

3.1.4 Different vehicle signalling and marking systems

It is necessary to adapt the vehicle to two traffic systems. Trams must be equipped with, among others, direction indicators, which are not required for rail traffic. On the other hand, the light marking of a railway vehicle needs to be adapted to railway requirements, among others, to properly mark the beginning and end of the vehicle in signal lights.

3.1.5 Durability in crash

There are no specific Polish requirements for endurance and durability requirements for tram frames. Thus the vehicles will have to meet the railway requirements. Railway requirements for maximum thrust longitudinal force are contained in PN-EN 12663-1:2010+A1:2014 depending on vehicle category. For P II passenger vehicles (train-sets and coaches), this power is 1500 kN. Increasing the structural strength of the vehicle frame is directly related to the weight gain of the vehicle. A track can be traversed by a vehicle with a maximum axle load of 100 kN at maxi-mum allowable tram weight.

3.1.6 Different track gauge

In Poland there is also a narrow gauge tram track width of 1000 mm. If it is necessary to adapt such a track to a two-system vehicle, it would be necessary to equip the track with a third rail (e.g. as it was done in Zwickau). In Poland, the narrow gauge tram tracks are found in Bydgoszcz, Torun, Elbląg, Łódź and Grudziądz.

3.1.7 Traffic control systems

In the case of trams the signal is coupled with the street signal, while on the railway tracks the traffic is driven from the dashboard and is based on the line and station lock systems. The solution is to adapt the two-mode vehicle to railway and tramway standards (rail traffic control equipment). In addition, the vehicle would have to be equipped with traffic safety devices - active dead man's switch, SHP and radio.

3.2 Organizational difficulties

3.2.1 Traffic intensity on the railway tracks and priority for the tram-train vehicle

It would be troublesome to add extra journeys on routes with heavy traffic. For example, in Karlsruhe, tram traffic jams occur in the morning and in the afternoon, where the capacity of tram routes is insufficient and the vehicles cannot be extended due to the regulations in force (in Poland the maximum tram length can be 65 m). All traffic control issues should be covered by appropriate instructions, regulations and internal PKP instructions. A two-system tram entering a railroad track would have to have a train number. There is also a large variation in the rules of tram and train traffic. Trams are driven using visibility – one behind the other, keeping a safe distance between each other. Trains, on the other hand, are managed on a "track section" basis, which means that only one vehicle can be pre-sent on a given section of the trail.

3.2.2 Driver or motorman?

The next organizational issue is the lack of coherence of traffic regulations on railway and tramway tracks. It will be necessary to train staff to navigate two types of roads. A person operating a two-mode vehicle on railway tracks should have a train driver's license with knowledge of rail traffic and traffic posts. On the other hand, a person running a tram in urban traffic should have the qualifications of a train operator. It will be easier to train drivers into operators due to the long duration of train drivers training. Alternatively, it is possible to make changes to the instructions on professional training of the PKP to include a provision on the two-system tram.

3.2.3 Passenger information

Adjustment of platform/tram stop information boards/passenger service (information boards showing waiting times for the nearest tram, ticket machines etc.). Introducing a uniform system of one time and periodical tickets valid for both rail and tram travel.

3.3 Legal difficulties

3.3.1 Approval processes for the vehicle to train and tram exploitation

In order for the vehicle to be able to move on tram tracks it must meet the conditions set out in the Regulation of the Minister of Infrastructure of 2nd of March 2011 on the technical conditions of trams and trolley-buses and the scope of their necessary equipment (Journal of Laws 2011 No. 65 item 344) and the Regulation of the Minister of Transport, Construction and Commerce of 28 May 2013, on type-approval of trams and trolleybuses (Journal of Laws 2015, item 38). On the other hand, a vehicle traveling on railway lines in Poland must obtain a permit from the Railway Transport Office for a free passage on railway tracks. This approval is granted on the basis of meeting the requirements set out in the Ordinance of the Minister of Infrastructure and Construction of 21 April 2017, on rail interoperability (Journal of Laws 2017, item 934) and related documents. These permits are made in two variants: TSI compliant or non TSI compliant. In the case of a vehicle conforming to the TSI and meeting the requirements specified therein, the authorized entity shall issue an EC Verification Certificate, under which the manufacturer creates an EC Declaration of Conformity. These two documents in turn are submitted to the Rail Transport Authority for authorization. In the case of a vehicle not compliant with the TSI, the requirements of standards and regulations published in the current Letter of the President of the UTK shall apply. Then, for a TSI-compliant part, the authorized body issues an EC Verification Certificate under which the manufacturer creates an EC Declaration of Conformity and for the non-TSI compliant entity the authorized body issues a Certificate of Conformity under which the manufacturer creates a Declaration of Conformity. On the basis of these documents, the Office of Rail Transport issues an indefinite permit.

3.3.2 Introducing new normative legislation

The implementation of the tram-train system in Poland will require the introduction of two coherent legislative systems for the design, construction and operation of rolling stock and infrastructure maintenance.

4 Comparison of basic requirements for trams and passenger rail vehicles

It is not possible to directly compare the requirements for tram and train traffic. These vehicles have completely different requirements backed by different standards and documents.

Examples of tests to be carried out to approve a vehicle (tram or train) for operation are:

- braking system tests;
- testing the acoustic properties;
- axle load testing;
- light and marking tests;
- tests of the combustible non-metallic materials used in the interior of the vehicle.

The approval requirements result from the documents listed in Table 1.

Table 1. Documents containing requirements for the admission of the tram and train to service.

Tram requirements	TSI conforming railway requirements	TSI non-compliant railway requirements
Braking system		
Journal of Laws 2011 No. 65 pos. 344; § 16-18	TSI 1302/2014 from 18.11.2014 and related documents	PN-EN 16452:2015-08, PN-EN 14198:2005, PN-EN 15734-1:2011, UIC 540, UIC 541-03, UIC 541-5, UIC 541-6, UIC 544-2, UIC 546, PN-K-88177:1998+Az1:2002 PN-EN 16334:2014-10, PN-EN 15612+A1:2011 UIC 541-1, UIC 612-1 UIC 544-1, UIC 543, UIC 567-1, UIC 567-2, PN-EN 14531-1:2016-02, PN-EN 15595+A1:2011,
Acoustic properties		
§ 3 pt. 8 and § 5 pt. 3 Journal of Laws 2011 No. 65 pos. 344	TSI 1304/2014 from 26.11.2014 and related documents	PN-EN ISO 3095:2013-12, ANX TM-2 PN-EN 15153-2:2013-06, UIC 644 (sound signalling)
Axle Loads		
§ 3 pt. 5 Journal of Laws. 2011 No. 65 pos. 344	TSI 1299/2014 from 18.11.2014 and related documents	PN-EN 14363:2016-04, ANX TM-1
Lighting and signalling		
Attachment 2 to Journal of Laws 2011 No. 65 pos. 344.	TSI 1302/2014 from 18.11.2014 and related documents	PN-EN 15153-1:2013-06 PN-EN 13272:2012 UIC 534
Properties of combustible non-metallic materials used in the vehicle interior		
PN-K-02511:2000	TSI 1302/2014 from 18.11.2014 and related documents	PN-EN 45545-2+A1:2015-12
Loading gauge		
PN-K-92008:1998/Ap1:1999	TSI 1302/2014 from 18.11.2014 and related documents	PN-EN 14033 (series), PN-EN 15273-2:2013-09, PN-EN 15746 (series), UIC 505-1, UIC 505-6, UIC 506, Regulation on the operation of railway vehicles (see Annex A-1); § 4

5 Legal barriers arising from the provisions on the use of railway infrastructure

5.1 Adjusting the height of the floor in a two-system vehicle to the edge height of the platform

The height of the tram platforms is smaller than the typical railway platforms. The possible solution is to rebuild the infrastructure or adapt the vehicle to the existing railway platforms by adding a deployable step. These elements should automatically hide when closing the door.

5.2 Rolling stock gauge – distance from the platforms

According to the tram regulations (Journal of Laws 2011 No. 65 item 344) the maximum width of the tram can be 2.65 m without mirrors (e.g. RegioCitadis manufactured by Alstom). It generates a problem of bigger distance between tram and platform. The solution is to adjust the existing infrastructure by reconstructing platforms or using the so-called track splices, or platforms adapted to handle two types of vehicles, each on a different platform side. This solution allows to use the stops by the classic tram by driving up to the very edge of the platform. In case it is not possible to rebuild the infrastructure, as in the case of different platform heights, a retractable ladder or ramp can be used. This will not only reduce the height difference, but also reduce the distance between the door and the distant platform.

6 Suggestions for alternative solutions for agglomeration communication by using trams

6.1 Park&Ride

An alternative method of reducing the load on public transport, without investing in new rolling stock and railway infrastructure, is to build park & ride parking spaces at tram or bus loops. This would allow a significant relief of traffic flows in the centre of cities. This solution is justified in cases where the suburban network is not significantly expanded and the frequency of interstate calls is not high. The idea of the park and ride system is to reach the city's outskirts by private transport and change to public transport for in city commute. This arrangement should include an estimate of the size of the car parks based on local conditions. These parking lots should be equipped not only with parking spaces for cars, but equally important is the provision of bike space.

6.2 Bus lanes

The designation of urban lanes for public transport buses will streamline bus journeys through the often crowded streets. This will allow the shortening of the passenger travel time by means of transport making it more competitive. Moreover, thanks to such a solution, it would be possible to move from the suburban areas quickly and without vehicle changes to reach the city centre. In the case where it is not possible to drive a separate bus lane in the centre of the city, a separate tramway track can be used (as far as its construction permits it) and treated as a bus and tram lane. It is also important to establish a priority system for bus crossings with traffic lights.

6.3 Railway ring

An alternative to the creation or adjustment of railway tracks for two-system trams is the creation of a perimeter railway ring, as it was done in the city of Ringbahn in Berlin. This railway bypass is designed to connect public transport lines (e.g. tram, metro) located on the outskirts of the city and downtown directly with inner city lines. In addition, it will allow for bypassing the city by freight trains.

7 Summary

It is undeniable that the development of public transport in agglomerations and metropolises is a necessity. To that end, all technically and economically viable means should be used. One of them is the use of the existing tram infrastructure in large cities in Poland, as has been done in some Western European countries. The favourable circumstance in most cases is the same track width of 1435 mm and level of development of power electronics, enabling the traction drive to be adjusted to different voltage levels. Admission of the tram to the railway tracks requires several construction projects, described above. However, it is important to overcome formal barriers related to the admission of a vehicle to rail and tram traffic, requiring the development and establishment of new technical standards for a tram-train vehicle. Railway Transport Office and Trans-port Technical Supervision will be involved in this stage, supported by experts from authorized scientific and rese-arch units.

References

1. T. Czauderna, *Próbne jazdy wagonu tramwajowego 105NT po torach kolejowych.* TTS 7-8 (1998)
2. W. Czyczuła, *Koncepcja zintegrowanego systemu transportu szynowego dla Krakowa.* TTS 9 (1999)
3. W. Czyczuła, *Koncepcja zintegrowanego systemu transportu zbiorowego dla Krakowa – dlaczego tramwaj dwusystemowy?* TTS 1-2 (2000).
4. W. Czyczuła, *Warunki techniczne jakim powinna odpowiadać infrastruktura kolejowa i tramwajowa ze względu na ruch pojazdów dwusystemowych.* TTS 1-2 (2000)
5. W. Czyczuła, J. Raczyński, *Pojazd dwusystemowy jako środek transportu regionalnego.* TTS 11 (2000)
6. J. Dąbrowski, *Dwusystemowe tramwaje – czyli tramwaj na torach kolejowych.* TTS 7-8 (1998)
7. J. Dąbrowski, *Rozważania nad koncepcją napędu tramwaju dwusystemowego w warunkach polskich.* TTS 9 (1999)
8. A. Drapała, *Pierwszy tramwaj dwusystemowy we Francji.* Transport miejski i regionalny 2 (2007)
9. G. Drechsler, *Light railway on conventional railway tracks in Karlsruhe, Germany.* Proceedings of the Institution of Civil Engineers – Transport Volume 117 Issue 2 (1996)
10. I. Gisterek, M. Krużyński, *Ocena możliwości i zasadności uruchomienia tramwaju dwusystemowego we Wrocławiu.* Transport miejski i regionalny 11 (2010)
11. A. Harassek, *Dwusystemowy tramwaj w Nordhausen.* TTS 6 (2004)
12. https://powietrze.poznan.wios.gov.pl/
13. http://warszawa.naszemiasto.pl/artykul/w-gdansku-tramwaje-beda-jezdzic-po-torach-kolejowych,16915,art,t,id,tm.html
14. C. Kraśkiewicz, W. Oleksiewicz W, *Tramwaj dwusystemowy – moda czy trend rozwojowy aglomeracyjnego transportu szynowego?* Logistyka nr 4 (2015)
15. C. Kraśkiewicz, W. Oleksiewicz, *Tramwaj dwusystemowy w Karlsruhe.* Logistyka 4 (2015)
16. C. Kraśkiewicz, W. Oleksiewicz, *Tramwaje dwusystemowe w wybranych miastach niemieckich.* TTS 12 (2016)
17. I. Merkisz, P. Lijewski, P. Fuć, J. Pielecha, Exhaust emission tests from agricultural machinery under real operating conditions. *SAE 2010 Commercial Vehicle Engineering Congress,* (SAE Paper 2010)
18. Ch. Muller, *RegioTram Kassel.* El-Eisenbahnigenieur 7/2006 (2006)
19. W. Wańkowicz, *Zagadnienia formalnoprawne a dwusystemowy Krakowski Szybki Tramwaj.* TTS 1-2 (2000)
20. I. Wieczorek, *Problemy eksploatacyjno-ruchowe związane z wprowadzeniem pojazdów dwusystemowych na sieć PKP.* TTS 1-2 (2000)

Effectiveness of urban transport modes

Agata Pomykala[1,*]

[1]Research Railway Institute, ul. Chłopickiego 50, 04-274, Warsaw, Poland

Abstract. Social expectations in terms of improving the conditions of life in cities whilst increase of mobility poses a difficult challenge for urban transport organizers. At the same time, conditions related to sustainable development policies with scarce resources and available reserves focus attention on transport optimization and reduction of environment harmful impacts. The paper presents an assessment of the efficiency of urban transport in terms of capacity, energy use, CO_2 emissions and costs, based on data on European medium cities in traffic conditions - peak (congestion) and off-peak traffic (no congestion). Compared were road vehicles: cars and buses of different propulsion, motorcycles and also rail vehicles: metro, tram/light rail and trains/heavy rail. The results are presented in tables or in graphs. The comparisons make it possible to identify the most efficient, in various scenarios, transport modes and may be helpful in deciding on issues related to the provision and organization of urban transport. In this article, cost evaluation omits external costs (although CO_2 emissions are taken into account) but includes fixed and variable costs associated with the use of vehicles (including fuel costs, maintenance, inspections, depreciation, insurance).In the preparation of the article information available in literature, publication and reports, particularly of the European Commission one was used.

1 Introduction

Nowadays, the necessity to solve problems related to urban and agglomeration transport is becoming more and more important. Cities generating significant benefits for the development of the economies of the regions at the same time face problems of high population density, congestion and excessive pollution. Already, 64% of all kilometres travelled relate to daily individual travels in the urban environment, and it is expected that their length will triple by 2050. It is anticipated that by 2050, the time spent in traffic jams by one inhabitant will be doubled and will be 106 hours [1]. Increasing avalanche mobility concerns cities to a greater extent than rural areas. Inhabitants of European cities already constitute more than 70% of Europeans and, according to forecasts, the number of inhabitants of urban areas by 2030 will exceed 80% of the European population, exacerbating related the problems [2, 3]. The congestion, whose cost in European urban areas is valued at EUR 100 trillion, in a particularly negative way in the area of cities affects the functioning of society and the economy [4]. Nearly 70% of road accidents occur in them, and in cities the most noticeable is the impact of transport on environmental pollution: 70% of pollutants emitted in transport are generated in them and 40% of total CO_2 emitted by road transport. [5]. Awareness of the negative impact of transport on the environment, including health and quality of life, but also climate change and the associated threat to biodiversity is reflected in the concept of sustainable transport and influences the emergence of local organizations interested in transport policy. The development of mobility in accordance with the principles of sustainable development is one of the most important challenges for the cities of the future.

Transport needs of residents of modern cities can be satisfied in various ways, using various means of transport: individual or collective. Already today, there are significant differences in this respect: for example, in Vienna, 75% of commuting to work is carried out using public means of transport, while in Palermo or Antwerp - less than 30% [6]. Decisions on the choice of means of transport are made by the citizens on the basis of various criteria, both objective and subjective, and the task of the city authorities is to create incentives to use the means of transport most conducive to the fulfilment of the development policy. Currently, it is important to take into account not only social expectations in terms of air cleanliness or health protection, but also sustainable urban development. On the one hand, it is expected to meet the ever-increasing mobility needs, including convenience, short travel time and low cost, and on the other: reduce nuisances related to excessive concentration of people and negative effects of transport on the natural environment and living conditions and ensure even growth socio-economic development. Decisions regarding transport systems in cities should take into account not only current transport needs but also, in accordance

* Corresponding author: a.pomykala@infotransport.pl

with the assumptions regarding the transport function as an integrator of socio-economic life, also issues related to spatial planning and shaping the functional and operational structure. Knowledge of the advantages and disadvantages of transport means is necessary especially in the context of the socially understood need to consider the concept of its sustainability. Analysis of the effectiveness of means of transport can be a starting point for the development of transport services in the city, including the social striving to protect and improve the quality of life. The aim of the article is to indicate the effectiveness of individual means of transport to facilitate decision making regarding their use and affecting the improvement of living conditions in the urban environment.

2 Effectiveness

The increasing importance of transport in social and economic life requires its evaluation not only from a technical and organizational point of view but also in terms of macroeconomic efficiency of the entire system, i.e. ensuring effective implementation of the need to move both goods and people. Due to the complexity of the relationship between this sector and the socio-economic system, there is a need for a holistic view that takes into account the functionality of the entire system and not only its separate elements. Assuming that the transport system is a system of technical, organizational and human means related to each other in such a way as to efficiently carry out the movement of persons or (i) loads in time and space [7] and taking into account the expectations regarding the optimization and sustainability of transport development, it is necessary to consider the issues of efficiency in its planning. Effectiveness is an ambiguous term used in various disciplines of science. This word derives from: Latin *effectīvus* „practical", equivalent to effect(us) "productive, effective," from *effect-*, stem of *efficere* [8]. Microeconomic efficiency focuses on the functioning of the business unit and its operations, including investment. It reflects the benefits of this activity. According to the principle of rational management, they should be adequate to the expenditures incurred, understood as the degree of consumption of production factors in the process of implementing an economic undertaking. The increase in efficiency can be implemented in two ways:

- maximization of effects, i.e. obtaining a higher level of effects at a constant level of expenditure,
- minimizing expenditures, i.e. achieving certain economic effects while reducing expenditures [9].

Efficiency in transport is understood as the mutual relation between the consumption of natural resources and the costs generated on the one hand, and the benefits resulting from their use on the other. Transport costs are most often understood as those incurred by users and carriers. It sometimes happened that external costs incurred by the general public, for example costs of accidents or environmental impact were omitted in the assessment of transport efficiency. With the development of the concept of sustainable transport, it is required to consider external costs in cost-benefit analyses.

Efficiency is best understood in terms of social and environmental integration and in contrast to economic, production and financial efficiency, regulated by the "invisible hand of the market" requires institutional adjustment of the market mechanism through sectoral policy actions integrated with environmental policy [10].

3 Factors affecting the choice of a means of transport

The choice of mode of transport is dictated by both economic and technical conditions. Analysing the factors that influence the planning of the mode of a city transport the problem was considered in the following aspects:

- accessibility of space (transport capacity for particular means of transport with the assumed occupancy of the area);
- time to move around the city in various relations (directions);
- environmental performance of the mode of transport (degree of environmental impact);
- unit cost of transport in the aspect of the entire system (total infrastructure, vehicle and external costs).

4 Characteristics of modes of transport

In the assessment of urban means of transport, listed in Table 1 the following parameters have been accepted for comparison:

- capacity – defined in terms of passengers per vehicle and passengers per hour in relation to infrastructure capacity;
- energy use – defined in terms of MJ per passkm;
- CO_2 emissions – defined in terms of CO_2 per passkm;
- cost – defined in terms of EUR per passkm.

Indicators have been calculated in relation to scenarios for a medium city (according to OECD 200,000-500,000 habitants) in the conditions of:

- congested peak traffic;
- un-congested off-peak traffic.

Table 1. Urban transport modes comparison [11, p.3].

Transport mode	Fuels
Passenger car	Petrol Diesel Gas (CNG, LPG) Hybrid Electric (petrol only) Battery electric
Motorcycle, Moped	Petrol
Bus: Midi, Large, Bus Rapid Transit (BRT)	Diesel Gas (CNG) Hybrid electric (diesel) Battery electric
Rail: Tram/Light Rail (LTR) Metro,Heavy rail	Electric Diesel (Heavy rail only)

The theoretical capacity, of each mode of transport, has been determined mainly on the basis of the physical infrastructure characteristics,' real capacity 'corresponds to the Community understanding of this concept in the context of urban transport, which means defining capacity as the number of passengers who are transported per unit of time transport, as shown in Tab.2, different factors, marked "+", affect their value.

4.1 Private cars

In the case of passenger cars, the theoretical capacity results from the vehicle's capacity and the capacity of the road infrastructure. The vehicle capacity assumed for the analysis is 5 passengers, and the capacity of the road infrastructure depends on the road characteristics. For the purposes of this task, the reference value for urban capacity in terms of vehicle traffic per hour was adopted at 4500 pass / hour (in acc. with the Technical Note 10 Transport Transport for London's Roads Task Force, defining the values for different types of city roads).

Table 2. Capacity of urban transport modes affecting factors (self-study, based on [11, p.5]).

Transport mode	Vehicle capacity [pass./veh]	Theoretical capacity			Actual capacity		
		Road infr. capacity [veh/h]	PCU factor [PCUs/veh]	Theoretical frequency of service [veh/h]	Occupancy rate [%]	Demand profile impact	frequency of service [veh/h]
Private car	+	+			+	+	
Motorcycle	+	+	+		+	+	
Bus	+			+	+		+
Rail *	+			+	+		+

* Rail: Tram, Metro, Light Rail, Heavy Rail

4.2 Mopeds and motorcycles

Mopeds and motorcycles are treated similarly to cars, however their transport capacity is limited to one or two passengers and they also take up less space on the road. Theoretical capacity is the result of the maximum number of passengers, road infrastructure capacity and coefficient of equivalence of vehicles in terms of PCU. As the assumed maximum number of passengers, two passengers were accepted for the motorcycle, and for the moped - one passenger. The reference value for urban capacity is the same as for cars, i.e. 900 PCU per hour. The vehicle utilization rate in the EU-28 is 1.1 passenger per motorcycle, i.e. around 55% of the vehicle's capacity. The average number of passengers on a motorcycle was estimated on the basis of the TRACCs database and statistics of selected countries [11 p. 7].

4.3 Buses

The following vehicle capacity was assumed for the calculation of the theoretical carrying capacity:

- Midi bus (lenght: 10 m): 60 passenger per vehicle
- Large bus (lenght: 12 m): 100 passenger per vehicle
- BRT(lenght: 18 m): 140 passenger per vehicle.

For the actual capacity, the service frequency was additionally taken into account.

Table 3. Road transport modes capacity comparison (self-study, based on [11, p.5-10]).

Transport mode	Time period	Theoretical capacity [pass/h]	Actual capacity [pass/h]
Car	Peak	4500	1530
	Off-peak		920
Motorcycle	Peak	3600	1980
	Off-peak		1180
Moped	Peak	1800	1800
	Off-peak		1080
Midi bus	Peak	1200	270
	Off-peak		20
Large bus	Peak	2000	600
	Off-peak		200
BRT	Peak	2800	1800

4.4 Rail

For urban rail transport capacity is theoretically a matter of vehicle infrastructure and capacity. In reality, the occupation is less than 100%, and the frequency of the service is usually lower than the theoretical maximum frequency. Therefore, in order to estimate the theoretical transport capacity, the frequency of services and vehicle capacity were taken into account, using the same approach as for road (bus) transport [11 p. 10-12].
With regard to vehicle capacity, the following:

- Tram/Light rail: 170 pass/veh;
- Metro: 550 pass/veh;
- Heavy rail: 1000 pass/veh.

Table 4. Rail transport modes capacity comparison (self-study, based on [11, p.12]).

Transport mode	Time period	Theoretical capacity [pass/h]	Actual capacity [pass/h]
Tram/LTR	Peak	3400	1150
	Off-peak		480
Metro	Peak	16500	5690
	Off-peak		2200
Heavy rail	Peak	15000	2250
	Off-peak		400

5 Transport capacity comparison

The comparison of urban modes transport in terms of transport capacity indicates (Fig. 1), taking into account the theoretical aspect, the advantage of metro and trains. The smallest capacity is characterized by medium-sized buses.

The actual situation is different due to realignment of vehicle occupancy and consideration of differences in peak and off-peak hours (Fig. 2).

https://doi.org/10.1051/matecconf/201818003003

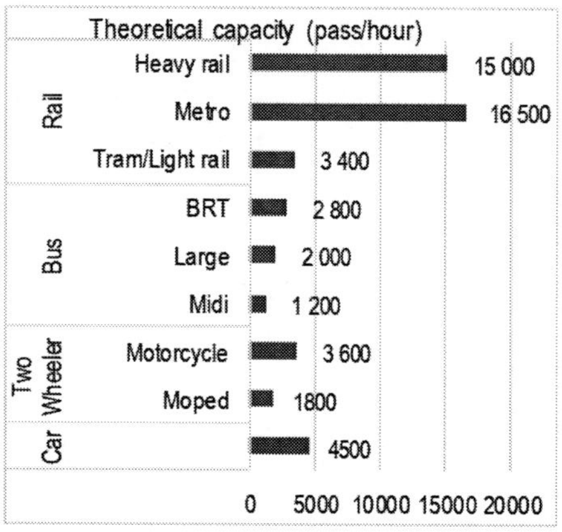

Fig. 1. Theoretical capacity (passenger/hour) in a medium city (self-study, based on [11, p.13]).

At peak times, collective transport provides greater efficiency than an individual, while in off-peak periods it is the reverse.

6. Energy consumption and CO_2 emissions comparison

Energy consumption and CO_2 emissions are closely related measures that are used to assess the environmental performance of transport. Average energy consumption calculated for four urban passenger transport means (cars, two-wheelers, buses and rail vehicles) was presented in megajoules per passenger kilometre (MJ / pkm), according to the WTW (well-to-wheel method, analysis from the source of energy carrier). This is representative data for transport systems throughout the EU. CO_2 emissions per kilometre were determined based on the results of energy consumption using coefficients of

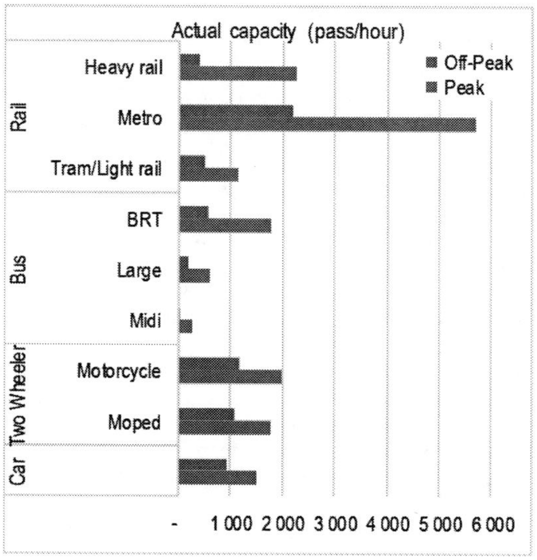

Fig. 2. Actual capacity (passenger/hour) in a medium city [11, p.16].

CO_2 emissions related to both direct (fuel combustion) and indirect (fuel production) emissions. Comparison of the average WTW CO_2 emission is shown in Fig.3

The energy consumption in individual means of transport depends mainly on the type of propulsion and speed of movement (off-peak or peak traffic). During peak hours, the estimated average speed is 30 km / h. Vehicle occupancy rates are assumed to be the same at both rush hour and away times, so they do not affect the results. Comparison of the average energy consumption by time period and modes of transport is shown in Fig. 4.

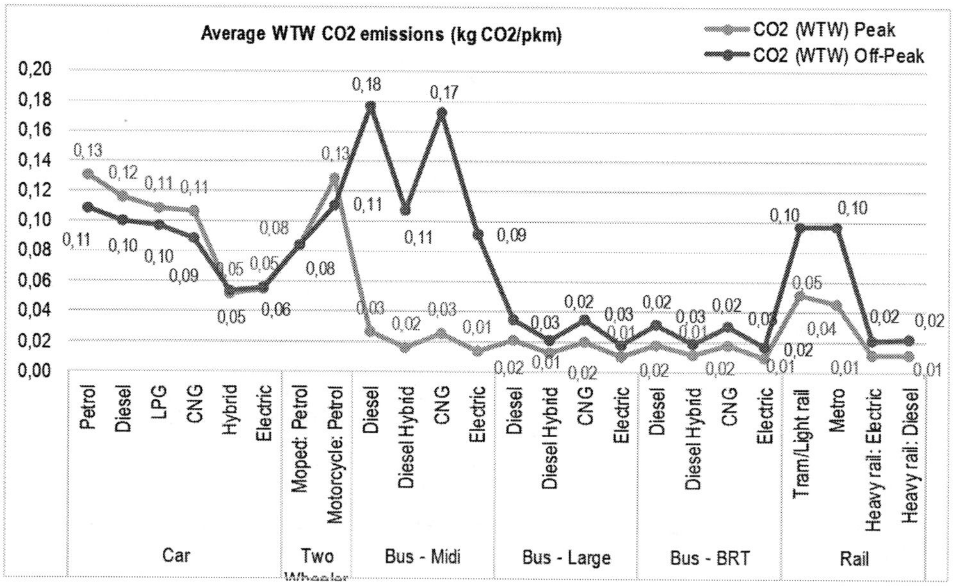

Fig. 3. Comparison of the average WTW CO_2 emission in medium city (self-study, based on [11 p.31]).

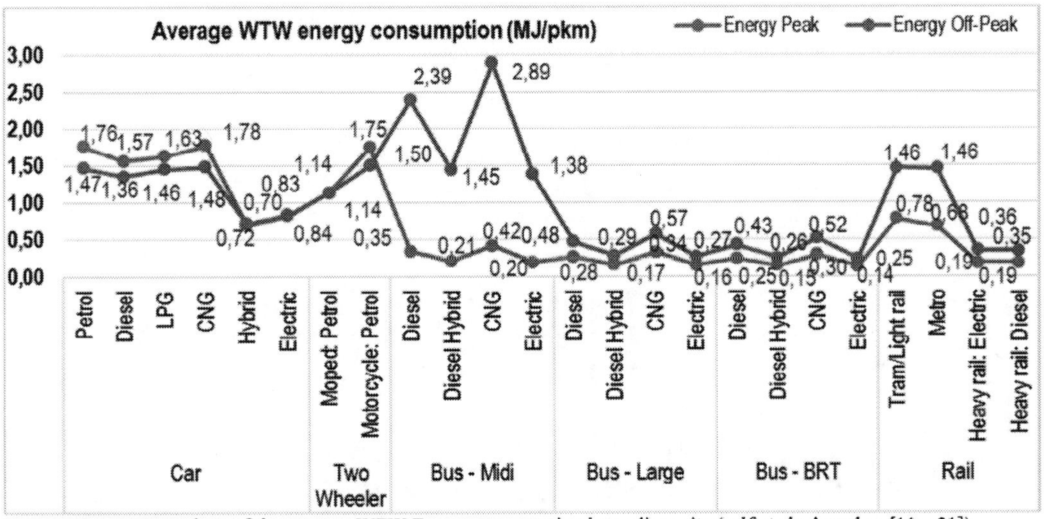

Fig. 4. Comparison of the average WTW Energy consumption in medium city (self-study, based on [11 p.31]).

7 Costs comparison

For compare of cost calculated the cost of transporting one passenger per passenger kilometre in different modes of urban transport [11, pp. 33-44]. The assessment of costs is related to their constituent elements considered in aspects:

- costs incurred by the customer (for public transport fares/ tariffs)
- costs incurred by the service provider - not included in the tariff
- social costs (external costs).

In this paper, the cost assessment omits external costs (although the comparison of CO_2 emissions is presented above), but takes into account fixed and variable costs related to the operation of vehicles (including costs related to fuel consumption and maintenance, inspections, depreciation, insurance).

7.1 Energy cost

The average energy cost in individual vehicle categories has been estimated based on the result of energy consumption estimation (with appropriate conversion factors) in combination with European averages. Energy costs for passenger cars in medium cities vary depending on the type of drive between 0.028 EUR / vkm for electric cars in peak hours (0.029 EUR / vkm - off-peak hours), hybrid cars - 0.044 EUR / vkm during peak hours and 0.045 EUR / vkm for off-peak time up to 0.111EUR / vkm in peak hours (0.093 EUR / vkm off-peak) for petrol cars [11, p.34]. The energy costs for single-track vehicles are respectively for motorcycles in rush hour - 0.071 EUR / vkm, off peak - 0.61 EUR / vKM, and for mopeds - 0.042 EUR / vkm regardless of the time of day. Other fixed and variable cost (not energy related)

The cost analysis includes other costs: purchase, maintenance, amortisation, ownership taxes and insurance those make up the costs of producing/delivering a service. For public modes of transport, additional costs of remuneration and fixed costs are included. The costs of providing transport services in cities depend on local conditions (e.g. commercial speed). For medium and large buses, the same costs of service delivery were assumed, and BRT buses were treated separately. For the public transport rail service, all elements that make up the cost of producing and delivering the service have been included (even if they are not covered by the tariffs), as well as for road. Tab. 5 contains information on the costs of service production broken down by modes of public transport.

Table 5. Average total production cost (self-study, on based [11 p.41, 43]).

Transport mode	Fuel type	Average total production cost [EUR/vkm]
Midi bus	Diesel	3.52
	Diesel hybrid	3.42
	CNG	3.52
	Electric	3.44
Large bus	Diesel	3.77
	Diesel hybrid	3.64
	CNG	3.77
	Electric	3.67
BRT	Diesel	4.55
	Diesel hybrid	4.36
	CNG	4.52
	Electric	4.35
Tram/light rail		5.7
Metro		14.2
Heavy rail		16.0

For comparison of the transport costs of one passenger per kilometre in different modes, both individual and public, both the theoretical point of view and the actual situation resulting from the occupancy of vehicles, the speed achieved and the capacity of the infrastructure were taken into account. A comparison of the theoretical costs of urban means of transport indicates two-wheeled vehicles as the most expensive (0.15-0.20 EUR / paskm) and trains as the cheapest (0.02 EUR / paskm) regardless of the traffic conditions. A comparison of actual costs gives a completely different picture that varies depending on the time of day (traffic situation). During peak hours, individual modes of transport are characterized by the highest average cost per passenger-kilometre (0.15 to 0.26 EUR / paskm). The lowest average cost is characterized by trains (EUR 0,02 / pkm), BRT buses and large buses (about EUR 0,04 / paskm). The average bus costs are also relatively low (0.07 EUR / paskm). In off-peak hours, the actual costs are doubled, the most expensive is the medium-sized bus, which costs up to 0.5 EUR / paskm. As a result, the difference between the cost of individual and public transport decreases, although the train is still the lowest cost.

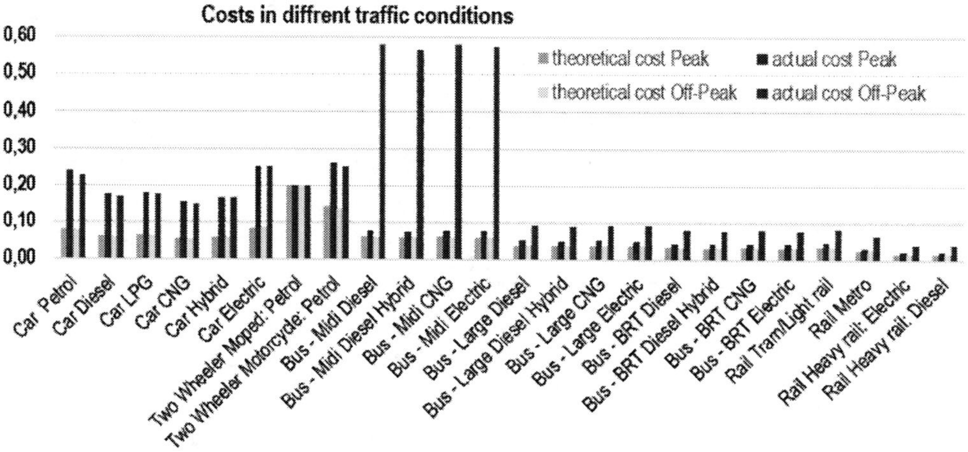

Fig. 5. Comparison of theoretical and actual costs in differential traffic conditions (self-study, based on [10 p. 46]).

8 Summary

The assessment of competitiveness of urban transport modes in the aspects of capacity, energy consumption, CO_2 emissions and costs was carried out on the basis of data on the medium size of European cities, taking into account changes resulting from the traffic volume, i.e. peak traffic (congestion) and off-peak traffic (without congestion). The paper were used the calculations and information contained in the report published in 2017 by the European Commission.

The effectiveness of public transport depends on a number of factors, such as infrastructure capacity and transport means capacity, the level of occupancy in vehicles and operation conditions. In addition, the assessment of the transport modes differs depending on what conditions and parameters are taken into account. Taking into account the different urban conditions, several key trends are noticeable:

1. In peak periods, capacity, costs and environmental performance per passenger in public transport is generally better than in case of individual transport. However, at off-peak hours, the effectiveness of the modes is more complex for them and its results may differ than for peak periods.
2. In general, in all conditions, costs are lower in public transport than in individual transport.
3. Overall infrastructure and transport means capacity for vehicles is the highest in rail transport systems. In the case of road transport, the capacity of individual and public transport (buses), in periods beyond the peak is comparable, and even due to the low level of occupancy rate in public transport, sometimes it is higher for cars.
4. Energy consumption and CO_2 emissions per passenger are generally lower for public transport during peak hours, although the metro and light rail systems are relatively energy-intensive. However, during off-peak periods, individual transport is characterized by lower energy consumption and lower CO_2 emissions per passenger.
5. In general, all transport modes play their part in an effective urban transport system. Public transport should be basic during peak hours, with rail transport providing the highest capacity for the most busy routes. To improve the environmental performance of public transport systems, electric and hybrid systems should be preferred. Individual transport modes play a special role in off-peak periods when there are not enough passengers to ensure a high rate of occupancy in public transport. Effectiveness data can be used by cities to optimize the organization of urban transport.

References

1. ARTHUR D. LITTLE & UITP, *The Future of Urban Mobility 2.0*, (January 2014)
2. *Urban Europe - statistics on cities, towns and suburbs - executive summary*, (Data extracted in February–April 2016), http://ec.europa.eu/eurostat/statistics-explained/index.php/Urban_Europe_-

_statistics_on_cities,_towns_and_suburbs_-_executive_summary [acces:15.08.17]
3. United Nations, *World Urbanisation Prospects: The 2014 Revision*, (New York, 2015)
4. T. Dyr, and M. Kozlowska, *Congestion costs in the European Union*, Technika Transportu Szynowego, tts, **7-8**, 18-23 (2017) [In Polish]
5. European Commission, *Urban mobility*, https://ec.europa.eu/transport/themes/urban/urban_mobility_en [access: 01.09.17]
6. European Commission, *Statistics on European cities*, http://ec.europa.eu/eurostat/statistics-explained/index.php/Statistics_on_European_cities [access: 01.09.2017]
7. M. Jacyna, *Modeming and evaluation of transport systems*, (Oficyna Wydawnicza Politechniki Warszawskiej, Warszawa 2009) [In Polish]
8. Douglas Harper, Online Etymology Dictionary (2010), http://www.dictionary.com/browse/effectiveness
9. T. Dyr, and P.R. Kozubek, Evaluation transport infrastructure investments financed by European Union funds, (Instytut Naukowo-Wydawniczy „SPATIUM", Radom, 2013) [In Polish]
10. B. Pawłowska, *Sustainable transport development against the background of contemporary socio-economic processes*, (Wydawnictwo Uniwersytetu Gdańskiego, Gdańsk, 2013) [In Polish]
11. European Commission, *Study on urban mobility – Assessing and improving the accessibility of urban areas. Annexe 3: Task 3 Report – Relative efficiency of urban passenger modes*, (Brussels, 2017)

MATEC Web of Conferences **180**, 03004 (2018) https://doi.org/10.1051/matecconf/201818003004
MET'2017

Evaluation of the efficiency of hybrid drive applications in urban transport system on the example of a medium size city

Marek Pawełczyk[1,*], and *Emilia* Szumska[2]

[1]Faculty of Management and Computer Modelling, Kielce University of Technology, Al. Tysiąclecia P.P.7, 25-314, Kielce, Poland
[2]Faculty of Mechatronics and Mechanical Engineering, Kielce University of Technology, Al. Tysiąclecia P.P.7, 25-314, Kielce, Poland

Abstract. Unconventional drive systems of the urban buses become more and more popular. The reason is clear: reduction of the harmful emissions as well as the fuel consumption reduction. Environmental aspects are becoming of the great importance. Local governments of large cities are "pushing" conventional buses away from city centres, trying to substitute them by the other more environmentally friendly urban vehicles. The goal of the contribution is to present a comparison of the selected types of the urban buses equipped with alternative drive systems under conditions of medium size city. The first part of the paper presents the methodology and construction of urban bus driving cycles prepared for Kielce city (Poland). The next part shows fuel economy and energy consumption obtained from simulation of urban buses powered by conventional diesel engine and hybrid electric drive under the developed driving cycles. The last part presents results of the Total Cost of Ownership (TCO) calculations estimated for selected urban buses.

1 Introduction

To evaluate the efficiency of the hybrid drive for urban public transport applications some most popular alternatives of the hybrid drives have been taken into account. The most attractive alternatives for the classical Diesel engine drive system are pure electric drives and hybrid drives. Pure electric autonomous vehicles are still very expensive although the price ratio of the conventional vehicles and electric vehicles has become more acceptable lately. Great progress has been taken place in the field of batteries offered, whose capacities and life time have increased significantly and the price has decreased. Recently, there has been a great deal of interest from the self-governments of the Polish cities manifesting itself in that they have already bought or are preparing to purchase a significant number of electric vehicles in Poland. More popular option is hybrid drive urban buses, which are offered in some different options by some Polish bus manufacturers. Among them the most typical are parallel and series hybrid vehicles, which have been chosen for the further consideration. Evaluation of the hybrid bud effectiveness has been conducted by use of the simulation model developed by the US National Renewable Energy Laboratory within the program "Partnership for a New Generation of Vehicles" (PNGV), co-financed by the US Department of Energy and the biggest American car manufacturers [4]. The model was developed in Matlab/Simulink simulation environment. The ADVISOR packages used to be able to download free of charge from the Program web site. The model can be used for alternative drive systems of the different types of the vehicles, including urban buses. To accommodate the model to local conditions of the medium-size city some driving cycles have been worked out.

2 Driving cycles

The driving cycle – Is a record of vehicle movement, represented as its speed vs. time. It does not provide detailed information about driving style, but illustrates the communication conditions of a given city and gives the basis for energy analysis of vehicle traffic [1, 2]. The first driving cycle was developed in the 50s of the 20th century in Los Angeles to investigate the impact of emissions on the formation of smog. The cycle consisted of 11 modules, including the acceleration phase, driving at constant speed, braking and staging [3]. Standardized road cycles are currently used to estimate emissions of harmful emissions and fuel consumption.

In the EU, the NEDC cycle is now in use, in the USA the FTP-75 cycle is used, in Japan there are J10 or J15 cycles, in India – it is IDC cycle. Standardized On-Road Tests (SORT) have been developed to standardize mileage fuelling procedures for buses and public transport. They include three types of tests [4, 5]:
- SORT 1 (Heavy Urban), which reflects the operating conditions in the centre of large cities (average speed 12 km/h);
- SORT 2 (Easy Urban Cycle), which reflects the operating conditions for a typical urban route (average speed is 18.0 km/h);
- SORT 3 (Easy Suburban Cycle), which reflects traffic conditions in the suburbs of large cities and on inner routes of smaller cities (average speed is 25.3 km/h).

* Corresponding author: m.pawełczyk@tu.kielce.pl

The complete cycle consists of a number of speed profiles that group basic cycles, with the following values: acceleration (decelerations), driving at constant speed, stopping time [2, 6, 7]. Standardized road cycles do not, however, reflect the actual communication conditions of a particular city. Therefore, in order to carry out an energy analysis of a particular vehicle and the emission of harmful exhaust particles, many researchers have undertaken to develop a driving cycle for specific cities [8-14]. This contribution presents the method of creating Kielce cycles of driving, which reflect the actual traffic conditions of urban communication vehicles in Kielce.

Two urban bus driving cycles have been worked out for the medium-size city (Kielce), the first one for a flat route (driving cycle I) and the second - for a route with a varied vertical profile (driving cycle II), both of them are based on the measurements conducted on the real bus routes in Kielce (Fig. 1).

Kielce driving cycle I reflects the route of line 13 that runs from the east to the west side of the city and has 39 bus stops. Its total length is 20.2 km. The Kielce II cycle was built on the basis of line 30, which runs from the north to the southern part of the city. There are 36 bus stops in the route and its length is 19.5 km.

Kielce is characterized by uplifting terrain, with the absolute height of between 260 and 400 m, therefore one of the route selection criteria was its vertical profile. Route 13 is characterized by a relatively flat vertical profile with a maximum slope of 1%. Line 30 runs along the hilly route and the highest slope of the route is 4%.

During the measurements of the Kielce cycles, the public transport bus of 12 meters in length. The vehicle was equipped with a Diesel engine of 209 kW (284 HP), its own weight was 10400 kg, while the maximum allowable weight was 18000 kg. Real-time traffic records were recorded using Kiestler GPS CDS Data Logger, installed by the authors on the vehicle. The data sent from the bus on-board GPS system to the ZTM Kielce (bus provider) monitoring system was also taken into account.

a)

b)

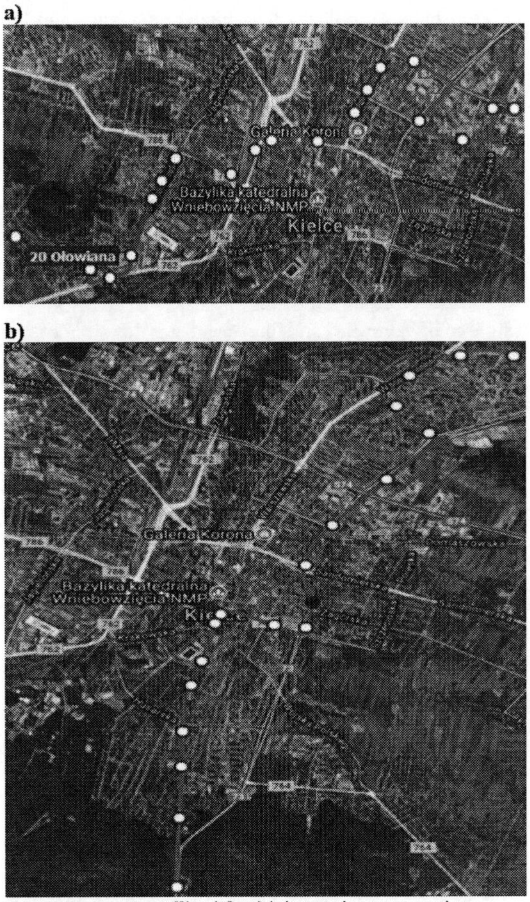

Fig. 1. Bus routes utilized for driving cycles construction: a) bus route No. 13; b) bus route No. 30.

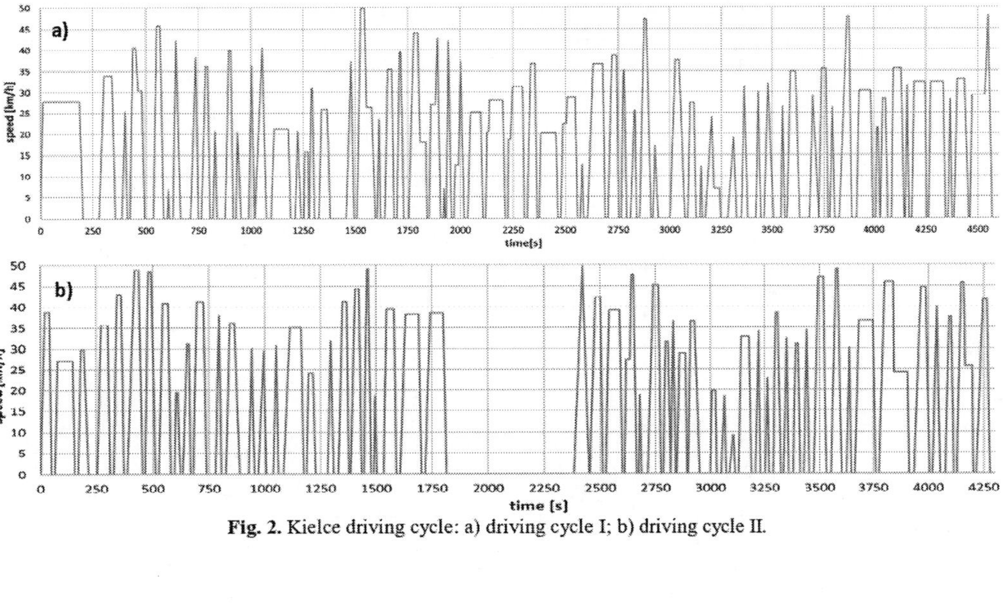

Fig. 2. Kielce driving cycle: a) driving cycle I; b) driving cycle II.

When compiling the actual driving cycles, the following vehicle performance parameters were selected: instantaneous speed, instantaneous acceleration, driving time, distance travelled, and vehicle momentary. On the basis of the available data, journeys with similar stop counts and average speeds and similar cycles were identified. Then one of them was selected for each of the routes studied. These journeys were taken as actual driving cycles, recorded as instantaneous velocity distributions as a function of time (Fig. 2).

The actual driving cycles were divided into short runs between the following stops. For each of them the average speed, acceleration and deceleration were calculated. and delay. The mean value of the real cycle velocity is equal to the average speed of the representative cycle, and the mean acceleration of the actual cycle is equal to the average acceleration of the representative cycle.

Both routes run in urban areas, so the maximum speed should not be higher than 50 km/h. The representative Kielce driving cycle I lasts 4568 s and its length is 20,25 km, its average speed is 15.95 km/h. This route is characterized by high variations in speed values and high variability of acceleration. The average acceleration is 0.55 m/s2 and the average deceleration is 0.67 m/s2. The total number of stops on the route was 66. The basic parameters of both cycles were presented in Table 1.

The driving cycle II is shorter by about 700 m and is characterized by a higher average speed (16.44 km/h). Average acceleration is 0.54 m/s2 and averagė deceleration is 0.81 m/s2. The route is also characterized by a smaller proportion of congestion, which results in fewer stops. In the middle of the cycle, according to the actual timetable for this line, there a stop lasting 568 s.

Most of the time of the two cycles were stops. For the driving cycle I it was 46%, and for the driving cycle II it was 52% of the total cycle time. Since the both worked out representative cycles show the urban bus driving cycles, the total stoppage times included stopping times at stops and stopping times due to traffic conditions. For the driving cycle I, the percentage of times of the two types of stops differed only 2% whereas for the driving cycle II the differences were much higher – total stop time caused by traffic conditions was 38% higher than total bus stop time.

Table 1. Basic parameters of the driving cycles.

Parameter	Driving test I	Driving test II
Test cycle time [s]	4568	4276
Cycle length [km]	20,25	19,53
Maximum speed [km/h]	49,89	49,79
Average speed [km/h]	15,95	16,44
Maximum acceleration [m/s^2]	1,34	1,24
Maximum deceleration [m/s^2]	1,49	1,66
Average acceleration [s/s^2]	0,55	0,54
Average deceleration [m/s^2]	0,67	0,81
Number of stops	66	55

3 The urban bus variants

Three types of vehicles have been chosen for simulations: the parallel hybrid bus, the series hybrid bus and the "classical" Diesel engine urban bus. The last one has been chosen to enable the comparison of the fuel consumption as the main parameter describing the bus usage efficiency.

Table 2. Technical parameters of the drive components of the analysed vehicles.

	SHEV1	SHEV2	PHEV1	PHEV2	Conventional
Engine: Maximum power: Engine displacement: Mass:	 140 kW 7,3 dm^3 476 kg	 170 kW 7,3 dm^3 476 kg	 190 kW 6,7 dm^3 404 kg	 200 kW 6,7 dm^3 443 kg	 205 kW 6,7 dm^3 443 kg
Electric motor: Power max: Mass:	 150 kW 82 kg	 180 kW 88 kg	 40 kW 48 kg	 55 kW 66 kg	-
Generator: Power max: Mass:	 150 kW 82 kg	 180 kW 88 kg	-	-	-
Energy storage device: Storage capacity: Nominal voltage: Mass:	 9,4 kWh 362 V 193 kg	 10,6 kWh 362 V 231 kg	 5,6 kWh 335 V 73 kg	 5,6 kWh 335 V 73 kg	-
Transmission: Mass:	BAE HDS200 31 kg	BAE HDS200 31 kg	ZF 5HP590 30 kg	ZF 5HP590 30 kg	ZF 5HP590 30 kg
Gear ratio of the main gear:	5,857	5,857	4,33	4,33	4,33

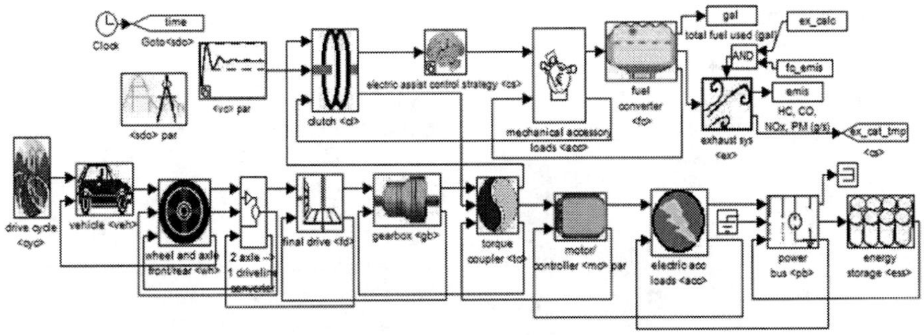

Fig. 3. Parallel hybrid vehicle simulation model – ADVISOR simulation environment.

4 The simulation results

The simulations have been conducted for two above mentioned driving cycles and three variants of the bus drive systems.

Simulation software, called ADVISOR, was implemented in Matlab/Simulink for simulating vehicles with different drive train configurations. As it was mentioned before it was developed by the American National Renewable Energy Laboratory (NREL). The program has built-in models of conventional, hybrid, parallel and electric vehicles. Fig. 3 presents a diagram of a vehicle model with a hybrid parallel drive. The program allows modifying the vehicle models by entering the vehicle data files, the characteristics and parameters of the drive components and the energy storage device. Different driving cycles can be chosen for the simulations and the user's driving cycle can be implemented [15-17].

The parameters of the vehicles chosen for the simulations were the Solaris 12 m bus parameters according to the producer data sheets. It was assumed that the maximum permissible total mass of each of the vehicles was 18 tons. The technical parameters of the drive components of the analysed vehicles have been shown in Table 2.

Authors have focused on the energy dependencies and fuel consumption issues. The simulation results presented below do not contain the tailpipe emissions, however in our opinion the emissions are proportional to the fuel consumption level.

The comparison of the two types of the hybrid buses shows that under conditions of the driving cycles taken into account the better fuel efficiency has been obtained for the parallel hybrid drive. The energy flow from the battery to the drive system and from the system to the battery (during regenerative braking) have been also studied.

Average fuel consumption for the driving cycle I has been shown on Fig. 4. The lowest average fuel consumption was recorded for vehicles equipped with a series hybrid system. These values were nearly 20% lower than for conventional vehicles. PHEV1 and SHEV2 hybrid vehicles have reported a 27% and 24% reduction in fuel consumption in comparison of a conventional drive vehicle. Average fuel consumption for the driving cycle II has been shown on Fig. 5.

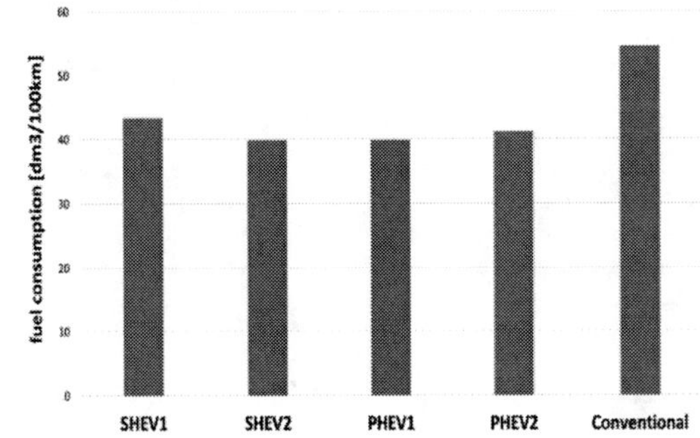

Fig. 4. Average fuel consumption for the driving cycle I [dm^3/100 km].

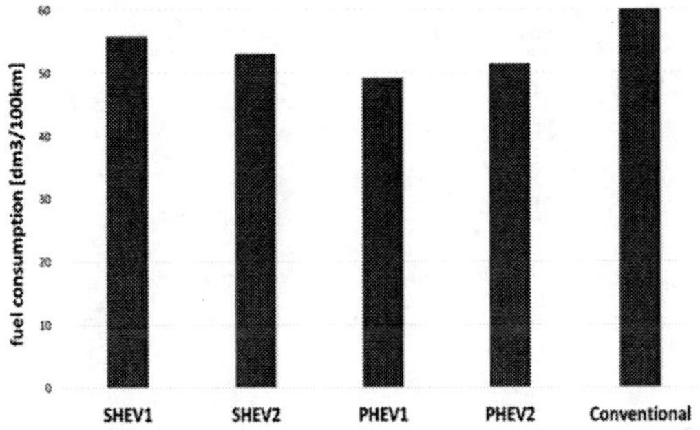

Fig. 5. Average fuel consumption for the driving cycle II [dm^3/100 km].

Fig. 6 presents amounts of energy coming from the fuel combustion and returned from the storage tank in total amount of energy consumed in the cycle. The value of the energy required to complete the driving cycle I by the SHEV1 vehicle is 6%, and by SHEV2 is 10% less than the energy consumed by the conventional vehicle. The energy returned from the energy storage device (ESD), consumed by the SHEV1 vehicle, is 15% of the total amount of the consumed energy. For the SHEV2 vehicle this ratio is 18%. Vehicles with hybrid parallel drives needed 10% less energy than conventional vehicles. The ratio of ESD energy required to complete the cycle by the PHEV1 vehicle is 18% and 16% for the PHEV2 vehicle.

The lowest values of average fuel consumption were recorded for vehicles with parallel hybrid drive. During the driving cycle II PHEV1 recorded 11% and PHEV2 7% consumed less energy than a classic-powered

https://doi.org/10.1051/matecconf/201818003004

vehicle. We should add that these amounts of energy were calculated for one driving cycle only, so to evaluate the real energy effectiveness SOC level for ESD must be taken into account.

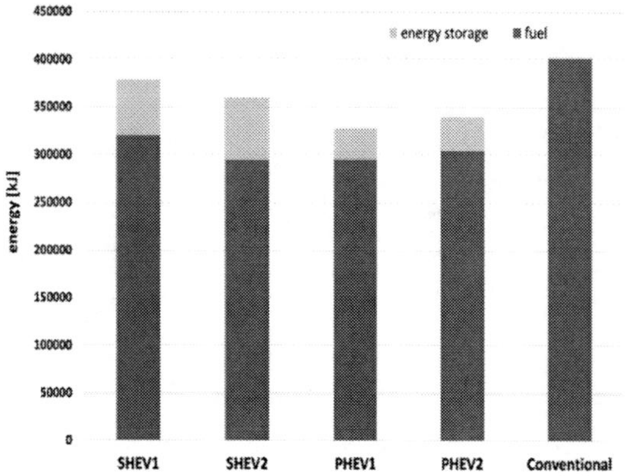

Fig. 6. Amounts of energy [kJ] coming from the fuel combustion (green) and returned from the storage tank (yellow) in total amount of energy consumed in the cycle I.

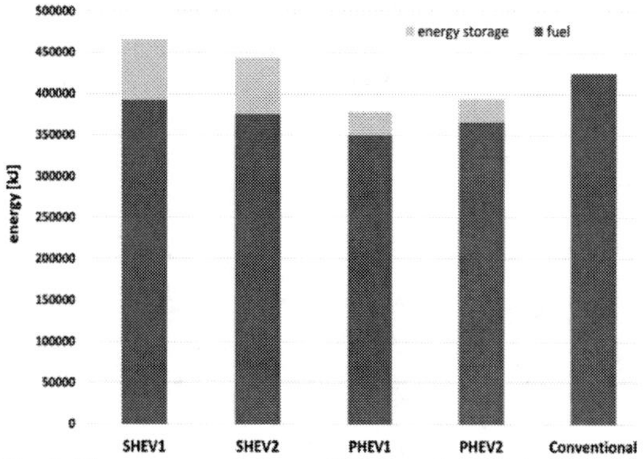

Fig. 7. The amounts of energy [kJ] from the fuel (green) and from the ESD (yellow) in total energy amounts consumed in both driving cycle II.

Our investigations show that hybrid vehicles require about 11% less energy than conventional vehicles. The estimated share of energy from batteries for parallel hybrid vehicles was 7%. For SHEV1 and SHEV2 these ratios were 16% and 15% respectively. The amount of energy from the fuel and from the ESD in total energy consumed in second driving cycle can be seen on Fig. 7.

For both driving cycles we have stated that the lowest energy values were observed for vehicles with parallel hybrid drive. For the driving cycle II, vehicles with series hybrid powertrain needed 19% more energy than for the driving cycle II. In the case of parallel hybrid vehicles, PHEV1 needed 13% and PHEV2 14% more energy for the driving cycle II than for driving cycle I. The share of the energy returned from the storage tank was higher for the driving cycle II. The biggest differences were observed for SHEV2.

5 The TCO (Total Cost of Ownership) evaluation

The TCO has been used for a vehicle total cost of ownership estimation the for a specific period of time. It includes all the expenses spent on fuel, insurance, maintenance, repairs, service, interest on loan payments as well as the losses incurred due to depreciation of the car at the end of the same period [18-23].

The life cycle costs, calculated for the three types of buses mentioned, including fuel costs, service costs, the battery exchange costs, show that the hybrid drive buses can be the cost effective alternative for a classical bus, even having taken into account the actual hybrid drives and batteries prices.

Total Cost of Ownership (TCO) is the sum of all costs of a vehicle, from its purchase, through use, to disposal. TCO analysis allows to evaluate the direct and indirect costs associated with vehicle usage and possession. In addition to the purchase cost, the total costs of living are included (fuel and/or energy costs, insurance, maintenance and repair costs).

The total cost of ownership can be written as follows:

$$K_{TCO} = K_Z + N_P \cdot (K_f + K_u) + N_W \cdot K_A \qquad (1)$$

where K_{TCO} – total vehicle cost [PLN], K_Z – vehicle purchase cost [PLN], N_P – vehicle operating period, K_f – cost of fuel [PLN/year], K_u – cost of maintenance [PLN/year], N_W – number of battery exchanges during the life of the vehicle, K_A – purchase cost of the battery [PLN].

Calculating TCO for a vehicle equipped with a hybrid drive some additional costs connected with the battery exchange must be taken into account. The vehicle life time is as a rule longer than the battery usage time, so the battery must be exchanged a few times. A very important factor, determining the cost of the purchase of a new vehicle, the battery exchange cost as well as the energetic effectiveness of the bus is a battery capacity.

Three city buses with a length of 12 meters were selected for the analysis. The annual mileage for each selected bus is 70 200 km. The life time of each bus is 18 years. The first one has a conventional combustion engine, which complies with the European Euro 6 emission standard. According to the producer data it has been assumed that the average fuel consumption of this vehicle is 43.3 dm^3/100 km. The second vehicle is the Solaris Urbino 12, equipped with a hybrid Eaton hybrid drive. The bus has a lithium-ion battery with a capacity of 3.6 kWh. Average fuel consumption is 33.7 dm^3/100 km. The third vehicle is the next-generation Solaris Urbino 12 Hybrid (launched in 2016), equipped with a hybrid BAE HybriDrive. According to the bus manufacturer data, average fuel consumption is 27 dm^3/100 km. The bus has a lithium-ion (LFP type) battery with a capacity of 11.6 kWh. It was assumed that the battery life time is 6 years. Therefore, the battery pack replacement is required two times during vehicle lifetime. The cost of lithium-ion battery is 900 PLN per kilowatt-hour. The following purchase prices were assumed: the conventional vehicle – 950,000 PLN, the parallel hybrid – PLN 1,200,000 PLN, the series hybrid – 1,300,000 PLN. According to [24] the average maintenance cost assumed for conventional bus – is 35 000 PLN per year, for hybrid buses – 48 000 PLN per year. The fuel price is 4.4 PLN per dm^3. It should be noted that the TCO analysis was carried out for current prices of fuel and lithium-ion battery per kWh. These prices are variable and it is difficult to predict their level in the future. It should also be noted that the changes of money value over time has not been taken into account in these calculations.

The conducted analysis has shown that the total cost of ownership of a conventional bus, referenced to 1 km, is PLN 3.12. The TCO of the parallel hybrid bus is 3.09 PLN/km, and the TCO for the series hybrid drive system is 7% lower – it has been estimated as 2.89 PLN/km. As it has been mentioned before the TCO has been calculated as the sum of fuel costs, maintenance costs and battery replacement costs. Hybrid buses have lower operating costs (TCO) than conventional vehicles. These differences are respectively about 10% for parallel hybrids and 22% for serial hybrids.

The costs of fuel accounts for the largest share of total vehicle ownership. In the case of TCO of a conventional drive vehicle, their share is 61%. For hybrid vehicles, the cost of fuel is lower – it is 48% for parallel hybrids and 41% for serial hybrids.

The share of the purchase cost in TCO of hybrid buses is more than 30%. They have a much higher production costs. Additional costs stem mainly from the use of electric drive components such as batteries, electric motors and electronic components for energy management and thermal balance. Electric auxiliaries improve energy efficiency, but since the market for these items is still small, their costs are quite high.

6 Conclusions

The results of investigations, presented in the paper, show, that hybrid buses are becoming an attractive alternative for urban public transport, especially in medium size cities. However, presented results have not taken into account the environmental aspects of the vehicles' operation. Hybrid buses offer better energy efficiency and fuel efficiency in comparison with conventional buses. They have also a significantly lower Total Cost of Ownership (TCO) then buses equipped with conventional drive.

The trends of reducing air pollution caused by exhaust gases emitted by the urban buses and the tendency to add to TCO some additional costs connected with the environmental effects will have a positive effect on evaluating the effectiveness of the utilizing hybrid drive technology. The authors have shown, that even taking to account only TCO proves a profitability of implementation hybrid buses in cities.

The purchase cost of buses equipped in hybrid drive is still much higher than for the conventional bus. It is possible that the difference in the purchase costs between hybrid and conventional buses will decrease in the future. The advantages of hybrids is that they do not require additional and expensive infrastructure and bring additional environmental and social benefits.

References

1. J. Peng, D. Pan, H. He, Study on the driving cycle construction for city hybrid bus, *Procedeengs of International Conference on Intelligent Systems Research and Mechatronics Engineering (ISRME 2015)*, China, 1-9 (2015).
2. U. Galgamuwa, L. Perera, S. Bandara, *Developing a general methodology for driving cycle construction: comparison of various established driving cycles in the world to propose a general approach*, Journal of Transportation Technologies, **5**, 191-203 (2015)
3. T.J. Loyson, J.R. Konworthy, P.I. Austin, P.W.G. Newman, *The development of driving cycle for fuel consumption and emissions evaluation*, Transportation Research, **20** (6), 447-462 (1986)
4. UITP 2004, *SORT – Standardized On-Road Test Cycles*. UITP – International Association of Public Transport (2004)
5. UITP 2009, *SORT – Standardized On-Road Test Cycles*. UITP – International Association of Public Transport (2009)
6. Z. Dai, D. Niemeier, D. Eisinger, Driving cycle: A new cycle-building method that better represents real-world emissions, *U.C. Davis-Caltrans Air Quality Project*, Department of Civil and Environmental Engineering, University of California, Davis (2008)
7. J. Peng, D. Pan, H. He, Study on the driving cycle construction for city hybrid bus, *Procedeengs of International Conference on Intelligent Systems Research and Mechatronics Engineering (ISRME 2015)*, China, 1-5 (2015)
8. E. Szumska, D. Zielińska, M. Pawełczyk, *Konstrukcja cykli jazdy komunikacji miejskiej w Kielcach (Construction of the Kielce Driving Cycles)*, Autobusy – Technika, Eksploatacja, Systemy Transportowe, **17**, 1372-1376 (2016)
9. P. Chauhan Boski, G. Joshi, P. Purnima, *Vehicular speed pattern analysis on access controlled urban roadway in Surat City*, Technical Research Organization India, **2**, 137-143 (2015)
10. E. Tzirakis, F. Zannikos, Urban Driving Cycles for Greece, Procedeengs of 19th International Transport and Air Pollution Conference (TAP 2012), Greece, 1-6 (2012)
11. J. Zhu, Q. Shi, J. Zhou, The city bus driving cycle construction, *Procedeengs of Second International*

Conference on Mechanic Automation and Control Engineering (MACE), Chiny, 2687–2690 (2011)

12. D. Covaciu, I. Preda, D. Florea, J. Timar, O. Câmpian, Analysis of a driving cycle proposed for Brasov City, *Proceedings of International Automotive Congress (CONAT 2010)*, Romania, 1-8 (2010)
13. X. Gao, B. Zhang, X. Xiong, M. Dong, H. Li, *Construction and analysis of the Dalian Driving Cycle*, International Journal of Control and Automation, **8** (6), 363-368 (2015)
14. S.H. Kamble, T. Mathew, G. Sharma, *Development of real-world driving cycle: Case study of Pune, India*, Transportation Research Part D: Transport and Environment , **14** (2),. 132-140 (2009)
15. T. Markel, A. Brooker, T. Hendricks, V. Johnson, K. Kelly, B. Kramer, M. O'Keefe, S. Sprik, K. Wipke, *ADVISOR: a systems analysis tool for advanced vehicle modeling*, Journal of Power Sources, **110**, 255–266(2002)
16. D. Chen, X. Li, L. Chen, L. Yang, F. Tian, D. Xu, *Research on simulation of the hybrid electric vehicle based on software ADVISOR*, Sensors & Transducers, **171**, 68-77 (2014)
17. http://adv-vehicle-sim.sourceforge.net
18. E. Szumska, M. Pawełczyk 2017, *Porównanie TCO autobusów miejskich z hybrydowym i konwencjonalnym układem napędowym (TCO comparison of the urban conventional and hybrid buses)*, Zeszyty Naukowe Politechniki Warszawskiej, Seria Transport, z. **118**, 277-285 (2017)
19. *Review of the Research Program of the Partnership for a New Generation of Vehicles: Seventh Report.* Transportation Research Board Washington, https://doi.org/10.17226/10180
20. B.M. Al-Alawi, T.H. Bradley, *Total cost of ownership, payback, and consumer preference modeling of plug-in hybrid electric vehicles*, Applied Energy **103**, 488–506 (2013)
21. L. Nurhadi, S. Boren, H. Ny, *A sensitivity of total ownership for electric public bus transport systems in Swedish medium sized cities*, Transportation Research Procedia **2**, 818-827 (2014)
22. B. Propfe, M. Redelbach, Cost analysis of plug-in hybrid electric vehicles including maintenance & repair costs and resale values, *Procedeengs of International Battery, Hybrid and Fuel Cell Electric Vehicle Symposium EVS26*, USA, 1-10 (2012)
23. G. Wu, A. Inderbitzin, C. Bening, *Total cost of ownership of electric vehicles compared to conventional vehicles: A probabilistic analysis and projection across market segments*, Energy Policy **80**, 196–214 (2015)
24. *Panorama et évaluation des différentes filières d'autobus urbains 2015,* ADEME - French Environment and Energy Management Agency, http://www.ademe.fr

Power supply of trams in Poland – current state and perspectives of development

Adam Szeląg[1,*], and *Tadeusz* Maciołek [1]

[1] Warsaw University of Technology, Institute of Electrical Power Engineering, Electric Traction Division, ul. Koszykowa 75, 00-662, Warsaw, Poland

Abstract. Sustainable development of agglomerations requires efficient and ecological transport systems, i.e. electrified transport. In order to achieve the required results of operation of electrified urban transport, especially energy efficiency, it is required to closely examine the existing power supply systems. The biggest share of urban traction systems in Poland belongs to tram systems (in 15 agglomerations and cities). Recent years have witnessed an increased interest in modernisation of the existing lines and construction of new ones. Enhancement of RAMS (Reliability, Availability, Maintainability, and Safety) and energy efficiency of supply systems is a necessary requirement, due to fact that new, modern tram rolling stock with higher power poses significant challenge for the existing, in many systems, old power supply infrastructure. Furthermore, due to low driving-range of autonomous vehicles equipped with batteries and a need of frequent charging of storage devices, catenary supplied urban transport will dominate in the areas of its use. In addition, it might be helpful in developing hybrid vehicles supplied both from a catenary and from energy storage devices (charged during run under catenary) on sections without catenary. The paper presents parameters characterizing tram power supply systems in Poland. The analyses carried out for many tram lines have shown that even at relatively low investments for modernisation of the tram power supply system, it is possible to obtain fast return (energy saving due to improvement of efficiency of recuperation and the resulting reduction of CO_2 emission). Other advantages of modernisations include: enhancement of standards in supply of modern trams with higher power and improvement of reliability due to the reduced risk of disturbances and damages.

1 Introduction

The interest in electrified rail transport systems has increased considerably in West-European countries at the end of the 20th century [4,8], and in Poland since 21th century. due to significant changes in environment transformed by humans (including high population concentration in cities and suburban areas with simultaneous urban congestion and environmental pollution caused by industry and combustion engine vehicles) This is practically the only effective solution for communication problems in large agglomerations.

The main directions of development of these systems can be described primarily as introduction of new technical solutions for rolling stock, power supply systems and infrastructure, which at the same time fulfil requirements of technical standards, comfort and ensure optimization of both economic and financial indicators [1,2,3,5,6,7].

2 Tram traction power supply systems in Poland

Currently in Poland, trams are operated in 15 cities (Warsaw, Łódź together with suburban lines, Silesia conurbation, Cracow, Wroclaw, Poznań, Szczecin, Gdańsk, Bydgoszcz, Toruń, Elbląg, Gorzów Wlkp., Grudziądz, Częstochowa and Olsztyn), and in some cities, trams were eliminated. The total length of tram single track in Poland equals over 1900 km, and it is the second longest network in the EU, after a tram network in Germany, which is almost 2 times longer.

After a long time of stagnation, currently in Poland, one may witness a period of development of tram lines and rolling stock, with a large share of funds coming from urban transport development programmes financed by the EU.

* Corresponding author: adam.szelag@ee.pw.edu.pl

Fig. 1. Urban transport systems in Poland: T-trams, TR-trolleybuses, WKD- Warsaw suburban light rail.

In Warsaw, a new long-term contract for supply of new low-floor trams, including bidirectional ones and units equipped with energy storage devices, is under preparation. Furthermore, intensive works have been carried out in order to modernise various types of already wear-out trams, especially popular 105N type trams, by equipping them with choppers, which does not require replacement of motors, or by replacing DC series motors with asynchronous motors supplied by inverters, and by introducing, at least partially, low-floor trams. However, such an approach imposes certain requirements for often outdated electric power infrastructure. In many cases, the selection of a location for a traction substation TS may be limited due to the lack of proper area (especially in city centres) or a distant location of points of connection to a public electrical grid. If substations are located too far one from another, a higher supply voltage may be required (e.g. 1.5 kV DC instead of 600 or 750 V DC). A solution may be to construct a larger number of substations with smaller dimensions and a simplified structure (e.g. container type), which should facilitate selection of a place suitable for a traction substation. In extreme cases one may seek for other solutions, such as: use of a rectifier with a controlled characteristic or leading long feeders with a large cross-section from the existing substation to a traction network at place, where a substation is required to be built. Not without significance is the cost of laying long cables. However, due to lack of accessibility of area in the cities as well as the prices of a land, every effort is made to limit the number of substations to a minimum. Another solution might include constructing an underground substation – it is quite typical under metro conditions, but it can be found in other systems as well (this solution has been applied, e.g. for supply of tram networks in the city centre of Poznan); however, such a solution increases the investment costs substantially.

The largest share among the Polish urban carriers using electric traction belongs to tram transport companies. One may observe a noticeable increase in interest in construction of new tram systems (return of trams in Olsztyn) as well as in modernisation of the existing systems. Improvement of reliability, availability and energy efficiency of power supply systems is necessary and of high priority. In recent years one could witness intensified purchase and modernisation of rolling stock thanks to the EU funds.

The structure of a power supply network of urban electrified transport may be very diversified.

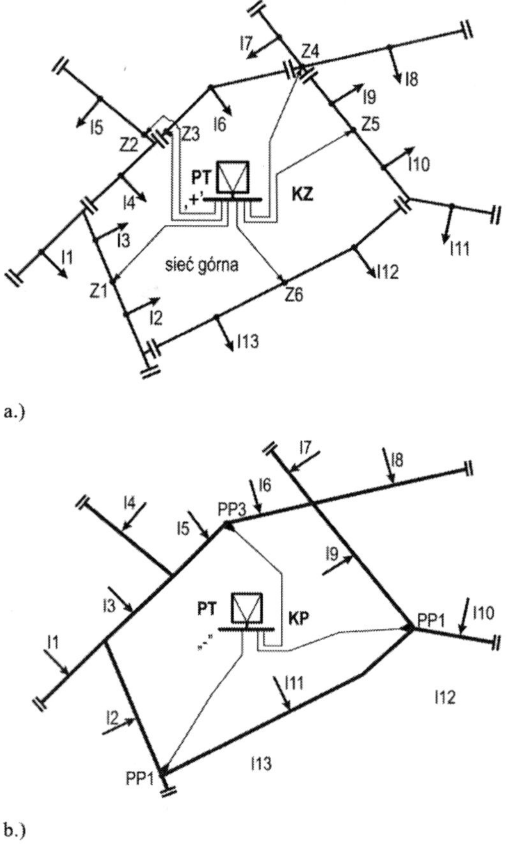

Fig. 2. A typical centralized scheme of a supply network TS a) supply network, b) return network.

It depends on [6]:

• dimensions of a transport network and power demand during peak hours;

• power and energy transmission from a local electrical grid;

• quality and quantity of energy needed by traction vehicles and auxiliary circuits.

Usually, a power supply network consists of several subsystems providing:

• traction current flow from traction substations to vehicles (supply and return network),

• AC low voltage supply and distribution for auxiliaries and lighting purposes;

• energy distribution of medium voltage lines (6, 15, 20 or 30 kV) supplying traction substations as well as auxiliary and non-traction need transformers.

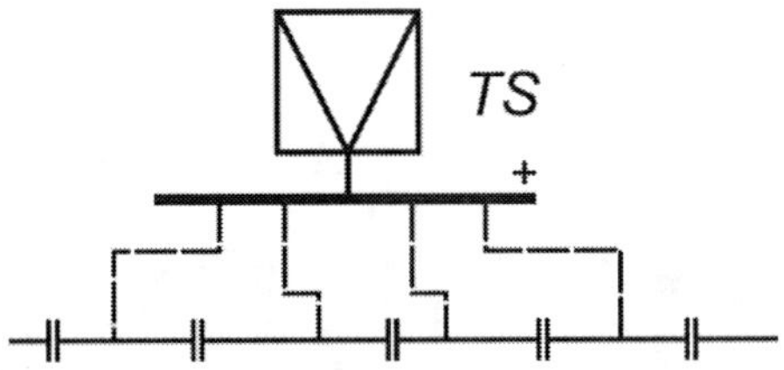

Fig. 3. A linear scheme of a network supplied by a traction substation TS.

There is still some considerable negligence in this area, especially as far as smaller systems are concerned. The upcoming years should change it. Apart from the purchase of rolling stock, it is necessary to make some investments in tram power supply systems. Because the structures of power supply systems were designed many years ago and were based on that time applicable standards. Currently, as outdated from a technical point of view, they are reconstructed and expanded using new technology and solutions.

On the basis of statistical data (obtained from the Polish Chamber of Urban Transport-IGKM) for 13 out of 15 tram systems that exist in Poland, Fig. 4 - 22 (tram systems are numbered from 1 to 13) the selected values characterising tram power supply systems are presented. These values area averaged for each tram transport power supply system and they refer to the year 2016. Unfortunately not all data were available from all tram systems, so in some diagrams appears value 0 – lack of data, mostly for a tram system numbered 7).

The values presented in the figures mainly show the structure of a power supply system. When comparing the values, one should take into account the fact that they are represented by average values and they do not reflect a local specific situation that is included in the division of transport into the central parts of cities and suburban areas, where transport service frequency is lower. For high loads in a power supply system a voltage drop in a catenary should be selected as a limitation while assessing operation quality, and for lower loads, a voltage drop in supply and return feeders. Based on statistics simplified in such a manner, it may be quite difficult to assess these power supply systems in terms of quality. In each case, such an assessment would require conducting a study on a power supply system, which should include a thorough analysis of losses and drops of voltage and loads for specific conditions [6]. The qualitative assessments should also include aspects of reliability, availability and impact on the environment and infrastructure.

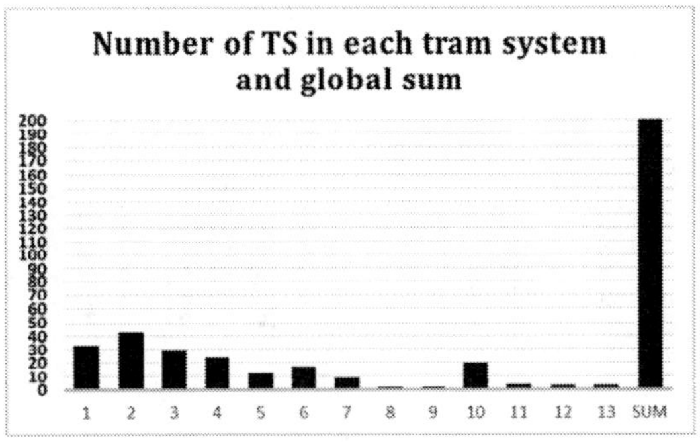

Fig. 4. The number of TSs in 13 tram systems in Poland and a sum of TSs in 13 tram systems, and a global sum.

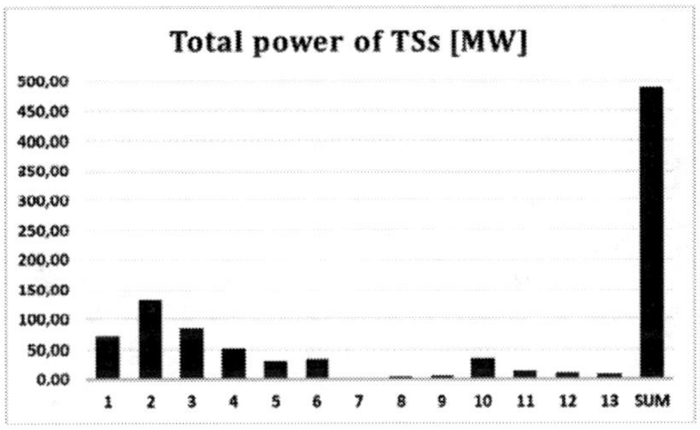

Fig. 5. Power installed in traction substations in 13 tram systems in Poland, and a global sum.

• the number of TSs in each of the 13 systems varies from 2 to 43 (Fig. 4), which gives globally over 200 TSs, with most of them being controlled remotely. Power installed in a TS exceeds 500 MW (Fig. 5), the number of rectifier groups used in all 13 tram systems is over 600 (Fig. 6), the average number of rectifier units installed in traction substations varies from 2.7 to 3.9 pieces, while the average number of rectifier groups in one TS is 3 (Fig. 7), and typically one group constitutes a reserve. It gives a spread of power in a range from 135 to 448 kW per 1 km of a network (Fig. 8),

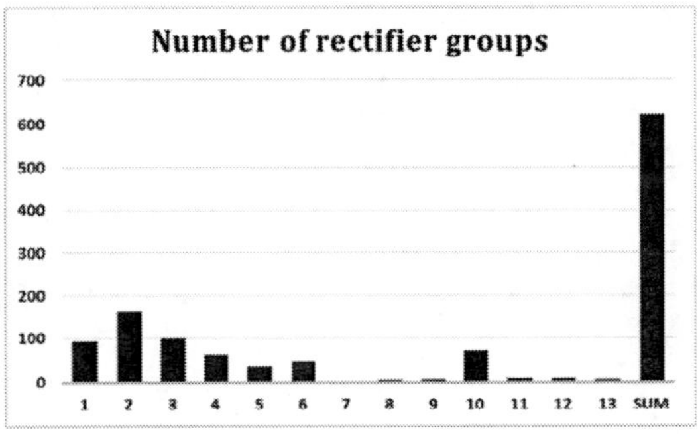

Fig. 6. The number of rectifier groups in a traction substation in 13 tram systems in Poland, a sum of TSs in 13 tram systems, and a global sum.

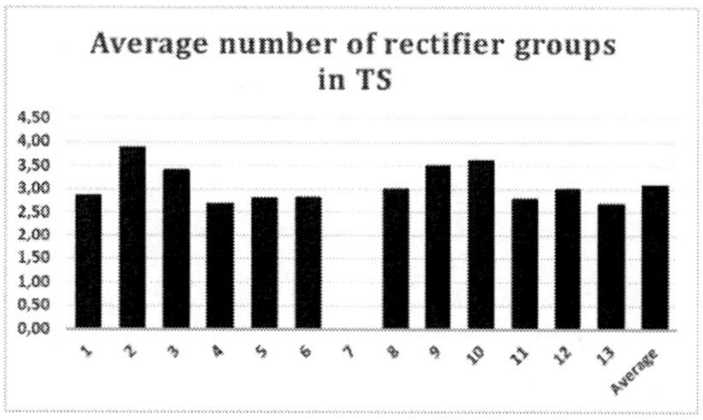

Fig. 7. The average number of rectifier groups in a TS in each of 13 tram systems in Poland and the averaged value for all 13 tram systems.

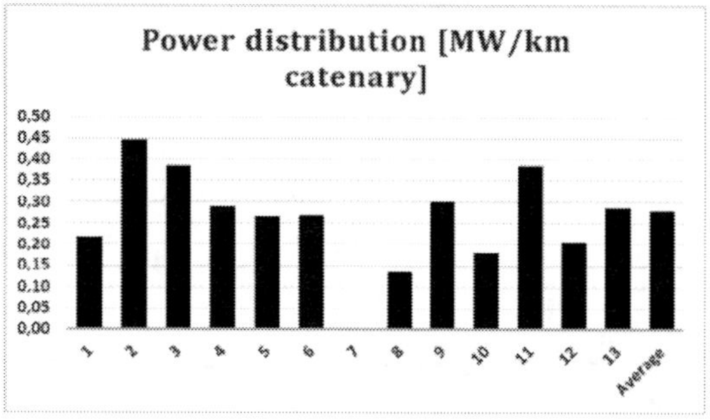

Fig. 8. Average power distribution installed in a TS per 1 km in each of 13 tram systems in Poland and the averaged value for all 13 tram systems.

Based on the presented data describing tram traction power supply systems in Poland, the following conclusions may be drawn:

• the central power supply system is a typical power supply system for tram traction in large urban agglomerations (Fig. 3), with the sizes of areas and powers of a substation characteristic for such a power supply system (Fig. 5), and large distances of network sections (Fig.9), supply cables (Fig. 10, Fig. 14) and return cables (Fig. 11), and a large number of power supply units (Fig.12) and return cables (Fig.13) per one traction substation.

• the length of feeders per 1 km of a network is between 0.39 and 1.94 km, and per a substation – from 5 to over 15 km with a number of power supply units in a substation varying from 4.5 to 10;

• length of return feeder cables per 1 km of a network is in the range from 0.27 to 1.56, and per a substation from 2.06 to 13.45 km (Fig. 17), with their number per a substation varying from 3.5 to 9.0 (Fig.13). A lower limit corresponds to a smaller number of installed units and longer sections of a catenary.

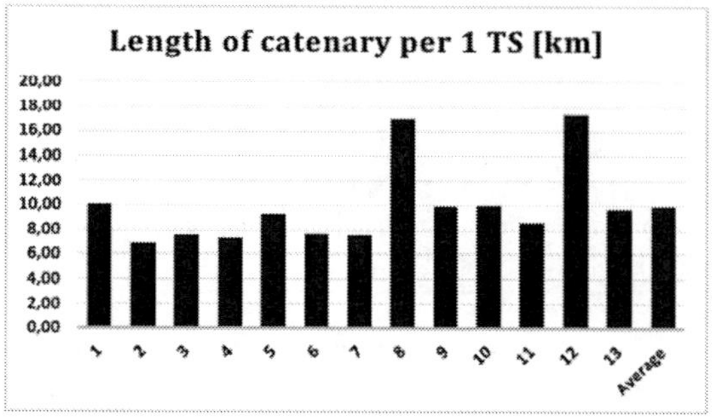

Fig. 9. The average length of a catenary supplied by 1 TS in each of 13 tram systems in Poland and the averaged value for all 13 tram systems.

• the average length of a catenary of a supply area of one substation is in the range between 6.9 and 17.35 km. Longer lengths occur in the cities with a smaller transport volume;

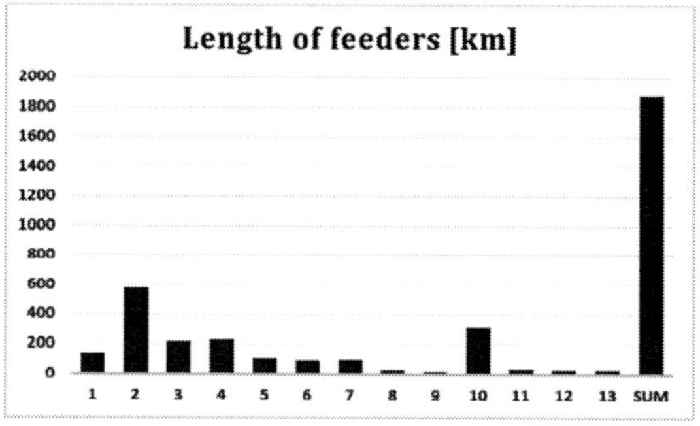

Fig. 10. The length of feeders in 13 tram systems in Poland, a sum of TSs in 13 tram systems, and a global sum.

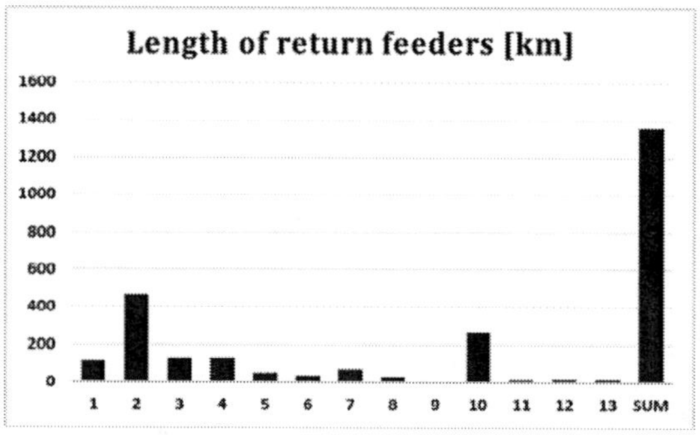

Fig. 11. The length of return feeders in 13 tram systems in Poland, a sum of TSs in 13 tram systems, and a global sum.

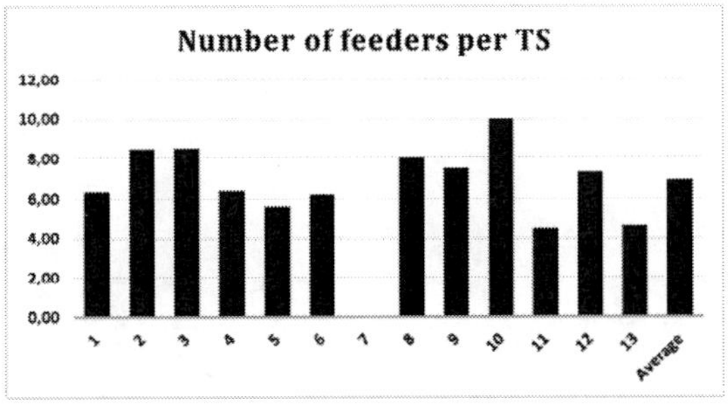

Fig. 12. The average number of feeders in one TS in each of 13 tram systems in Poland and the averaged value for all 13 tram systems.

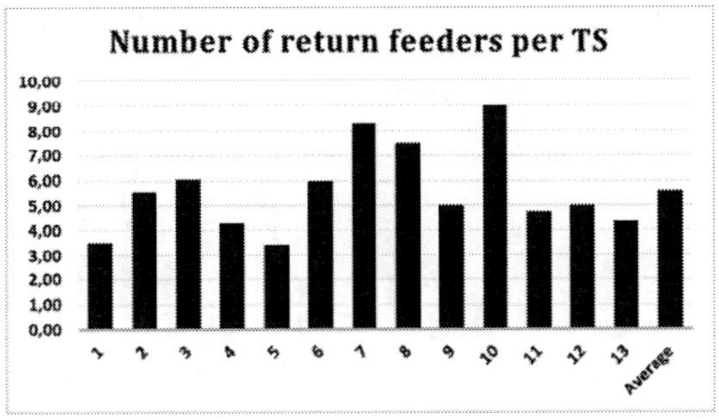

Fig. 13. The average number of return feeders in one TS in each of 13 tram systems in Poland and the averaged value for all 13 tram systems.

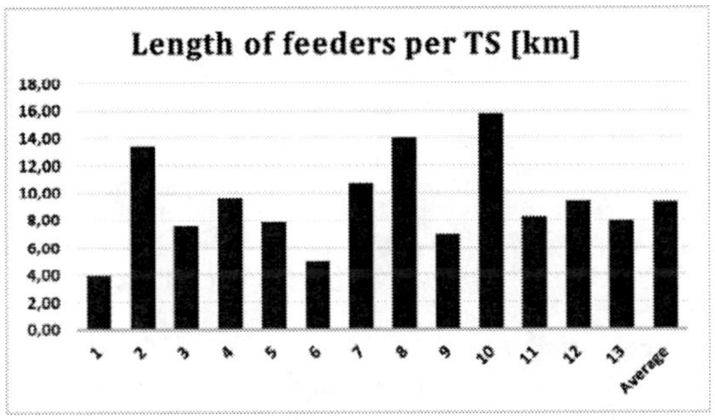

Fig. 14. The length of feeders per 1 TS in 13 tram systems in Poland and the averaged value for all tram systems.

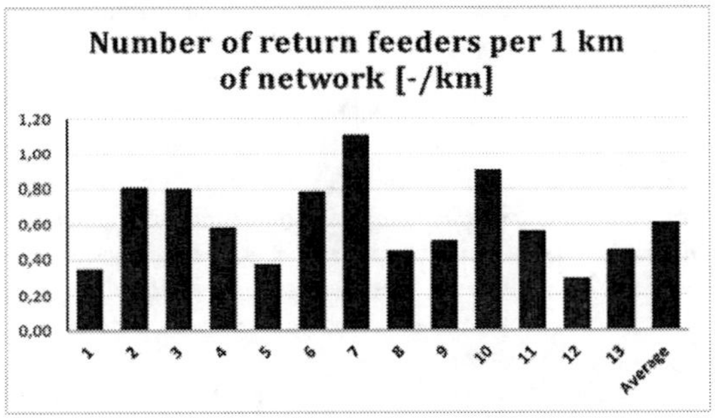

Fig. 15. The number of return feeders per 1 km of a network in each of 13 tram systems in Poland and the averaged value for all 13 tram systems.

• the number of power supply units per 1 km of a network is from 0.42 to 1.23 (Fig. 15), which means that a section length of a network supplied by 1 power supply units is between 0.815 and 2.37 (Fig. 16).

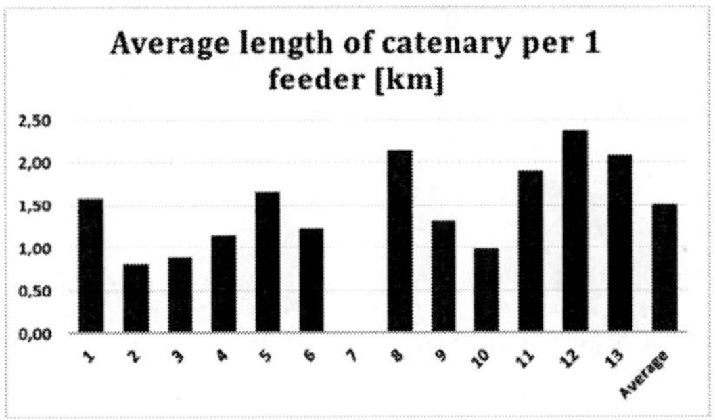

Fig. 16. The average length of a catenary per 1 feeder in each of 13 tram systems in Poland and the averaged value for all 13 tram systems.

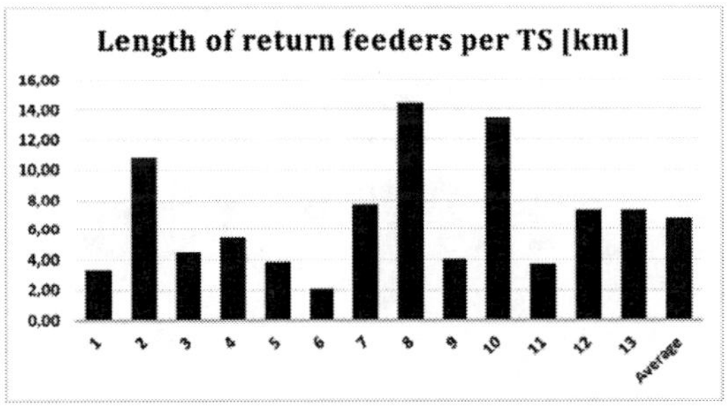

Fig. 17. The length of return feeders per 1 TS in 13 tram systems in Poland and the averaged value for all tram systems.

• longer sections appear in the cities with a lower traffic volume. The average number of return cables per 1 km is in the range from 0.35 to 0.9, which means that a catenary length per 1 return cable is from 1.1 to 3.47 km;

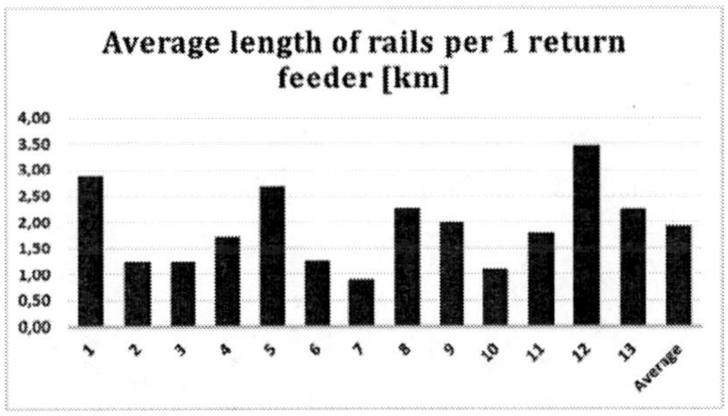

Fig. 18. The average length of rails per 1 feeder in each of 13 tram systems in Poland and the averaged value for all tram systems.

 https://doi.org/10.1051/matecconf/201818003005

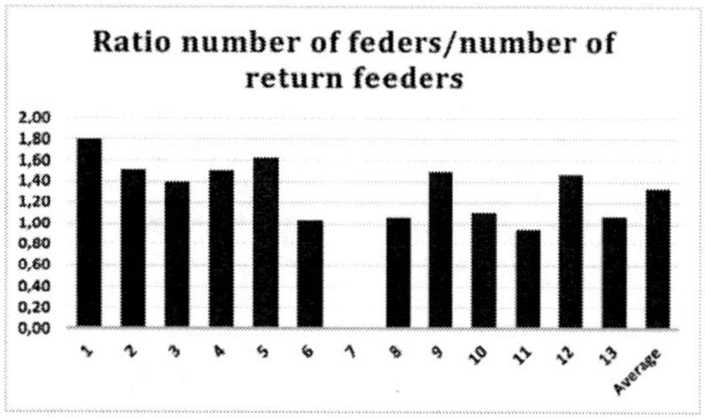

Fig. 19. The ratio of the number of feeders to the number of return feeders in 13 tram systems in Poland and the averaged value for all tram systems.

• only in 1 system there is more return cables than supply cables, while in the remaining the majority belongs to supply cables; the ratio of supply cables to return cables is from 1.04 to 1.8 (Fig. 18);

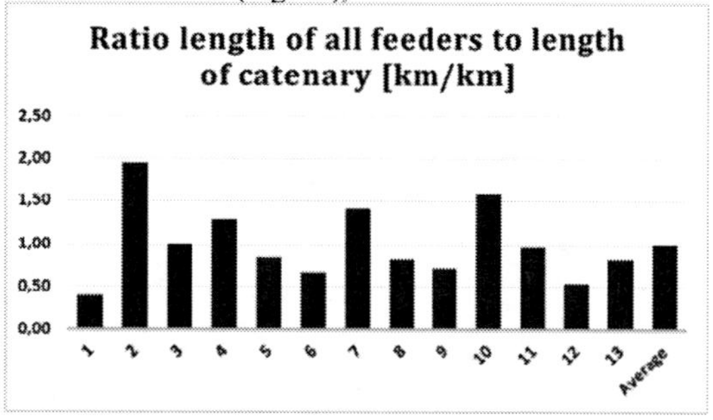

Fig. 20. Ratio of length of all feeders to length of catenary in 13 tram systems in Poland and the averaged value for all tram systems.

• properly selected locations of return points of a rail network for a central power supply system require that the ratio of supply cable lengths to return cable lengths have a value not exceeding 1.4 (half of the systems fulfil this requirement - Fig. 19);

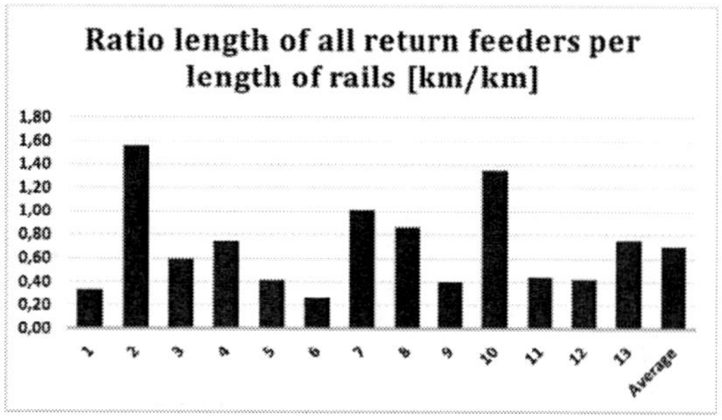

Fig. 21. The length ratio of all return feeders to the length of rails in 13 tram systems in Poland and the averaged value for all tram systems.

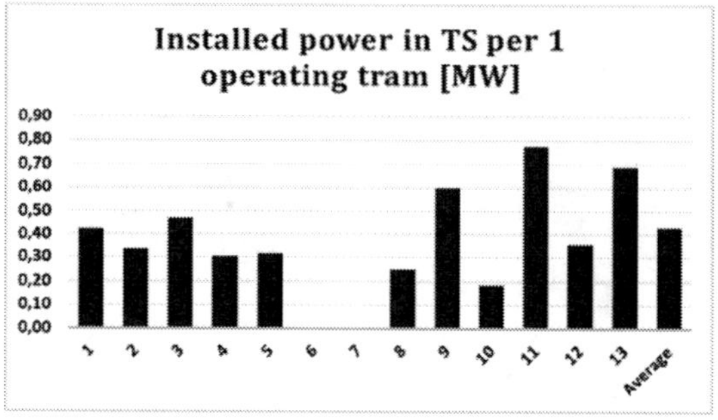

Fig. 22. The average value of installed power in TSs per 1 operating tram in each tram systems in Poland and the averaged value for all tram systems.

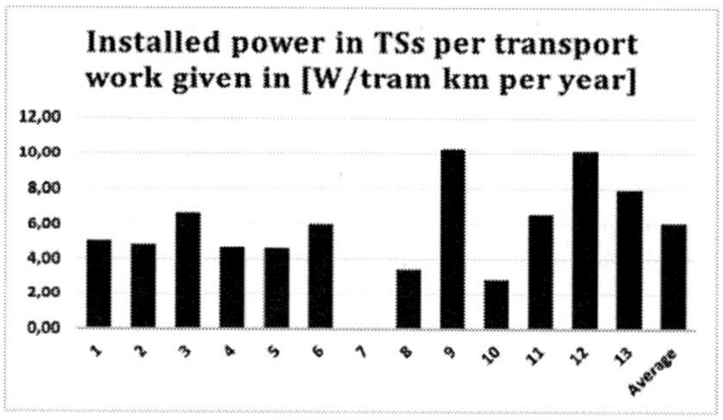

Fig. 23. The value of installed power [W] in TSs per one-year transport work of trams [in tram km] in each tram system in Poland and the averaged value for all tram systems.

• in many tram systems, the lengths of a cable network are larger than those of a catenary. Fig. 20). This is a consequence of application of central power supply systems and multiple unit traction substations. The ratio indexes of supply cable lengths to the length of a catenary (Fig. 20) and return cables to the length of a rail network (Fig. 21) provide an overview on the extent of the area of substation supply, but also allow for assessment of power and energy losses and danger of stray currents at known values of load;

• power installed in traction substations per one km of a line (Fig. 8) and the ratio of power installed in traction substations to average annual number of wagons being operated (Fig. 22) constitute an indicator of use of supplying devices. Depending on a system, power installed in a supply system per an operated tram-train ranges from 187 to 776 kW during a year, while installed power per train-km per year ranges from 2.8 to 10.3 kW (Fig. 23). The lower the powers are, the greater is the use of power installed in substations for execution of transport tasks. A unitary power recalculated to the average number of cars in peak hours (including heating and auxiliaries) defines the capacities of a power supply system or – with known average power of vehicles – a limit capacity of a power supply system.

The last indicator should be additionally recalculated for each substation separately.

3 Conclusions

1. Introduction of modern rolling stock poses challenge for the existing electric power infrastructure of urban traction, which is present in the majority of large cities (except for Białystok, Kielce, Radom, Zielona Góra and Opole) (Fig. 1).
2. Sustainable development of major urban centres requires efficient ecological urban transport, i.e. electric transport [8]. However, one should not forget about the fact that Poland has practically no own resources of liquid fuels. Therefore, the strategic role of electric traction cannot be omitted [2].
3. In order to achieve the required effects of operation of electrified urban traction systems, especially in terms of energy savings, one should thoroughly analyse the operational conditions of the existing power supply systems. In some cases technical infrastructure solutions for tram networks existing in Poland are as old as over 100 years, and even after further modernisation, they are still not suitable for supply of rolling stock with a higher power and energy recuperation [1,3]. This implies the need for a detailed analysis of operational conditions of traction electric power systems due to specific requirements of these vehicles. In most cases, it will be necessary to modernise particular sections of power supply and of substations in order to both provide reliability of power supply system operation and achieve an appropriate level of RAMS (*Reliability, Availability, Maintainability and Safety*) for the target share of rolling stock with recuperation; while on the other hand, increase in effectiveness of the use of regenerative braking and decrease in energy consumption.
4. Analyses conducted for power supply systems of many traction substations [7] proved that at even with considerable low expenditures on the modernised power supply systems one may obtain quick results: an average estimated return on investment amounted to several years of energy savings and, as a result, reduction of CO_2 emission. Furthermore, due to modernisation investments, it is possible to achieve effective power supply of modern rolling stock and improve reliability of power supply by decreasing the probability of disturbances and failures. The conducted research [7] includes the following modifications to the existing power supply system in terms of increase of effectiveness of regenerative braking:
   – cross-connection via a non-polarised high speed breaker (HSB) (2-point power supply within a traction substation);
   – a longitudinal connection – permanent, without a HSB (2-point power supply within a traction substation);
   – longitudinal connection – permanent, without a HSB (2-side power supply) between 2 traction substations;
   – insulator displacement;
   – displacement of a return feeder connection;
   – displacement of a supply feeder connection;
   – supply feeder extension;
   – installation of an energy storage device in a tram;
   – installation of an energy storage device in a traction substation;
   – installation of an energy storage device outside TS, for instance at the end of supply section.
5. The projects related to electromobility pose challenge, but at the same time constitute an opportunity for development of public electric transport in Poland. Due to a still limited extent of electric autonomous vehicles and the need of frequent charging of energy storage devices, a urban traction network will dominate within the areas of its use, and within the area where a trolleybus system is operated, it will provide an opportunity for hybrid vehicles, namely autonomous and catenary supplied – to broaden the area of their use by charging during a run, and next operating on the sections without a catenary.

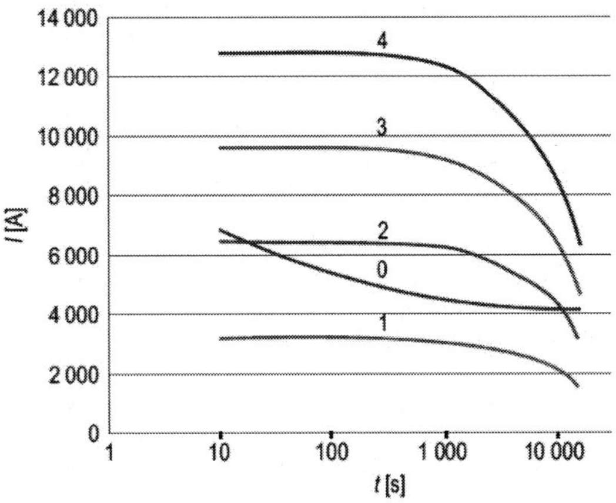

Fig. 24. Results of simulation of a TS current *I* (curve 0) and curves of load current capacity of a TS with 1,2, 3 and 4 rectifiers with 1600 A rated current and V IEC class of overload capacity.

On the other hand, considerable power reserve in traction substations under conditions of normal load (typically in a TS 1 rectifier unit with a rated power of 800-1000 kW could be switched-off and kept as a cold reserve – Fig. 24) and no loads during night-time (installed power in tram systems is above 500 MW). This amount of installed power justifies examining the possible use of traction substations (or certain points of a traction catenary) as an energy source for supply of fast charging points (1 rectifier set could supply up to 4 points with 200 kW power for electric busses) or even creating smart-grids.

References

1. S. Abe, H. Fuimori and T. Ito, DC feeding system suitable for rolling stocks with regenerative braking system utilizing thyristor rectifier, *Proceeding of Fifth European Conference on Power Electronics and Applications, 1993 IET, Brighton, UK,* 168-173 (1993)
2. G. Bajno, Influence of energy regeneration on the power supply of the Filtrowa and Annopol traction substations in Warsaw Trams [Wpływ rekuperacji energii na układ zasilający podstacji trakcyjnej

„Filtrowa" i „Annopol" w Tramwajach Warszawskich], Proceeding of 11th International Conference *"Modern Electric Traction"* MET'2013, Warsaw, Poland,(2013) [in Polish]

3. J. Czucha, K. Karwowski, M. Mizan and P. Pazdro, *Efficiency of recuperative electrodynamics braking energy in urban transport [Efektywność odzysku energii hamowania elektrodynamicznego w komunikacji miejskiej]*, Przegląd Elektrotechnicz, **10(80)**, 1016-1019 (2004) [in Polish]
4. M. Bartłomiejczyk, Practical application of in motion charging: Trolleybuses service on bus lines, Proceeding of 18th International Scientific Conference Electric on Power Engineering (EPE), Kouty nad Desnou, Czech Republic, 1-6 (2017)
5. M. Bartosik, W. Kamrat, M. Kaźmierkowski, W. Lewandowski, M. Pawlik, T. Peryt, T. Skoczkowski, A. Strupczewski and A. Szeląg, Power sources and safety of electrical energy distribution in Poland [Bezpieczeństwo elektroenergetyczne dla pokoleń], Przegląd Elektrotechniczny, **8(92)**, 268-282 (2016) [in Polish]
6. A. Majchrzycki and K. Tuliński, Energy regeneration – a key for optimalization of finances or additional technical problems [Rekuperacja energii – klucz do optymalizacji kosztów czy dodatkowe problemy techniczne], *Electric traction. Scientific researches. Monography of the 2nd congress of Polish electric power industry [Trakcja elektryczna. Prace Naukowe. Monografia II Kongresu Elektryki Polskie]*, edited by A. Szeląg, (COSIW, SEP, Warsaw 2016) [in Polish]
7. Ch. Schreyer, Ch. Schneider, M. Maibach, W. Rothengatter, C. Doll and D. Schmedding, *External costs of transport. Update study. Final Report*, (Zurich/Karlsruhe 2004)
8. A. Szeląg, *Improvement of the electrical effectiveness of railway transport [Zwiększanie efektywności energetycznej transportu szynowego]*, Railway Transport Tecgnology [Technika Transportu Szynowego], 12, (2008) [in Polish]
9. A. Szeląg, T. Maciołek, Z. Drążek and M. Patoka, *Aspects of effectiveness and energy saving in the process of modernization of power supply of tram traction [Aspekty efektywności i energooszczędności w procesie modernizacji układow zasilania trakcji tramwajowej]*,Pojazdy Szynowe, **3**, 34-42 (2011) [in Polish]
10. A. Szeląg, Z. Drążek, T. Maciołek and M. Patoka, *Study of power supply of tram traction in Warsaw Trams PLC [Studium układu zasilania trakcji tramwajowej Tramwajów Warszawskich sp. z o.o.],* Research of Electric Traction Division of Warsaw University of Technology on the order of Warsaw Trams PLC [Praca Zakładu Trakcji Elektrycznej PW na zlecenie Tramwajów Warszawskich sp. z o.o.], Warsaw (2010-2011) [in Polish]
11. M. Yasunami, *Transportation Systems: Their Technical Trends*, Hitachi Review, **4(40)**, (1991)

https://doi.org/10.1051/matecconf/201818004001

Analysis of wayside energy storage devices for DC heavy rail transport

Petru Valentin Radu[1,*], and *Zbigniew* Drazek[1]

[1] Warsaw University of Technology, Institute of Electric Power Engineering, Electric Traction Division, ul. Koszykowa 75, 00-662, Warsaw, Poland

Abstract. The proposed paper presents the possibility of using the wayside energy storage devices (WESD) for the DC Heavy Rail Transport treating the design, costs and payback time. Moreover a case study comparison for the use of wayside energy storage devices on the heavy transport at the supply voltage of 3.3kV DC is presented. A method of sizing the energy storage devices using vehicle characteristics, traction power supply and running timetable is presented. The paper also presents the cost analysis for the most commonly used energy storage devices and the payback time.

1 Introduction

Nowadays most of the traction vehicles have the ability of regenerative braking. The energy resulted from the braking process is used by other vehicle located in the vicinity of the braking vehicle. If the receiving vehicle is not able to receive the entire energy then it is lost on braking resistors to keep the line voltage under the limits. Different types of energy storage devices are proposed for storing the resulted braking energy like: batteries, supercapacitors, flywheels, superconducting energy storage systems or a combination of them [1, 2, and 3]. Beside the main scope of saving energy the energy storage devices can improve the train traction characteristic, the line voltage and can provide with longer distances between the traction substation supplies [4, 5].

The energy storage devices are widely used in the light transport and there are many research papers and studies detailing their design, control, effectiveness and payback time. The maximum power required for the light transport vehicles (metro, tram) is around 1.5MW compared with heavy rail trains that can be up to 6MW [6]. In theory the heavy rail trains could save more energy if we consider the power needed for powering them and the longer braking distances. If wayside energy storage devices are used, there is no restriction and constraints regarding the size and location allowing us to freely design those according with the power demanded.

2 Applications of wayside energy storage devices

The lead acid battery has been used for more than 150 years [7]. The first lead acid battery used for the electrified transport was called "battery post". They were used in Japan from 1912 to 1927 on Shin-etsu Line. The battery was installed at the Maruyama and Yagasaki traction substation. The space required for battery installation was 20% bigger than that for substation electrical equipment. The installed battery capacity was 1332 Ah and rated voltage 624V. The scope of battery use was to suppress the voltage drop [8].

A Ni-MH battery produced by Kawasaki Heavy Industry has been successfully used for wayside light rail application. The batteries Kawasaki Gigacell used for Osaka Subway improved the minimum line voltage from 703V to 718V and reduced the voltage spikes from 939V to 854V [9].

In table 1 is presented examples were the batteries Kawasaki Gigacell was used [10].

Table 1. Kawasaki Gigacell applications.

Location	Year of installation	Battery capacity	Installation scope
New York Subway	2010	367kWh	Verification tests
Osaka Subway	2011	205kWh	Power savings
Washington D.C. Subway (WMATA)	2012	385kWh	Verification tests
Tokyo Monorail, Shinagawa Substation	March 2013	203kWh	Power savings
Osaka Subway	2013	204kWh	Power savings
Sapporo Subway	2013	204kWh	Power savings
Tokyo Monorail, Tamagawa Substation	March 2014	203kWh	Power savings

* Corresponding author: petru.radu@ee.pw.edu.pl

Lithium ion batteries produced by Hitachi was successfully tested on Myodani substation of the Seishin-Yamate Line of the Kobe city Subways in 2004 by Kobe Municipal Transportation Bureau. The rated power of the system was 1000 kW and the rated energy 37.4 kWh. The role of energy storage device was to absorb regenerative power. In parallel with the storage devices, the invertor equipment was used to check if they could operate together and no interference was noticed. It was noticed that the installation of the system can save more than 10% of the energy for power consumption [8].

In 2011, in South Korea on the Seoul Metro Line number 9 a stationary energy storage device was installed produced by Hitachi with lithium ion battery at three substation 903 (0.3MW inverter), 909 (1MW storage system) and 921 (1MW storage system). The estimation for the annual power savings on the substation 921 was 510MWh. During the operation in the first month the energy saved was 94 MWh which surpassed the initial expectations [11].

First stationary energy storage device on electrified railway using a flywheel was introduced in 1988 in Japan. The stationary storage device was installed at the Zushi substation by Keihin Electric Express Railway. The device was able to convert regenerative energy to mechanical energy using a generator motor and converter and back to electrical energy. The total power of the system was 2000kW and was able to store up to 25kWh energy. The energy saved by the storage devices was believed to be around 12% [8].

3 Simulation of a DC heavy rail transport

Figure 1 gives the assumed line profiles including the stations and substations.

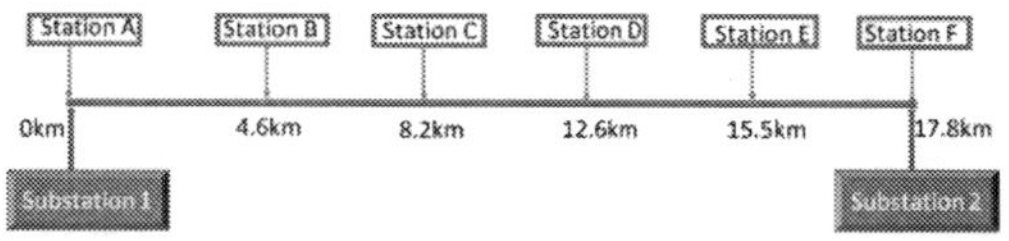

Fig. 1. Simulation line profile.

In table 2 the assumed specification is presented for the vehicle and the line.

Table 2. Simulation parameters.

Parameter	Unit	Quantity
Track length	km	17.8
Catenary and rail resistance	Ω/km	0.07
Substations internal resistance	Ω	0.21
System Voltage	kV DC	3.3
Number of stops	-	6
Number of trains	-	1
Train mass	t	120
Maximum train speed	km/h	120
Average train speed	km/h	80
Average train acceleration / deceleration rate	m/s^2	0.8

If we consider a train running from Station A to Station F the train voltage, current and power simulation results are presented in figure 2. Train simulation was provided in the train simulation program prepared by Electric Traction Division of the Warsaw University of Technology [12, 13].

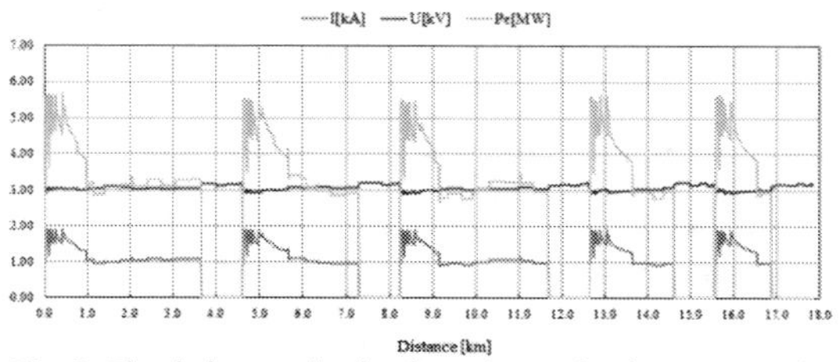

Fig. 2. Simulation results for the train running between Station A and Station F.

The results of the simulation for a train travelling from Station A to Station F are presented in table 3.

Table 3. Simulation results for the train running between Station A and Station F.

Stations	Track length [km]	Total Energy used [kWh]	Energy used per km [kWh/km]	Average power [MW]
A-B	4.6	132.2	28.73	2.86
B-C	3.6	106.3	29.52	2.79
C-D	4.4	121	27.5	2.77
D-E	2.9	83.7	28.86	2.65
E-F	2.3	68.1	29.6	2.6
Total	**17.8**	**511.3**	**28.72**	**2.73**

Figure 3 shows the train movement diagram used for operation calculations for a 5 minute headway and 1 minute stop at each station.

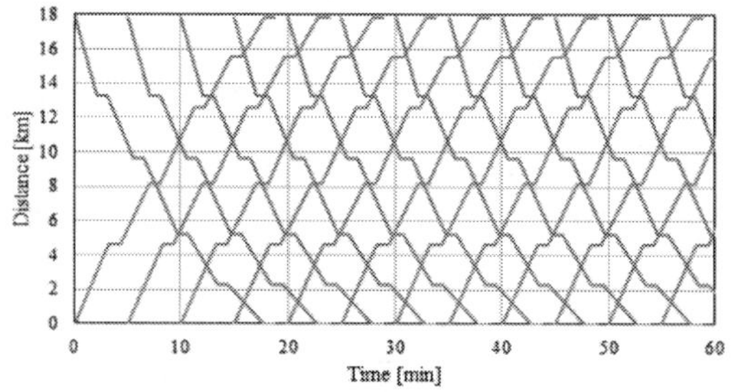

Fig. 3. Assumed train movement diagram for 5 minute headway.

From figure 3, the train operation was calculated for one year and presented in table 4.

The average train speed for the simulation was considered 80km/h with a train mass of 120t. From figure 4 we can conclude that the wayside energy storage system should be capable of absorbing minimum energy of 38.4 MJ (10.7kWh). The charging / discharging current for the WESD is considered 1kA. If we consider the WESD power 3MW and the absorbed energy of 40MJ the charging time for the WESD is 20s

(40MJ/3MW=13s). For 13s charging time and 2C current the depth of discharge (DOD) is 0.7% (13*2/3600=0.7%), for 3C current the DOD is 1.1% (13*3/3600=1.1%) and for 5C current the DOD is 1.8 % (13*5/3600=1.8%).

Table 4. Assumed train operation for a calendar year.

Time	Headway	Cycles per Day	Total cycles
Peak hours (6-9AM and 4-7PM)	5 minutes	144	52560
Off - peak hours and weekends	15 minutes	128	46720
Between 1AM and 3PM	No train services	-	-
Total			**99280**

4 Calculation of the energy that can be saved

The Polish transport company SKM provided test of braking with regeneration on 3.3kV DC trains manufactured by Polish company Newag S.A in April 2012 [14]. The measurements was taken in train with the energy meter LE3000 Elester. Table 5 presents the measurement results provided by SKM.

Table 5. Measurements results provided by SKM.

Train Type	EZT19 WE	EZT27 WE	EZT35 WE
Max power	3.5 MW	4MW	4MW
Train mass	190t	190.5t	197t
Maximum train speed	130km/h	160km/h	160km/h
Average train speed	42km/h	41km/h	36.8km/h
Track length	Approx. 35.5km	Approx. 23km	Approx. 29km
Stations	Pruszkow – Sulejowek Miłosna	Warszawa Zachodnia – Sulejówek	Warszawa Zachodnia (Platform 8) - Legionowo Piaski
Total energy used	362.2kWh	302.9kWh	329.7kWh
Total energy saved	129.9 kWh	59.4kWh	121.5kWh
Percent energy saved	35.31%	19.58%	37%
Average energy used per km	10.2kWh/km	13.17kWh/km	11.37kWh/km
Average energy saved per km	3.1kWh/km	2.58kWh/km	4.19kWh/km

The total energy that can be recovered by the WESD can be calculated with the formula:

$$E_{WESD} = E_{Breake} - E_{Aux} - E_L \quad (1)$$

where:
E_{WESD} - potential energy that can be recovered by the storage device [kWh],
E_{Breake} - total energy resulted from the braking vehicle depending on the mass, speed and brake force [kWh],
E_{Aux} - energy recovered by the braking vehicle and used for auxiliary supply (lighting, ventilation, heating etc.) [kWh],
E_L - energy line loses [kWh].

If we consider that the traction vehicle loses only mechanical energy and we have only kinetic energy (E_{Breake} = E_k) we can use the following formula to determine the energy resulted from braking:

$$E_k = m \times V^2 / 2 \quad (2)$$

where:
V - the vehicle velocity in [km/h],
m - the vehicle mass in [kg],
E_k - the kinetic energy in [kWh].

Figure 4 shows estimated braking energy for different vehicle mass and velocity calculated according with formula (2).

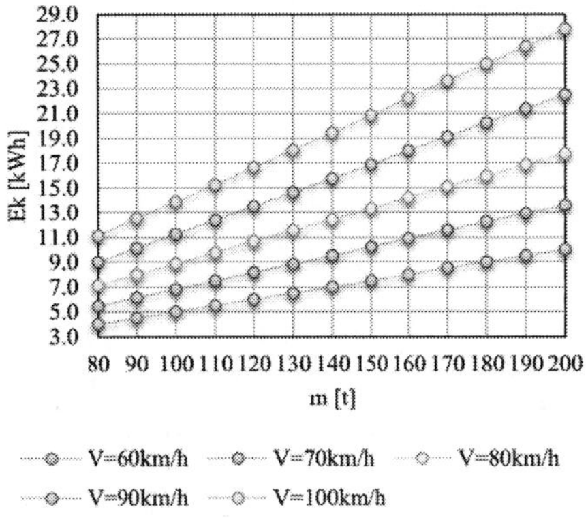

Fig. 4. Braking energy in kWh.

The power needed for the auxiliary train equipment can be around 40kW depending on the weather conditions (cold or hot), period of the day (day or night). We assume the energy recovered by the braking vehicle to be around 3kWh.

Depending of the location of the WESD and the braking current we can have different catenary losses.

$$\Delta E_L = R \cdot \int_0^t I^2 dt \quad (3)$$

where:
ΔE_L - energy losses in wires [W] function of time t [s],
R - resistance of the line [Ω],
I - the load current [A].

If we consider the energy loses in catenary around 10% of the braking energy 10.7 kWh according to figure 5 we have approximately 1.1kWh catenary loses for a maximum distance of 17.8km between the braking train and the WESD.

According to figure 4 and figure 5 the potential energy that can be recovered by the storage device is approximately 6.6kWh (10.7 kWh - 3 kWh - 1.1 kWh = 6.6kWh) considering that there is no other train to recover the braking energy which is around 60% from the braking energy that can be recovered.

The recovered braking energy can take values from 5% to 30% from the total consumed energy depending on the receptivity of the other trains to consume the energy. The energy for a train to travel from Station A to Station F is 511kWh. For the worst case scenario where the recovered energy is 5% we have 25kWh potential recovered energy divided between the catenary losses, auxiliary train equipment and the WESD.

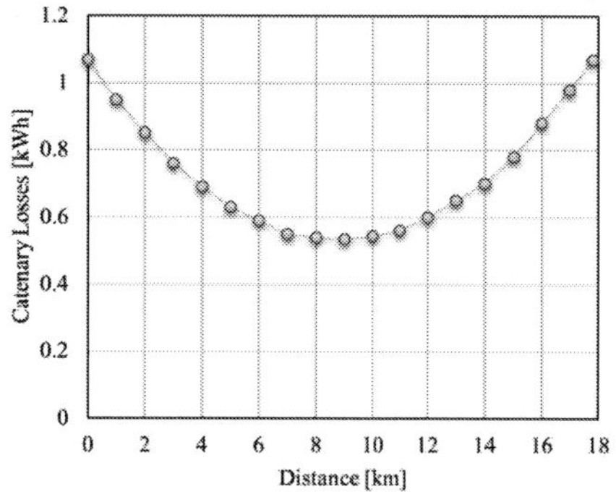

Fig. 5. Catenary loses for different location of the WESD in kWh.

From the above results the WESD can recover around 15kWh (60% from 25kWh) for a train travel between Station A and Station F. For 99280 cycles per year in both directions, theoretically we can save up to 1490MWh/year.

The kWh in Poland costs around 0.09 US $ for industry [15] which results on a potential saving of 134 000 US $ per year. For a 10 years calendar life of WESD system we can calculate the money saved as 1.34 million US $ (0.134million US $ * 10years = 1.34 million US $).

5 Designing of the energy storage devices

To directly connect the energy storage devices we need to design them to the voltage line to avoid electrical separation.

5.1 Lead acid batteries

The Hitachi LL1500-WS battery [16] has 2V and 1500Ah per cell with a recommended maximum discharging current of 900A, a maximum of charging current of 450A and energy capacity of 3kWh (1500Ah * 2V / 1000 = 3kWh). To connect the WESD to the 3.3kV we need 1650 batteries (2V * 1650batteries = 3.3kV) in series. The energy capacity of the unit is 5MWh (3kWh * 1650batteries = 5MWh). Because the maximum discharge current is 900A the power rating for the unit is limited to 3MW (3300V * 900A / 1000 = 3MW).

The number of cycles given by Hitachi for a 70% DOD is 4500 cycles. For 0.7% DOD number of cycles is 450 000 cycles and for 1.1% DOD number of cycles is 286 000 cycles. If we consider the calendar life of the battery to be 10 years, the number of cycles should be sufficient because on peak hours the braking energy is transferred between trains and the DOD is much smaller than calculated. For a DOD of 1.1% the maximum energy that can be saved by the WESD is 55kWh.

The price given by Hitachi Chemical for the battery is approximately 300 US $ per kWh [17]. If we consider the battery management system to cost 0.5 million US $ the WESD cost is around 2.5 million US $ (for 1650 batteries).

5.2 Nickel-metal hydride (Ni-MH) batteries

The Kawasaki Gigacell battery [10] has 36V (30cells * 1.2V = 36V) and 150Ah per battery with a charging/discharging recommended maximum current of 750A (5C) and energy capacity of 5.4kWh (150Ah * 36V / 1000 = 5.4kWh). To connect the WESD to the 3.3kV we need 91 batteries (36V * 91 batteries = 32730V) in series. The energy capacity of the unit is 491,4kWh (5.4kWh * 91batteries = 491,4kWh). The power rating for the unit at a 5C maximum charge/discharge rate is 2.46MW (3276V *750A / 1000 = 2.46MW).

If we want to capture all the braking energy we need to design the WESD capable of absorbing 40MJ. For 2 units connected in parallel we have 1.5kA maximum current at a 5C maximum charge/discharge rate per unit. The energy capacity of the system is 983kWh (491.4kWh * 2 = 983kWh).The power rating for the system at a 5C maximum charge/discharge rate is 4.9MW (3276V * 750A * 2 / 1000 = 4.9MW).

The number of cycles given by Kawasaki for a 0.33% DOD is 2 million. For 0.7% DOD number of cycles is 943000 cycles and for 1.8% DOD number of cycles is 367000 cycles. If we consider the calendar life of a battery as 10 years for our system the number of cycles should be sufficient because on peak hours the braking energy is transferred between trains and the DOD is much smaller than calculated. For a DOD of 1.8% the maximum energy that can be saved by the WESD is 17.7kWh.

For a battery price of 25000US $ and 0.5 million US $ cost assumed for the battery management system the cost of the WESD is around 5.1 million US $ (for 182batteries).

5.3 Lithium Ion (Li-ion) batteries

The Hitachi CH-75-6 battery [18] has 22.2V (3.7Vcell * 6cells = 22.2V) and 75Ah per pack with a charging/discharging recommended maximum current of 3C (225A) and energy capacity of 5kWh (225Ah * 22.2V / 1000 = 5kWh). To connect the WESD to the 3.3kV we need 148 batteries (22.2V * 148batteries =

3.3kV) in series. The energy capacity of the unit is 4.95MWh (5kWh * 148batteries = 740kWh). The power rating for the system at a 3C maximum charge/discharge rate is 4.9MW (3286V * 225A / 1000 = 739kW).

If we want to capture all the braking energy we need to design the WESD capable of absorbing 40MJ. For 4 units connected in parallel we have 900A maximum current at a 3C maximum charge/discharge rate for unit. The energy capacity of the system is 3MWh (740kWh * 4 = 3MWh). The power rating for the system at a 3C maximum charge/discharge rate is 3MW (3286V * 225A * 6/1000 = 3MW).

The number of cycles given by Hitachi for a 75% DOD is 4000 cycles. For 0.7% DOD number of cycles is 428000 cycles and for 1.1% DOD number of cycles is 273000 cycles. The calendar life of the battery given by Hitachi is 10 years the number of cycles should be sufficient because on peak hours the braking energy is transferred between trains and the DOD is much smaller than calculated. For a DOD of 1.1% the maximum energy that can be saved by the WESD is 33kWh.

The price given by Hitachi Chemical for the battery is approximately 900 US $ per kWh [17] (4500 US $ per module) excluding the battery management system that can be assumed to cost 0.5 million US $. If we consider a cost of 7000 US $ per module the WESD cost is around 4.8million US $ (for 592batteries).

5.4 Supercapacitors (SC)

The Maxwell BMOD0063 P125 B08 supercapacitors [19] has 125V (48cells), 140A continuous current and with a charging/discharging recommended maximum current of 1900A and energy capacity of 144Wh (3Wh * 48cells = 144Wh). Because the ESR can increase up to 100% during 10 years of service we have to reduce the continuous current to 100A. To connect the WESD to the 3.3kV we need 27 supercapacitors (125V * 27supercapacitors = 3.3kV) in series. The energy capacity of the unit is 3.9kWh (144Wh * 27supercapactiors = 3.9kWh). The power rating for the system at a 100A charge/discharge current is 0.38MW (3375V * 100A / 1000 = 0.38MW).

If we want to capture all the braking energy we need to design the WESD capable of absorbing 40MJ. For 10 units connected in parallel we have 1kA continuous current and a maximum charge/discharge current of 19kA. The energy capacity of the system is 39kWh (3.9kWh * 10 =3 9kWh).The system power rating for 1kA continuous current is 4.9MW (3276V * 1000A * 10 / 1000 = 3.3MW).

The number of cycles for charge/discharge given by Maxwell is 1 million times at 25° C with a live period of 10 years. The maximum energy that can be saved by the WESD is 39kWh.

The price for the BMOD0063 P125 B08 is approximately 5000 US $ not including the converter and the voltage equalizing circuit which can be estimated to cost around 0.5 million US $ the WESD cost is around 1.9million US $ (for 270 supercapacitors).

5.5 Flywheel

The Vycon VDC flywheel [20] has input voltage regulated between 400V to 600V, 500kW output power and can store up to 7.5MJ.

To connect the WESD to the 3.3kV we need 6 modules (600V * 6modules = 3.3kV) in series.

If we want to capture all the braking energy we need to design the WESD capable of absorbing 40MJ. For 6 units connected in parallel we have 45MJ (6units * 7.5MJ = 45MJ) and 3MW power.

The Vycon VDC flywheel has 20 year operational live with low maintenance (equipped with magnetic bearing).

The price for the flywheel is assumed 1 million US $ / MW and 0.5million US $ for the control circuit which give a total cost of the WESD 3.5 million US $.

6 Conclusion

In the article an example was presented on how to calculate the train braking energy. Simulation was provided with scope of calculation the energy needed for a train to run on an example line. The main cost for the WESD is the energy storage medium. The design and cost estimation was made for a period of 10 years calendar life given by the manufacturers for the batteries.

For a 10 years period the energy that can be saved for the example provided is 15 GWh which is 1.34 million US $. For 10 years life time the cost for the WESD with battery is 2.5 million US $ for lead acid, 5.1 million US $ for Ni-MH, 4.8 million US $ for Li-ion, 1.9 million US $ for WESD with SC and 3.5 million US $ for WESD with flywheel. When calculating the total cost of the WESD the installation, building premises, cooling and heating or maintenance were not taken into account; these factors will considerably increase the total costs.

The revised WESD from the example provided does not save enough energy to justify the installation by the customer. The most probable WESD that will be used on electrified heavy rail transport in the future will be with SC because it is the most cost effective.

References

1. P. Radu and A. Szelag, *A Cuk converter integrated with lead-acid battery and supercapacitor for stationary applications,* Proceeding of IEEE 18th International Scientific Conference on Electric Power Engineering (EPE), Kouty nad Desnou, Czech Republic, 1-6 (2017).
2. N. Ghaviha, J. Campillo, M. Bohlin, E. Dahlquist, *Review of Application of Energy Storage Devices in Railway Transportation,* The 8th International Conference on Applied Energy – ICAE, Energy Procedia, **105**, 4561 - 4568 (2017).
3. P. Radcliffe, J. S. Wallace, L. H. Shu, *Stationary Applications of Energy Storage Technologies for Transit Systems,* IEEE Electrical Power & Energy Conference, 1-7 (2010).
4. T. Ratniyomchai, S. Hillmansen, P. Tricoli, *Optimal Capacity and Positioning of Stationary Supercapacitors for Light Rail Vehicle Systems, International Symposium*

on Power Electronics, Electrical Drives, Automation and Motion, IEEE, 807-812 (2014).

5. H. Kobayashi, T. Saito, K. Kondo, *A Study on a Method to Design Energy Capacity of Wayside Energy Storage Devices in DC-electrified Railway Systems,* Power Electronics and Applications EPE ECCE-Europe, 1-9, (2015).
6. J. G. Yu, Martin P. Schroeder, D. Teumim, *Utilizing Wayside Energy Storage Substations in Rail Transit Systems – Some Modelling and Simulation Results,* APTA Rail Conference, 1-8 (2010)
7. P.T. Moseley, D.A.J. Rand, *In celebration of the sequicentennial of the lead-acid battery,* Journal Of Power Sources, **195**, 4423, (2010).
8. A. Okui, S. Hase, H. Shigeeda, T. Konishi, T.Yoshi, *Application of Energy Storage System for Railway Transportation in Japan,* Proc. Int. Conf. Power Electronics Conf. IPEC, Sapporo, Japan, 3117–3123 (2010).
9. M. Ouattara, J. G. Yu, *WMATA Energy Storage Demonstration Project*, FTA Report 0086 Federal Transit Administration, http://www.fta.dot.gov/research (2015).
10. http://www.khi.co.jp/english/gigacell/. Retrieved on 2017-06-18.
11. H. Takahashi, K. Aoki, A. Maoka, Y. Ik Kim, *Current and Future Applications for Regenerative Energy Storage System,* Hitachi Review, **61**, **7**, 336-340 (2012).
12. A. Szelag, T. Maciolek, Z. Drazek, M. Lewandowski, *Conditions for the effective use of traction substations with possible return of trains regenerative braking energy into a power network in a 3 kV DC supply system,* Logistyka **3**, 4683-4692 (2015).
13. A. Szelag, *The influence of the voltage in the 3kV DC* traction *network on the energy and traction parameters of supplied vehicles,* Wyd. INW Spatium, Radom, 158, (2013).
14. http://www.skm.warszawa.pl/dane-techniczne-pojazdow.html, Retrieved on 2017-06-18.
15. http://ec.europa.eu/eurostat/statistics-explained /index.php/Energy_price_statistics, Retrieved on 2017-06-18.
16. H. Kodoma, *Electrical energy storage devices & systems,* Hitachi Chemical Technical Report, **57**, 6-15 (2015).
17. M. Sawato, Credit Suisse, Asia Pacific / Japan, Equity Research, Specialty Chemicals (Chemicals/Textiles (Japan) / Market Weight, Hitachi Chemical, 4217 (2015).
18. M. Yoshida, T. Hirasawa, K. Sumiya. *Electrical energy storage devices & systems,* Hitachi Chemical Technical Report, **55**, 4-8 (2013).
19. http://www.maxwell.com/images/documents/125vmodule _ds_1014696-7.pdf, Retrieved on 2017-06-18.
20. http://www.ampower.com.tw/wp-content/ uploads/2014/pdf/brochure-VDC%20flywheel.pdf, Retrieved on 2017-06-18.
21. A.J.C. Schmeitz, *A Semi-Empirical Three-Dimensional Model of the Pneumatic Tyre Rolling over Arbitrarily Uneven Road Surfaces*, PhD thesis, TU Delft (2004).

 https://doi.org/10.1051/matecconf/201818004002

Dual Active Bridge (DAB) DC-DC converter for multilevel propulsion converters for electrical multiple units (EMU)

Marek Adamowicz [1,2,*], *Zbigniew* Krzemiński [2,3], and *Paweł* Stec [4]

[1]Gdansk University of Technology, Faculty of Electrical and Control Engineering, ul. Narutowicza 11/12, 80-233, Gdansk, Poland
[2]H. Cegielski-Energocentrum Sp. z o.o., ul. 28 Czerwca 1956 r. 223/229, 61-485, Poznań, Poland
[3]MMB Drives Sp. z o.o., ul. Maszynowa 26, 80-298, Gdańsk, Poland
[4]PESA Bydgoszcz S.A., R & D Department, ul. Zygmunta Augusta 11, 85-082, Bydgoszcz, Poland

Abstract. Semiconductor power devices made from silicon carbide (SiC) reached a level of technology enabling their widespread use in power converters. Two different approaches to implementation of modern traction converters in electric multiple units (EMU) have been presented in recent years: (i) 3.3-kV SiC MOSFET-based three-level PWM inverter with regenerative braking and (ii) 6.5-kV IGBT-based four-quadrant power electronic traction transformer (PETT). The former has successfully reached optimized dimensions and efficiency but still requires a bulky line frequency transformer for multisystem applications. The latter characterizes inherent galvanic isolation from AC traction, which is realized by cascaded system of power electronic cells containing medium frequency transformers (MFT). The downsizing of the 6.5-kV IGBT-based cells is, however, problematic. The present paper proposes a different approach, that involves the use of a fast switching 1.2-kV SiC MOSFETS. The SiC-based PETT proposed in the paper is dedicated first for the DC traction. For multi-system application the input voltage of the proposed PETT can be adjusted using weight-optimized adjusting autotransformer. Thanks to utilization of fast-switching SiC-based power modules the weight and size of the power electronic cells can be optimized in a convenient way.

1 Introduction

Modern semiconductor devices allow realization of traction power converters with minimized weight and size, and high efficiency [1]. Silicon carbide (SiC) semiconductor material is treated as a direct successor to silicon (Si). The high breakdown electric field, low specific on-resistance, high operating frequency and high junction temperature of SiC transistors are beneficial for the efficiency, power density, specific power and reliability of traction power converters. SiC *Metal Oxide Field Effect Transistors* (SiC MOSFETs) enable lower system costs by providing the ability for increasing power density and frequency of operation, thereby reducing the volume and weight of the power converter [2]. To increase the number of applications, that SiC MOSFETs can serve, manufacturers of these devices have extended their capability to higher voltages [3], [4]. A megawatt-scale traction inverter using 3.3 kV SiC MOSFETs, made by Mitsubishi company, has been successfully used in Japanese EMU trains powered from 1500V DC traction. The three-level PWM inverter with regenerative braking using 1.5 kA SiC MOSFETs has been applied to Shinkansen train, providing a 2.5 kV AC voltage for the group of 305 kW power, self-cooling traction motors, operating in parallel.

The 3.3 kV SiC-based power modules have over 70% less switching losses than Si-based GTO thyristors used earlier. As it was reported in [5], the use of SiC modules in N700 Shinkansen bullet train reduced the size and weight of its inverter system by approximately 55% and 35%, respectively, compared to inverters using GTOs.

In the 3kV DC traction network, either 6.5kV IGBT-based two-level inverters or multilevel inverters using 3.3.kV switches can be used. Multilevel inverters characterize lower dv/dt slew rates than two-level inverters and can operate at higher resulting frequency without augmented electromagnetic interference (EMI) emission. Multisystem EMUs operated in Europe within different traction systems: 3kV DC, 25kV/50Hz and 15kV/16.7Hz have on-board power distribution systems, which are reconfigured on-line [6] by using high voltage switchboards and multi-winding traction transformers. Three main goals of multi-system EMU design are selected. They are time to reach full speed operation of the train (acceleration improvement), energy saving and weight and dimension reduction of the electrical equipment. The limiting factor for the EMU train's acceleration can be overcome by increasing the number of powered axles, which, however, binds with increasing cost of the system. More serious limitations apply to electrical equipment downsizing problem. The low

* Corresponding author: marek.adamowicz@pg.edu.pl

frequency transformer (LFT) is the heaviest and bulky constituent of the multi-system rolling stock and it represents the worst efficiency between different main components of traction drive. Traditionally, LFT occupies 12% of the weight and volume of rail vehicles. Nevertheless it is 82% smaller than its infrastructure on-ground counterpart. Today, one the most important evolution to be realized on traction transformer is the power electronics on-board traction transformer (PETT) [7], [8]. The power electronics on-board transformer combine several high voltage AC-AC converters connected in series and directly connected to catenary, several high-frequency transformers (HFT) providing galvanic isolation and several rectifiers, each connected to the secondary side of one high frequency transformer, which supplies usual DC-link or several independent DC-links used by other on-board converters. On-board PETT provides control functionality which are far beyond conventional LFT capabilities. Modular approach can cope with any grid voltage by increasing the number of cascaded cells. Cascaded full-bridge topology is actually the most popular solution for constructing PETT prototypes. However the connection of full-bridge converters at the high-voltage side, e.g. 25kV or 15kV is still a challenge. Unlike full-bridge converters, the three-level neutral point clamped (NPC) converter [5] does not provide modularity. Bi-directional three-level NPC converters cannot be scaled to 15kV or 25kV voltage level, except for using of direct series connection of power semiconductors [9].

The world's first PETT installed in the Ee 933-Type shunting locomotive fed from 15-kV/16.7-Hz traction has shown in [10]. The PETT system described in [10] is constructed from isolated AC-DC-DC cells, which includes resonant DC-DC converters. Each power-electronic cell is made of two types of power modules: high voltage and low voltage. Application of different voltage level subassemblies causes increased complexity of the system. The high voltage PEBBs consist of bulky 6.5kV IGBTs connected to 3.6kV dc-link, while the faster, low voltage 3.3kV IGBTs are connected to dc-link of the propulsion converter. Very slow speed of high voltage IGBTs limits the operating frequency of HFTs. The PETT system in [10] needs, at minimum, N = 8 cells. Selection of 6.5-kV IGBTs for converter design also entails unfavourably high cost of capacitors, inductive components, protection devices and the whole system.

The present paper aims the development of another type of isolated modular traction converter for multi-system EMUs [11]. The essence of the proposed approach is to apply 1.2kV SiC power modules to construct small-size and fast-switching PETT operating at full DC traction voltage and partial AC traction (25kV or 15kV) voltage.

For AC traction application the inherent galvanic isolation of the proposed converter need not be duplicated by bulky line-frequency transformer. Instead, a weight-optimized autotransformer can be used for adjusting the input voltage of the converter. First of all, the basic topology which can be directly connected to DC traction grid is presented. For multi-system applications the whole system of PETT with LF autotransformer is proposed. In Section II the very recent achievements in the power electronic transformer technology are described. Section III shows a SiC-based Dual Active Bridge (DAB) DC-DC converter as the key component of SiC-based PETT. Experimental results of the proposed SiC-based DAB converter are given in section IV. Discussion and conclusions are covered in section V.

2 Recent achievements in power electronic transformer (PET) industrial applications

Electrical drives consume 64% of all electricity used in industry. Use of frequency converters with electronic speed control providing recuperation of brake energy into the system leads to save energy and increase efficiency. Many medium voltage (MV) variable speed drive applications need one or two LF transformers matching voltage levels in the system. In the oil industry, typical offshore installations of megawatt scale electric submersible pumps (ESP) have step down transformer that supplies low voltage variable speed drive (VSD) and step up transformer providing MV for the high horsepower (above 1000 HP) ESP motor. With the power electronic transformer (PET), supply line voltage up to 6.6kV does not need to be stepped down, furthermore is no requirement for a multi-tap step up transformer as the VSD can provide the required medium voltage required for the ESP motor. The removal of LF transformers has reduced the space and weight of the surface electrical equipment, which has a significant influence on the size and load bearing capabilities of new installations [12].

Application of PET topology to industrial drives leads to several advantages such as: (*i*) power scalability, (*ii*) ability to shape output voltage with significantly reduced harmonic content and, hence, operation with small size passive filter and (*iv*) ease of maintenance and inspections. In [12] the benefits of application of PET in the oil industry has been presented. The typical state-of-the-art surface equipment containing two low-frequency transformers occupies significant deck space, of about 10 m^2. The PET-based VSD doesn't need any additional transformers and makes it possible to obtain advantageously smaller footprint. Hence, it gives possibility to upgrade the surface equipment in terms of higher power or possibility of installing additional drive on the same used surface.

The first PET-based VSD has been installed in Saudi Aramco fields in 2014 [12], [13]. The overall scheme of IGBT-based 3.3kV/3.3kV PET is shown in Fig. 1 [13]. The PET topology features isolated dual active bridge (DAB) DC-DC converter technology [14] with a PWM multilevel cascaded full-bridge rectifier at the input and PWM multilevel cascaded full-bridge inverter at the output of the system [15]. Each phase leg of the PET from Fig. 1 consists of 3 cascaded cells. The full-bridge converters in Fig. 1. are constructed from 1.7kV IGBTs and DAB DC-DC converters utilize 1:1 HFTs. The use of universal modules on the primary and secondary side reduces complexity of the system and makes easier maintenance of the system and reduces production costs when compared to 15kV PETT from [10].

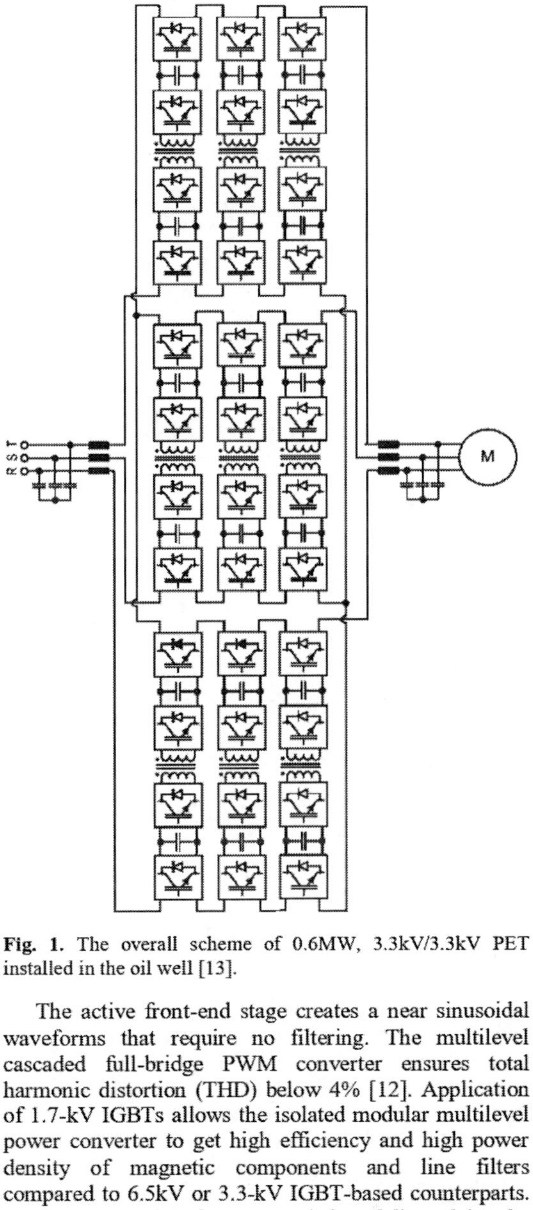

Fig. 1. The overall scheme of 0.6MW, 3.3kV/3.3kV PET installed in the oil well [13].

The active front-end stage creates a near sinusoidal waveforms that require no filtering. The multilevel cascaded full-bridge PWM converter ensures total harmonic distortion (THD) below 4% [12]. Application of 1.7-kV IGBTs allows the isolated modular multilevel power converter to get high efficiency and high power density of magnetic components and line filters compared to 6.5kV or 3.3-kV IGBT-based counterparts. The cleaner, spike free power being delivered by the PET converter along the long cable to the asynchronous motor installed from 2km to 5km underground helps to ensure the longest possible ESP run life by minimizing the degradation of the ESP electrical system insulation [12]. The PET converter having NEMA-4 enclosure to protect against windblown dust and water, occupies twice smaller deck space than existing LV drives supplied from line frequency transformers and is capable of uninterrupted operation at high summer temperatures of up to 55°C.

Application of fast switching 1.2-kV SiC MOSFETs, enables the PET-based variable speed drives to obtain improved functionality. Fig. 2. Shows SiC-based successor of PET from Fig. 1. The increase in frequency of HFTs from 8 kHz to 25 kHz allows noiseless operation of the PET and HFTs downsizing. The 1MW, 4kV/4kV PET, shown in Fig. 2, has been constructed from 1.2-kV SiC MOSFETs and consists of six cascaded cells in each phase.

Fig. 2. Air-cooled 4kV/4kV PET, constructed from 1.2kV SiC MOSFET power modules (*Photo courtesy of MMB Drives*)

3 Novel autotransformer-based PETT for multi-system EMUs

Replacement of bulky 6.5kV IGBTs by SiC-based counterparts in the PETT system from [10] will not completely eliminate the problem of limited frequency of HFTs. High-voltage SiC devices suffer from high dv/dt and di/dt slew rates required to limit the switching losses. Fast 1.2kV SiC-MOSFETs give more freedom and the ability to design a noiseless device with less EMI emissions. First of all the DC-AC PETT for 3kV DC traction is proposed. The overall scheme of the compact traction converter combining multilevel propulsion inverter and auxiliary power converter in the same housing is shown in Fig. 3. The proposed PETT topology consists of: (*i*) filter inductor, (*ii*) cascaded connection of isolated DC-DC converters including input filter capacitors and HFTs operated with high frequency, above 20kHz, (*iii*) the three phase propulsion inverter realized as cascaded full-bridge multilevel converter and (*iv*) integrated auxiliary converter.

The propulsion inverter features a cascaded full-bridge multilevel DC-AC converter topology. Each phase of the propulsion inverter consists of three full-bridges connected in series. A cascaded topology having 3 isolated input sources in each phase provides 7 levels to synthesize the AC output voltage waveform. This 7-level voltage waveform enables significant reduction of harmonics in the synthesized output current feeding the traction motor. The high frequency isolated dual active bridge (DAB) converters galvanically separate the propulsion inverter and auxiliary power converter from traction grid. In the PETT from Fig. 3 the DC traction voltage u_{dc} is divided between five DC-link capacitors of DAB DC-DC converters.

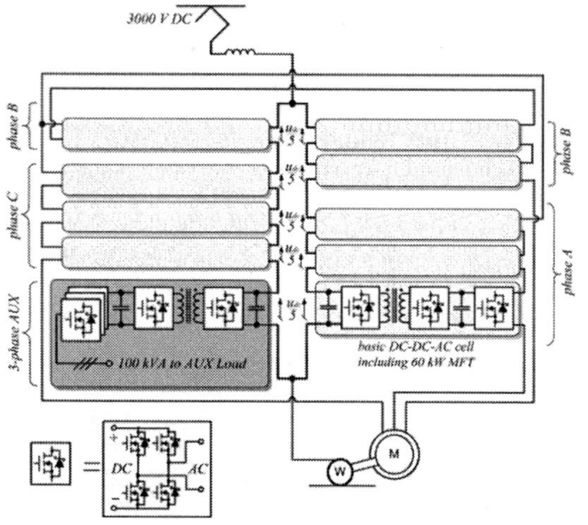

Fig. 3. The proposed DC-AC PETT

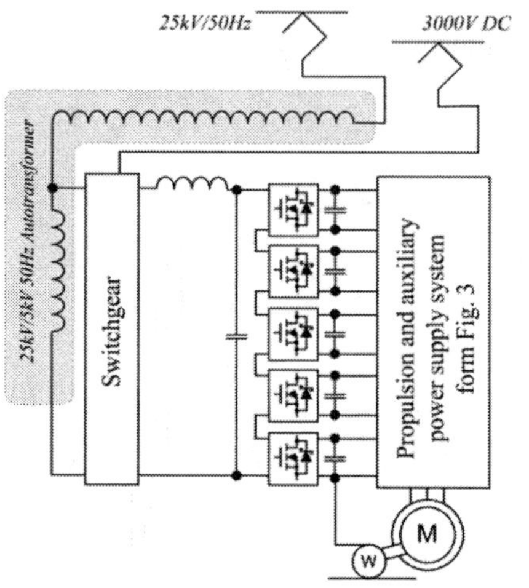

Fig. 4. Functional scheme of the proposed autotransformer-based traction converter

One fifth of the actual DC traction voltage is connected to each DAB converter. Five DAB DC-DC converters connected in series, step up or down the input voltage voltages during voltage transients in DC traction line. Two sets of DC-DC-AC converters operate within the compact traction power converter. As the propulsion inverter use 9 isolated power cells, the one remaining isolated DAB DC-DC converter serve as isolating stage for the 3x400V auxiliary power converter. The concept the traction converter shown in Fig. 3 can be extended to multi-system version. The replacement of the isolated DC-DC-AC cells with four quadrant isolated AC-DC-DC-AC counterparts is therefore required. The inherent galvanic isolation of the traction converter from Fig. 3 can be utilized in multisystem converter for galvanic isolation from AC traction grid. To cope with the problem of connection of PETT with the AC traction voltage (15kV or 25kV), a low-frequency autotransformer, shown in Fig. 4, is proposed. According to the proposed concept, the four quadrant isolated AC-DC-DC-AC cells, connected in cascade, form the three-stages of the PETT: (*i*) the single phase multilevel cascaded full-bridge rectifier connected to the voltage-adjusting autotransformer, (*ii*) isolated DC-DC stage and (*iii*) three phase multilevel cascaded full-bridge propulsion inverter integrated with auxiliary power converter.

The concept of the autotransformer-based traction converter for the multisystem EMUs is shown in Fig. 5.

The advantages of the proposed system are: expected lower weight of the LF autotransformer when compared to HFT, lower costs, increased reliability and flexibility and preferably lower complexity when compared to existing PETT [10]. The design and optimization of the line frequency adjusting autotransformer is out of scope of the present paper and will be described in further publications.

4 SiC-based Dual Active Bridge converter as a key component of PETT

The key component of the proposed PETT, which is the isolated DAB DC-DC converter (Fig. 5), has been investigated experimentally.

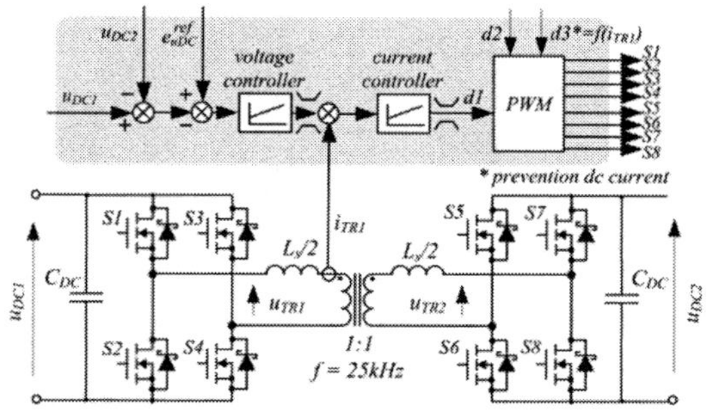

Fig. 5. Isolated dual active bridge DC-DC converter under investigations.

The 400kW, 2340V asynchronous traction motor has been assumed for further investigation. According to the maximum power of 500kW, the nominal power of the designed DAB DC-DC converter is about 60 kW. The investigated DAB converter from Fig. 5 has been built from CAS120M12BM2 power modules characterizing continuous drain current of 138A at 90 Celsius degrees. The DAB DC-DC converter enables bi-directional power flow which is controlled using phase shift control. The DC voltages: u_{DC1} on primary side and u_{DC2} on secondary side are converted into high frequency rectangular pulses, with constant or modulated pulse width.

The transferred power depends on the mutual phase shift d_1 ratio between primary and secondary voltages u_{DC1} and u_{DC2}. For pure phase-shift control, when the voltage pulse duty cycle is fixed at 50%, the transferred power is described by the following relation:

$$P = \frac{u_{DC1} \cdot u_{DC2}}{2 \cdot f \cdot L_s} \cdot d_1 \cdot (1 - d_1) \qquad (1)$$

where f =25kHz denotes switching frequency of the SiC MOSFETs. Fig. 8 shows the output power as a function of the phase shift d_1. For dual phase-shift control voltage pulse width can be adjusted second phase shift parameter d_2 between pair of switches S_i, S_{i+2} (Fig. 5), $i = 1, 2, 5, 6$. Phase shift d_2 modulates the voltage pulse width and can be used for shaping of the transformer current (loss minimization) and for prevention from the core saturation.

Fig. 6 shows the characteristic waveforms of DAB converter: the secondary transformer voltage u_{TR2}, secondary transformer current i_{TR2} (inverted) and voltage of S_2 with half load (30kW). The commercially available high frequency 100kW transformer (oversized) with N87 UIIU/93/245/30 core has been used in the first experiment. Two identical series inductors LS/2=10μH consisting of ferrite cores and Litz wires were connected to the transformer to adjust the inductance value Ls which affect the power flow. A significant ringing can be observed in the transformer voltage and current. It can be deduced that their magnitude will increase with increased power. For the second experiment an optimized high frequency transformer has been designed within the present project. Litz wire coils, N87 core material and insulation resin with optimized temperature conductivity coefficient have been used in the design. Fig.7 shows characteristic waveforms of 60 kW DAB converter at half load: secondary transformer voltage u_{TR2}, secondary transformer current i_{TR2} and S_1 switch voltage. The ringing in transformer voltage and current have been reduced significantly. The measured dv/dt value during SiC transistor switching in Fig. 7b is 3.5kV/μs and the basic frequency of the SiC transistor voltage oscillations is 30 MHz. Fig. 9 and Fig. 10 show the results of the temperature test. The investigated DAB converter operated with full load of 60kW.

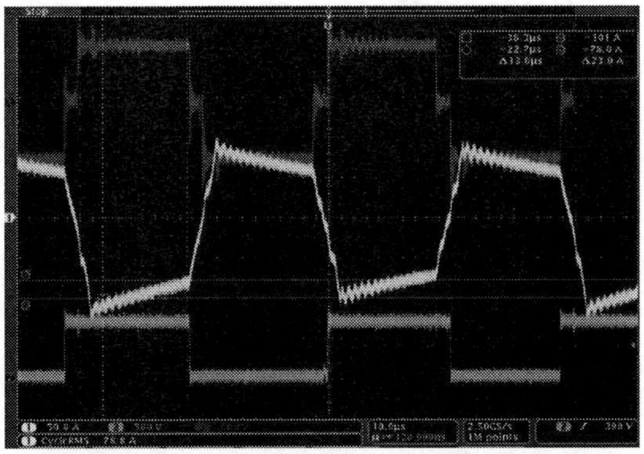

Fig. 6. Characteristic waveforms of DAB converter with commercially available high frequency transformer at half load (30kW).

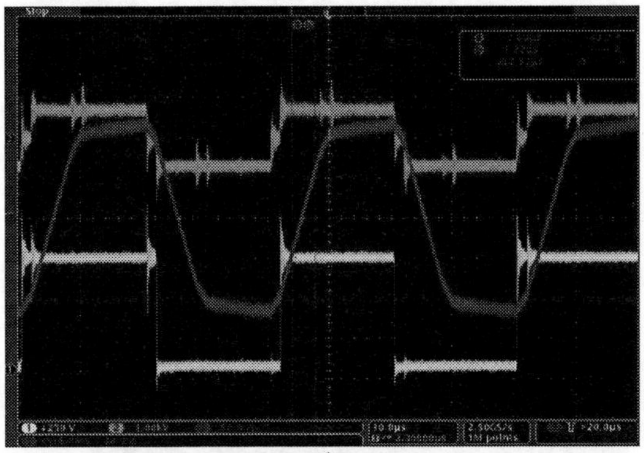

a)

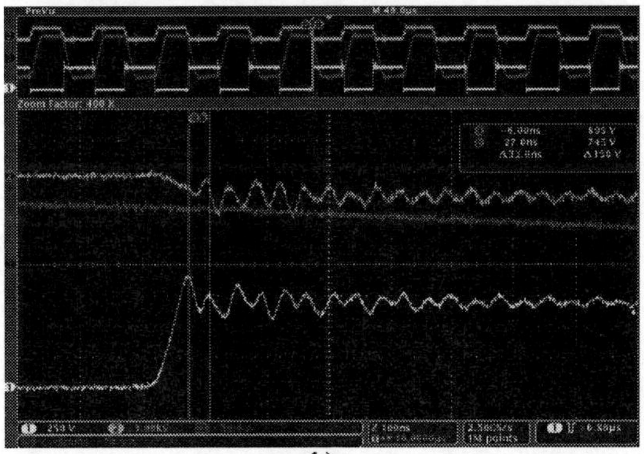

b)

Fig. 7. Characteristic waveforms of DAB converter with 60kW HFT optimized for the project, at half load (a) and transients during SiC switching (b).

5 Conclusions

The concept of novel autotransformer-based PETT for multi-system EMUs was proposed. The investigated SiC-based PETT features noiseless operation and galvanic isolation realized with the use of high-frequency transformers. Special attention was given on isolated DAB DC-DC converter, which provides power flow control with high efficiency.

The investigated isolated DAB DC-DC converter, constructed from 1.2kV SiC MOSFETs demonstrates 60kW transferable power using 25 kHz switching frequency. The measured temperature of heatsinks did not exceed 55 Celsius degrees during constant load operation, which is a promising result when taking into account application in roof-mounted container.

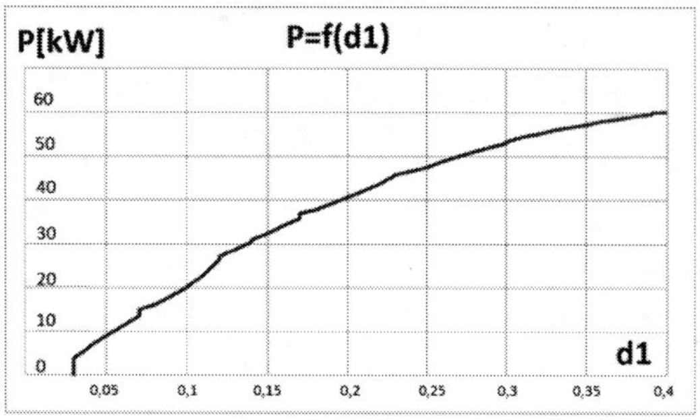

Fig. 8. Output power in function of phase shift d1.

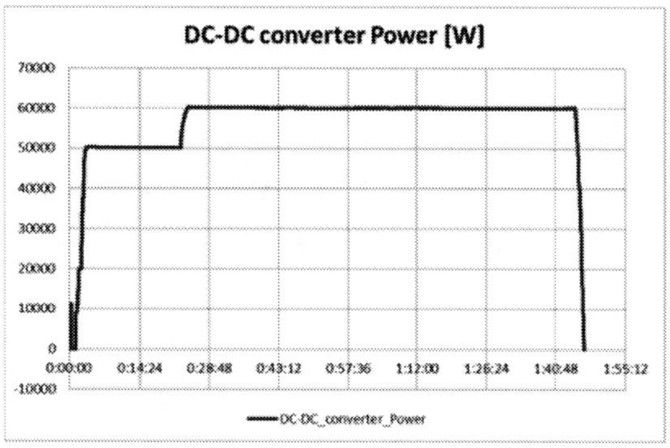

Fig. 9. Output power during temperature test.

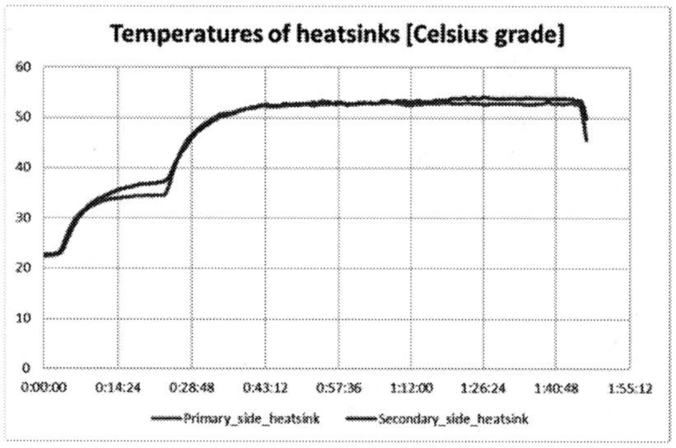

Fig. 10. Temperatures of DAB heatsinks.

The next step of investigations should be detailed EMC analysis on compliance with railway standards.

This scientific work is financed from resources for science as a Research Project of the National Centre for Research and Development under agreement POIR.01.02.00-00-0193/16-00.

References

1. Drofenik U., Canales F., European Trends and Technologies in Traction, *Proceedings International Power Electronics Conference IPEC,* 1043–1049, (2014).
2. M. K. Das, *SiC MOSFET Module Replaces up to 3x Higher Current Si IGBT Modules in Voltage Source Inverter Application*, Bodo's Power Systems, **2**, 22-24, (2013), www.bodospower.com
3. Pala V., van Brunt E., Casady J., *Simplifying power conversion with medium voltage SiC MOSFETs*, Compound Semiconductor, **3**, 44–47, (2016),
4. Negeshi T., Tsuda R., Ota K., Iura S., Yamaguchi H., 3.3-kV All-SiC Power Module for Traction System Use, *Proceedings Int. Conf. PCIM Europe*, 51-56, (2017)
5. Mitsubishi electric corporation news releases No.2942, *Mitsubishi Electric Installs Railcar Traction System with All-SiC Power Modules on Shinkansen Bullet Trains*, June, 25 (2015)
6. *BORDLINE® CC1500 MS. High power multi-system propulsion converter,* Product data sheet, ABB (2012)
7. Vetterli C., Suranyi A., *Concept et potentiel d'économie d'énergie d'un transformateur de traction électronique "PETT"*, Rapport annuel, ABB Sécheron SA (2013)
8. Chamaret A.P., Frugier D., Future of on-board traction transformer for rolling stock, *Proc. 11th World Congress of Railway Research*, 1-6, (2016)
9. Huber J. E., Kolar J.W., *Optimum Number of Cascaded Cells for High-Power Medium-Voltage AC–DC Converters*, IEEE Journal Emer. Select. Top. Power Electr., **5**(1), 213-232, (2017)
10. Zhao C., Dujic D., Mester A., Steinke J.K., at al.: *Power Electronic Traction Trans-former–Medium Voltage Prototype*, IEEE Transactions on Industrial Electronics, **61**(1), 3257–3268, (2014)
11. Adamowicz M., Krzemiński Z., *Multi-system locomotive feeding system*, Polish Patent No PL 226553 B1
12. Orlowski T., Radcliffe Alan R., Application of New Generation Variable Speed Drive Solutions To Reduce Footprint and Operating Cost, Proceedings SPE Middle East Artificial Lift Conference, 1-9, (2016)
13. Adamowicz M., Krzemiński Z., *Medium voltage (MV) modular multilevel converters*, Automatyka Elektryka Zakłócenia, **5**(3), 56-71, (2014) in Polish, available online.
14. Inoue S., Akagi H., *A Bidirectional Isolated DC–DC Converter as a Core Circuit of the Next-Generation Medium-Voltage Power Conversion System*, IEEE Transactions On Power Electronics, **22**(2), 535–542, (2007)
15. Lewicki A., *DC-link voltage balancing in cascaded H-Bridge converters*, Archives of Electrical Engineering, **63**(3), 439-455, (2014)

Experimental verification of visual method for measuring displacements of contact line elements

Jacek Skibicki[1,*]

[1]Gdansk University of Technology, Institute of Electrical Engineering, Division of Electrical Engineering in Transport, ul. Narutowicza 11/12, 80-233, Gdańsk, Poland

Abstract. The increase of rail vehicles speed, as well as the increase of their power, puts high demands on the power delivery system for traction vehicles The most critical point in the vehicle's power supply circuit is the contact between the current collector and contact wires. Ensuring a reliable co-operation of the current collector and contact line, requires technical development of the contact line and current collector construction, as well as improvement of diagnostic systems of these components. This is translated into the need to develop new measurement methods, which allow for more complete definition of the parameters of these elements of traction vehicle power supply system. Each new measurement method requires verification, performed by calibrating or comparing the results with another measurement method which has already been tested. The article presents results of experimental verification of a visual method for measuring the displacement of contact wires, with a proven method based on short range laser rangefinders and calliper as reference measurement.

1 Introduction

Diagnostics of elements of the traction vehicles power supply can be divided into diagnostics of the contact line itself, diagnostics of current collectors and monitoring of proper cooperation of these elements. Diagnostic stands of the contact line are constructed in the form of mobile diagnostic wagons or other vehicles [1, 2]. Measurements of contact line parameters are often performed by using special measurement current collectors [1]; visual, contact-less methods, where the measuring system does not affect the contact line are also known [3-6]; sometimes measurements performed with the use of a collector are supplemented and supported by visual techniques [7, 8]. In contrast, diagnostics of current collectors is usually performed with the use of stationary diagnostic stands, where static characteristics of the collector and suspension of the slipper are tested [9]. Evaluation of the technical condition of strips is carried out using 3D visualization devices [10-13]. On the other hand, correctness of cooperation between the collector and the contact line can be monitored with the use of stationary measurement points which help to determine the correctness of the collector's adjustments based on displacements of the contact line caused by its impact when a vehicle passes through the diagnostic stand. In order to measure contact line displacements, short range laser rangefinders or other optical devices are most commonly used. They make it possible to evaluate movements of the contact wire in the vertical axis [14, 15].

The proposed method based on a stationary 2D camera and advanced image analysis allows for the measurement of the position (displacement) of contact wires in a contact line, or of other elements, in a two-dimensional plane. For monitoring the cooperation of the current collector and the contact line, measuring displacements of contact wires in a two-dimensional plane gives more possibilities, compared to the existing stands, which allow only for measurements in the vertical axis. In addition to determining the condition of adjustment of the collector's clamping force, it is possible to evaluate, to some extent, the technical condition of the slipper strip.

Each new measurement method requires verification, performed by calibrating or comparing the results with another measurement method which has already been tested. This article presents the results of experimental verification of the above-mentioned measurement method for displacements of contact wires in the catenary network with a verified method using short range laser rangefinders.

2 Subject of measurement

The subject of measurement is the displacement of the contact wire of contact line, caused by external force coming from the current collector or from a different source. The displacement range determined on the basis of previous measurements was set at ± 10 cm in both axes, with the possibility of exceeding it by 2 cm in each direction, to allow for recording of displacements deviating from the typical ones. Frequency of oscillation

* Corresponding author: jacek.skibicki@pg.edu.pl

 https://doi.org/10.1051/matecconf/201818004003

courses of the contact line does not usually exceed a few Hz [14]. To represent the course of changes correctly, it is assumed that the sampling rate of the signal should not be less than 100 Hz.

3 Methodology of measurement

In the first stage, comparative measurements of oscillations of the contact wire of contact line caused by external force were performed with the use of a laboratory test stand, whose diagram is shown in Fig. 1.

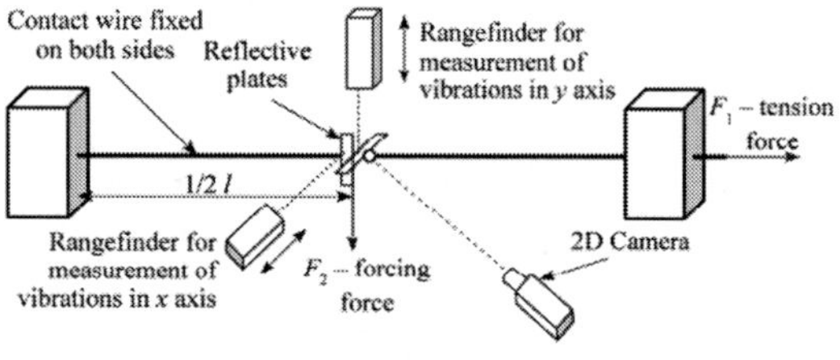

Fig. 1. Diagram of stand used for comparing results obtained with the use of two measurement methods.

A section of typical contact wire type Djp100 used by the PKP, with the length of $l = 28.8$ m, was fixed on both sides and tensioned with the force of $F_1 \approx 10$ kN. Mid-way along the length l, the forcing force F_2 was applied and, after its peak removal, oscillatory displacement of the contact wire was recorded using both measurement methods at the same time.

The second stage of the comparison was to test which method gives more accurate results, i.e. the results with a lower level of measurement uncertainty. For this purpose, measurements were performed at the stand presented in Fig. 2.

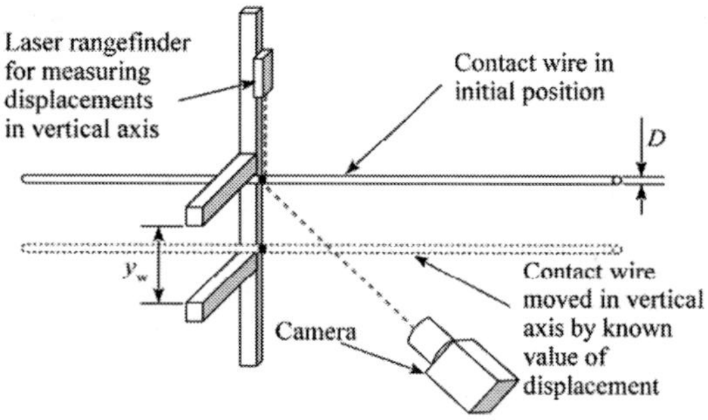

Fig. 2. Stand used for comparing accuracy of measurement methods.

The contact wire was moved by a known value in the vertical axis, and this displacement was recorded using a laser sensor and the camera system. The results were compared by way of a reference measurement performed with a calliper. In all the cases, the measurement uncertainty analysis was conducted, and all the indirect measurements were made with the use of high quality laboratory instruments. Visual measurements were made for the configuration of the stand where the uncertainty analysis showed that the highest measurement accuracy would be achieved.

Detailed information on each measurement is given below.

4 Measurements performed with the use of laser rangefinders

The measurements were performed with the use of short range laser rangefinders. Due to torsional fluctuations of the contact wire, it was necessary to use a third rangefinder to compensate for those fluctuations. The principle of compensation is presented in Fig. 3.

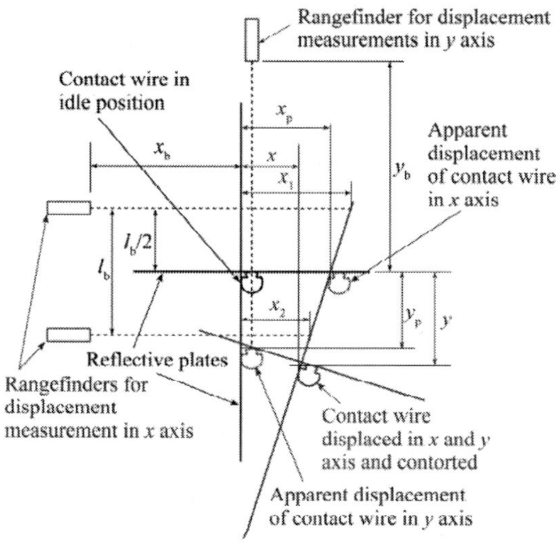

Fig. 3. Compensation of torsional vibrations with the use of a third laser rangefinder.

In the system presented in Fig. 3., in the idle state, the contact wire location is determined as the indications x_b and y_b. If no torsion compensation was applied, then, after the wire had been moved from the idle position, the measurement system would have indicated its displacement with the value of x_p and y_p, which are significantly different from the actual displacement of x and y. The use of an additional, third laser rangefinder in the horizontal axis gives the possibility of compensating the indications, so that, despite the torsion of the wire, the result is consistent with its actual displacement. For the presented geometrical layout, the dependence involving displacements in the horizontal axis x is as follows:

$$x = \frac{\frac{x_1 + x_2}{2} - y_p \cdot \left(\frac{x_1 - x_2}{l_b} \right)}{1 + \left(\frac{x_1 - x_2}{l_b} \right)^2} \tag{1}$$

For the vertical axis, the following formula has been derived:

$$y=y_p+\left[\frac{\frac{x_1+x_2}{2}-y_p\cdot\left(\frac{x_1-x_2}{l_b}\right)}{1+\left(\frac{x_1-x_2}{l_b}\right)^2}\right]\cdot\left(\frac{y_1-y_2}{l_b}\right) \quad (2)$$

where: x_1 – change of the indication of the first rangefinder measuring the displacement in the horizontal axis x; x_2 – change of the indication of the second rangefinder measuring the displacement in the horizontal axis x; y_p – the apparent displacement of the contact wire in the vertical axis y (change of the indication of the rangefinder measuring the displacement in this axis); lb – the distance between the rays emitted by the rangefinders recording the displacement in the horizontal axis.

During measurements, the electrical signals from the three rangefinders were recorded using an oscilloscope and then processed in a spreadsheet, in order to obtain the desired measurement results. Displacements in the vertical axis were recorded by a SICK rangefinder type OD250-150P142, whose measuring range is 100 to 400 mm. Two rangefinders type OD130-50P142, also made by SICK, with the measuring range from 80 to 180 mm, worked in the horizontal axis. A Tektronix TDS 540 oscilloscope, whose recording speed was set at 400 samples per second, was used to record the results.

5 Measurements performed with the use of a visual system

A visual system based on a high-speed 2D imaging camera, together with a system for recording advanced analysis of the image was used for measurements. The configuration of the visual measurement stand is presented in Fig. 4.

For the spatial configuration of the stand shown in Fig. 4, the displacement of the contact wire in the horizontal axis x is defined by the dependence:

$$x=\frac{(k-F)\cdot x'\cdot\cos\beta}{\cos\alpha\cdot(F\cdot\cos\beta-y'\cdot\sin\beta)-x'\cdot\sin\alpha} \quad (3)$$

Similarly, for the vertical axis y the formula is:

$$y=\frac{(k-F)\cdot(y'+x'\cdot\sin\beta\cdot\tan\alpha)}{F\cdot\cos\beta-y'\cdot\sin\beta-x'\cdot\tan\alpha} \quad (4)$$

The measurements were performed with the use of Basler's 2D imaging camera, type acA 2040-180 kc, with the following basic technical specifications:

- sensor resolution: 2046x2046 px (4 Mpix);
- sensor size: 11.26x11.26 mm;
- maximum recording speed: 180 fps.

The camera was equipped with a Jupiter 9 2/85 lens, whose rated focal length is f = 84.944 ± 0.078 mm (all the uncertainties presented in the article are standard uncertainties).

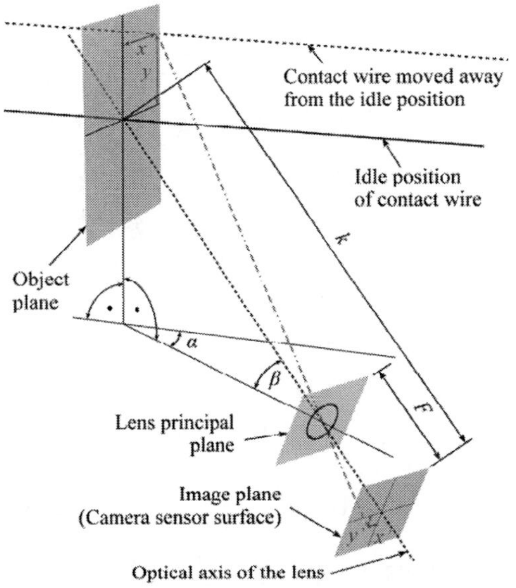

Fig. 4. Compensation of torsional vibrations with the use of a third laser rangefinder Configuration of the measurement stand, where: x, y – displacement of the contact wire in the horizontal and vertical axis respectively; x', y' - displacement of the image of the object on the camera sensor in the horizontal and vertical axis respectively, k – the distance between the central point of the object plane and the image plane, F – the distance between the lens principal plane and the camera sensor (plane of the image); α – the angle between the vertical plane in which the contact wire lies and the vertical plane in which the optical axis of the lens lies; β – the angle of inclination of the optical axis of the lens in relation to the level.

The measurements were performed for the following stand configuration:

- distance k = 2742.46 ± 0.24 mm;
- distance F = 87.751 ± 0.083 mm;
- angle value for α = 52.750 ± 0.029°;
- angle value for β = 26.80 ± 0.28°.

The measurements were performed at the maximum recording speed of the camera, which is slightly higher than the declared one, i.e. 187 fps. The recorded images were subjected to advanced processing in LabView in order to obtain the measurement results for x' and y', based on which, in accordance with the dependencies (3) and (4), the target measurement result, i.e. the displacement of the contact wire in the horizontal and vertical axis, was determined.

6 Measurements results

6.1 Comparative measurements for contact wire oscillation

A fragment of an exemplary course of displacement of the contact wire of the catenary network in the horizontal axis x is shown in Fig. 5. Similar measurement performed for displacements in the vertical axis y are shown in Fig. 6.

 https://doi.org/10.1051/matecconf/201818004003

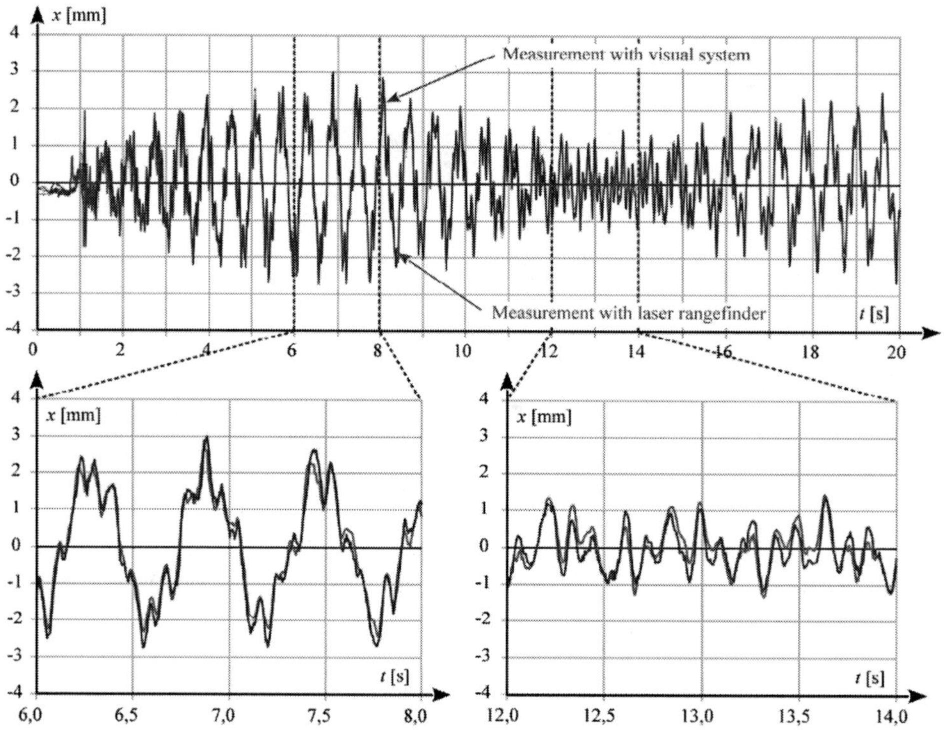

Fig. 5. Comparison of the results of measurements of the catenary network contact wire displacement in the horizontal axis *x*, performed with a laser rangefinder system and a visual system.

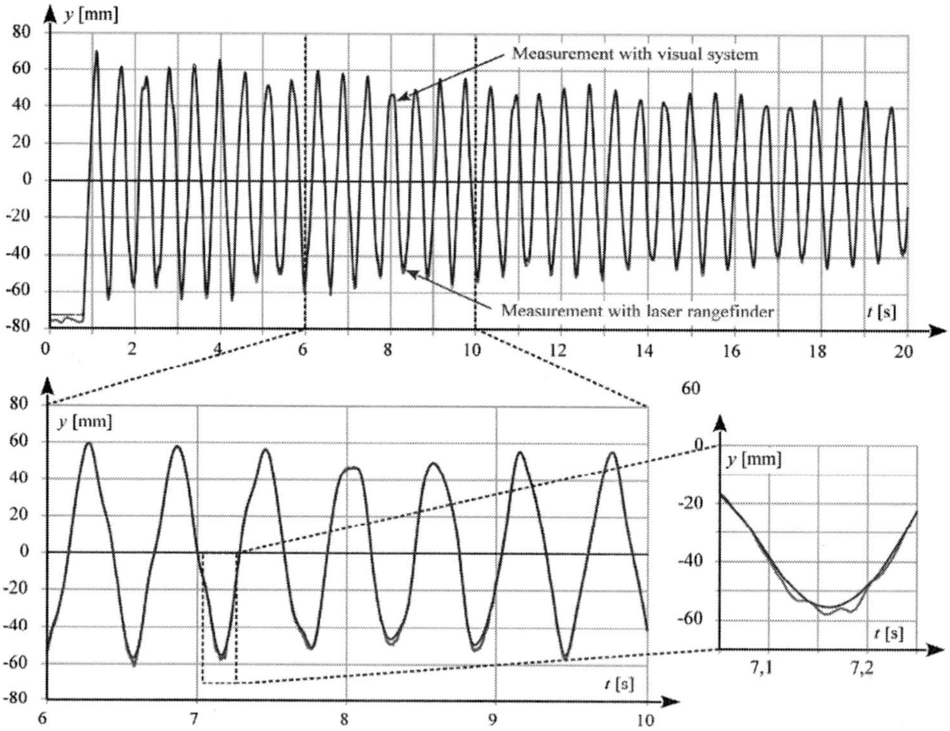

Fig. 6. Comparison of the results of measurements of the catenary network contact wire displacement in the vertical axis *y*, performed with a laser rangefinder system and a visual system.

The measurement results show good convergence of the measurement results obtained with the use of both methods. For measurements of displacements in the horizontal axis, the differences in instantaneous values practically do not exceed 0.5 mm (the mean squared normalised error MSE is equal 0,048 mm), and the nature of the vibrations is reproduced in the same way in both cases. In the case of measurement of displacements in the vertical axis, there are differences of several millimetres in the results – MSE is equal 2,3 mm (see Fig. 6 – enlargement), visible in the lower folding of the oscillation waveforms. This difference is caused by the mechanical characteristics of the reflective plates attached to the contact wire, which began to vibrate slightly in this position. It is visible in Fig. 6, where it can be observed that the trajectory of the measurement performed with the use of the visual system is smooth and the laser measurement is subject to additional oscillations.

6.2 Comparative measurements for contact wire oscillation

The measurement results presented in p. 6.1. show good convergence of the results obtained with both measurement methods and thus demonstrate the suitability of the two methods for performing the said measurements. However, they do not provide an answer to the question which method is more accurate, i.e. which results have a lower level of measurement uncertainty. To answer this question, measurements were performed at the stand shown in Fig. 2.

6.2.1 Calliper reference measurement

The result obtained with the use of a calliper was considered to be the reference measurement. For the calliper measurement, the value of the contact wire displacement is given by the formula:

$$y_{pw} = y_w - D \tag{5}$$

where: y_w – the distance between the contact wire vertical movement limiters; D – wire diameter.

Values y_w and the diameter of the contact wire D were measured using a Yato YT-7201 digital calliper with calibration uncertainty of 0.02 mm for the measurements below 100 mm and 0.03 mm for larger distances. A series of 10 measurements was performed for each of the values.

6.2.2 Measurement performed with the use of a laser rangefinder

For the measurements performed with a laser rangefinder the same rangefinder was used as in the measurement of oscillation courses, i.e. the OD250-150P142 model made by SICK. The current source of the rangefinder was connected to a reference resistor with the resistance of R = 100 Ω. The voltage signal was recorded by a Hioki recorder, model 8853. The whole procedure involved the performance of 10 series of measurements. 500 measurements at the top position of the wire and then 500 measurements at its bottom position were recorded in each series. After averaging and calculating the difference, the displacement value was obtained for the given series. An example measurement is shown in Fig. 7.

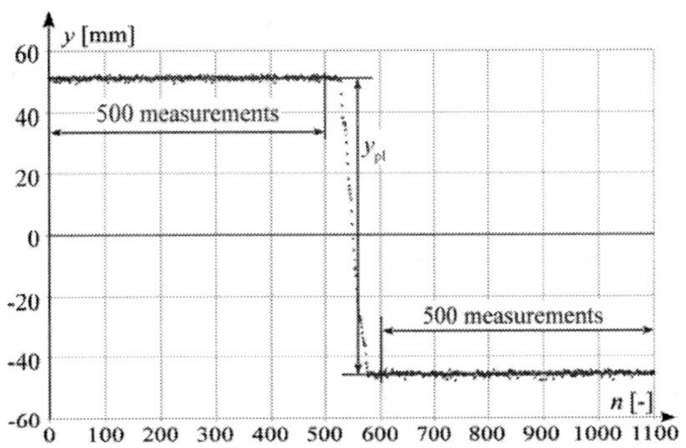

Fig. 7. Exemplary measurement performed with a laser rangefinder, where: n – number of measurements.

6.2.3 Measurement performed with the use of a visual system

The analysis of measurement uncertainty showed that for visual measurements with the spatial configuration given in Fig. 4, the highest accuracy (the lowest level of measurement uncertainty) was obtained for the distance $k \approx 15$ m [16]. To maintain the reproduction scale, a Sonnar 4/300 lens with a 2x type K-6B teleconverter was attached to the camera. The total rated focal length of such an optical system was f = 559.12 ± 0.22 mm. For such an optical system the other characteristics were:

- distance k = 14549.46 ± 0.87 mm;
- distance F = 582.43 ± 0.23 mm;
- angle α = 42.7200 ± 0.0057°;
- angle β = 5.290 ± 0.067°.

For visual measurements, 10 measurement series were also performed, in an identical manner as for the laser rangefinder.

6.2.4 Comparison of measurement results

The results for a series of 10 measurements, along with standard uncertainties, are presented in Table 1. Due to multiple measurements, both type A and B uncertainties were taken into account.

Table 1. Comparison of measurement results.

Calliper measurement y_{ps} [mm]	Laser rangefinder measurement y_{pl} [mm]	Visual system measurement y_{pk} [mm]
97.245 ± 0.023	96.7 ± 9.9	97.206 ± 0.066

As it can be seen, the differences in the results between the measurements performed with the laser rangefinder and the visual system, as well as the reference measurement, fall within the limits of measurement uncertainties. However, the considerably higher level of measurement uncertainty obtained when using a laser rangefinder is an interesting issue. It is a direct result of the manufacturer's declared uncertainty of calibration of the instrument, amounting to ±3% of the maximum distance of the measurement scope, which gives the calibration uncertainty of Δy = 12 mm for the rangefinder used. The level of uncertainty obtained with visual measurements is three times higher than that of the reference measurement performed with the use of a calliper, and two orders of magnitude smaller than in the case of a laser rangefinder.

7 Summary and conclusions

The conducted experiment has proved the suitability of the visual measurement system for the analysis of vibrations and displacements of the contact line contact wire, or of its other elements. The obtained results are identical in quality to the existing ones, coming from a short range laser rangefinder measuring system, yet they are far superior in terms of accuracy. When using the visual system, the measurement uncertainty is two orders of magnitude smaller than in the case of laser rangefinders. Such a difference is a completely new measurement quality, unattainable when using the methods applied so far. Furthermore, the measurement of displacements in the bi-axial plane for the realization variant with the rangefinders is possible only in tests conducted in laboratory conditions. For field measurements on the catenary line, the laser rangefinders can only be used for measuring vertical displacements, which greatly limits the functionality of such measurements. Measurements of horizontal dislocations are practically impossible due to the difficulty in mounting the reflective plate and the rangefinders in such a way that they do not interfere with the current collector gauge. Such limitations do not occur during measurements performed with the visual system. Furthermore, the visual system does not require any additional elements to be mounted on the contact wire, so the mechanical parameters of the network itself are not changed. There are also no problems with simultaneous measurement of the displacements of the contact wire and the catenary, or both wires and the catenary, which, in the case of laser rangefinders, requires doubling or tripling the number of sensors. The disadvantages of the visual system include higher costs, the need for very precise positioning, and greater sensitivity to changes in the level of external lighting. However, as has been shown by the results of the analysis, in situations where high measurement accuracy is required, the visual system is irreplaceable.

References

1. Z. Giętkowski, K. Karwowski, M. Mizan, *Diagnostics of contact line.* (Wydawnictwo PG, Gdańsk, 2009) [In Polish]
2. S. Kusumi, K. Nezu, H. Nagasawa, *Overhead Contact Line Inspection System by Rail-and-Road Car.* RTRI Vol. **41** (4), 169-172 (2000)
3. Rav.M. Galiulin, Rish.M. Galiulin, J.M. Bakirov, V.V. Bydanov, D.R. Bogdanov, Optoelectronic measurement of a contact net geometric parameters in railways. *IFAC Proceedings Volumes, Volume 33, Issue 20,* 143-147 (2000)
4. H. Hofler, M. Dambacher, N. Dimopoulos, V. Jetter, Monitoring and inspecting overhead wires and supporting structures. *IEEE Intelligent Vehicles Symposium/Parma*, 512-517 (2004)
5. Z. Liu, W. Liu, Z. Han, A High-Precision Detection Approach for Catenary Geometry Parameters of Electrical Railway. *IEEE Trans. Instrum. Meas.*, 1-11 (2017)
6. J. Skibicki, M. Bartłomiejczyk, *Analysis of measurement uncertainty for contact-less method used to measure the position of catenary contact wire, performed with the use of Monte Carlo method.* Measurement **97**, 203-217 (2017)
7. I. Aydin, M. Karaköse, E. Akin, A robust anomaly detection in pantograph-catenary system Based on Mean-Shift Tracking and Foreground Detection. *IEEE International Conference on Systems, Man, and Cybernetics.* 4444-4449 (2013)
8. C.J. Cho, Y. Park, *New monitoring technologies for overhead contact line at 400 $km \cdot h^{-1}$.* Engineering **2**, 360-365 (2016)
9. Z. Giętkowski, K. Karwowski, M. Mizan, P. Pazdro, *Review of research and developement on the technical diagnostics in the field of electric traction.* Pomiary Automatyka Kontrola **12**, 66-70 (2003)
10. L. Jarzebowicz, S. Judek, 3D machine vision system for inspection on contact strips in railway vehicle current collectors. *International Conference on Applied Electronics (AE).* (2014)
11. S. Judek, L. Jarzebowicz, Algorithm for automatic wear estimation of railway contact strips based on 3D scanning results. *International Conference and Exposition on Electrical and Power Engineering (EPE 2014).* 724-729 (2014)
12. S. Judek, L. Jarzebowicz, *Wavelet transform-based approach to defect identification in railway carbon contact strips.* Elektronika ir Elektrotechnika, Vol. **21**, No 6, 29-33 (2015)
13. E.D. Stefano, E. Ruffaldi, C.A. Avizzano, Automatic 2D-3D vision based assessment of the attitude of a train pantograph. *IEEE International Smart Cities Conference (ISC2)*, 1-5 (2016)
14. K. Karwowski, M. Mizan, D. Karkosiński, *Monitoring of current collectors on the railway line.* Transport **0**, 1-9 (2016)
15. C.A.L. Vázquez, M.M. Quintas, M.M. Romera, Non contact sensor for monitoring catenary-pantograph interaction. *IEEE International Symposium on Industrial Electronics.* 482-487 (2010)
16. J. Skibicki, *The issue of uncertainty of visual measurement techniques for long distance measurements based on the example of applying electric traction elements in diagnostics and monitoring.* Measurement **113**, 10-21 (2017)

Modal analysis of railway current collectors using Autodesk Inventor

Andrzej Wilk[*,1], *Sławomir* Judek[1], *Krzysztof* Karwowski[1], *Mirosław* Mizan[1], and *Paweł* Kaczmarek[1]
[1]Gdansk University of Technology, Faculty of Electrical and Control Engineering, ul. Gabriela Narutowicza 11/12, 80-233, Gdańsk, Poland

Abstract. The paper presents the results of modal analysis of railway current collector type 160EC. In the first place, the analysis was carried out analytically for a simplified two lumped mass pantograph model. Then numerical analysis was conducted in the Autodesk Inventor (AI) on the prepared multibody model using the AI modal analysis algorithm, which is based on the finite element method (FEM). Model elements which are most relevant for attaining a correct representation of vibration properties when using AI modal analysis were indicated. The influence of selected parameters of modal analysis algorithm on results accuracy was investigated, e.g. the FEM mesh density. The natural frequencies and shapes of the first few vibration modes are shown. The results indicate that the frequencies of natural vibrations of the moving pantograph components are within the range of up to tens of hertzs. The possible use of results of pantograph modal analysis at the design stage and in the operation phase was also discussed.

1 Introduction

The development of high-speed railways requires that modern current collectors, apart from the correct static characteristics, must be characterized by special dynamic properties to provide uninterrupted contact between the contact wire and the contact strips of the slipper [1-4]. The pantographs' dynamic properties depend on the accepted design solutions, e.g. detailed construction of drive system, but also on the adjustment of parameters e.g. inertia of components, friction parameters, aerodynamic resistance, etc. [5-8]. It is expedient to analyse and pre-evaluate the construction of the pantograph and the proper selection of its parameters already at the design stage.

Modern methods of current collectors designing use software packages of CAD, CAM and CAE (Computer Aided Design / Manufacturing / Engineering) type. Recently, software manufacturers strive to integrate the CAD/CAM/CAE processes into one complex program, enabling the implementation of optimization procedures. Among a number of software packages that integrate these processes, one of the more popular is Autodesk Inventor (AI) [9]. The AI environment provides convenient geometric designing of models with respect to manufacturing technology and the possibility of parameterization of executive documentation, which is an advantage in the case of a product intended for manufacturing in many variants. AI software allows to providing static and dynamic structural strength analysis of a device, as well as its frequency- and modal analysis [10]. The AI environment allows efficient and comprehensive modelling and analysis of even very complex three-dimensional (3D) structures.

Modal analysis is used both in theoretical and experimental studies to analyse dynamic properties of technical objects - especially mechanical ones. The theoretical modal analysis uses a physical model of the object including its mass, stiffness and damping properties, taking into account their spatial distributions, which leads to the determination of mass, stiffness and damping matrices. Elements of these matrices are the coefficients of a system of normal differential equations of motion. The superposition principle of a linear dynamic system enables the transformation of these equations into a typical eigenvalue problem. Its solution called a modal data of the system is obtained in the form of a set of natural frequencies of simple harmonic motions, mode shapes and damping factors, which allow to predict the behaviour of an object in case of any disturbance of its equilibrium.

The results of modal analysis are used for: structural modifications, diagnostics of structure condition, synthesis of control device in active systems for vibration suppression, and verification and validation of numerical models such as finite element models [11-14]. Currently, many commercial simulation software suits based on the finite element method (FEM) allow the creation of discrete models of almost any linear dynamic structure, which remarkably enlarges the capacity and range of theoretical modal analysis.

2 Modal analysis of pantograph two degrees of freedom

The 3D model of a pantograph in interaction with overhead contact line, which was created in Autodesk Inventor, is presented in Fig. 1a. Alternatively, Fig. 1b

* Corresponding author: andrzej.wilk@pg.edu.pl

shows the simple linear, viscous-damped, lumped parameter model of the current collector, as a more commonly used version in pantograph dynamic simulations [15, 16, 10]. It is a relatively simple conventional lumped mass model consisting of two equivalent masses, connected by springs and damping elements, representing the main components of pantograph, such as panhead, and arms. The model with one degree of freedom (DOF) is not considered here because the response frequency is relatively low. Even more a simple one-mass model with one-DOF, sporadically used to simulate selected phenomena such as waveform of contact force or contact wire uplift, is not considered here because it does not reflect too many significant physical quantities. Also, its response frequency is relatively low.

a)

b)

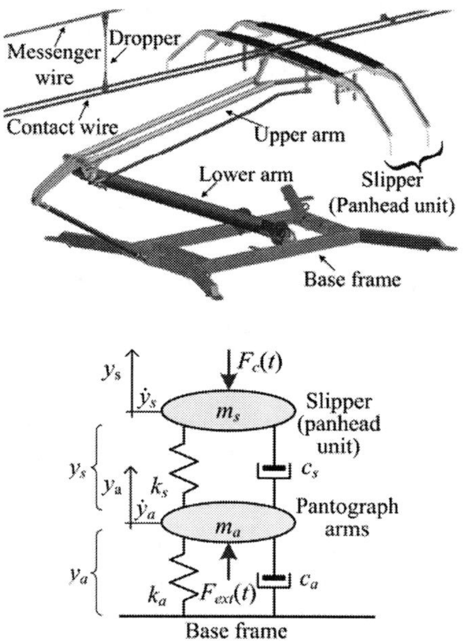

Fig. 1. Current collector in interaction with overhead contact line: a) 3D CAD sketch excluding drive system; b) equivalent lumped two-mass model.

The pantograph model (Fig. 1b) is a two-DOF system which can be described, omitting gravity forces, by an equation of motion in matrix notation as [5, 13]:

$$\begin{bmatrix} m_a & 0 \\ 0 & m_s \end{bmatrix}\begin{bmatrix} \ddot{y}_a \\ \ddot{y}_s \end{bmatrix}+\begin{bmatrix} c_a+c_s & -c_s \\ -c_s & c_s \end{bmatrix}\begin{bmatrix} \dot{y}_a \\ \dot{y}_s \end{bmatrix}+ \\ +\begin{bmatrix} k_a+k_s & -k_s \\ -k_s & k_s \end{bmatrix}\begin{bmatrix} y_a \\ y_s \end{bmatrix}=\begin{bmatrix} F_{ext} \\ -F_c \end{bmatrix} \quad (1)$$

where: m_a, c_a, k_a, m_s, c_s, k_s represent the equivalent masses, dumping coefficients and stiffness of element of the model for pantograph arm and slipper respectively, F_{ext} is the static lift force, F_c is the contact force.

Since the static lift force F_{ext} is constant and does not affect the vibration characteristic of the pantograph, F_{ext} can be neglected. The change of the contact force F_c and interaction of overhead contact line are also omitted. The modal analysis was performed for the pantograph only.

Passing to matrix notation the equation (1) becomes:

$$\mathbf{M}\ddot{\mathbf{y}}+\mathbf{C}\dot{\mathbf{y}}+\mathbf{K}\mathbf{y}=\mathbf{F} \quad (2)$$

where: **M**, **C** and **K** are the mass, damping and stiffness matrices respectively; $\mathbf{y}$, $\dot{\mathbf{y}}$ and $\ddot{\mathbf{y}}$ are the corresponding displacement, velocity and acceleration and **F** represents forces applied to the pantograph.

In case of an undamped and unforced system the matrices and vectors from (2) become:

$$\mathbf{M}=\begin{bmatrix} m_a & 0 \\ 0 & m_s \end{bmatrix},\quad \mathbf{C}=\begin{bmatrix} 0 & 0 \\ 0 & 0 \end{bmatrix},\quad \mathbf{K}=\begin{bmatrix} k_a+k_s & -k_s \\ -k_s & k_s \end{bmatrix},$$

$$\mathbf{F}=\begin{bmatrix} 0 \\ 0 \end{bmatrix},\quad \mathbf{y}=\begin{bmatrix} y_a \\ y_s \end{bmatrix}. \quad (3)$$

As a result, the notation of equation (2) reduces to:

$$\mathbf{M}\ddot{\mathbf{y}}+\mathbf{K}\mathbf{y}=\mathbf{0} \quad (4)$$

Assuming a solution in the form of: $\mathbf{y}(t)=\mathbf{Y}e^{i\omega t}$, which leads to: $\ddot{\mathbf{y}}(t)=-\omega^2\mathbf{Y}e^{i\omega t}$, it can be expressed as:

$$[\mathbf{K}-\omega^2\mathbf{M}]\mathbf{Y}=\mathbf{D}(\omega)\mathbf{Y}=\mathbf{0} \quad (5)$$

where: **Y** is a nonzero two-element vector that represents the amplitudes of the motions of the point masses, and $\mathbf{D}(\omega) \overset{def}{=} [\mathbf{K}-\omega^2\mathbf{M}]$ is called the dynamic matrix.

Equation (5) states the free vibrations eigenproblem for an undamped two-DOF system. For nontrivial solutions $\mathbf{Y}\neq 0$ the determinant of the dynamic matrix must become zero:

$$\det\mathbf{D}(\omega)=\det[\mathbf{K}-\omega^2\mathbf{M}]=0 \quad (6)$$

This forms the characteristic equation for free undamped vibrations. For a two-DOF system, the left side of (6) has a form of quadratic polynomial of variable ω^2, which yield two roots: ω_1^2 and ω_2^2.

Assuming a set of typical values of pantograph equivalent masses $m_a = 37.12$ kg, $m_s = 19.87$ kg and stiffness $k_a = 50$ N/m, $k_s = 16000$ N/m, the equation (6) takes a notation of:

$$\det\begin{bmatrix} 16050-37.12\omega^2 & -16000 \\ -16000 & 16000-19.87\omega^2 \end{bmatrix}= \\ =737.6\omega^4-912833.5\omega^2+800000=0 \quad (7)$$

The roots of equation (7) are real and nonnegative numbers:

$$\omega_1^2=0.877\,(\text{rad/s})^2 \text{ and } \omega_2^2=1236.74\,(\text{rad/s})^2 \quad (8)$$

called as the undamped natural circular frequencies.

Vectors $\mathbf{Y}_1(\omega_1)$ and $\mathbf{Y}_2(\omega_2)$ are mathematically called as eigenvectors:

$$\mathbf{Y}_1 = \begin{bmatrix} 1 \\ 1.0011 \end{bmatrix} \text{ and } \mathbf{Y}_2 = \begin{bmatrix} 1 \\ -1.8661 \end{bmatrix} \quad (9)$$

Note that in the first mode (ω_1) the masses oscillate in phase while in the second one (ω_2) they move opposite to each other. The visualization of an eigenvector as a motion pattern is called a mode shape of the system.

Normalization criteria are often used in the analysis i.e. the highest absolute value of $\mathbf{Y}_i$ is searched, and then $\mathbf{Y}_i$ is divided by it. Normalized eigenvector is called $\mathbf{\Phi}$, for assumed parameters of two-DOF pantograph model it is as follows:

$$\mathbf{\Phi}_1 = \begin{bmatrix} 0.9989 \\ 1.0 \end{bmatrix} \text{ and } \mathbf{\Phi}_2 = \begin{bmatrix} 0.5359 \\ -1.0 \end{bmatrix} \quad (10)$$

After orthogonalization and orthonormalization it is possible to write modal equation of motion for generalized coordinates [13] as:

$$\mathbf{M_g}\ddot{\boldsymbol{\eta}} + \mathbf{K_g}\boldsymbol{\eta} = \mathbf{0} \text{ or short } \ddot{\boldsymbol{\eta}} + \mathbf{K_g}\boldsymbol{\eta} = \mathbf{0} \quad (11)$$

where: $\mathbf{M_g} = \mathbf{\Phi}^T\mathbf{M}\mathbf{\Phi} = \mathbf{I}$ is generalised mass matrix, that reduces to the identity matrix $\mathbf{I}$ hence in short notation it may be omitted, $\mathbf{K_g} = \mathbf{\Phi}^T\mathbf{K}\mathbf{\Phi} = diag[\omega_i^2]$ is generalised stiffness matrix, and $\boldsymbol{\eta}$ is generalised coordinates vector.

Because the matrices $\mathbf{M_g}$ and $\mathbf{K_g}$ are diagonal, equation (11) uncouples into two homogeneous, ordinary second order differential equations already in canonical form. Once solutions η_a and η_s are available, they can be combined via the superposition procedure to get the physical response:

$$\mathbf{y}(t) = \mathbf{\Phi}\boldsymbol{\eta}(t) \quad (12)$$

To solve this equation, its initial conditions in modal coordinates are needed. For analysed example of two-DOF pantograph model the solution in generalised and physical coordinates for exemplary initial conditions $\mathbf{y_0} = \begin{bmatrix} 0 \\ 0 \end{bmatrix}$, $\mathbf{v_0} = \begin{bmatrix} 1 \\ 0 \end{bmatrix}$ are respectively:

$$\boldsymbol{\eta} = \begin{bmatrix} 5.2486\sin(0.936t) \\ 0.1024\sin(35.167t) \end{bmatrix} \quad (13)$$

$$\mathbf{y} = \mathbf{\Phi}\boldsymbol{\eta} = \begin{bmatrix} 0.6950\sin(0.936t) + 0.0099\sin(35.167t) \\ 0.6957\sin(0.936t) - 0.0185\sin(35.167t) \end{bmatrix} \quad (14)$$

$$\dot{\mathbf{y}} = \mathbf{v} = \begin{bmatrix} 0.6508\cos(0.936t) + 0.3492\cos(35.167t) \\ 0.6516\cos(0.936t) - 0.6516\cos(35.167t) \end{bmatrix} \quad (15)$$

Fig. 2 shows waveforms of displacements $\mathbf{y}$ and velocities $\mathbf{v}$ for the two DOF model of current collector (see: Fig. 1b).

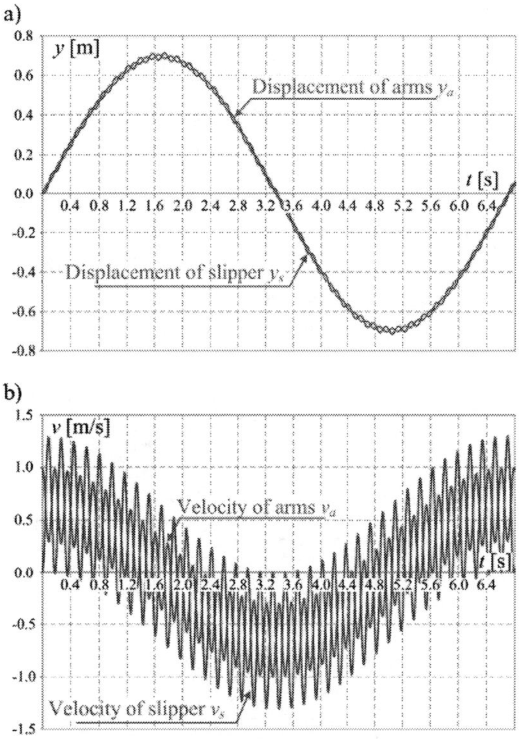

Fig. 2. Response waveforms for analytical modal analysis for lumped mass model of pantograph for arms and slipper respectively: a) displacements y_a, y_s; b) velocities v_a, v_s .

In the case of the damped system, using (1) and (2), the equation (11) will take the following form:

$$\ddot{\boldsymbol{\eta}}(t) + \mathbf{C_g}\dot{\boldsymbol{\eta}}(t) + diag(\omega_i^2)\boldsymbol{\eta}(t) = \mathbf{f}(t) \quad (16)$$

where: $\mathbf{M_g}$ is reduced to the identity matrix $\mathbf{I}$ therefore it was omitted in the notation above, $\mathbf{K_g}$ becomes a diagonal matrix with squared frequencies stacked along its diagonal, $\mathbf{f}(t)$ is called a modal forces vector, the modal matrix $\mathbf{C_g}$ generally will not be a diagonal matrix. For such case, it is not possible to decoupling of modal equation of motion.

The above example shows that modal analysis for a highly simplified current collector model can be carried out even analytically at relatively low time, especially in the case of omitting the damping phenomena, but it provides little information – in practice it is only possible to determine the basic vibration frequencies of the main components of the pantograph in this way.

3 Computer modelling of current collectors in Autodesk Inventor

In order to perform a comprehensive simulation analysis of the pantograph using AI, it is necessary to develop its basic physical model. The main operations at model creating and at modal analysis of the current collector in this environment are presented in Fig. 3.

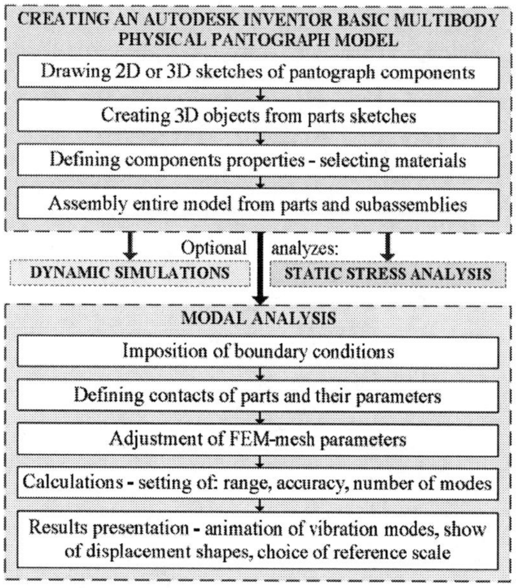

Fig. 3. Flowchart of pantograph modal analysis using Autodesk Inventor.

The process of modelling is typical in AI, although it should be noted, that some stages may be difficult because e.g. such materials as carbon for contact strips are not included in program library. For proper reflection of dynamic properties at modal analysis it is necessary to determine relevant parameters of those materials such as density, elasticity, Young's modulus etc. The correct determination of the types of mechanical contacts between pantograph parts together with the adoption of adequate values of contact parameters especially for mutually moving parts is of particular importance. In modal analysis AI allows only to define two types of contacts: bonded and spring, for the latter it is possible to independently set the normal and tangential stiffness. This process requires a lot of engineering experience, as these values affect the results of the collector eigenvibrations frequencies. Another important task is to set the parameters of FEM-mesh, especially in places of spring contacts, because its automatic generation can cause significant inaccuracy of analysis results. The program allows selective mesh densification in areas indicated by the user.

Figure 4a shows the prepared basic physical model of the pantograph with its main components marked. Figure 4b presents an enlarged part of the suspension system of the pantograph head. For obtaining representative results in modal analysis, this subassembly must be modelled with relatively high precision because it has a significant influence on the dynamic properties of the current collector. Particularly essential is the precise elaboration of the suspension springs model, wherein apart from the proper indication of material and dimensions, it is important to correctly place them in the nests and also appropriately set the density of the FEM-mesh, what is shown on Fig. 4c.

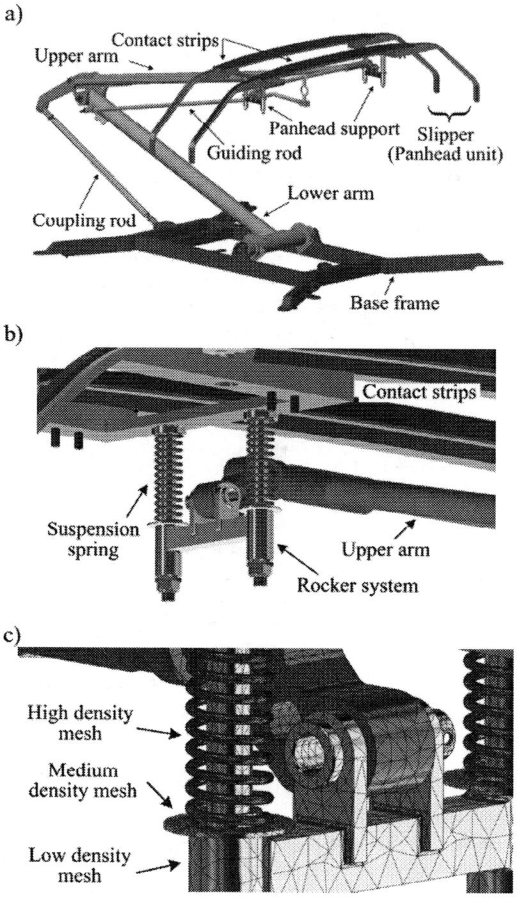

Fig. 4. Autodesk Inventor model (excluding drive system) of 160EC railway current collector: a) entire pantograph with marked main parts; b) slipper suspension unit; c) enlarged fragment of suspension unit with superimposed FEM-mesh.

4 Finite element modal analysis results

Based on the created AI model of the current collector, the modal analysis was performed. In the first stage, the study concerned only the slipper unit together with its suspension system, which was separated from the rest of the device. Among other, the aim was to determine the right FEM-mesh density distribution based on the comparison of the obtained frequency of slipper's basic vibration mode with the results derived from the simplified two-mass model analysis (see Section 2) and from the laboratory experiment. Fig. 5 shows the displacement shapes and frequencies of the first three vibration modes of the slipper unit. The vibrations represented by modes 1 and 2 are of major importance in the interaction between the pantograph and the overhead contact line – note that their frequencies are very close. It can be stated that the oscillation of the upper mass in the equivalent lumped two-mass model is mainly a superposition of these two modes of vibration, and therefore its frequency should be within the interval determined by two obtained values. Achieved results fulfill this condition.

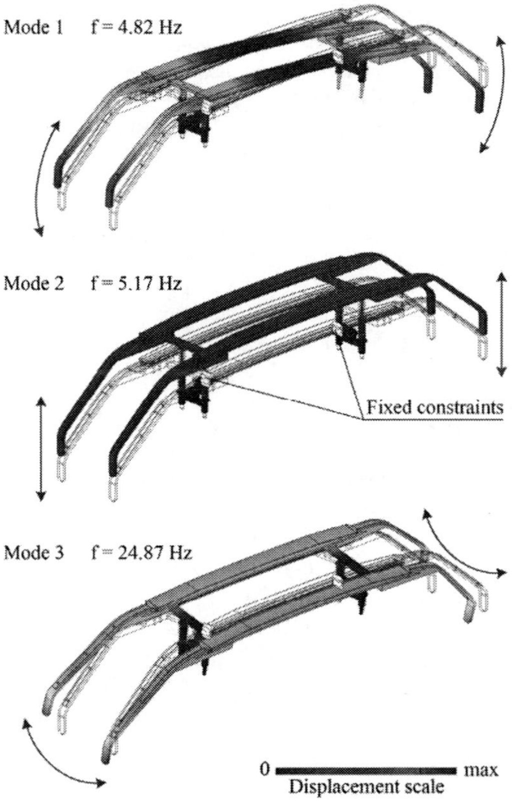

Fig. 5. Displacement shapes and corresponding frequencies of the first three vibration modes of the slipper unit obtained by AI modal analysis.

If the slipper is treated as a lumped mass, then the natural frequency can be calculated as:

$$f = \frac{1}{2\pi}\sqrt{\frac{k}{m}} = \frac{1}{2\pi}\sqrt{\frac{16188}{16.06}} = 5.05\,\mathrm{Hz} \qquad (17)$$

where k is equivalent stiffness of suspension springs, m is the sum of the masses of the components of the slipper. Obtained values correspond to mode 2 presented in Fig. 5.

Subsequently, an analysis was performed for the entire pantograph, but without the drive system, for the selected uplift height within the normal operating range. The results for vibration modes with frequencies up to about 25 Hz are shown on Fig. 6. It should be noted that in modal analysis of the AI the graphically presented displacements are relative, i.e. they are scaled in relation to their highest value in a given vibration mode. This way they do not reflect the real absolute magnitude of deviations. On the Fig. 5 and Fig. 6, the fixed constraints i.e. the locations of the rigid mounting of the analyzed device are indicated, which constitute the boundary conditions for modal analysis.

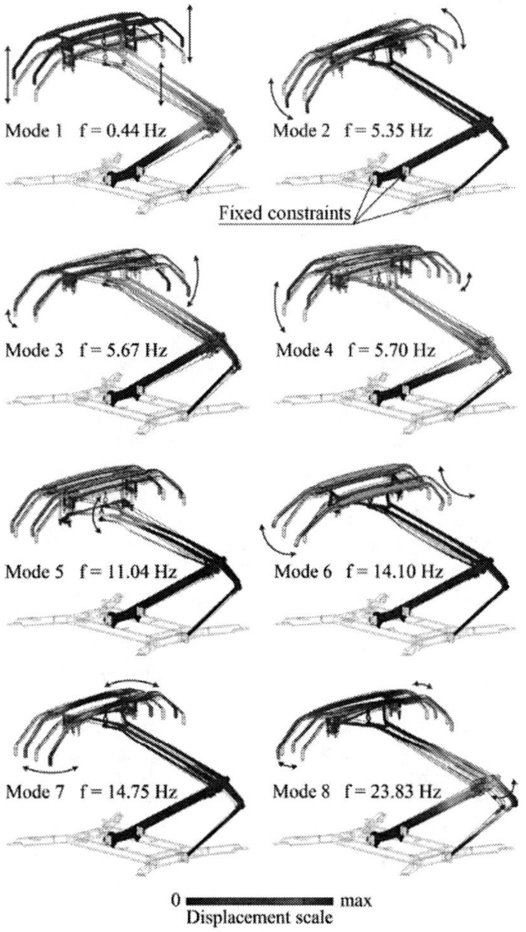

Fig. 6. Displacement shapes and corresponding frequencies of the pantograph eigenvibration modes with frequencies up to 25 Hz obtained by AI modal analysis.

It can be seen that the frequency of the collector first vibration mode is consistent in range magnitude with the lower frequency obtained from the analytical calculations of the two-mass model (see Section 2). It should be emphasized, however, that the created AI pantograph model does not cover its drive system yet – the work in this field continues, hence the difference between both values is noticeable. On the other hand, the frequencies of the main vibration modes of the slipper unit – modes No 2, 3, 4 – have not changed significantly with respect to results given above for the unit separated from pantograph arms.

Looking at the pantograph vibration modes obtained by modal analysis, some of them associated with spring connection of the components are of practical importance in studying the dynamic interaction between the current collector and the overhead contact line, but the other result from the inevitable existence of clearance in articulated joints or from the limited rigidity of the long components e.g. lower and upper arm or slipper contact strips. The latter modes may be more interesting rather at the design stage than in the operational phase, while the first group gives information that can be used at the exploitation check of the collector's technical condition.

5 Conclusions

Performed modal analysis indicates that the frequencies of the excited natural vibrations of the moving pantograph components are within the range of up to tens of hertzs. Higher frequencies generally involve harmonic distortions of components or vibrations caused by a slight clearance in moving links. Such marginal phenomena have a negligible effect in normal operation of the pantograph. The comparison of the results obtained for main vibration modes of the full-bodied AI-model with the results obtained for the simplified model, e.g. with two degrees of freedom, allows for the adjustment or verification of equivalent parameters of the latter. Simple lumped parameters models with a small number of degrees of freedom are often used in simulation procedures for assessing the quality of dynamic co-operation between overhead lines and current collectors – as required by certification procedures for newly developed or renovated overhead contact lines. The use of a precise 3D-multibody FEM-model in this case is not justified and would lead to significant complications and increased workload.

The results presented in this article are of preliminary nature - currently the work is focused on completing the AI pantograph model including its propulsion system. Future study will concern the impact of changes in the most important parameters of the current collectors, related to the wear and deterioration of individual components, and will also focus on the frequencies and mode shapes of the eigenvibrations of the device. This may be a premise to use the results of modal analysis in the technical diagnostics of pantographs.

References

1. J. Jiang, Z. Liu, and X. Lu, Optimization of the pantograph parameters based on matching performance between pantograph parameters and dropper interval, *Proceedings of the 35th Chinese Control Conference*, 9742–9747 (2016)
2. J.-P. Massat, C. Laurent, J.-P. Bianchi, and E. Balmès, *Pantograph catenary dynamic optimisation based on advanced multibody and finite element co-simulation tools*, Vehicle System Dynamics, **52**, 338–354 (2014)
3. T.X. Wu, and M.J. Brennan, *Active vibration control of a railway pantograph*, Proc. Inst. Mech. Eng. Part F: Journal of Rail Rapid Transit, **211**, 117–130 (1997)
4. T.X. Wu, and M.J. Brennan, *Basic Analytical Study of Pantograph-catenary System Dynamics*, Vehicle System Dynamics, **30**, 443–456 (1998)
5. J. Ambrósio, F. Rauter, J. Pombo, and M.S. Pereira, *A Flexible Multibody Pantograph Model for the Analysis of the Catenary–Pantograph Contact*, in: K. Arczewski, W. Blajer, J. Fraczek, M. Wojtyra, (Eds.), Multibody Dynamics: Computational Methods and Applications. Springer Netherlands, Dordrecht, 1–27 (2011)
6. J.P. Bianchi, E. Balmès, G.V. des Roches, and A. Bobillot, Using modal damping for full model transient analysis. Application to pantograph/ catenary vibration, *Proc. of the Int. Conf. on Adv. Acoustics and Vibration Eng. ISMA 2010, Leuven Belgium,* 1167-1180 (2010)
7. C. Zhao, N. Zhou, H. Zou, R. Li, R., and W. Zhang, Comparison of dynamic characteristics of different pantograph models, *Proceedings of the 35th Chinese Control Conference*, 10216–10221 (2016)
8. N. Zhou, W. Zhang, and R. Li, *Dynamic performance of a pantograph-catenary system with the consideration of the appearance characteristics of contact surfaces*, Journal of Zhejiang University-SCIENCE A, **12**, 913–920 (2011)
9. AUTODESK Inventor, http://www.autodesk.pl/ products/inventor/overview
10. A. Wilk, K. Karwowski, S. Judek, and M. Mizan, A new approach to determination of the two-mass model parameters of railway current collector, *12th Int. Conf. Modern Electrified Transport MET'2015, Trogir, Croatia*, 164-170 (2015).
11. S. Judek, and L. Jarzebowicz, *Wavelet Transform-Based Approach to Defect Identification in Railway Carbon Contact Strips*, Elektronika Ir Elektrotechnika, **21** (6), 29-33 (2015)
12. Karwowski, K., Mizan, M., Karkosiński, D., *Monitoring of current collectors on the railway line*, Transport, **33**, 177–185 (2018)
13. P. Lengvarský, and J. Bocko, *Theoretical Basis of Modal Analysis*, Am. J. Mech. Eng., **1** (7), 173–179 (2013)
14. G. Santamato, M. Solazzi, and A. Frisoli, A Detection Method of Faults in Railway Pantographs Based on Dynamic Phase Plots, *World Acad. Sci. Eng. Technol. Int. J. of Mech. and Mechatr. Eng.*, **10** (8), 1474-1485 (2016)
15. Commission Regulation (EU) No 1301/2014 of 18 November 2014 on the technical specifications for interoperability relating to the 'energy' subsystem of the rail system in the Union (2104)
16. EN 50318, *Railway applications. Current collection systems. Validation of simulation of the dynamic interaction between pantograph and overhead contact line* (2003)

MATEC Web of Conferences **180**, 04005 (2018) https://doi.org/10.1051/matecconf/201818004005
MET'2017

Traction induction motor

Jakub Bernatt[1,*], *Stanisław* Gawron[1], *Tadeusz* Glinka[1], and *Artur* Polak[1]

[1] Institute of Electrical Drives and Machines KOMEL, Rozdzienskiego Ave. 188, 40-203, Katowice, Poland

Abstract. The paper presents a traction induction motor with an innovative rotor winding design. The winding of the rotor is almost identical to the armature winding in the dc traction motors, with the exception of commutator. Winding is double layered. The bars are insulated. Along the iron core bars are secured with wedges and in the winding outhang with bandages. The bars of the bottom layer in the winding outhang part near the iron core are bent radially in the direction of the shaft axis. On the winding outhang of the bottom layer an inner bandage is applied. On the winding outhang of the top layer an outer bandage is applied. All the ends of the bottom and top layers bars, on both sides of the rotor's core, are connected together by ferrule. There is a ring placed on the shaft or on the coil support on one side of the rotor's core. It is best if this ring is made of copper. The ring is characterized by mp tongues, which are distributed evenly along its circumference (p is number of winding pole pairs and m is number of rotor winding phases). Tongues are connected to equally uniformly placed ferrules. Bandages are made of carbon fibre or glass fibre tape.

1 Rotor winding of induction traction motor

Induction motors for electric drives of rail vehicles are equipped with squirrel-cage rotor winding. These motors are supplied from inverters and used for traction drives of rail vehicles such as tram cars, metro cars, electric multiple units (EMUs), electric railway locomotives and mine locomotives. Drives using induction motors are characterized by high power efficiency and good control properties in two zones of speed control: one, where load torque is constant and the other, where load power is constant. Motors rated up to 100 kW (tram motors) are fitted with cast aluminium rotor winding. Larger motors are equipped with rotor winding made up of copper bars short-circuited with end rings. We may recall here

LK-450 X6 motors driving EMUs EN 57. One of the drawbacks of bar-type rotor winding is its high failure rate. After several years of service, the junction between winding bars and end ring cracks. This kind of damage is due to fatigue. The motors built into bogies are coupled with the wheels of traction vehicle via mechanical gearbox. Vibrations of drive wheels and tolerances of toothed gear are transmitted to the winding end-rings. This causes fatigue cracking of junctions between bars and end-rings. The authors of this paper became acquainted with the problem of end-ring breaking off during service tests run on railway vehicles under normal running conditions.

Fig. 1. Rotors of LK-450 X6 motor: brand-new one (upper photo) and with broken junctions between bars and end-ring (lower photo).

* Corresponding author: jakub.bernatt@komel.katowice.pl

This problem is described in the literature of the subject, such as publication [2]. The old drive systems of rail vehicles used dc motors, which are characterized by much lower efficiency and high failure rate of commutators and brush holders. However, damages caused by ripping of bandages from coil outhangs were rare. Innovative design of rotor winding of induction motor is presented in the paper as well. The winding is manufactured in a manner similar to that of armature winding in dc traction motors, excepting the commutator and brush holding apparatus.

2 Operating conditions of induction traction motors

In this section we present results of investigating operating conditions of asynchronous LK450X6 traction motors, installed in EMU type EN57 AKM No. 1582. Tests were run in 2014. The ratings of induction LK-450 X6 motors are as shown in Table 1.

Table 1. The ratings of induction LK-450 X6.

Power continuous rating/one-hour rating	250/300 kW
Rated voltage at 50 Hz frequency	2340 V
Rated/maximum frequency	50/120 Hz
Rated/maximum current	78/160 A
Rated/maximum rotational speed	987/2400 rpm
Rated/maximum torque at 50 Hz frequency	2419/4400 N·m
Break-down torque at 50 Hz	6600 N·m
Efficiency	≥ 0.94
cos φ	≥ 0.84
Cooling	Forced
Ingress protection rating	IP22
Ambient temperature	up to 40°C

Motor car is equipped with four motors: two in the front bogie (M1 and M2) and two in the rear bogie (M3 and M4). Two motors are supplied from a common inverter – see Fig.2. Each motor, via a mechanical gearbox, drives one bogie axle, to which the drive wheels are attached. Motors common to one bogie drive two different axles; these axles are connected via the drive wheels and rails. Therefore, this drive system may be classified as parallel operation of two motors supplied from a common inverter and driving a common system of mechanical axles. Motor pairs M1 and M2, M3 and M4, if mounted casually, may differ in parameters within the allowable tolerance range set out in standard [3].

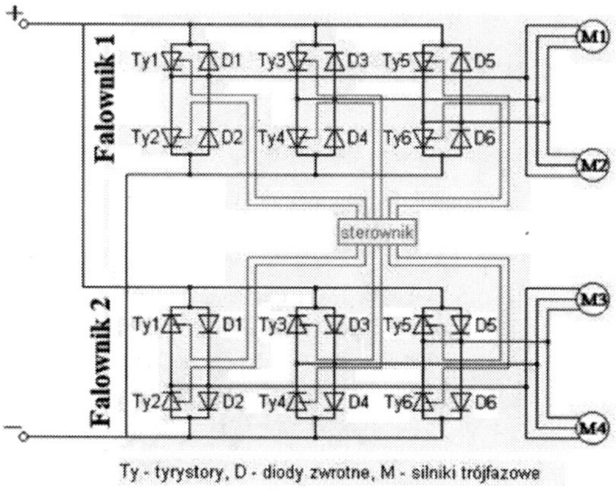

Fig. 2. Electrical scheme of LK-450 X6 motor supply - motor car.

Examples of current RMS-value waveform of M1 and M2 currents are shown in Fig.3. Differences in instantaneous values of RMS current are as high as 100%; this proves that motors have not been selected (matched) properly. When train is braked (Fig.4), the differences in motor loadings are even greater.

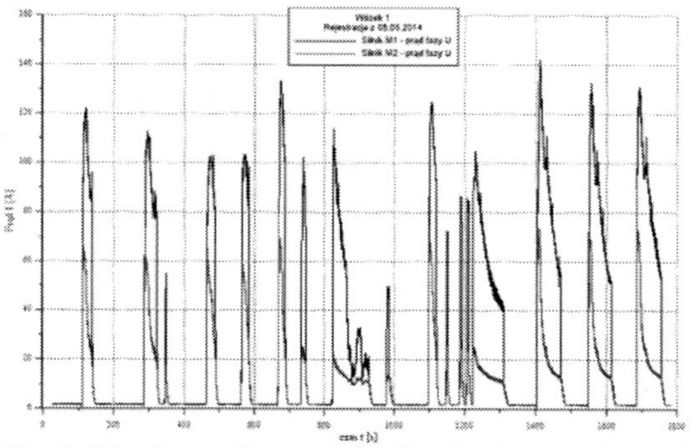

Fig. 3. Waveforms of currents (RMS values), motors M1 and M2, during train run.

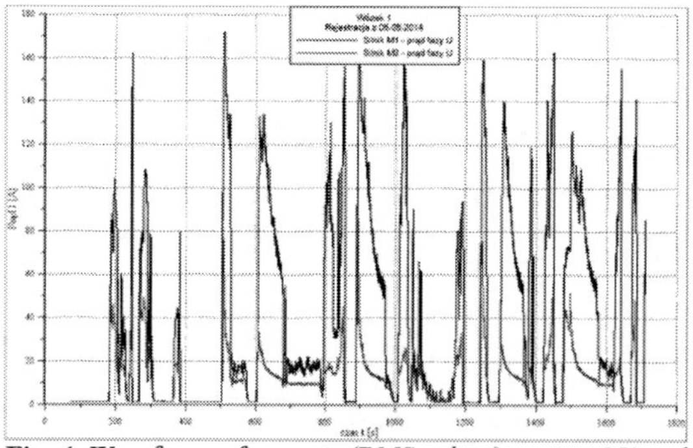

Fig. 4. Waveforms of currents (RMS values), motors M1 and M2, during train braking with braking energy recuperated.

 https://doi.org/10.1051/matecconf/201818004005

Figs. 5 and 6 show similar waveforms for motors M3 and M4 of bogie #2. The differences in motor loading are much smaller. The motors for parallel operation must be selected (matched) in such a way, that their parameters should be nearly the same. If motors are properly matched, then difference in load currents should not exceed 10%.

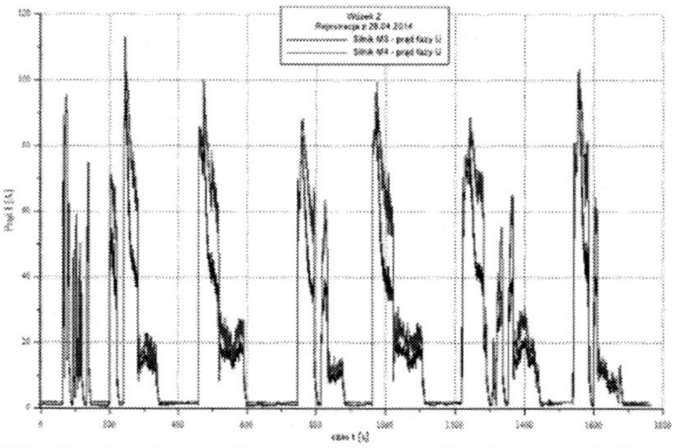

Fig. 5. Waveforms of currents (RMS values), motors M3 and M4, during train run.

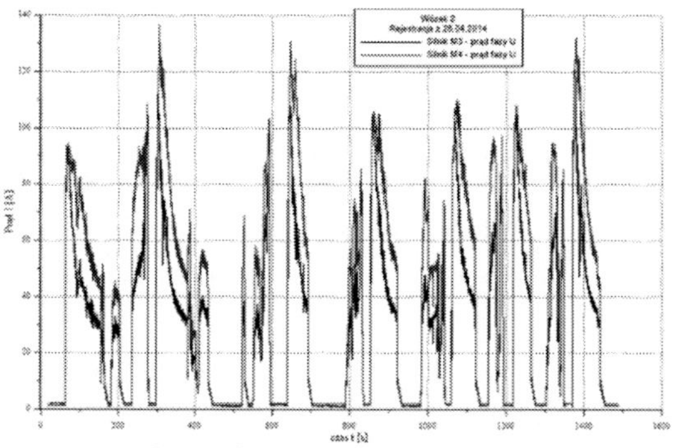

Fig. 6. Waveforms of currents (RMS values), motors M3 and M4, during train braking with braking energy recuperated.

Proper selection of motors may be achieved by using:

- no-load curves $I_0 = f(U)$ or simply I_0 at U_N and
- short-circuit curve $U_z = f(I)$ or just u_z at I_N.

The curves or parameters I_0 and u_z are measured by the manufacturer during product tests and are set out in the motor machine card; they are known to the service teams assembling motors into bogies.

Vibrations transmitted to motors by the bogie constitute another problem. Motors are attached to the bogie frame and mounted (by axial slide bearing) to the drive wheels' axles; this results in transmitting wheel vibrations to the motor. Standard PN-EN 61373 (IEC61373) specifies requirements as to tests of railway vehicle equipment subjected to vibrations and shocks, arising from the peculiarities of railway operation environment [4]. Maximum RMS-values for vibration accelerations, in accordance with Appendix A.3 of the Standard (category 3 – equipment mounted at wheelset axle), are equal to:

- vertical axis 43 m/s2 (c. 4.3 g);
- lateral axis 39 m/s2 (c. 3.9 g);
- longitudinal axis 20 m/s2 (c. 2.0 g).

According to these requirements, the LK450 X6 motor during operation may be subjected to vibrations with maximum acceleration equal to 43 m/s^2 (c. 4.3 g). LK 450 X6 motor was tested by the manufacturer (complete tests), with vibrations equalling 50 m/s^2 (directions x, y, z) and test results were positive.

Vibration acceleration measurements of LK-450 X6 motor were carried out during train run. Example of vibration acceleration waveform is given in Fig.7. The tests show that motors during normal operation were subjected to vibration accelerations as high as 150 m/s^2.

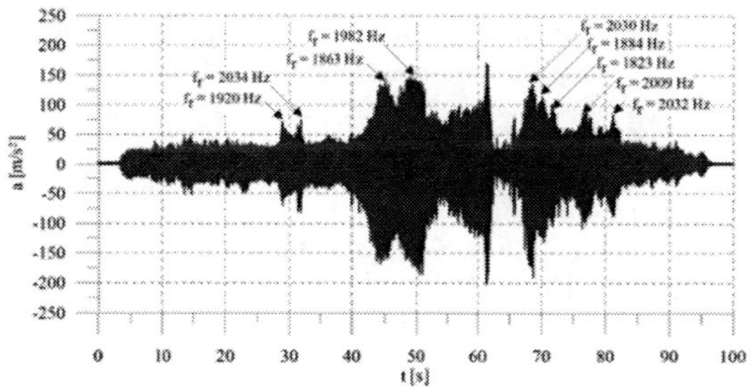

Fig. 7. LK-450 X6 motor vibration acceleration waveform, recorded on 23.04.2014.

LK-450 X6 motors fulfil the standard requirements [4]; however, vibrations acting upon the motors during normal operation attain values three times as high as values set out in the standard.

The third factor adversely influencing motor durability is backlash in the mechanical gear. Transmission of driving torque from motor shaft to bogie drive axle is achieved via mechanical gear. Small toothed gear (pinion) is mounted in the motor shaft, while main gear is attached rigidly to the car drive axle. Toothed gear is characterized by backlash. Induction motor supplied from inverter generates variable components of electromagnetic torque, apart from the constant component. This is a natural effect. These variable components of electromagnetic torque, when load at shaft is small and constant component of electromagnetic torque is likewise low, may result in appearance of negative instantaneous values of total electromagnetic torque. In the backlash range the motor's rotor will accelerate and decelerate. The main gear (which is rigidly fixed to the drive wheels' axle), will tend to maintain a constant rotational speed, on account of high car inertia. The backlash between gear teeth will produce clatter and mutual abrasion of gear teeth, which in turn will enlarge the backlash. When rotor accelerates and gear teeth come into contact, a sudden (surge) deceleration and braking of rotor will take place. A derivative of rotor' angular speed generates dynamic torque, which acts upon constructional elements of the rotor, including end-rings short-circuiting rotor bars. Shearing torque affecting the junction between bars and ring is equal to $J_p(d\omega_m/dt)$, where J_p is moment of inertia of the end-ring, and ω_m is rotor's angular speed. This torque causes fatigue cracking of junction between rings and winding bars. The backlash (clearance) should be eliminated from torque transmission path, and toothed gear and the motor should be elastically unsprung from the drive wheels and entire car mass.

3 Innovative design of rotor winding

Innovative design of rotor winding is shown in Figs. 8-12 [1]. Fig.8 demonstrates the longitudinal cross-section of the rotor, lower and upper layer winding bars are shown in Figs. 9 and 10, respectively, the end-rings with tongues are given in Fig.11 and junction of winding to the end-ring is illustrated in Fig.12.

Rotor winding of traction induction motor is two-layered. Turns are insulated and winding is fixed in the slot part with wedges, and coil outhangs are protected with special cloth ("bandages"). Winding bars in the lower winding layer (2.2), in the outhang part near the laminations (1), are radially bent towards shaft axis (8) and are circumferentially formed as shown in Fig.9. Specially formed outhangs of the lower layer bars (2.2) are placed on coil supports (6), which in turn are fixed to rotor laminations (1), or lie upon the shaft (8). Inner protective cloth ("bandage") (3.2) is wound around outhangs of the lower layer winding (2.2). All bar ends of the upper layer winding (2.1) and lower layer winding (2.2) are connected by ferrules at both sides of rotor laminations (1). Coil outhangs of the upper layer winding (2.1) are covered with outer protective cloth ("bandage") (3.1). A ring (5) (copper if possible) is placed either at shaft (8) or at coil support (6), on one side relative to laminations. This ring is distinguished by uniformly spaced tongues (5.1), mp in number (p is number of winding pair poles; m is equal to number of rotor winding phases). Tongues (5.1) are connected with the ferrules, which are also uniformly distant one from the other along the circumference. The ring (5) may be insulated from the shaft (8) by insulation sleeve (7); however, it may be also placed directly upon the shaft (8). It is beneficial to attach the ring (5) to the lateral surface of coil support (6), instead of placing it upon the shaft, since in such case the tongues (5.1) joining ring (5) to ferrules may be shorter. The easiest way of attaching end-ring (5) to the coil support (6) is by using screws. The tongues (5.1) may be formed by e.g. casting (together with the ring); they may also be welded (or thermally welded) or soldered to ring (5). Number of rotor slots should fulfil the condition that quotient Q/mp is an integer number (Q is number of slots). Winding with number of pole pairs p = 2 and 36 slots is shown in Fig.12. The ferrules connecting the upper bars (2.1) to lower bars (2.2) produce a closed ring out of winding. Connecting any mp ferrules (symmetrically placed along the circumference) to the end-ring (5), provided that m =3, generates a 3-phase winding with delta arrangement, with number of parallel groups equal to p. Ring (5) short-circuits the winding. The coil outhangs are covered with two layers of protective clothing: outer (3.1) and inner (3.2) "bandages". The cloths (3.1) and (3.2) are best manufactured from carbon fibre tape such as NEOXECOVER or glass fibre tape or composite tape (plastic material strengthened with fibre). Bending of coil outhangs (2.2) of the lower winding layer towards the shaft axis (8) creates space for the inner "bandage" (3.2), which protects the coil outhangs from deformation due to centrifugal forces.

Traction induction motor with rotor manufactured in accordance with above described design does not possess end rings; hence, the reason for high failure rates of induction traction motor is eliminated.

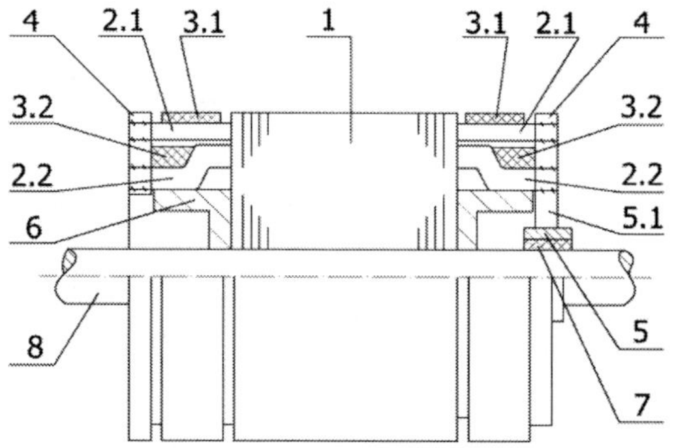

Fig. 8. Longitudinal cross-section of rotor.

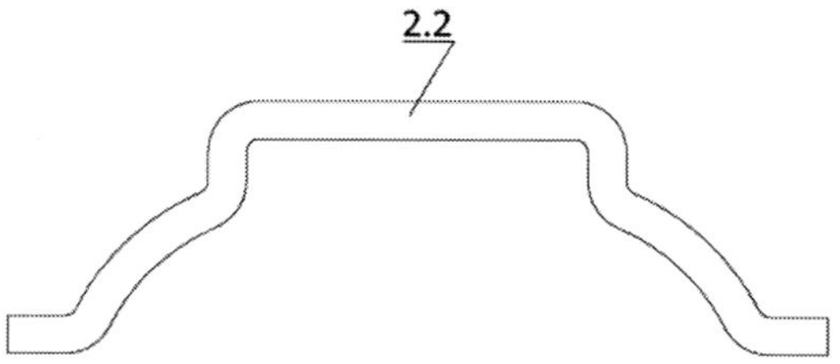

Fig. 9. Bar of winding's lower layer.

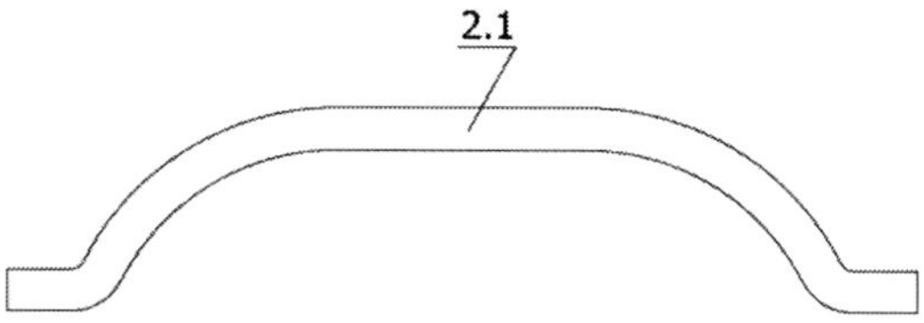

Fig. 10. Bar of winding's upper layer.

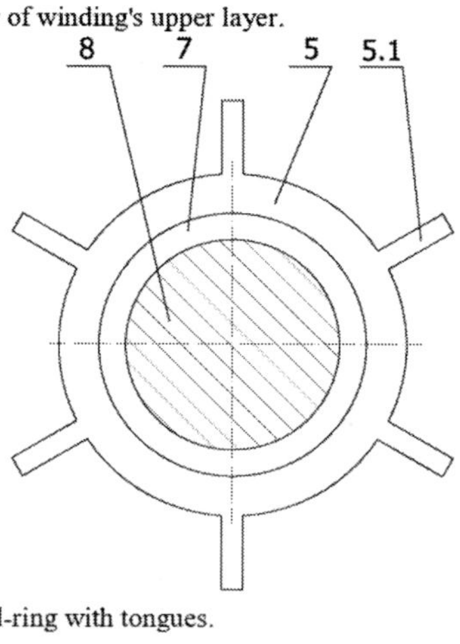

Fig. 11. End-ring with tongues.

https://doi.org/10.1051/matecconf/201818004005

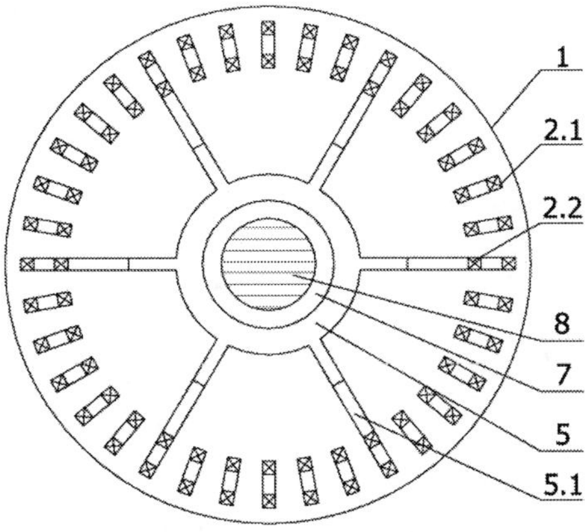

Fig. 12. Connection of winding to end-ring.

4 Conclusions

Replacing dc drive systems present in EMUs EN57 with asynchronous drives lowers energy consumption by several per cent, which is beneficial. The unfavourable effect is high failure rate of utilized LK-450 X6 induction motors. The rotor winding is often damaged due to cracking in end-rings which short circuit winding bars. Observation and testing of motor currents and vibrations in running trains has made it possible to ascertain reasons of this phenomenon. These are:

- non-uniform loading of motors supplied from one inverter, this is caused by improper selection (matching) of motors,
- vibrations transmitted to motor from bogie,
- backlash in mechanical gear.

We have proposed innovative design of rotor winding. The winding is m-phase, best if it is three-phase and connected into delta arrangement. The winding ends are short-circuited with the ring attached either to coil support or mounted directly on the motor shaft.

References

1. J. Bernatt, S. Gawron, T. Glinka, W. Poprawski, *Uzwojenie wirnika silnika indukcyjnego trakcyjnego (Rotor winding of traction induction motor – in* Polish), *Patent application P.418971*, 03.01.2016.
2. R.N. Lal Ganesh, *Development of reliable designs of squirrel cage rotors for traction motors used in locomotives working in ardnous operating conditions*, IET Electric Power Applications, **vol 4**, 487-499 (2010)
3. Polish Standard PN-EN 60349-2 (IEC 60349-2), *Trakcja Elektryczna – Maszyny prądu przemiennego zasilane z przekształtników elektronicznych (Electric traction – Rotating electrical machines for rail and road vehicles*, (2011)
4. Polish Standard PN-EN 61373 (IEC 61373), *Zastosowania kolejowe. Wyposażenie taboru kolejowego. Badanie odporności na udary mechaniczne i wibracje (Railway applications – Rolling stock equipment – Shock and vibration tests)*, (2011)

MATEC Web of Conferences **180**, 05001 (2018)
MET'2017
https://doi.org/10.1051/matecconf/201818005001

Analysis of the influence of train timetable on energy consumption on the metro line

Sławomir Barański [1,*], and *Marcin* Steczek [1]

[1] Lodz University of Technology, Institute of Electrical Power Engineering, Faculty of Electrical, Electronic, Computer and Control Engineering, ul. Stefanowskiego 18/22, 90-924, Łódź, Poland

Abstract. The latest technical solutions provide new opportunities for reducing energy consumption in electrified transport systems. The basic solution is to eliminate vehicles based on resistor start-up system and replace them with vehicles equipped with semiconductor based converter propulsion system (choppers and voltage source inverters). This type of drive also allows to recover energy (recuperation) from train braking, which is repeated very often case of metro lines. The problem is to find the receiver for this energy while the vehicle is braking. Large opportunities are also available when using various types of energy storage devices: stationary installed on the side of traction line and mobile ones used on-board the traction vehicles. Choosing the type of energy storage device for operating conditions, their capacity and the location of installation are complex optimization processes. This type of solution involves significant financial input, which reimbursement is spread over a long period of time. Therefore, it seems rational to seek solutions with limited financial outlay. This paper presents a solution analysis of adjusting the timetable and time-shifts between vehicle streams on the metro line to increase the efficiency of energy recovery of regenerative braking vehicles. This operation will reduce the amount of energy consumption by vehicles from the traction substations. The results presented in the paper were obtained by computer simulation of operation on the first line of the Warsaw subway, assuming that there was no energy storage devices and traction substations were operating without recuperation of the energy to the power system.

1 Introduction

Electric means of urban transport are characterized by relatively low power of used vehicles and high volume of traffic. Under such conditions, the use of recuperative braking in used vehicles can achieve its highest efficiency, up to 40% [1, 2]. The metro system is characterized by the highest regularity of traffic due to complete separation from the road traffic. This makes the traffic in the metro system easier to control due to its predictability, in comparison with e.g. the tram traffic. This can provide more potential for being able to affect some parameters that influence the efficiency of the braking vehicle's electrical energy recovery. In a system without subsystem energy recovery or possibility of storage in the reservoirs, the problem solution comes down to the proper synchronization of braking vehicles with vehicles that accelerate. The problem of energy consumption by electric means of transport is a determinant of their profitability and therefore requires continuous research [3]. In [4] it is reported that systems ATR (Automatic Train Regulation) and AOC (Adaptive Optimal Control) are being developed. These systems are designed to maximize the throughput of subway lines and minimize energy consumption.

Implementation of mentioned algorithms, however, requires full cooperation with both, the signalling system and whole structure of additional devices that support the work of entire the system. Another method of reducing energy consumption may be the optimization of the driving technique for a single train [5, 6]. In the article, a method of limiting energy consumption in the metro system is presented, based on the regulation of time intervals between vehicles entering the metro line. The main advantage of the proposed method is that, for implementation, it does not require any additional financial investment in infrastructure. Some aspects of the reduction of the energy consumption with low investment costs in the underground transportation was described in [8] but the technique proposed in this paper was not taken into account. The proposed method can be used on older lines that have been in operation for a long time, where no future investment in infrastructure is expected.

2 Simulation object description

The subject of analysis is the metro line I in Warsaw, section A1 Kabaty - A11 Politechnika. The test section consists of eleven stops and six traction substations. Traction substations operated in this section do not provide the possibility of reclaiming the energy recovered by vehicle's recuperative braking in the electrical system (no inverters at the substation). In examined configuration energy storage is not possible, in result, it must be used by

* Corresponding author: slawomir.baranski@p.lodz.pl

another vehicle at the time of its recovery. Otherwise it will be dispersed in braking resistors of a braking vehicle. Electricity is supplied to vehicles by means of a third rail. The line uses longitudinal sectioning with one cross connection as part of the section at the point where the third rail is supplied with power. The significant equipment used on the analysed line, from the point of the research, is the system of automatic speed limitation SOP-2. Mentioned system, developed at the Electrical Traction Department of the Technical University of Lodz [7], belongs to the group of Automatic Train Protection (ATP) systems, whose primary task is to ensure safe train operation by the driver or ATO (automatic train operation) system. The operation of this system significantly influences train traffic, especially in the case of very dense traffic, by limiting the maximum travel speed. The maximum vertical profile of the route is 25‰ while the accepted length of the platforms is 120 m.

3 Simulation software description

The software allows to simulate the operation of the metro line, the following simplifications have been adopted: traffic is carried out with the equal traction vehicles, there are no branches of the line, the traffic resistance assumed for the simulation calculations take into account travelling through tunnels.

The unitary resistance of motion of the modelled vehicle was described using the following equation:

$$w = w_0 + \frac{1}{k}\left[\frac{v}{10}\right]^2 + i \qquad (1)$$

where:

- i - profile of the track,
- k - coefficient of motion resistance ,
- v - velocity.

The w_0 is the coefficient dependent from the state of the vehicle. When the vehicle is under influence of pulling or breaking force the w_0 equals 2 (w_0=2). When traction and breaking force are equal zero then w_0 can be described with following formula:

$$w_0 = 2\cdot\left(1+\frac{gl}{g}\right) \qquad (2)$$

where:

- gl - weight of motor car,
- g - weight of whole vehicle.

The available acceleration for the vehicle for specified operating point can be calculated:

$$a = \left(\frac{f}{g} - \frac{w}{1000}\right)\cdot\frac{9.81}{\alpha} \qquad (3)$$

where:

- f - tractive force,
- w - unitary motion resistance,
- α - coefficient of rotating mass.

The operation of the simulation software consists in determining, at the assumed time step (uploaded as one of the output data), the locations of the trains, the voltages on their current collector, the electrical currents they take from the traction network and the load currents of the traction substations. If the value of the time step is sufficiently small, the above instantaneous values of voltages and currents can be considered as the average values for the period of the calculation step.

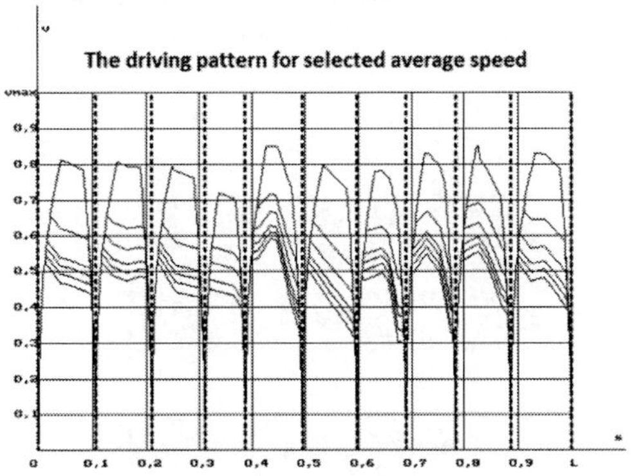

Fig. 1. Determining the energy-efficient theoretical drive with fixed average speed.

In the model adopted for the simulation of the line work, three basic functional blocks can be identified: the timetable block, the theoretical pass block and the electrical power calculation block, where the last two blocks work with a specific "feedback loop" - the theoretical pass block provides data to the electrical power calculation block (the location of trains on the line and currents taken from or off the network by them), the electrical power calculation block forwards data to theoretical pass block, with the information concerning current values of voltages on the e.m.u. current collector.

The timetable block controls the movement of the trains on the section of the line for which the simulation is carried out. Block accomplishes above mentioned task by fulfilling four basic functions by :

- positioning trains on the line at the beginning of the simulation;
- allowing new trains at the beginning of the line at the correct time;
- controlling traffic safety by ensuring adequate distances between trains on the route, depending on the current train velocities;
- withdrawing trains from traffic at the end of the line.

From the point of view of constructing the timetable block line in the model, the way the SOP system works out the permissible speeds for particular trains is very important. This system assumes the division of lines into fixed block lengths of different distances, while the permissible speed of the train at a given distance depends on the number of free block lengths before the train and amounts to: 87 km/h at four or more free block lengths, 76 km/h at three block lengths, 60 km/h at two and 37 km/h with one free block lengths in front of the train. When the preceding block is busy, the train is stopped. The model assumes that the delayed train will always try to reduce the delay by extending the driving distance under current, if its delay is not too considerable. In case of any delay of more than half of the scheduled interval between trains, which is already a significant disturbance in the uniformity of traffic on the line and which is difficult to compensate by extending the drive under current, the timetable for the whole line is rebuilt in such a way that most delayed train becomes the reference point for the new timetable - in this new layout, the remaining trains will reduce the speed accordingly, thereby ensuring that the trains are evenly distributed on the route. Figure 2 shows a simplified algorithm for the simulation program.

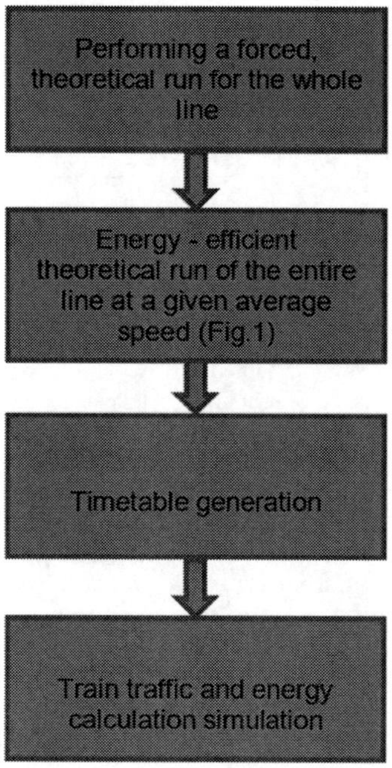

Fig. 2. Subsequent and simplified simulation steps.

After the data upload, that describes the line and the trains running on it, software performs a forceful run of the entire line in order to determine the possible average speed of passing through the line. Subsequently, after inputting given, average speed value, the theoretical energy-efficient passage is determined, that allows the given average speed to be reached with minimal energy consumption (Fig. 1). For executed run, a timetable, prepared with the given parameters, is being generated, and accordingly, software starts the simulation of train traffic on the line, at the same time performing the calculations to determine energy consumption during the simulation.

4 Simulation results

The maximum achievable train speed on the line at 30 s stop at the station was 39.5 km/h. However, a passage with such an average speed is only possible at relatively high intervals between successive trains. For a shorter succession time, the ATP system limits the maximum speed, hence the achievable average speed is also lower. As the traffic simulation was conducted for 90 s and 120 s, the average speeds were assumed to be on 30 to 36 km/h level. The cumulative results of the simulation are presented in the 3D graph in Fig. 3 and Fig. 4.

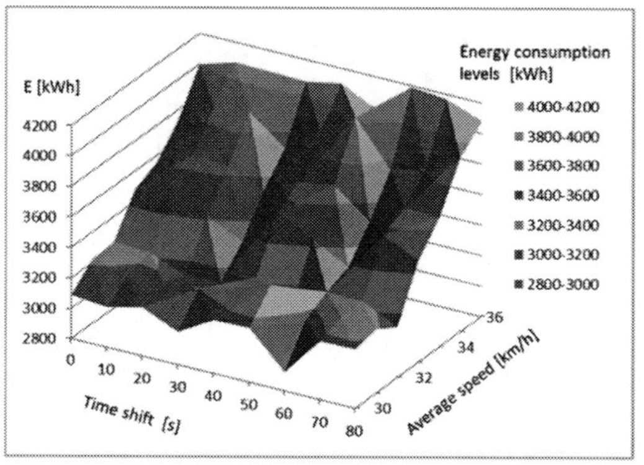

Fig. 3. Energy consumption for passages with a 90 seconds succession, with a change in the average speed and time shift between traffic directions.

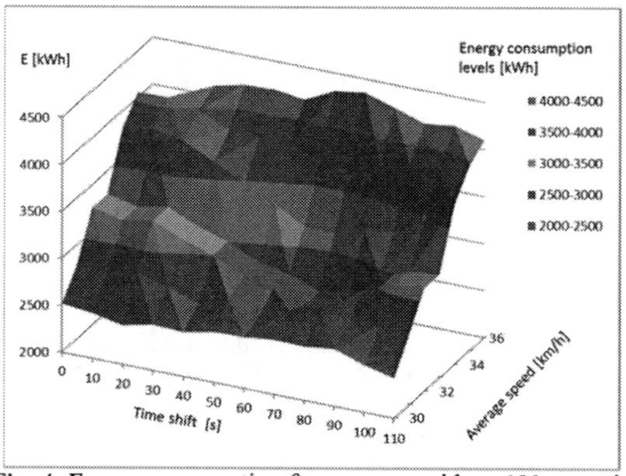

Fig. 4. Energy consumption for passages with a 120 seconds succession, with a change in the average speed and time shift between traffic directions.

The next figures shows the separated line graphs of energy consumption during driving with a succession time of 90 and 120 seconds for the average speed of the train running from 30 to 36 km/h and the change of the shift time between the traffic directions on the line from 0 to 80 s (Fig. 5) and 0 to 110 s (Fig. 6).

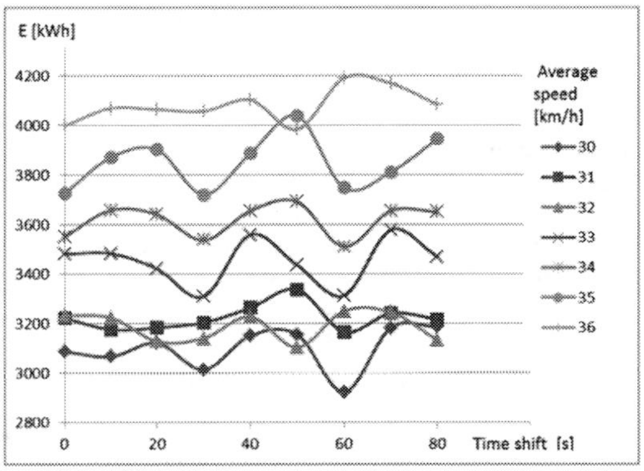

Fig. 5. Graph of energy consumption for traffic on the line with the succession time of 90 seconds.

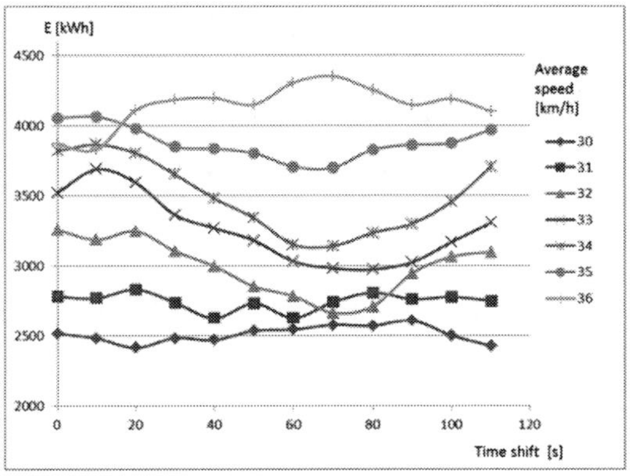

Fig. 6. Graph of energy consumption for traffic on the line with the succession time of 120 seconds.

With the average speed increase, evidently, the energy consumption increases. However, a very interesting result of the conducted simulations is that the energy consumption also depends on the shift of the traffic in both directions. There are noticeable differences reaching in some cases from about 8 % (Fig. 7) to 19 % (Fig. 8) presents selected, characterized by the highest variability, percentage charts of changes in energy consumption depending on the traffic shift.

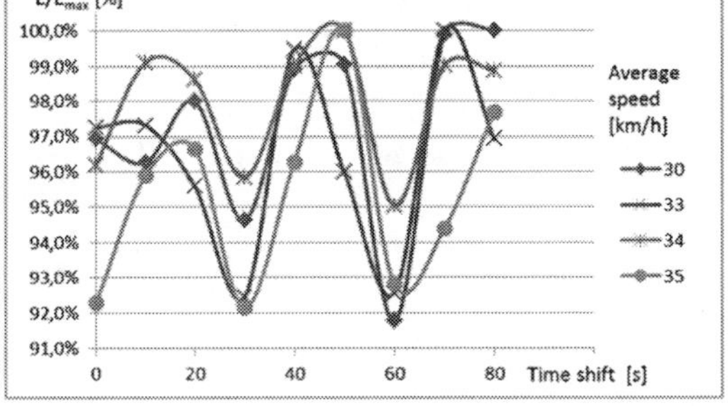

Fig. 7. Relative energy changes for line traffic with succession time of 90 sec.

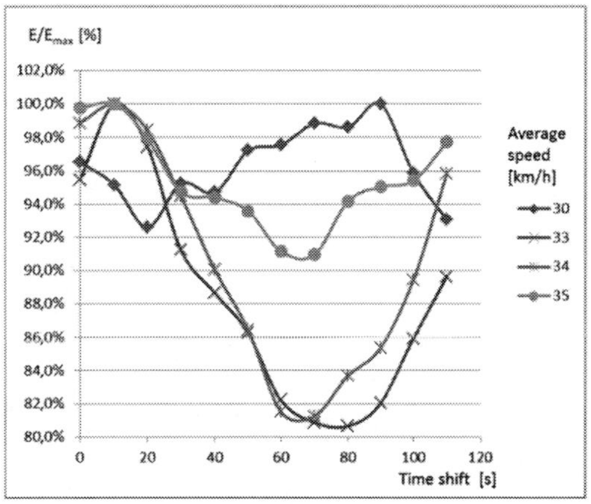

Fig. 8. Relative energy changes for line traffic with succession time of 120 sec.

The relative, minor changes in energy consumption, for the remaining average speeds and slightly different distribution of its minima, result from the fact that, at these speeds, even with changes in the travel time shifts on the line between driving directions, a sufficient number of motion situations corresponding to simultaneous recuperation braking and start up within the same electrical power section, cannot be found.

5 Conclusions

The article describes the case of the metro line, where modern trains, that allow energy recuperation during braking, are used. However traction substations on this line are not able to recuperate to the power network, nor are they equipped with energy storage units. On such a line of energy savings during recuperation will arise only if within the same power section, at the same time, one of the trains will be braking while the other will start up. The simulation results presented in the article indicate that it is possible to achieve savings in the traction energy consumption on the metro line by organizing traffic on the line based on adjustment of the time shift of deploying trains into traffic. Obviously, the savings achieved in above mentioned way are lower than in the case of systems enabling the return of energy to the power utility grid or equipped with energy storage units. However, these savings are achieved practically without investment, only by changing the timetable of trains.

The results presented in this paper were obtained with the omission of electricity consumption by metro trains for non-traction needs. Thus the relative energy savings are expected to be slightly lower than presented. Presented results are only valid for specified metro line. Simulation results are influenced by such factors as location of stations, vertical profile of line, traffic control system etc. For every metro line the calculations should be provided individually with mentioned factors taken under consideration. However, it should be expected that there exists time shift between vehicles moving in opposite directions which gives energy savings without additional investments in line infrastructure.

References

1. R. Barrero, J. Mierlo, X. Tackoen. *Energy Savings in Public Transport.* IEEE Vehicular Technology Magazine (2008)
2. W. Gunselmann. *Technologies for increased energy efficiency in railway systems.* in Proc. EPE 2005, Dresden. [CD-Rom] (2005)
3. A. Szelag, Z. Drążek, T. Maciołek. *Elektroenergetyka miejskiej trakcji elektrycznej.* Instytut Naukowo-Wydawniczy „Spatium" (2017)
4. W.S Lin, J.W. Sheu. *Optimization of Train Regulation and Energy Usage of Metro Lines Using an Adaptive-Optimal-Control Algorithm.* IEEE Transactions on Automation Science and Engineering, vol. 8, no. 4, pp.855-864 (2011)
5. N. Zhao, C. Robert, S. Hillmansen, G. Nicholson. *A Multiple Train Trajectory Optimization to Minimize Energy Consumption and Delay.* IEEE Transactions on Intelligent Transportation Systems, vol. 16, no. 5,pp.2363-2372 (2015)
6. A. Szelag, P. Gorczyca. *An Algorithm of a Tram Running Cycle Forecasting for Purposes of Energy Consumption Calculations.* 8th International Conference "Modern Electric Traction in Integrated XXIst Century Europe" Warsaw (2007)
7. H. Karbowiak, S. Barański, K. Bergiel. *Polskie systemy SOP bezpiecznego prowadzenia pociągu na liniach metra.* TTS Technika Transportu Szynowego, nr 1-2 s.52-56 (2014)
8. J. Czucha, K. Karwowski, M. Mizan, P Pazdro: *Efektywność odzysku energii hamowania elektrodynamicznego w komunikacji miejskiej.* Przegląd Elektrotechniczny. R. 80, nr 10/2004, s. 1016–1019 (2004)

MATEC Web of Conferences **180**, 05002 (2018) https://doi.org/10.1051/matecconf/201818005002
MET'2017

Effect of the supply voltage fluctuations on the efficiency of the technique of shaping the current spectrum generated by VSI traction drive

Marcin Steczek [1,*]

[1] Lodz University of Technology, Institute of Electrical Power Engineering, Faculty of Electrical, Electronic, Computer and Control Engineering, ul. Stefanowskiego 18/22, 90-924, Łódź, Poland

Abstract. The issue of interfering influence of traction vehicles on railway traffic control systems is a serious problem posing a threat to railway traffic safety. This issue is gaining in importance as more and more vehicles equipped with traction inverters are put into service. Current harmonics generated by traction vehicles equipped with voltage source inverters can be controlled in such a way that they do not exceed the limits imposed by railway operators. One of the current harmonic control techniques, proposed by the authors, is the use of a combination of SHE (Selective Harmonic Elimination) and SHM (Selective Harmonic Mitigation) techniques to control the traction voltage source inverter. In this article, the author present the results of investigations on the influence of supply voltage fluctuations on the efficiency of shaping techniques of the current harmonic spectra taken from the DC network by VSI traction drive system, using the SHE and SHM modulation techniques. Regarding the fact that in 3 kV DC electric traction systems, fluctuations of the vehicle's supplying voltage are accepted in a wide range, it is necessary to determine the influence of voltage changes on the proposed modulation techniques. The research object is a simulation model of traction drive which consisting of an asynchronous motor fed by a voltage source inverter. The simulation model was verified by the means of measurements performed at a low scale laboratory stand. Analysis of the tested techniques when supplying voltage is fluctuating, allows to evaluate their usefulness in real operating conditions. The results obtained for SHE and SHM techniques were compared with SPWM modulation.

1 Introduction

The modern traction vehicles are commonly equipped with power electronics devices to provide fluent and dynamic control over torque of traction motors. This type of devices are known to be the significant source of current harmonics which may occur as the distortion for railway signalling system. That is the reason why railway operators provides limits for admissive amplitudes of current harmonics generated by traction vehicles. For example for Polish Railways (PKP) this limits are formulated in [2]. Fig.1 ('Limits Icat') presents mentioned above limits only for frequency band 1300-3100 Hz. This frequency band is known to be critical because of high level of current harmonics generated by traction drives and relatively low attenuation of vehicle's input filter. This was the motivation for the author to take this frequency band under consideration.

For purposes of this work 'Limits-Icat' were recalculated on the DC-link side of a low pass gamma-type input filter as the 'Limits-If' (Fig. 1). The current I_{cat} can be easily recalculated to current I_f (Fig.2), using formula (1) where impedances of the choke $Z_{Lf}(j\omega)$ and the capacitor $Z_{Cf}(j\omega)$ are taken under consideration. This formula is correct only for high frequencies where impedance of input filter can be assumed linear:

$$I_f(j\omega) = I_{cat}(j\omega) \cdot \left[\frac{Z_{Lf}(j\omega)}{Z_{Cf}(j\omega)} + 1 \right] \tag{1}$$

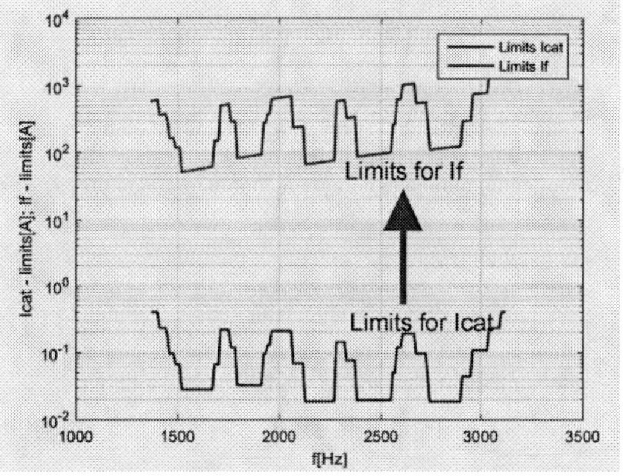

Fig. 1. Limits for catenary (I_{cat}) and recalculated for DC-link (I_f) current harmonics, generated by a traction vehicle in the frequency band between 1300 and 3100 Hz;

One of the opportunities to reduce the level of distortions generated by vehicles is to apply multilevel

* Corresponding author: msteczek@p.lodz.pl

traction converters instead of two-level ones [1] [4]. However, replacement of two-level traction inverters with three-level do not solve finally the problem of distortions generated by traction vehicles. The suitable control algorithm is required.

2 Problem description

In previous works [7, 8, 11] the Author presented the possibility of application of selective harmonic elimination (SHE) and selective harmonic mitigation (SHM) techniques to eliminate or control chosen current harmonics generated by traction drive to fulfil limits from Fig. 1. In mentioned analysis the DC traction power supply system was assumed to be modelled as perfect voltage source because voltage fluctuations were not in the scope of the conducted research. In this paper the Author took under consideration the influence of supply voltage fluctuation on efficiency of SHE and SHM techniques in traction current harmonics control.

The issue presented in this paper is to prove that change of supplying voltage can be compensated with change of modulation index of voltage waveform generated by traction inverter with sustained control of chosen current harmonics bellow assumed limits. The permissible changes of voltage in 3 kV DC traction system are set between 2000 and 3600V [9]. The task is to change the supplying voltage U_{DC} in the range set by the standard and change modulation index in such a manner to keep traction drive in one operating point described by the constant value of inverter's voltage basic component (U_{out}). Afterwards the corrections to harmonic control strategies must be implemented to keep current harmonics below assumed limits.

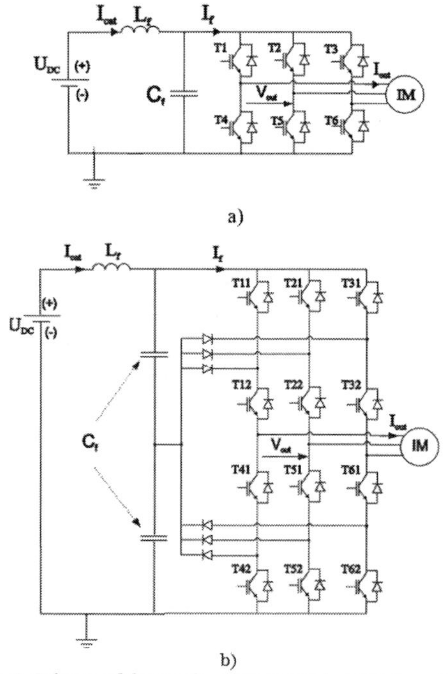

Fig. 2. Schema of the traction VSI, a) two level b) three-level.

3 Simulation model

The traction multi-drive simulation models were implemented using the Matlab Simulink environment. The two-level and three-level NPC VSI topology were chosen (Fig. 2). Simulation models of 500 kW induction traction motors were implemented. The measurement verification of simulation models used in this work was conducted on low scale laboratory model based on DSpace 1104 platform and presented in [7, 8, 11]. Simulation results were carried out for quasi-steady states of the analysed drive system with usage of time domain transient simulation option. The supply voltage was modelled as the perfect DC voltage source with variable voltage magnitude U_{DC}.

4 Selective harmonic elimination - SHE

The SHE modulation is one of the non-carrier based techniques. It was described for the first time in 1960s [10] and it was developed by Patel and Hoft in [5] [6]. When voltage harmonics are not eliminated but their values are set on certain values the technique is called Selective Harmonic Mitigation (SHM) [3]. SHM technique is still taken into account in the latest research and industrial applications [12, 13]. Both techniques (SHE and SHM) are based on direct determination of switching angles (time points) for VSI's transistors by solving the set of nonlinear equations, developed from the Euler's coefficients of Fourier series (2):

$$f(\omega t) = a_0 + \sum_{n=1}^{\infty}[a_n \sin(n\omega t) + b_n \cos(n\omega t)] \quad (2)$$

where: a_0, a_n and b_n – coefficients described by the Euler's formulas.

In this paper the quarter-wave symmetry of voltage waveform is assumed. In this case for two-level inverter coefficients a_0 and b_n are equal zero and a_n, can be described as follows:

$$a_n = \begin{cases} \dfrac{4 \cdot U_{DC}}{2n\pi}\left[1 + \sum_{i=1}^{N}(-1)^i \cdot 2\cos(n \cdot k_i)\right] & for\ odd\ n \\ 0 & for\ even\ \ n \end{cases} \quad (3)$$

Therefore for three-level inverter a_n, can be described as follows:

$$a_n = \begin{cases} \dfrac{2 \cdot U_{DC}}{2n\pi}\left[1 + \sum_{i=1}^{N}(-1)^i \cdot \cos(n \cdot k_i)\right] & for\ odd\ n \\ 0 & for\ even\ \ n \end{cases} \quad (4)$$

where: N- number of switching angles, k - switching angle.

Using N switching angles in quarter period N-1 voltage harmonics can be eliminated (or controlled) from the spectra of inverter's output voltage ($V_{out}(f)$). Regarding [7, 8, 11] using this technique, DC side current harmonics spectra can be controlled indirectly to match it to the assumed limits. In next part of this paper the case of U_{DC} variation for one inverter's operating point will be taken under consideration.

5 Assumptions

In this work one operating point of the two and three-level traction inverter was taken under consideration with different values of supply DC voltage (U_{DC}). The parameters of analysis starting point are presented in Table 1.

Table 1. Operating point parameters.

Inverters fundamental frequency (f_{fal})	**60 Hz**
Modulation index (M)	0.7
Voltage fundamental component (U_{out})	1200 V
Motor phase current (I_{ph})	215 A
Supplying voltage (U_{DC})	3400 V
Number of switching angles (N)	5

The simulations were provided for one inverter with one motor but the limits for I_{cat} current [2] were divided by 4 ('lim/4') for case of vehicle driven by 4 motors and by 8 ('lim/8') for case of double traction units. The limits were recalculated od DC-link current I_f regarding equation (1). The inductance and capacity of gamma filter were assumed as follows: L_f = 8.7 mH, C_f=6.5mF The aim of the study is to select a control strategy to meet the most restrictive 'lim/8' limits. for operating point from Table 1 the elimination of the lowest voltage harmonics was chosen (5th, 7th, 11th and 13th) On Fig. 3 it is shown that this control strategy is effective for 3400V U_{DC}. All current harmonics ale below assumed 'lim/8' limits. In next part of this paper a variation of U_{DC} will be implemented.

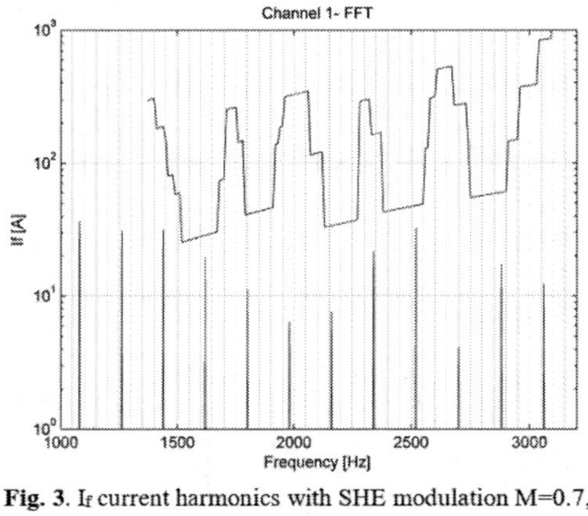

Fig. 3. I_f current harmonics with SHE modulation M=0.7, U_{DC}=3400V.

To prove the advantage of SHE over different modulation techniques the SPWM technique was implemented for the same operating point of the model.

Fig. 4 present I_f current harmonics and it can be observed that for SPWM technique current harmonics exceeded assumed limits.

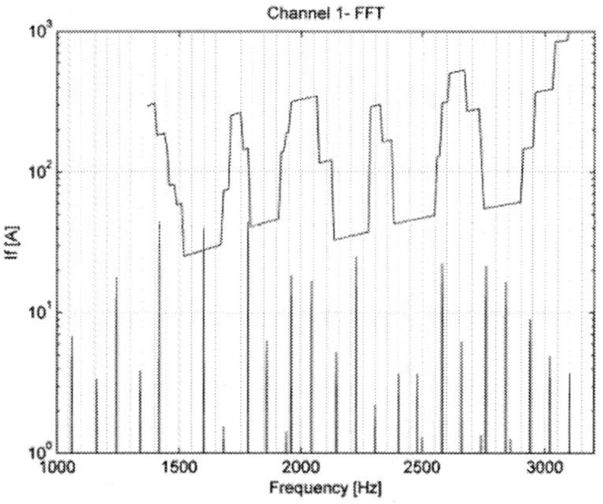

Fig. 4. I_f current harmonics with SPWM modulation M=0.7, U_{DC}=3400V.

6 Simulation results for three-level inverter

In this section the simulation results for three-level NPC inverter will be presented. In the first place the SHE strategy will be the same for all values of supplying voltage U_{DC}, only modulation index will be changed to keep constant value of VSI's output voltage basic component U_{out} = 1200V to keep traction drive in the same operating point. That means 5th, 7th, 11th and 13th voltage harmonics will be eliminated and modulation index will change according to values from Table 2.

Table 2. Modulation index correction regarding U_{DC} variations.

	U_{DC} [V]	M [-]	V_{out} voltage harmonics [-]			
			5th	7th	11th	13th
OP1	3600	0.66	0	0	0	0
OP2	3400	0.70	0	0	0	0
OP3	3200	0.75	0	0	0	0
OP4	3000	0.80	0	0	0	0
OP5	2800	0.85	0	0	0	0
OP6	2600	0.92	0	0	0	0
OP7	2400	1.00	0	0	0	0

Fig. 5 presents the values of I_f current harmonics for U_{DC} values from Table 2. It can be observed that for points describes in Table 2 as OP4, OP5, OP6 limits 'lim/8' are exceeded by harmonic 1620 Hz. The conclusion is: for this point, correction of control strategy is required.

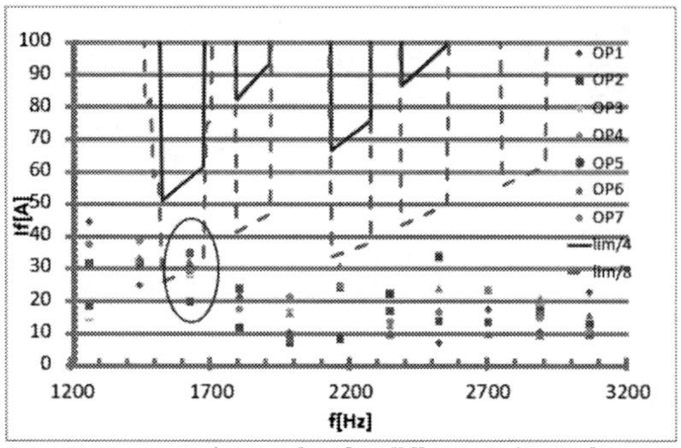

Fig. 5. I_f current harmonics for different values of U_{DC} and modulation index M (Table 2)-for three-level inverter.

For cases where limits were exceeded the SHE strategy was replaced by SHM with suitable values of mitigated harmonics. The pattern of harmonics control is presented in Table 3. Fig. 6 presents I_f current harmonics for U_{DC} values from Table 3 with revised control strategy to meet limits 'lim/8'.

Table 3. Revised voltage harmonics control strategy.

	U_{DC} [V]	M [-]	V_{out} voltage harmonics [-]			
			5th	7th	11th	13th
OP1	3600	0.66	0	0	0	0
OP2	3400	0.70	0	0	0	0
OP3	3200	0.75	0	0	0	0
OP4	3000	0.80	0	0	0.11	0.13
OP5	2800	0.85	0	0	0.11	0.13
OP6	2600	0.92	0	0	0.11	0.15
OP7	2400	1.00	0	0	0	0

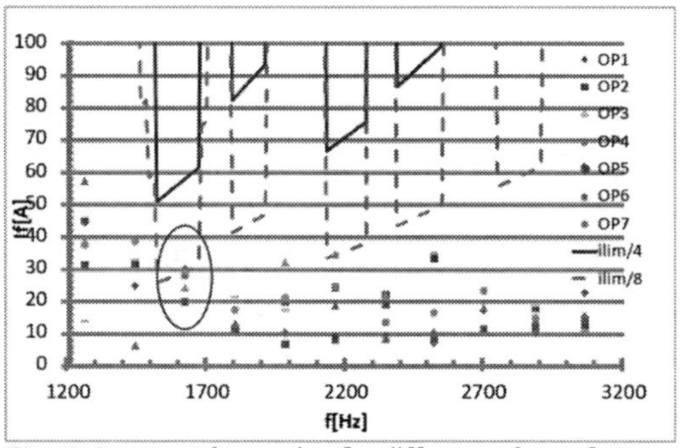

Fig. 6. I_f current harmonics for different values of U_{DC} and modulation index M (table 3))-for three-level inverter.

The following conclusion can be formulated: for chosen operating point of traction inverter the combination of SHE and SHM techniques can provide positive results in full range of supply voltage U_{DC}. Increasing of chosen voltage harmonics in SHM technique brings disadvantage in the form of increased THD of motor phase current and what follows increment of losses. The change of THD is presented in Fig. 7. It can be observer that maximum THD increment is 0.4 % for point OP4 and OP6, and it is relatively low in comparison to the benefits of meeting the compatibility requirements.

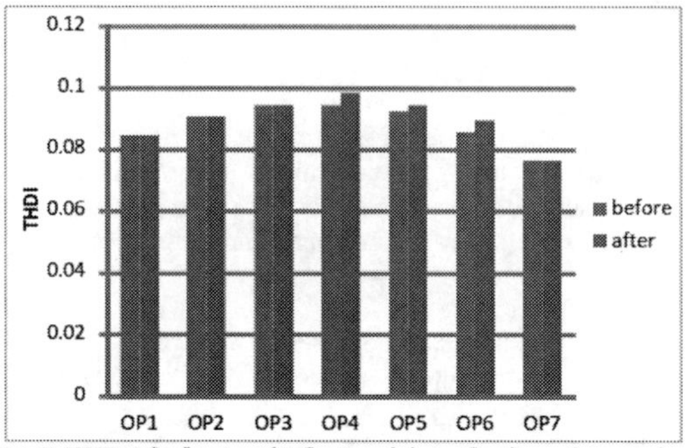

Fig. 7. THD I 'before' and 'after' revision of voltage harmonics control strategy for three-level inverter.

7 Simulation results for two-level inverter

In this section the results for two-level inverter will be presented. The approach to the problem will be similar to that presented in Section 6. Firstly the inverter's operating points from Table 4 will be analysed, taking into account the change of the supply voltage U_{DC} without correction of the harmonic elimination strategy. The results from this stage of work are presented in Fig. 8. It can be observed that current harmonic of 2160 Hz frequency exceeds the assumed limits for most operating points that were taken under consideration.

Table 4. Modulation index correction regarding U_{DC} variations

	U_{DC} [V]	M [-]	U_{out} voltage harmonics [-]			
			5th	7th	11th	13th
OP1	3600	0.66	0	0	0	0
OP2	3400	0.70	0	0	0	0
OP3	3200	0.75	0	0	0	0
OP4	3000	0.80	0	0	0	0
OP5	2800	0.85	0	0	0	0
OP6	2600	0.92	0	0	0	0
OP7	2400	1.00	0	0	0	0

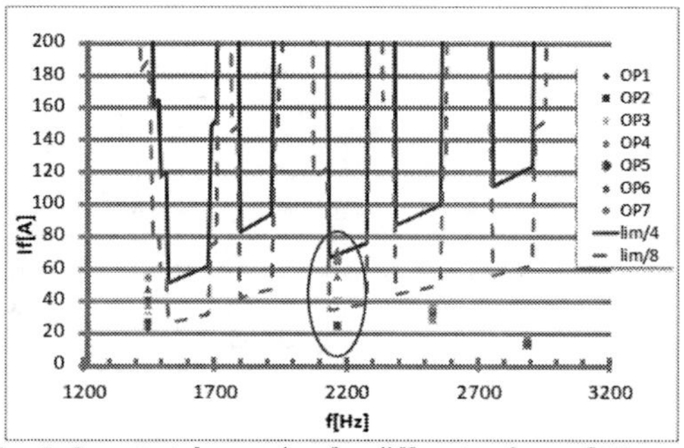

Fig. 8. I_f current harmonics for different values of U_{DC} and modulation index M (Table 4)-for two-level inverter.

The second step is to adjust the strategy of harmonics control to prevent exceeding the limits. The new strategy is described in Table 5. Fig. 9 presents that assumed limits are not exceeded using revised strategy.

Table 5. Revised voltage harmonics control strategy.

	U_{DC} [V]	M [-]	U_{out} voltage harmonics [-]			
			5th	**7th**	**11th**	**13th**
OP1	3600	0.66	0	0	0	0
OP2	3400	0.70	0	0	0	0
OP3	3200	0.75	0	0.1	0	0
OP4	3000	0.80	0	0.1	0	0.15
OP5	2800	0.85	0	0.1	0	0.1
OP6	2600	0.92	0	0.1	0	0.22
OP7	2400	1.00	0	0.1	0	0.1

The disadvantage of this method is negative impact on motor phase current THD (Fig. 10). The THD increase can reach even 2%. That means the losses in motor will increase as well. However the compatibility issues have priority, thus the proposed method is effective and ready for implementation in traction drives.

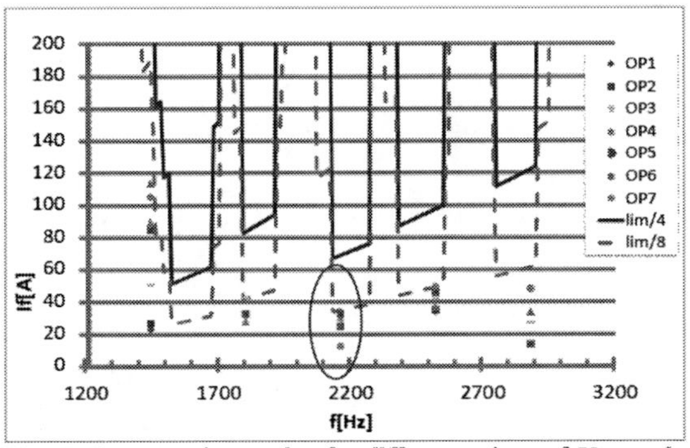

Fig. 9. I_f current harmonics for different values of U_{DC} and modulation index M (table 5))-for two-level inverter.

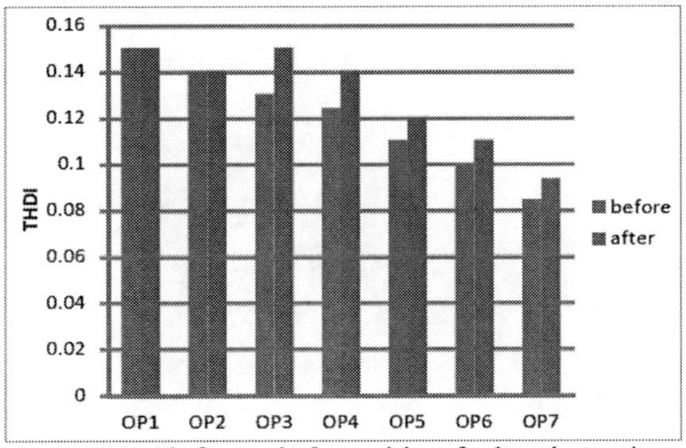

Fig. 10. THD I 'before' and 'after' revision of voltage harmonics control strategy for two-level inverter.

8 Conclusions

The variations of supplying voltage is a usual phenomenon in the electric traction power supply system. It is essential to take it under consideration during development of new traction drive control strategy. In this paper the Author presented the example of suggested corrections which must be implemented in SHE and SHM techniques to compensate the effect of DC voltage changes in traction drives. It was presented that using of proposed techniques it is possible to find solution which fulfil assumed restrictions regarding current harmonics in full range of DC voltage variations. The analysis were provided for both: two and three-level traction VSI. For every operating point of the VSI the different values of the supplying voltage must be taken under consideration. It is time consuming process but calculations are provided offline only once and implemented for whole lifetime of the drive. Presented work will be essential input in developing fully functional traction drive system based on SHE and SHM techniques.

References

1. H. Bernhard, A. Mariscotti, D. Wuergler. *Recommendations for the calculation of the total disturbing return current from electric traction vehicles.* IEEE Transactions on Power Delivery 19.3 (2004): 1190-1197 (2004)
2. A. Białoń, A. Kazimierczak, J. Furman, Ł. Zawadka, D. Adamski, P. Pajka, K. Ortel. *Defining limits and levels of disturbances in signalling and control railway systems (in Polish).* Railway Institute, Warsaw, Report 4430/10 for PKP PLK S.A. (2011)
3. J. Napoles, J. I. Leon, R. Portillo, L. G. Franquelo, M. A. Aguir. *Selective Harmonic Mitigation Technique for High-Power Converters.* IEEE Trans on Ind. Electron., vol. 57, no. 7, pp.2315-2323 (2010)
4. A. Ogusola, A. Mariscotti. *Electromagnetic Compatibility in Railway. Analysis and Management.* Lecture Notes in Electrical Engineering., Berlin Heidelberg Springer-Verlag, pp. 1-528, ISBN 978-3-642-30281-7 (2013)
5. H. S. Patel, R. G. Hoft. *Generalized techniques of harmonic elimination and voltage control in thyristor inverters: part I - harmonic elimination",* IEEE Trans. Industry Applications, vol. IA-9, pp.310-317. (1973)
6. H. S. Patel, R. G. Hoft. *Generalized techniques of harmonic elimination and voltage control in thyristor inverters: part II-voltage control techniques.* IEEE Trans. Ind. Applications, vol. IA-10, pp.666-673 (1974)
7. M. Steczek, A Szeląg. *Modification of Selective Harmonics Elimination Method for Effective Catenary Current Harmonics Reduction.* IEEE Int. Conf. Electrical Drives and Power Electronics (EDPE), pp. 394-401 (2015)
8. M. Steczek, P. Chudzik, A. Szeląg. *Combination of SHE and SHM - PWM techniques for VSI DC-link current harmonics control in railway applications.* IEEE Transactions on Industrial Electronics,

vol.64, issue 10, DOI 10.1109/TIE.2017.2694357 (2017)

9. A. Szeląg, Z. Drążek, T. Maciołek. *Elektroenergetyka miejskiej trakcji elektrycznej*. Instytut Naukowo-Wydawniczy SPATIUM, Radom, ISBN 978-83-62805-42-6, page 59 (2017)
10. F. G. Turnbul. *Selected harmonic reduction in static dc-ac inverters*. IEEE Trans Communication and Electronics, vol. 83, pp. 374-378 (1964)
11. M. Steczek, A. Szeląg, D. Chatterjee. *Analysis of disturbing effect of 3 kV DC supplied traction vehicles equipped with two-level and three-level VSI on railway signalling track circuits.* Bulletin of the Polish Academy of Sciences Technical Sciences, Vol. 65, No. 5, 2017 DOI: 10.1515/bpasts-2017-0071 (2017)
12. M. Sharifzadeh, H. Vahedi, R. Portillo, L. G. Franquelo, K. Al-Haddad. Selective harmonic mittigation Based self-elimination of triplen harmonics for single-phase five-level inverters. IEEE Transactions on Power Electronics, earlu access DOI 10.1109/TPEL.2018.2812186 (2018)
13. A. Moeini, H. Zhao, and S. Wang, "A Current Reference based Selective Harmonic Current Mitigation PWM Technique to Improve the Performance of Cascaded H-bridge Multilevel Active Rectifiers," IEEE Transactions on Industrial Electronics, vol. 65, pp. 727 - 737, (2018)

MATEC Web of Conferences **180**, 05003 (2018) https://doi.org/10.1051/matecconf/201818005003
MET'2017

Electrified lines entering tunnels in Poland, safety aspects of the running with fire

Marek Pawlik[1,*]

[1]Railway Research Institute, ul. Chłopickiego 50, 04-275, Warsaw, Poland

Abstract. Potential fire on board of the passenger train creates risk for passengers health and passengers life. Using non-flammable materials for producing Electric Multiple Units (EMUs) is not considered to be sufficient as passengers are carrying different types of materials as their luggage and belongings. In case of fire on board of the train, the EMU should stop in a place which is appropriate for passengers evacuation and fire fighting. Not all places along railway lines are considered to be appropriate for such purpose. Especially stopping in tunnels is considered inappropriate. Meanwhile the amount of tunnels on the Polish railway network is growing significantly. This requires application of the Technical Specification for Interoperability regarding Safety in Railway Tunnels (TSI SRT) and set of fire standards EN 45545. Using those requirements is associated with putting tunnels and rolling stock into service under railway interoperability directive 2008/57/WE. Additional analyse is however necessary from operational safety point of view due to growing amount of tunnels and growing diversity of rolling stock. Paper proposes conducting risk analyse in accordance with common safety method for risk evaluation and assessment based on a dedicated European Standard EN 50533 which defines requirements regarding running capability in case of fire on board of the rolling stock. Paper analyses applicability of that approach for group of tunnels and rolling stock fleet comprising different types of rolling stock including Electric Multiple Units.

1 Introduction

EN 50533 standard [7] distinguishes three types of fire and defines reference fire (with which the train should be able to continue running). It also defines assessment model, which is to be applied for assessing trains and their critical functionalities important for running capability in case of fire. The critical functionalities are especially those ones linked with ensuring appropriate traction and braking necessary for reaching safe place appropriate for passengers evacuation and fire fighting. Quoted standard is generally dedicated for assessment of new trains, as it defines specific technical measures, compliance with which allow the 'presumption of conformity' with Commission Regulation (EU) 1303/2014 [2] concerning technical specification for interoperability relating to 'safety in railway tunnels'.

Despite that, the key challenge in relation to running with fire via tunnels in case of Poland is not to verify new trains, but to verify respective overall safety which is changing due to opening of many tunnels on the Polish railway network taking into account not only new but also existing trains, which are in operation usually longer than twenty years.

2 Operational safety

The 10-by-10 safety matrix defined by author in previous publications [10, 11] show, that operational safety as well as activities in case of emergency are associated with ten technical safety aspects T1-T10, out of which two, namely T4 and T9, also have security dimension. The technical aspects which have to be considered are: T1 - malfunctioning safety, T2 - construction safety, T3 - electrical safety, T4 - preventing unauthorized access and fire, T5 - wheel/rail interaction safety, T6 - control command and signalling (active safety systems), T7 - power supply/signalling interaction safety, T8 - safety of the operational rules and staff competences, T9 - preventing panic, and T10 - IT support for safety. Not all of them are associated with running with fire on the board of the EMU in the context of growing amount of the electrified lines entering tunnels in Poland. However not only T4 which is directly dedicated to fire hazards have to be considered, but also relevant requirements which can be seen as associated with T1, T2, T3, T6, T8 and T9.

* Corresponding author: mpawlik@ikolej.pl

https://doi.org/10.1051/matecconf/201818005003

3 Preventing unauthorized access and fire (T4)

On one side from fire risk point of view it is important to prevent unauthorised access to railway tunnels. This is not directly associated with protection against fire itself, however from the fire hazard point of view it is important to verify that in that respect appropriate measures are defined and applied. On the other side preventing unauthorised access to passenger trains, which are entering tunnels in commercial operation is irrelevant. Passengers can enter trains carrying different types of materials as their luggage and belongings including also flammable materials and products. As unauthorised persons one can point passengers without appropriate tickets, but luggage fire-checks never takes place.

The amount of materials which are carried by passengers is very limited in comparison to materials used for passenger train construction. Taking that into account national railways since many years were defining fire requirements for construction materials. Common approach in that respect was necessary for interoperability purpose since adoption of the interoperability directive [1].

The long and challenging works on common European fire requirements for rolling stock materials resulted finally with EN 45545 standard [5]. This works were conducted with an important and long involvement of Polish experts from Railway Institute (IK) and were disseminated by Institute via publications [8, 9] and organization of the international fire conferences, which take place every second year (e.g. III International Conference on Modern Trends of Fire Protection in Rolling Stock, which took place in Warsaw in 2016). The already agreed standard [5], which sets obligatory requirements for interoperable rolling stock (as it is quoted by European regulation) and non-interoperable rolling stock (as it is quoted by Polish regulation [4]) is still under on-going standardisation works. European Committee for Standardisation (CEN) working parties undertake, together with IK experts, works focused on improving set of seven parts of the EN 45545standard:

- Part 1 – general requirements
- Part 2 – fire behaviour of materials and components
- Part 3 – fire barriers resistance requirements
- Part 4 – fire safety in railway rolling stock design
- Part 5 – electrical equipment fire safety
- Part 6 – fire control and management systems
- Part 7 –flammable liquid and gas installations.

Smoke properties, toxicity of combustion products, flame propagation, and flammability are defined taking into account [8]:

- weight of material and dimensions of elements,
- distance from the place of potential ignition, and
- hazard level due to vehicle type & train category.

Technical Specification for Interoperability regarding '**safety in railway tunnels**' (TSI SRT) [2] defines requirements not only for prevention pointing EN 45545, but also for mitigation, escape and rescue. Those three aspects are associated with ensuring measures which are necessary for activities, which have to be possible in case of emergency.

4 Risk assessment

Judgement in relation to mitigation, escape and rescue for new type of rolling stock and for new tunnels is performed on the basis of the TSI SRT [2]. This leads to fully safe situation after modernisation of tunnels and replacement of old rolling stock by new one. This requires however many years and therefore it is proposed to use risk assessment methodology for verifying fire safety for existing and new tunnels as well as existing and new rolling stock.

4.1 EN 50533 based malfunctioning safety (T1)

EN 50533 standard is dedicated for verification of running capability in case of fire on board of the train.

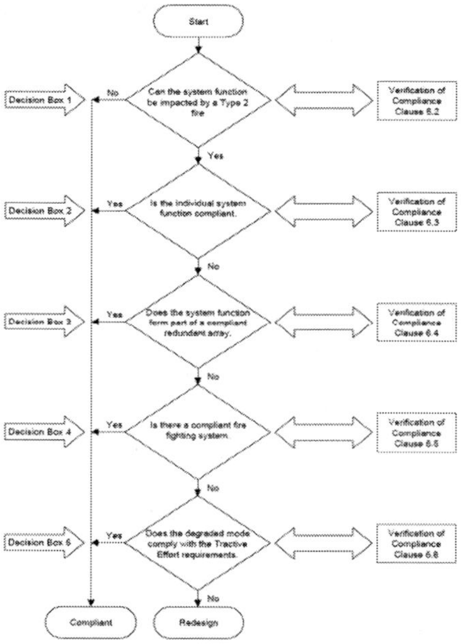

Fig. 1. Decision box approach to running capability. [7]

Figure 1 shows decision box approach defined in EN 50533 Standard for assessing running capability in case of fire on board of rolling stock. Shown algorithm is intended to be used separately for all functionalities, which are identified as critical (especially those ones linked with ensuring appropriate traction and braking necessary for reaching safe place appropriate for passengers evacuation and fire fighting). It is composed by five decision boxes defined in respective chapters of the EN 50533 Standard.

It is important to understand, that these steps are generally performed to prove that all such individual functions would be ensured in case of type 2 fire. The type 2 fire is defined as a *"fire which, due to its size*

 https://doi.org/10.1051/matecconf/201818005003

and/or location, presents a risk of serious injury and/or threat to life and for which it is reasonably practicable to define running capability requirements". Running capability for individual critical functionalities is assessed on the basis of a type 2 reference fire defined as a "*fire which has the power vs. time profile of Ignition source 5 of CEN/TS 45545-1; 75kW for 2 minutes followed by 150 kW for 8 min*". Three safety measures are foreseen to be used when individual functionality is not compliant itself: moving device performing such functionality to another place in the train, adding second and possibly third device independent from the first one performing the same functionality and located separately to create so called 'redundant array' and adding appropriate fire fighting systems ensuring required running capability.

4.2 Ensuring control command operation in case of fire and inhibiting passenger break (T6)

Ensuring correct operation of the control command on-board equipment is an example of functionality which has to be assessed. Fire cannot affect neither performed functions nor fail-safety and SIL 4 safety integrity of respective equipment used for receiving electronic movement authorities and continuous verification of running in compliance with authorities.

Special function is foreseen for new trains and for the European Train Control System ETCS to ensure inhibiting passenger break when as a result of use of such break by passenger train would stop inside the tunnel – in a place inappropriate for evacuation and possibly required fire fitting.

4.3 Basic hazard log

All critical identified hazards associated with malfunctioning of critical functionalities, lack of functioning of critical devices associated with critical functionalities as well as loss of fail-safety or SIL4 related functions have to be seen as hazards and put into hazard log. An example hazard log is shown on Figure 2.

Such hazard log prepared in accordance with risk evaluation and assessment regulation [3] is legally required for individual trains, which are being modernized. Application of EN 50533 in case of new types of EMUs and other new types of vehicles can be performed directly according 'decision box' approach.

4.4 Complementary safety aspects (T2 T3 T8 T9)

Type 2 fire cannot affect construction safety (T2). Especially it cannot affect within 8 minutes, from the beginning of fire: housings, seals, devices and cables mounting racks, etc.

Type 2 fire cannot affect electrical safety (T3). Especially it cannot cause within 8 minutes neither loos of power supply nor overvoltage. It cannot cause short-circuits as well as presence of electric arch.

All operational rules (T8) necessary in degraded modes of operation taking place in case of type 2 fire have to be defined and communicated to operational and railway rescue staff. The staff competences and health have to be regularly verified.

All types of equipment foreseen to be used for preventing panic (T9) by e.g. staff communication in emergency and public address systems as well as evacuation ways and doors together with emergency lighting have to be ensured and regularly verified.

	device malfunctioning				device not functioning			affected device fail-safety or SIL		
Identified hazards										
Associated function										
Associated device										
Hazard description										
Hazard reasons										
Remarks										
Potential consequences										
Existing safety means										
Frequency level *)										
Severity category *)										
Risk acceptance *)										
Additional safety means										
Responsibility split **)										
Frequency level *)										
Severity category *)										
Risk acceptance *)										

Fig. 2. Step by step approach to hazards (own elaboration):
*) as defined in EN 50126 [6],
**) as defined in regulation 2013/402 [3].

4.5 Complementary security aspects (T4 and T9)

Moreover it is a must to ensure security especially in relation to prohibiting unauthorised access and protecting all critical places [T4] like control rooms, interlocking rooms, technical compartments, cabinets, etc. They have to be protected not only against vandalism and usual crime (against theft) but also against intentional fire ignitions and terrorist attacks which can result with fire.

Also rules defining split of staff responsibilities and procedures dedicated for preventing panic [T9] have to be defined, known and used.

4.6 Extended hazard log with sensitivity verification for network running capability assessment

Complementary safety and security aspects show, that some activities are not connected with fire verification of individual EMUs and vehicles forming trains.

Fire relevant fact have to be considered e.g. that amount of electrified lines entering tunnels on the Polish network is growing significantly, some tunnels are old and their reconstruction is not foreseen, new and old tunnels are used in commercial operation by new Electrical Multiple Units and existing EMUs as well as other trains including cargo ones.

It is proposed to use extended hazard log for assessing running capability in case of fire in the Polish network scale. It can be done by defining network

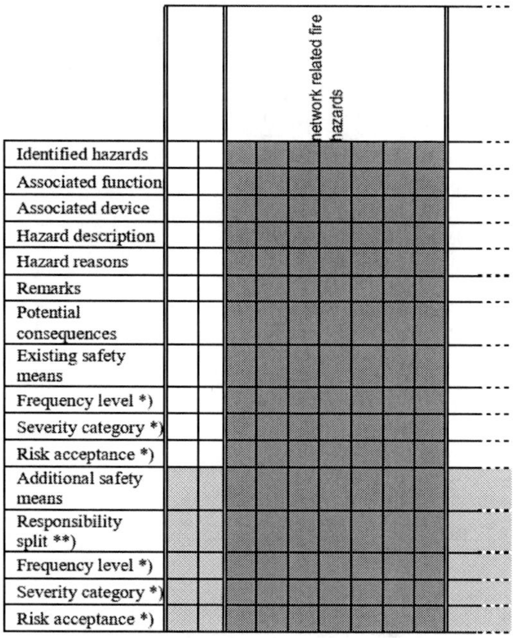

Fig. 3. Proposed network approach to tunnel fire hazards. (own elaboration)

related fire hazards, representing most sensitive combinations of trains and tunnels. It is proposed to take into account:

- existing old railway tunnel near Tunnel station on the railway line nr 8 constructed in the nineteen century (1882–1884) used by passenger and freight trains operated by different railway undertakings on the line between Warsaw and Cracow,
- recently constructed (2008-2012) new railway tunnel already in operation, which is leading to railway station Warszawa – MPL Okęcie at the Warsaw airport used by passenger EMUs operated by two regional railway undertakings (Mazovian KM and Warsaw SKM),
- railway tunnel presently under construction over four tracks between Gdańsk Główny and Gdańsk Śródmieście, being constructed under full railway operation on long distance tracks between Warsaw and Gdańsk (Pendolino trains and freight trains with containers and with tank vehicles carrying flammable and explosive liquids going to the Gdańsk Seaport) and on local tracks between Gdańsk Śródmieście and Gdynia (old and new EMUs fulfilling and not fulfilling requirements related to running capability in case of fire.

It is proposed not only to consider different types of trains – new and existing operated since many years but also to take into account:

- presence of two trains in tunnel both in case of running on different trucks in opposite directions and in case of a train following another train which is also in a tunnel,
- pure passenger traffic with fire on board of one train as well as situations in which passenger train is in a tunnel together with freight container train with type 2 fire on board of the train.

Protection against fire of the tank vehicles carrying flammable and explosive liquids cannot be ensured – such fire is considered to be EN 50533 type 3 fire. However [redacted] fires which can take place in case of [redacted] locomotives and [redacted] Multiple Units have to be taken into account. Lack of protection against fire in case of tank vehicles carrying flammable and explosive liquids does not mean that there is no need for minimising fire risk in case of dangerous goods – all trains and vehicles carrying dangerous goods have to be loaded, operated and unloaded in accordance with the COTIF (Convention concerning International Carriage by Rail) RID (Regulations concerning the International Carriage of Dangerous Goods by Rail) requirements.

Defining extended hazards for Polish case is therefore not only possible but also reasonable. The main representative hazards can be defined on the [redacted] tunnels and trains which are in regular operation over those tunnels. Such analyse would allow defining probably presently required additional safety means and the ways they would affect frequency levels and severity categories used for risk acceptance.

Such analyse should however be verified by sensitivity check. Defined additional safety means may ensure enough protection in case of scheduled traffic. Sensitivity analyse should verify whether they would ensure protection in case of reorganizations of traffic which takes place in case of operational disturbances. It is probable, that as a result some operational restrictions may be defined for resolving operational disturbances. For instance some old tunnels may be closed for some types of freight in normal operation and used for such purpose only in case of operational disturbances. Such use may be subject of individual decisions and may be forbidden to take place together with passenger traffic served by old trains or by all passenger trains.

5 Conclusions

It is reasonable and possible to conduct risk based analyse of safety aspects of the running with fire on the board of the EMU in the context of growing amount of the electrified lines entering tunnels in Poland. Such analyse may result with necessary technical safety measures and/or safety restrictions for operation.

References

1. Directive 2008/57/EC of the European Parliament and of the Council of 17 June 2008 on the interoperability of the rail system within the Community (EU OJ L 191/1 of 18 July 2008) with further changes (which is to be superseded by Directive (EU) 797/2016)
2. Commission Regulation (EU) No 1303/2014 of 18 November 2014 concerning the technical specification for interoperability relating to 'safety in railway tunnels' of the rail system of the European Union (EU OJ L 356/394 of 12 December 2014)
3. Commission Implementing Regulation (EU) No 402/2013 of 30 April 2013 on the common safety method for risk evaluation and assessment and repealing Regulation (EC) No 352/2009 (EU OJ L 121/8 of 3 May 2013) with changes
4. Office of Railway Transport President List of the relevant national technical rules and normative documents, which are being applied for achieving fulfilement of the essential requirements relevant for railway system interoperability, Office of Railway Transport, Warsaw 19 January 2017 (Lista Prezesa Urzędu Transportu Kolejowego w sprawie właściwych krajowych specyfikacji technicznych i dokumentów normalizacyjnych, których zastosowanie umożliwia spełnienie zasadniczych wymagań dotyczących interopera-cyjności systemu kolei, Urząd Transportu Kolejowego, Warszawa 19 stycznia 2017 r.)
5. EN 45545:2013 Railway applications – Fire protection on railway vehicles –Requirements for fire behaviour of materials and components – all parts (from EN 45545-1 to EN 45545-7)
6. EN 50126-1:1999 Railway Applications - The Specification and Demonstration of Reliability, Availability, Maintainability and Safety (RAMS) (which is to be superseded by FpprEN 50126-1:2017)
7. EN 50533:2012 Railway applications - Requirements for running capability in case of fire on board of rolling stock
8. J. Radziszewska-Wolińska, "Development of Requirements for Fire Protection of Rolling Stock in Poland and ITS Comparison with EN 45545", Railway Reports volume 57/2013 issue 160, Railway Research Institute, Warsaw 2013, ISSN 0552-2145
9. J. Radziszewska-Wolińska, "Revision Process of EN 45545", Railway Reports volume 58/2014 issue 164, Railway Research Institute, Warsaw 2014, ISSN 0552-2145
10. M. Pawlik, "Bezpieczeństwo kolei w kontekście powiązań pomiędzy dyrektywami o bezpieczeństwie kolei i o interoperacyjności kolei, analiza z punktu widzenia zarządzania bezpieczeństwem" ("Railway safety in the context of relationships between railway interoperability directive and railway safety directive, analyse from the point of view of safety management"), Przegląd Komunikacyjny (Transport Review), 09/2016, ISSN 0033-22-32
11. M. Pawlik, "Safety, security and cybersecurity in railway operation", Safety and Reliability – Theory and Applications, 2017 Taylor & Francis Group, London, ISBN 978-1-138-62937-0

Integrated power supply system for station equipment of rail traffic control

Ryszard Mielnik[1], and *Zofia* Wróbel[2,*]

[1]Cracow University of Technology, Institute of Electromechanical Energy Conversions E-2, ul. Warszawska 24, 31-155, Kraków, Poland
[2]PKP Polish Railway Lines JSC, Railway Lines Establishment in Rzeszów, ul. St. Batorego 24, 35-005, Rzeszów, Poland

Abstract. Railway Traffic Control Systems (RTCS) provide a safe, reliable and efficient movement of rolling stock on railway networks. RTCSs are classified as critical process systems. The power supply system is required to provide a high level of power reliability so that a station equipment of an RTCS safely executes the rail traffic process. The article describes typical solutions of power supply of station equipment of RTCSs. Analysis of these solutions led to the development of assumptions and structures for the integrated power supply system of station equipment. Useful tools for the formal modeling of such systems can be Petri Nets. These nets are a graphical and formal tool for modeling, formal analysis, and design of discrete event systems. The basic element of the integrated power system which is described in the article is the Automatic Reserve Switching system (ARS system). The ARS controller Marked Petri Net (ARS controller MPN) that has been developed by the authors is presented here. The net was subject to reduction resulting in an ARS controller hierarchical simplified MPN. The article also presents the practical application of the ARS controller built using the hardware platform and graphical programming environment LabView National Instruments (NI).

1 Introduction

Nowadays measurement and control digital systems (MCD systems) are commonly used in industrial processes [1]. These systems are built on the basis of hardware platforms such as PLC's (Programmable Logic Controller), industrial computers or embedded systems. The functionality of MCD systems depends on software applications as implemented by particular users. Thus, the quality of the action of this system is influenced not only by the hardware platform, but also by its user's software. The need to ensure high reliability of the software and the reliability of the entire MCD system is required especially in the field of the so-called critical applications. In these applications malfunction or failure of MCDs during their operation may lead to abnormal functioning of supervised process resulting in the loss of life and material waste. Ensuring the required quality of an MCD system during the design and prototyping process requires the use of appropriate tools and methods to formalize its functions and actions. Suitable instruments which can be used for this purpose are Petri Nets. They are graphic and formal tools for modeling, analysis and design of discrete event systems. The model represented by such nets allows for the analysis of the characteristics of the system's behavior, and its evaluation in each phase of the system's life cycle.

The aim of this publication is to present the opportunities for using Petri Nets in modeling MCD systems in critical applications [2, 3, 4, 5, 6], i.e. the integrated controller of the Automatic Reserve Switching (ARS) device of the power supply system of Railway Traffic Control Systems (RTCS). Here, the ARS controller Marked Petri Net is presented. This publication also shows the practical implementation of such ARS controller. It has been implemented using the hardware platform of the family of controllers NI CompactRIO and a software application implemented in the National Instruments graphical environment LabView [6, 7, 8].

2 Power supply system for Railway Traffic Control Systems

2.1 Railway Traffic Control Systems

Railway Traffic Control Systems (RTCS) provide safe, reliable and efficient means for the movement of rolling stock on railway networks [2]. Therefore, RTCSs are classified as critical equipment and systems [4, 5, 6, 7]. The history of RTCSs is dated back to the end of the 19th century when the industrial revolution introduced rolling stock for the transport of goods and people. Initially, railway traffic safety devices were simple, and the main role in the decision making and execution process was in the hands of human operators. Currently, mounted station RTCS systems are relay and computer based systems or computer based systems only [2, 3]. In relay and computer based systems, the security is ensured in the relay layer, and the setting functions are performed by a computer control panel [2]. The

* Corresponding author: zwrobel@prz.edu.pl

development of information and computer technology has led to the construction of secure computer systems. For systems of this type, PLCs or computers with "2 of 2" or "2 of 3" voting systems are used to provide the required level of security [2, 3, 7]. The use of relay and computer based systems or computer based systems requires the provision of adequate reliability and quality of power supply [9]. Therefore, new, more suitable power supply systems are introduced so that they can meet strict sets of requirements.

2.2 Power supply systems for RTCS

The current requirements for the design and manufacture of power supply systems for RTCS systems are contained in Chapter 13 of Management Order No. 1/2014 of PKP (Polskie Koleje Państwowe) Polish Railways. [9]. The general requirements (§ 72) of this chapter state that the primary sources of power supply for station RTCS systems are power grids that meet the requirements for:

- regulations for the construction of power equipment,
- principles of protection against overvoltage,
- principles of shock protection.

The block diagram of power supply system's general structure intended for a RTCS computer station is shown in Figure 1 [7, 9, 10, 11, 12].

The use of relay and computer based systems or computer based systems requires the provision of adequate reliability and quality of power [9]. Therefore, new, more suitable power supply systems are introduced, so that they can meet the increased requirements.

2.3. ARS controller

Due to the above, based on the analysis of the requirements and technical solution of the currently produced power supply systems of station RTCSs a development of an integrated ARS controller was proposed. This controller integrates power components and contains:

- measurement systems,
- logic of the ARS system operation,
- diagnostics system with data transfer interfaces.

The measuring system checks the correctness of the power supply of 230/400 V AC +/- 10%. If the measured voltage is not within the given measurement range, then a logic signal is generated to the controller's system logic block. Based on logic rules and set priorities, this system turns on or off relays that turn on voltage for station equipment. Information about the status of the power elements is transferred to the control panel and the remote diagnostics system. The location of this controller in the power supply system is shown in Fig. 2.

2.4. Application steps of ARS controller

Based on the analysis of current solutions [2, 10, 11, 12] and the requirements for power supply systems of station RTCS [9], a functional model of a controller with a Petri Net has been developed. Based on the developed net and the use of the NI LabView graphical programming language, a dedicated software application has been developed. In parallel to the design and implementation of the software, several assumptions have been made regarding the hardware platform of the controller. NI CompactRIO platform has been selected as the development platform. The next step of the ARS controller implementation is the integration of the software application developed for the hardware platform. Following the integration and positive activation of the controller, functional tests were performed on the basis of a previously developed research program. This enabled for the verification of assumed assumptions for the ARS controller.

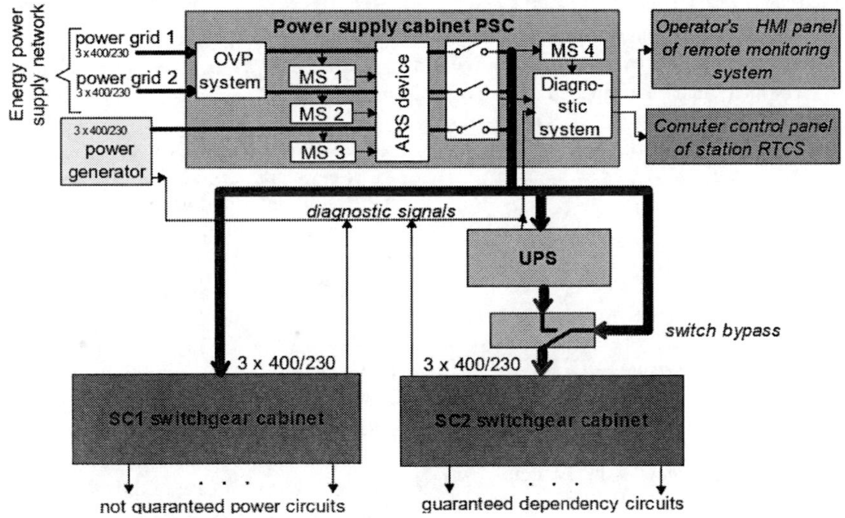

Fig. 1. General structure of a power supply system for the computer station RTCS's [7].

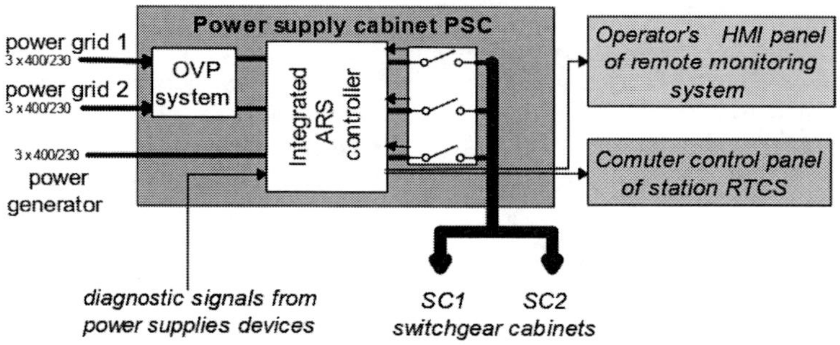

Fig. 2. Structure of the station RTCS power supply system with an ARS controller.

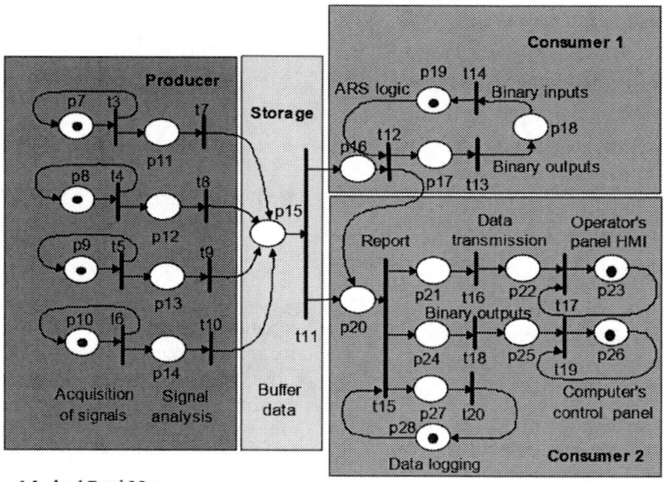

Fig. 3. The ARS controller Marked Petri Net.

3. ARS controller's Petri Net

3.1. Petri Nets

In 1962 Carl A. Petri developed based on graph theory based tool called Petri Nets (PN) to model discrete systems [13-17]. As a result of long-term development of the theory of these nets a number of classes of PNs have emerged for which numerous applications have been established [4, 5, 6, 7, 8, 13, 14]. During the development phase of assumptions and design requirements as the graphical tool for modeling of complex systems Petri Nets allow for a good and clear communication between designers and customers.
Therefore, *Marked Petri Net (MPN)* [15, 17]:

$$N = (P, T, A, M_0) \qquad (1)$$

where the following conditions are met:

$N = (P, T, A)$ is the PN,

$M_0 : P \rightarrow \mathbb{Z}_+$ it is a function defined on the set of places called *initial markings* of the N net.

In this class of nets places have an unlimited capacity for tokens, but arcs can carry only one token. The execution of transition involves eliminating of individual tokens from the input places and adding the individual tokens to the exit places of this performed transition. In the class of *Generalized Petri Nets (GPN)* [15, 17] it is possible to eliminate or add to one place more tokens in a single transition by assigning appropriate weights to the arcs. By introducing the function of limiting capacity of the places for tokens to the GPN, one of the most common net classes in literature [15, 17] called *Places and Transitions Petri Net (PTPN)* is obtained [15, 17]. The PNPT can easily replace a GPN which has the same properties. Accordingly, the term PTPN often refers to GPN in the literature [15, 17]. It can be stated that the MPN class is a special case of GPN and PTPN in which there is an unlimited token capacity and arcs carry only one token (the weight of the arcs is 1).

3.2. ARS controller Marked Petri Net

Based on the features of the ARS integrated controller power supply system of station RTCS, a Marked Petri Net (ARS controller MPN) has been developed for the ARS controller. This MPN in which the capacity of the places are unlimited and only one token can pass through the arc is a special case of the GPN and PTPN class of a PN. The developed MPN for the ARS controller can – highlighted in Figure. 3. It includes

- Producer process for generating tokens
- Storage process (token buffer)
- Two consumer processes for absorbing tokens.

Figure 4 shows the hierarchical simplified MPN for the ARS controller. This net was developed as a result of the hierarchy of the ARS controller MPN from Figure. 3, using the bottom–up method of creating [17]. The resulting hierarchical simplified MPN is easier to analyze. This net corresponds to the producer-consumer model system. Classes of this type net are well known in literature [17]. One of the important features of this net is its liveliness. This means that the net does not get jammed, and every producer's token is consumed by consumer processes [17]. Fig. 5 shows the block-diagram of the ARS controller with the macro transition MT1 from the hierarchical simplified Marked Petri Net.

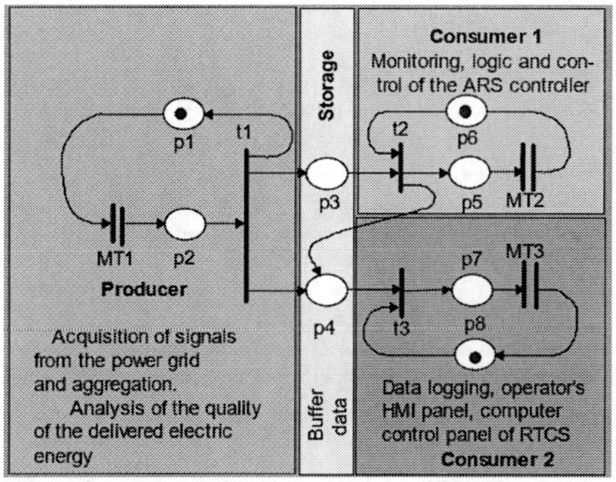

Fig. 4. The ARS controller hierarchical simplified Marked Petri Net [7].

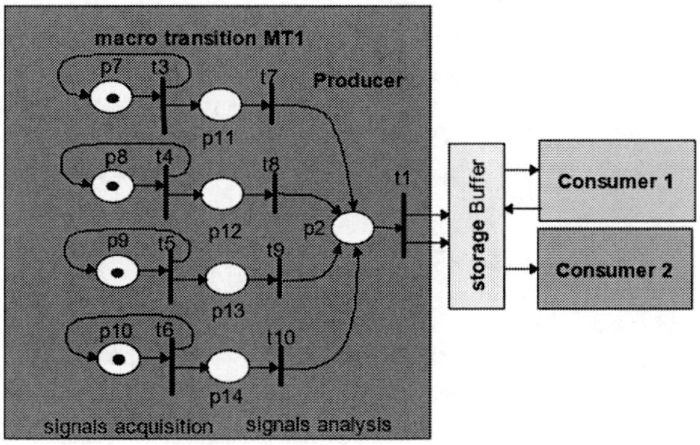

Fig. 5. The ARS controller block`-diagram with the macro transition MT1.

4. Practical implementation of the ARS controller

4.1. Hardware platform of ARS controller

The ARS controller hardware platform was configured using NI CompactRIO components [8]. It is a modular real-time platform. It meets the stringent requirements of environmental standards approved by the Management of the Polish Railway Lines. This platform consists of:

- NI cDAQ-9135 cassette with processor module equipped with: Atom processor 1.33 GHz, 32 GB RAM, 8 slots, Linux RT real-time operating system, interface to HMI panels,
- NI PS power supply module,
- 4 pieces of analog input modules NI 9244,
- 2 pieces of digital inputs/outputs NI 9375.

4.2. Software application of ARS controller

The National Instruments LabView graphical software environment [6, 8, 18] was used to implement the ARS controller software application. This environment incorporates numerous tools for a rapid software development. The main advantage of this environment is that the execution of the program is based on the data flow between the nodes of the program. These include simple functions, complex functions, VI Express functions and SubVI functions developed by users. Executing the node function will only occur if the node inputs will have data. This requirement is similar to the mechanism of triggering transitions in Petri Nets. This environment also includes programming structure patterns. One of them is the producer-consumer pattern. It transmits data from the producer to the consumer through the data buffer so that no data is lost. The pattern of this structure was used in the software application of the ARS controller. In the application the authors also used the previously mentioned mechanism of building own subprograms (SubVI). For the analysis of the measured signals Electric Power tool was used in which functional blocks are used to analyze the quality of electricity. DAQmx functions driver have also been used for the acquisition of measured signals. Software in the LabView environment is developed in two windows. The first one, called the Front Panel, is responsible for the communication between the operator and the system. In the second window, called the Block Diagram, the program code of the application is placed.

5. Conclusions

Following the integration of the hardware platform and the application software as well as the launch of the ARS controller, functional laboratory tests of the controller were performed. Laboratory tests were carried out on the developed controller. Three regulated 230/400 V AC power supplies with a wide range of output voltage regulation were connected to the controller. These studies were made on the basis of a previously prepared research program. After minor modifications to the controller's application software it can be concluded that the ARS controller's operation is correct. This tested controller and the documentation were provided to a manufacturer of the power supply system of the RTCS stations. The controller is currently undergoing field testing. Once completed it will be possible to apply for its admission for the use of the integrated ARS controller in RTCS stations within the Polish Railway Network.

References

1. W. Winiecki, *Organization of computer measuring systems,* Warsaw University of Technology Publishing House, Warsaw, 2006) [In Polish]
2. M. Dąbrowa–Bajon, *Basics of railway traffic control* (Warsaw University of Technology Publishing House, Warsaw 2002) [In Polish].
3. A. Lewiński and Perzyński, *New solutions of control computers in railway traffic control systems on the example of SSP,* (Conference materials, Transport in the 21st century, Warsaw 2001) [In Polish]
4. R. Mielnik, *Petrii Net as a tool for modeling a microprocessor measurement-control system using in critical application, (*Proceedings of the IMEKO-TC7 Symposium, Measurement Science of the Information Era, Cracow 2002).
5. R. Mielnik, *Petri Net in the design of metering and control systems for discrete events in critical applications,* (Collective work edited by A. St. Jagiełły - Electrical engineering in traction applications, Monograph no. 450, Cracow University of Technology 2014) [In Polish]
6. R. Mielnik, *Synthesis of controller for railway – level crossing devices using Petri Nets and state machine,* (Technical Transactions, Electrical Engineering, Issue 1-E (2), Cracow University of Technology, 2016).
7. R. Mielnik, *Synthesis of control devices of ARS power suplly railway network usig Petri nets and LabView,* (Proceedings the XVII Conference Electric Traction, Semtrak'2016, Zakopane 2016, Poland) [In Polish]
8. National Instruments website www.ni.com, 2017
9. *Technical guidelines for the construction of railway traffic control devices, Ie-4 (WTB-E10),* (Polish Railway Lines, Warsaw, 2014) [In Polish]
10. M. Kornaszewski, *Backup power supply systems for railway traffic control devices*, Journal "Logistyka", Łódź, 2013) [In Polish]
11. *Album of power devices: Supply system for SZUS type control devices, (*KZA Kraków, 1995) [In Polish]
12. *Technical guidelines for the construction of railway traffic control devices, Ie-4 (WTB-E10),*, (Polish Railway Lines, Warsaw, 2014) [In Polish]
13. David, R., and Alla, H.: *„Discrete, Continuons, and Hybrid Petri Nets"* (Springer, 2005)
14. B. Hruz. and M.C. Zhou, *Modeling and Control of Discrete-Event Dynamic Systems with Petri Nets and other tool*, (Springer, 2007)
15. P. H. Starke, J. Żurek, *Petri Nets: basics, applications, theory*,. PWN, Warsaw, 1987)
16. T. Szmuc, M. Szpyrka, *Formal methods in software engineering of real-time systems,* WNT Warsaw 2010.) [In Polish]
17. M. Szpyrka, *Petri Nets in modeling and analysis of concurrent systems,* WNT, Warsaw 2008) [In Polish]
18. Z. Wróbel and R. Ziemba, *Simulation of atmospheric surges in the systems of non-traction power lines supplying railway traffic control devices*, (Technical Transactions, Electrical Engineering, Issue 1-E/2009, nr 15, Cracow University of Technology, 2009) [In Polish].

The influence of coefficients for the terms of a controller on generation of braking curve

Jan Anuszczyk[1], and *Andrzej* Gocek[1,*]

[1]Lodz University of Technology, Institute of Electrical Engineering Systems, ul. Stefanowskiego 18/22, 90-924, Łódź, Poland

Abstract. This paper presents the theory about generating a braking curve and the analysis of the influence of the braking controller parameters on generation of the braking curve of the train. It has been demonstrated that the developed algorithms can be used to verify the effectiveness of the braking controller and the adjustment of the terms, and the change of these algorithms affects the shape of the generated braking curve of the train. The developed simulator cooperating with braking control system, which is reflecting the influence of braking function of the applied system used in the subway, has been analysed in terms of the impact of the parameters on the generation of braking curve. The outcomes has been presented and commented. Authors have also shown the results of the tests in the terms of an impact of a selected controller on the generation of braking curve.

1 Theory of automatic train braking – the theoretical braking curve

Target braking (docking) is used mainly for underground lines in order to stop the train on the platform in case the length of the train and subway platform is virtually the same. Target braking may be called an automatic process of stopping the train on the station. That kind of braking is a very important part of operating the train as it requires stopping the train on a platform of a specified length. In case of closed platforms, where entire platform is screened from the track by a wall with automatic platform-edge doors, the train stopping at metro station has to be very accurate and needs to implement the principle of "door to door" braking and is not allowed to open the marginal door beyond the platform area. Target braking of the train on the train stations should provide such siting of the train (after stopping) relative to the platform, so that all the doors of the train allow passengers to get out directly on the platform. Precise train stopping is also very important from the point of view of passengers. While the train stops in the platform the situation that the train stops in the incorrect place and a passenger could get stuck in the gap between the platform and the train cannot be allowed. Automatic braking must start at the correct distance from the stopping point and requires proper regulation of the braking force, which shall secure stopping in the required point. Target braking is one of the cases of essential braking (duty). It is basing on the gradual reduction of the actual speed of the train V_{rz} to the value of allowable speed limit V_d=0, which shall be obtained at the point of restricting the speed x_d, without exceeding the allowable deceleration of braking a_h [1, 2, 4]. Automatic target braking (docking) of the train occurs during stopping the train at the platform. Stopping should be very accurate, and the actual place of stopping the train x_{rz} on the platform, should meet the following condition:

$$x_{rz} = x_d \pm c \tag{1}$$

where *c* is the stopping tolerance.

The theoretical braking curve should take into account the desire to provide possibly high capacity of arterial railway line, it is advisable to brake with the maximal deceleration a_{hmax}. The limitation here is a feasible braking force F_{hmax} ensuring sufficient wheels adhesion with rails. The maximum allowable braking force is a function of the actual speed of the vehicle and the profile of track and operating conditions of the line. The optimal braking process ensures gradual increase of deceleration up to a maximum value (phase I), braking with the maximum and constant deceleration (phase II) and the gradual decrease of deceleration to zero v_g=0 (phase III). One may distinguish two basic types of theoretical braking curve: (a) Three-sector braking curve, which takes into account I, II and III phase of braking; (b) Two-sector braking curve, which takes into account I and III phase of braking. The parameters necessary to develop a braking curve are: the coordinate of a point where velocity reduction started x_g, the maximum deceleration of braking a_{hmax} and the maximum value of the derivative of deceleration of braking $|\dot{a}|_{hmax}$. Obtaining of an appropriate braking curve, in other words the dependency $v_h(l_h)$ may be performed substantially in three ways:

- by storing typical waveforms $v_h(l_h)$ in memory chips,
- every-time generation of braking curve using computing devices,
- every-time generation of braking curve for phase I and III and using of the recorded curve in phase II.

Figure 1 shows an example of generated theoretical curves of braking $v_h(l_h)$. Additionally for the actual speed

* Corresponding author: andrzej.gocek@gmail.com

before braking process started $v_{rz}(s)$, equal to v_0=80 km/h, there indicated on this curve three phases of the braking process, which were discussed above. At the point of overrunning the braking curve $v_0(l_{hmax})$ there starts the process of reducing the speed to the boundary value of the speed v_g=0 km/h. The curve of actual speed before braking $v_{rz}(s)'$ (marked with gray – the dotted-intermittent) illustrates the case of starting braking from v'_0=40 km/h. Due to the allocation of the setup for generation of braking curve one may distinguish two groups of solutions:

- a generation in the vehicle with the transmission to the vehicle of output data for calculations, or
- a generation by the trackside equipment with the transmission of the results of calculations to the vehicle.

The choice of the location for the setup for generating braking curve depends on the range of automation of driving the vehicle, and especially depends on the amount of transmitted information and the type of data transmission equipment. It should be emphasized that the working conditions of the data transmission system between the track and the vehicle are very difficult, especially due to strong interference in the transmission channel. Hence the need to strive to minimize the required number of transmitted information [1, 2, 4].

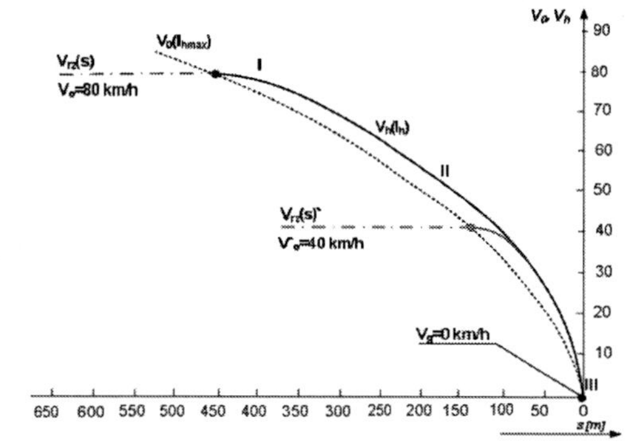

Fig.1. The theoretical curves of braking and braking start for the speed of braking start v_0=80 and 40 km/h.

1.1 Generation method of braking curve

The primary task of a brake control system is to obtain in the point of speed restriction x_d, permissible speed V_d with certain restrictions. To conduct this process the following data have to be obtained: constant parameters – capacity of electrodynamic and mechanical braking a_h, coordinates of the stopping points x_z, vertical and horizontal line profile m_{zp} given in promilles and variable parameters – actual train speed V_{rz}, actual train stopping position x_{rz}, permissible speed V_d, train load and any external interferences z – also including elementary resistance to motion. The executing system has to perform the following tasks (in the given order):

1) determine x_0 point for starting the braking process,
2) determine the actual train position x_{rz} and begin to measure the actual train speed V_{rz},
3) calculate the braking speed V_h,
4) calculate in real time the braking force F_h and control the braking process.

The steps from enumeration *1-3* are measurement and computing operations. Therefore, these steps are not affected by the integrated train braking system.

The above-mentioned actions in no. 4 have a control-execution nature and are related to the integrated braking system [1, 2, 4]. The structure of this system is shown in Fig. 2. The above-mentioned calculation structure may be easily calculated in a generic way in relation to a simplified computational model, and consists of the following tasks: *(a)* determination of initial braking point x_0, *(b)* calculation of the distance l_h – between the front of the train x_{rz} and the point of speed restriction x_d, *(c)* generation of the theoretical braking curve $V_h(l_h)$.

The tasks of the control-execution structure are based on the calculation of the difference Δv between braking speed V_h and the actual train speed V_{rz}, next the determination of the braking force F_h and the execution of the braking process (electrodynamic and mechanical braking phases). Braking devices are implementing the braking process with a specified value of braking deceleration a_h. The parameters which are permanently supplied to the system are dependent on the type of the train, applied the braking system and the weight of the train. Actual speed $V_{rz}(x_{rz},t)$ is a function of the place where the train is located x_{rz} and time t.

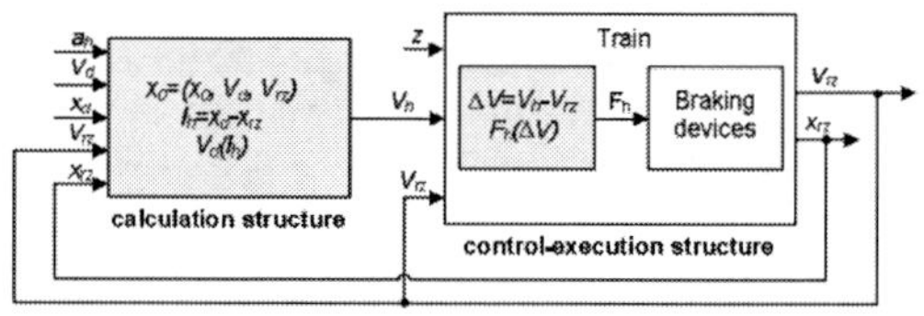

Fig. 2. Structure of the automatic braking system.

The place where the train is positioned on the track $x_{rz}(t)$ is a function of time. The train is the source of the values mentioned above. The allowed speed depends on the situation of the movement and the parameters of the rail setup on the metro line, and it depends on the location of points of speed restriction x_g on the metro line as well as on the time t. The source of this value is trackside equipment related to railway traffic control devices. The points of restrictions (speed change) are constant in time. The parameters required to calculate the theoretical curve of braking depend on the speed of the braking $V_h(l_h)$, which is affected by the actual - current distance between the vehicle and the point of speed limitation $l_h=x_g - x_h$ and deceleration a_{hmax}. The starting point of braking is dependent on the actual speed of the train V_{rz}, terminal speed V_g and technical capacity of the braking train. In this step, it becomes necessary to take into account the load of the train, which is strictly dependent on the load caused by passengers. For the proper implementation of target braking the control-executive system requires provisioning of the curve of braking speed $V_h(l_h)$, and the actual speed of the train $V_{rz}(x_{rz},t)$. The signal responsible for the implementation of the braking process is the braking force F_h, dependent on the braking speed $V_h(l_h)$ and the actual speed $V_{rz}(x_{rz}, t)$.

2 Simulation studies of the selected cases

2.1 Methodology of determining the braking precision

The outcome of the created simulator relies on the implementation of the controlling system, which corresponds to a layout used in the underground to generate docking stage, it takes into account [1]: the train movement equations, profile i, resistance to motion W, the combination of the traction characteristics $F_{ham}(v)$, delay of the controller, line spacing, wire loops lengths, wire loops crossings, permitted line speed, restriction of increase of braking deceleration, the derivative of acceleration with respect to time, weight of the train, train filling and electrical controlling signals used in the process of precision braking.

The simulation model communicates with a braking controller, which is based on a discrete PID regulator, in a connection time t_{kom}=50 ms. The braking controller reflects the operations of the on-board device embeddable on a real underground train, which cooperates with the system which realizes docking.

Depending on the signal from the controller, the simulator implements one of the available driving modes (steady ride, free run or braking). Next, the program checks the initial parameters of train physics, line profiles, the drive characteristics (F_{ham} – the braking force of traction machines), parameters of braking system, train weight, etc. The algorithm of the software detects the number of a wire loop (*loop detection*) and sets the value of the distance to the stopping point. Depending on the selected driving mode, the script performs calculations of resistance to motion, profile and the distance during braking. The calculations are performed with a time step $t_h=t_h+\Delta t$, to a train stop ($v=0$). The value of the total braking force $F_h=F_{ham}+W$ is being calculated during the braking process. For the following steps, the program calculates the actual braking distance s, speed v and the actual deceleration a_h on the basis of total braking force F_h, resistance of motion W and profile i. The actual train speed v and the distance s are the input data supplied to the regulator. The functional scheme of this process is shown in the Figure 3.

2.2 The influence of coefficients for the PID terms of a controller on generation of braking curve

The description of the authorial tool developed to perform the objective research and the applied target braking steering systems of the train with AC and DC drive were extensively discussed in papers [1, 2, 4]. The developed simulator cooperating with braking control system PID, which is reflecting the influence of braking function of the applied system used in the subway, has been analysed in terms of the impact of the parameters on the generation of braking curve. In particular, the effect of changing parameters of the terms of a controller to offset of the braking distance was analysed.

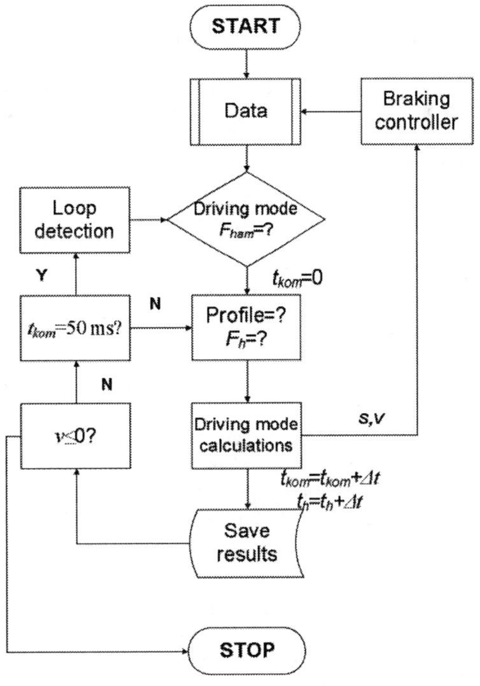

Fig. 3. A general block diagram of the simulation model – functional scheme.

For this purpose four sets with values of terms of parameters of controller were tested - K_p, T_i, T_d. Considering the proposed sets of terms of parameters of the controller, using the developed target braking simulator, the calculations were carried out for several theoretical rides of a train, for which an infringe of braking path was scheduled. The results of calculations are presented in Table 1, where the accuracy of stopping (D_{ham}) is given from the point of required stop. In a situation where the train passes by a stopping point the result has the sign "minus".

Table 1. The results of calculation of an infringe of braking path of a train for different controller parameters.

	Set 1	Set 2	Set 3	Set 4
The accuracy of stopping the train D_{ham} (infringe of braking path) [m]	-1,10	-3,41	1,34	0,38

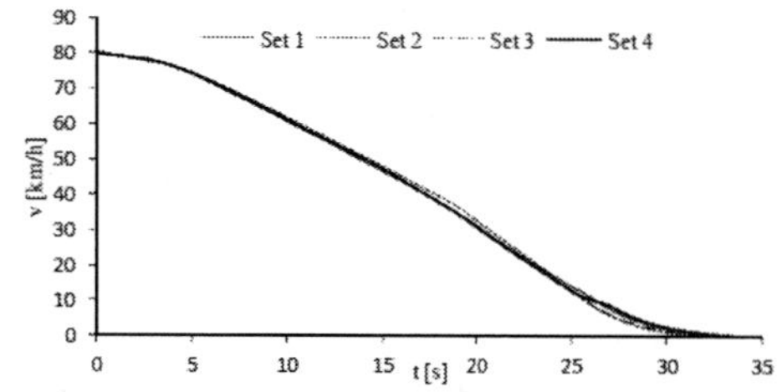

Fig. 4. Trajectories of braking speed depending on the adopted controller parameters (parameters of terms of controller); the degree of filling st=1,0; speed of initiation of braking v_p=80 km/h, the complex route profile.

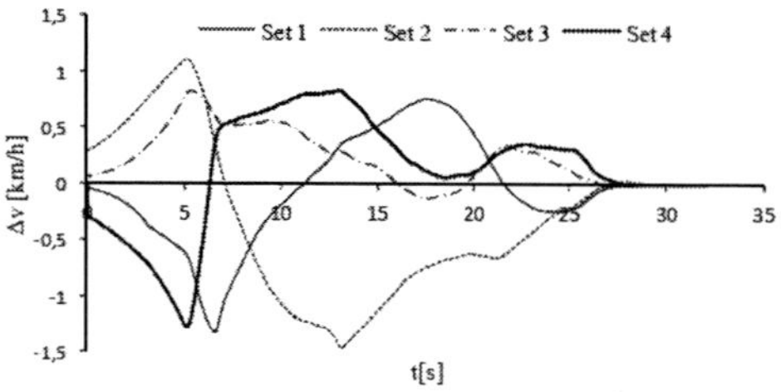

Fig. 5. The suppressed oscillatory course for adopted sets of terms of controller parameters; the degree of filling st=1,0; speed of initiation of braking v_p=80 km/h, the complex route profile.

After discerning literature in this field [3, 5, 6, 7, 8, 9, 10] a deficiency has been found, to expand the research of the influence of coefficients for the terms of a controller on generation of braking curve, the authors have considered two types of implemented controller – PID and PI, using as a base the set 4 of the parameters of terms of controller. Results are presented below in the form of 3D graphs.

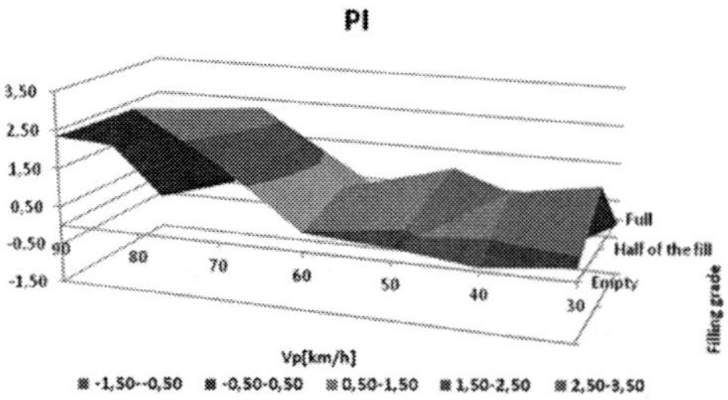

Fig. 6. Simulation for PI controller (T_d=0), T_p=100 ms, speed of initiation of braking – v_p, the degree of filling the train by passengers – filling grade, accuracy of stopping – D_{ham}.

Performed simulations show that a PI controller allows to improve static accuracy in range of medium and low speeds (30-60 km/h), for speeds above 60 km/h, braking accuracy is considerably reduced (comparing with outcomes for PID controller shown in the Figure 7). So it is justified to use a PID controller, this type of controller is typically used for objects subjected to rapid-change disturbances, what occurs in the braking process, and PID controller allows to shorten the adjustment time.

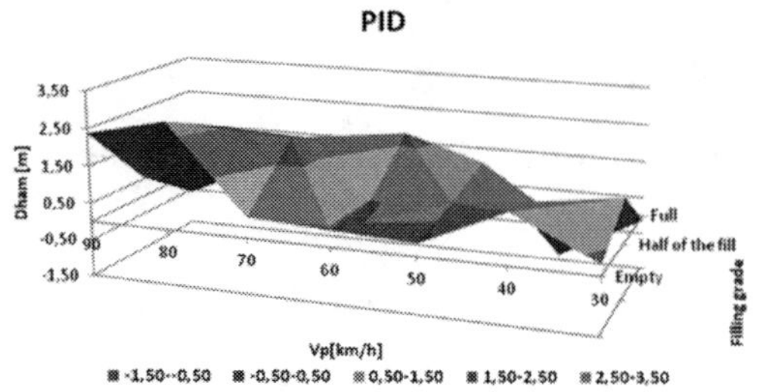

Fig. 7. Simulation for PID controller (parameters of terms of controller – set 4), T_p=100 ms, speed of initiation of braking – v_p, the degree of filling the train by passengers – filling grade, accuracy of stopping – D_{ham}.

PI controllers allow for the elimination of slow-moving disturbances, which translates into a zero deflection fixed and a longer regulation time. The PID controller enables shorter times of duplication than PI regulator without fear of the formation of oscillation in the closed system (shown in the Figure 5), so it eliminates the effects of interference more quickly.

3 Conclusions

The developed simulator [1, 2, 4] enables modelling the shape of the computational curve of braking speed, through the ability to change electrical parameters of the controller operating in the system of target braking, which is illustrated by characteristics shown in Figure 4.

As it can be seen in Figure 5, which presents the trajectories of braking curve f(t)=Δv during the braking process, the smallest oscillations of the concerned sets of terms of parameters of regulator are for the Set 4. All trajectories are characterized by suppressed oscillations. From simulation calculations it may be noticed that the change of terms of parameters of a controller affects the course of target braking process, and consequently the value of the infringe of the train's braking path.

As the results show, the PID controller enables better stopping accuracy than a PI controller. The developed simulator [1] cooperating with the brake adjustment block can be used to study the influence of electrical parameters on the process of target braking and may be used for selection and verification of parameters for braking controller. The analysis deduces that changing the electrical parameters of the regulator affects the process of target braking, and consequently influences the value of infringe of the braking path of the train and that the developed algorithms can be used to verify the effectiveness of the braking regulator.

References

1. Anuszczyk J., Gocek A.: The influence of operating parameters on braking precision of metro trains. *Conference Proceedings of XII International Conference MET2015 Modern Electrified Transport*, Trogir, Croatia, (2015), pp. 1-7.

2. Barański S., Gocek A., Steczek M.: The computer modeling of target braking of underground trains taking into account different types of traction drives. *Logistics - the science* ***Vol. 4/2015***, (2015), (in polish).
3. Domínguez M., Fernández-Cardador A., Cucala A. P., Pecharromán R. R.: Energy Savings in Metropolitan Railway Substations Through Regenerative Energy Recovery and Optimal Design of ATO Speed Profiles. *IEEE Transactions on Automation Science and Engineering*, **Vol. 9, No 3**, July, (2012).
4. Gocek A., Steczek M.: Simulation studies of target braking accuracy of an underground train driven by an AC motor. *Technical Magazine, Electrical Issue* ***1-E (2)***, pp. 67-77, (2016).
5. Gao Bing , Dong Hairong, and Zhang Yanxin: Speed Adjustment Braking of Automatic Train Operation System Based on Fuzzy-PID Switching Control. *2009 Sixth International Conference on Fuzzy Systems and Knowledge Discovery*. China, (2009).
6. Kawashima H., Komura F., Honma K., Iwasaki A., Takaoka T., Tashiro K., Monma N.: A Total Railway System Simulator and its Application to the Evaluation of Energy Saving Train Operation. *Wessex Institute of Technology Conference, COMPRAIL 92*, Washington, (1992).
7. Mincardi R., Savio S., Sciutto G.: Models and tools for simulation and analysis of metrorail transit systems. *COMPRAIL 94*. Italy, (1994).
8. Pniewska B.: Follow-up control of the traction vehicle at the scheduled timetable. *Phd Thesis. Technical University of Warsaw*, Warsaw, (1995), (in polish).
9. Zhao N., Roberts C., Hillmansen S., Western P., Chen L., Tian Z., Xin T., Su S.: Train Trajectory Optimization of ATO Systems for Metro Lines. *2014 IEEE 17th International Conference on Intelligent Transportation Systems (ITSC)*. Qingdao, China, (2014).
10. Zhu X., Liu X.: The Modelling of Test Systems of Automatic Train Operation (ATO) in Urban Rail Transit Based on LABVIEW. *International Conference on Computer Application and System Modelling (ICCASM 2010)*. China, (2010).

The use of the PN-EN 62305-2 standard to the risk assessment of losses in the equipped object in station- devices of the railway traffic controlling

Zofia Wróbel[1,*]
[1]PKP Polish Railway Lines JSC, Railway Lines Establishment in Rzeszów, ul. St. Batorego 24, 35-005, Rzeszów, Poland

Abstract. The valuation of the risk of losses makes easy the correct estimation of resulted threats as a result of atmospheric discharges and harms connected to them. Qualifying the need of lightning protection use for the object, we ought to take into account the risk R for the object and for the public services devices. In every case of examining risks we ought to fulfill: the identification of components RX forming the risk, counting the identified components of the risk RX, the entire risk R, identify the tolerated risk RT and compare the risk R with the tolerated value RT. As a result of such analysis we can come to a decision about the use of protection resources. In the report was realized the risk analysis of devices damages for a chosen object - the signal box with installed station - relaying devices of the railway traffic controlling with a computer adjustable control panel. From the realized analysis results that DEHN Risk Tool is a useful tool for the risk analysis of lightning losses in buildings, making possible the improvement of the calculations process, especially in the range of the choice of proper protection resources.

1 Introduction

The object damage risk analysis as atmospheric discharges and damages connected to them result, with regard to the random occurrence, are difficult. Therefore the decision about necessity and effectivity of resources usage of the lightning and surge protection and the choice of the efficient protection method, should be preceded by the risk valuation due to statics damages. The report presents the risk analysis of the railway traffic devices damages due to statics for the chosen object - a section inspector auty room with installed computer devices of controlling the railway traffic (rtc) and in the risk analysis, a fragment of the track section in which statics appeared, was taken into account. The risk results analyses was obtained by the chosen software DEHN Risk Tool according to the standard EN 62305 - 2:2008. Protection against lightning. - Part 2: Risk management., were represented [1].

2 The risk of lightning losses

The risk R is defined (according to the standard PN-EN 62305-2:2008 pt. 3.1.31 [1]) as the probable average value of one year's loss (people and goods). In consequence with the lightning influence, related to the object entire value (people and goods) subjected to the protection and the risk analysis give the threat estimation appears in consequence with earth lightning discharges in the object or the service – device. So it takes into account not only the threat of lively creatures and material losses due to the direct lightning stroke, but also the earth nearby discharges influence[2] and the influence of these discharges on the building installations, or connected to it devices in connections led and induced form, is also significant [3, 4]. On the ground of the performed risk analysis we can come to a decision about the protection resources use in order to minimalize losses in the object or the service - device, so the analysis performing gives the possibility of the protection resources selection that best fit the given object type, its equipments and the way it is used.

The risk assessment also makes possible the qualification of the proper lightning protection level for the object (LPL - Lightning Protection Level) and the complex conception creation of the protection before LEMP (Lightning Electromagnetic Impulse) with the use of shielding. In the performed analysis, the chosen object was a Signal box in Łancut (Łn) along with the attached computer rtc devices and the: feeding line, mast, two telephone lines, three railway automatic level crossings and two signals at the farthest point [5, 6, 7].

* Corresponding author: zwrobel@prz.edu.pl

3 The object description and its environment parameters

In the delimitation of the risk (PN EN 62305-2:2008 [1]) an essential parameter is the density of lightning discharges to-earth *Ng*. It determines the number of direct lightning discharges to-earth by km^2 a year. For areas, where an object is found - a signal box Łn (Fig. 1) according to lightning discharges to-earth density maps, the value *Ng* carries out 2.7 of lightning discharges by km^2 a year [8].

The object measurements in an essential way decide about the direct lightning stroke threat, and basing on them collecting areas for direct and indirect lightning strokes are qualified. For the Signal box Łn the following measurements were accepted: length - L_b: 14 m, width - W_b: 10 m, height - H_b: 8 m and the object highest point - H_{pb}: 16 m. For given above object measurements, the following collecting areas were calculated:

- direct discharges: 7 238 m^2,
- indirect discharges (nearby the object): 208 489 m^2.

A factor defining the number of possible direct and indirect lightning strokes is the environment surrounding the object. The relative position of the signal box is definite by the coefficient C_{db} which its value carries out 1. For the lightning discharges to-earth density, considering the size and environment of the object, we ought to expect the following frequencies:

- direct lightning strokes into the object: $N_D = 0.0049$ strokes / year,
- indirect strokes nearby the object: $N_M = 0.558$ strokes/ year.

The risk analysis performed by DEHN Risk Tool [9]. The programme is adapted to appoint the risk for different countries, including Poland according to the standard PN-EN 62305-2:2008. Protection against lighting Part 2. Risk Management [1].

4 The analysed object lightning protection zones

The zoning can be made taking into account such criteria as: the kind of the ground or floors, fire-proof barriers, spatial shielding, the internal systems arrangement, existing or expecting protection resources losses. For the analysis, the object - a signal box Łn that was divided to the following lightning protection zones (LPZ) i.e. zones for which the lightning electromagnetic environment was qualified. Limits of the zone LPZ are not always physical limits of the objects (eg. walls, floor and ceiling) [10]:

- LPZ 0B - object protected against an indirect stroke results:
  Z1-outside the building (signal box),
  Z2-the aerial mast.
- LPZ 1 - the internal zone of the protected object:
  Z3 - section inspector auty room,
  Z4 - the repair workshop room.
- LPZ 2 - the room/device in LPZ 1 of the screen proprieties:
  Z5-the relay room

Zones of the lightning protection in the standard are defined as:

LPZ 0_B: Protection against the direct stroke. The threat of a part of the thunder current and its entire electromagnetic field. Other internal systems are subject to the activity of a part of the failure current.

LPZ 1: The failure current is limited by its partition and by surge protected devices (SPD) at the bounds of the zones. The electromagnetic field is dead-beat by spatial shielding.

LPZ 2...n: Current impulses are still limited by their partition and by additional SPD at the bounds of the zones. The electromagnetic field is in general dead-beat by additional spatial shielding.

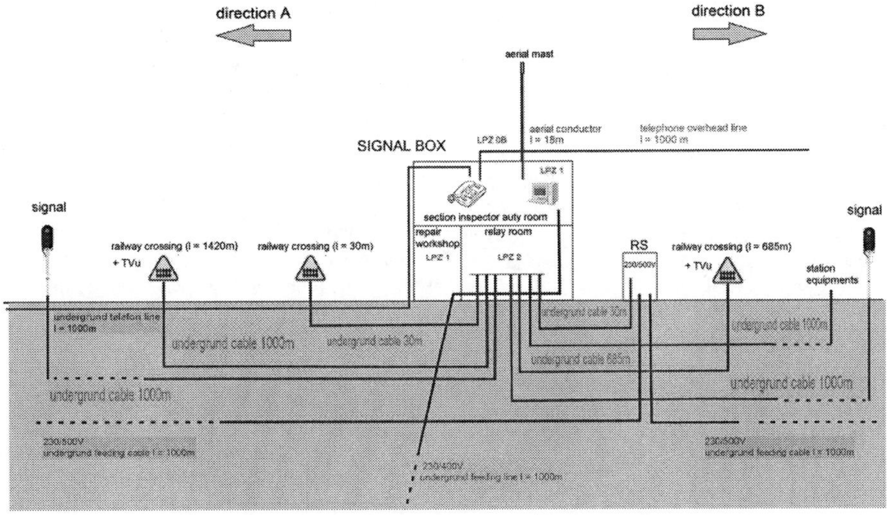

Fig. 1. Accepted to analysis object model Łn with connected devices [5]

5 Power supply cable

In the analysis were taken into account the entering lines to the Signal box and going out of it [10]. Conduct pipes are not taken into account, if they are connected to the main rail earthing. If they are not earthed then they are ought to take into account in the risk analysis (requirements concerning compensations of potentials [11]). In the risk analysis for the control room building (signal box) Łn the following lines were taken into account:

- electric AC 230/400 V,
- telephone 1 – underground cable,
- telephone 2 - aerial line,
- signal line - railway crossing (685 m),
- the aerial conductor,
- feeding line – voltage box (RS),
- signal line - railway crossing 1 (30 m),
- signal line - railway crossing 2 (1420 m),
- feeding line – signal 1 (10600 m),
- feeding line – signal 2,
- feeding line – station equipments,
- feeding cable – automatic line block system (230/500 V).

For each line the following parameters were qualified: the kind of the line (overhead/underground), the length of the line (outside of the building), the environment, the connected object to the line, the type of the internal cabling (shielded/unshielded) and the least voltage holdings out equipments (the strength of receiving devices).

5.1 The electric line AC 230/400V

For the analysed line, was accepted that it is buried in a ground of 500 Ωm resistivity. It was taken into account that the object is isolated – with no other objects nearby. The environment of the object is suburban (the height of buildings is less than 10 m). In the analysis was taken into account a transformer only for the service - device. The length of the line outside the building carries out 1000 m. Areas of assembling were qualified for the electric line as:

- The assembling area of discharges into the line: 21824 m^2,
- The assembling area of discharges striking nearby the line: 559 017 m^2.

The voltage holdings out equipment connected to the electric line AC 230/400 V, was definited for the zone Z1 and Z3, Z4 as 2.5 kV < Uw <= 4 kV and for the zone Z5 - as 1.5 kV < Uw <= 2.5 kV. The arrangement of lines in the building considering the zones: Z1, Z3, Z4 and Z5 was accepted as unshielded cable - lack of lofting to avoid the loop.

5.2 The underground telephone line 1

For this line, was accepted that it is an underground telephone line at a ground resistivity of 500 Ωm. It was taken into account that the object is isolated - lack of other objects nearby. The environment of the object is suburban (the height of buildings is less than 10 m). The length of the line outside the building carries out 1000 m. In the analysis, was taken into account a transformer only for the service - device. Appointed assembling areas are:

- The assembling area of discharges into the line: 21 824 m^2,
- The assembling area of discharges srtiking nearby the line: 559 017 m^2.

The voltage holdings out equipment connected to the telephone line 1 was definited for the zone Z1 as 1.0 kV < Uw <= 1.5 kVand Z3 as 1.5 kV < Uw <= 2.5 kV. The arrangement of lines in the building was qualified considering the zones: Z1 and Z3 as unshielded cable - lack of lofting to avoid the loop.

5.3 The overhead telephone line 2

For this line, was accepted that it is an overhead line at a ground resistivity of 500 Ωm. It was taken into account that the object is isolated - lack of other objects nearby. The environment of the object is suburban (the height of buildings is less than 10 m). The length of the line outside the building carries out 1000 m. In the analysis, was taken into account a transformer only for the service - device. Appointed assembling areas are:

- The assembling area of discharges into the line: 35 136 m^2,
- The assembling area of discharges srtiking nearby the line: 1 000 000 m^2.

The voltage holdings out equipment connected to the telephone line 1 was definited for the zones Z1 and Z3 as 1.0 kV < Uw <= 1.5 kV. The arrangement of lines in the building was qualified considering the zones: Z1 and Z3 as unshielded cable - lack of lofting to avoid the loop.

5.4 Signal line – railway crossing 1

The line was marked as buried in a ground of the resistivity 500 Ωm. It was taken into account that the object is isolated - lack of other objects nearby. The environment of the object is country-side (the height of buildings is less than 10 m) and in the analysis, was taken into account a transformer only for the service - device. The length of the line outside the building carries out 685 m. Basing on this the following assembling areas were qualified for the line:

- The assembling area of discharges into the line: 23 796 m^2,
- The assembling area of discharges striking nearby the line: 685 000 m^2.

The voltage holdings out equipment connected to signal - ride line 1, was definited for the zone Z1 as 1.0 kV < Uw <= 1.5 kV. The arrangement of lines in the building was qualified considering the zone Z1 as unshielded cable - lack of lofting to avoid the loop.

5.5 The aerial conductor

For this line, was accepted that is an overhead line at a ground resistivity of 500 Ωm. It was taken into account that the object is isolated - lack of other objects nearby. The environment of the object is suburban (the height of buildings is less than 10 m) and in the analysis, was taken into account a transformer only for the service - device. The length of the line outside the building carries out 18 m. The appointed assembling areas are:

- The assembling area of discharges into the line: 0 m^2.
- The assembling area of discharges striking nearby the line: 18 000 m^2.

The voltage holdings out equipment connected to the aerial conductor, was definited for the zones Z2 and Z3 as 1.0 kV < Uw <= 1.5 kV.and the arrangement of lines in the building for the aerial conductor was qualified considering the zones: Z2 and Z3 as unshielded cable - lack of lofting to avoid the loop.

5.6 Feeding line – railway crossing 2 and feeding line – voltage box (RS)

The line was marked as buried in a ground of the resistivity 500 Ωm. It was taken into account that the object is isolated - lack of other objects nearby. The environment of the object is suburban (the height of buildings is less than 10 m) and in the analysis, was taken into account a transformer only for the service - device. The length of the line outside the building carries out 30 m. Basing on this the following assembling areas were qualified for the line:

- The assembling area of discharges into the line: 134 m^2.
- The assembling area of discharges striking nearby the line: 16 771 m^2.

The voltage holdings out equipment connected to signal - ride line 1, was definited for the zones Z1 and Z5 as 1.0 kV < Uw <= 1.5 kV. The arrangement of lines in the building was qualified considering the zones: Z1 and Z5 as unshielded cable - lack of lofting to avoid the loop.

5.7 Feeding line – railway crossing 3 and feeding line – station equipments

The line was marked as buried in a ground of the resistivity 500 Ωm. It was taken into account that the object is isolated - lack of other objects nearby. The environment of the object is country-side (the height of buildings is less than 10 m) and in the analysis, was taken into account a transformer only for the service - device. The length of the line outside the building carries out 1420 m. Basing on this the following assembling areas were qualified for the line:

- The assembling area of discharges into the line: 21 824 m^2,
- The assembling area of discharges striking nearby the line: 559 017 m^2.

The voltage holdings out equipment connected to signal - ride line 1, was definited for the zones Z1 and Z5 as 1.0 kV < Uw <= 1.5 kV. The arrangement of lines in the building was qualified considering the Z1 and Z5 as unshielded cable - lack of lofting to avoid the loop.

5.8 The line signal – signal 1 and 2

For these lines was accepted that they are lines buried in a ground of the resistivity 500 Ωm. Each of the lines considered as an isolated object - lack of other objects nearby. The environment of the object is country-side and in the analysis was taken into account a transformer only for the service - device. The length of the line outside the building carries out 1000 m. The assembling areas carry out :

- The assembling area of discharges into the line: 21 824 m^2,
- The assembling area of discharges striking nearby the line: 559 017 m^2.

The voltage holdings out equipment connected to the signal line was definited for the zones Z1 and Z5 as 1.0 kV < Uw <= 1.5 kV. The arrangement of lines in the building for signal line - signals 1 and 2 was qualified considering the zones: Z1 and Z5 as unshielded cable - lack of lofting to avoid the loop.

5.9 Feeding line – automatic line block system (230/500 V)

For these lines was accepted that they are lines buried in a ground of the resistivity 500 Ωm and each line was considered as an isolated object - lack of other objects nearby. The environment of the object is country-side.and in the analysis was taken into account a transformer only for the service - device. The length of the line outside the building carries out 1000 m. The assembling areas carry out :

- The assembling area of discharges into the line: 21 824 m^2,
- The assembling area of discharges striking nearby the line: 559 017 m^2.

The voltage holdings out equipment connected to the signal line – feeding cable automatic electric block system was definited for the zones Z1 and Z5 as 1.0 kV < Uw <= 1.5 kV and the arrangement of lines in the building for signal line - signals 1 and 2 was qualified considering the zones: Z1 and Z5 as unshielded cable - lack of lofting to avoid the loop.

6 The fire risk

The fire risk determines an important criterion at qualifying the class of the lightning protection (LPS - lightning protection system) for the building. The gradation of the fire risk is based on values of specific fire load and it is defined in analysed standard [9]. The fire load is fixed by expert of fire precautions or defined after consultation with the owner of the building or with his insurance - firm. The fire risk for the given object Łn was qualified as a usual (common) one. Considering the number of persons being in the object Łn for zones Z1 and Z2 was assumed the lack of a special threat, and for zones Z3, Z4 and Z5 was fixed for a low level of panic (for no more than 100 persons).

7 External spatial shielding

Shielding spatial extinguishes the electromagnetic field inside the building, that appears at hits of the thunder into the building or close to it. This causes limitation of inducing surges in internal installations. This way is created a net of equalizing connections in which are taken into account all conductible parts of the building and internal systems. External or internal spatial shielding contains only the shielding part of the building structure. In the analysis was accepted the lack of shielding.

8 The valuation of the risk and the choice of protection resources

In the analysis, the risks R1, R2 and R4 were marked. For the risk R1 of the loss of human life for persons outside and inside the signal object Łn was fixed so that in the case of no-protection, the tolerated risk RT carries out 1E-05, and appointed in DEHN Risk Tool the risk R1 carries out 2.23E-06. With using protection, the risk R1 dropped to the value 9.5E-07 which is considerably less than the tolerated value. The risk R2 (the loss of public service), for the signal box Łn was fixed so that the tolerated risk RT carries out 1E-03. The appointed risk R2 carries out 1.32E-02 with no-protection that is to say the admissible level was 12.2 times exceeded. Whereas at using protection resources the risk R2 dropped to the value 2.7E-04.

The risk analysis R4 is performed to reduce the economic losses value taking into account the present and the target state. The result of these calculations is the cost of chosen protection resources with reference to the building and economic reason values. In marking were taken into account the following parameters: i - the interest rate: 1.1 %, at - the amortization time: 10 years, a - the depreciation rate 10 % and m - exploitation rate 1.1 %. To qualify the costs of the building, were taken into account: one year's cost of the building (C_B) carrying out 23 800 zloty, the content cost (C_C) 2 618 697 zloty and the cost of systems in the object (C_S) 130 000 zloty. Single costs of protection resources was accepted as 10 000 zloty. As a result of the risk R4 estimation was received the entire cost of losses on account of the thunder in the case of the lack of protection resources C_L: 17 412 zloty/year. The cost of deciduous losses C_{RL} on account of the thunder in the case of the presence of chosen protection resources carries out 437.7 zloty/year. The yearly cost of chosen protection resources CPM in the accepted amortization period of 10 years carries out 0 zloty/year. Yearly savings with using chosen protection resources carry out 16 974.6 zloty/year. So the use of chosen protection resources is well-founded economically.

Reducing the existing risk to an accepted level is obtained through the selection of suitable protection means. The selection of these resources determines a part of the risk management for the signal box Łn and is proper only with reference to this object. In the final part of the programme printout were placed protection resources, that should be used to assure a proper protection for the object with a zoning [5, 8].

9 Conclusions

From the performed analysis results that every used resource of the protection has an influence on its protection level. Properly well-chosen surge protection devices should be conditioned on the least profitable of connection arrangements, that can appear in given installation. Especially important is the use of suitable protection resources, including the protection of non-resident lines led to the control room (signal box) building and joint to computer devices. These circuits are most subject to damage due to surges of atmospherical origin.

The aim of further analyses is the realization of the risk assessment of lightning losses , in the new edition of the standard PN-EN 62305-2:2012 in which were introduced several of essential changes with relation to the edition from 2008r. Interesting will be so, how the introduced changes can shift themselves on results and the risk analysis assessments of lightning losses in specific cases of buildings, and the choice of required level and protection resources against the lightning electromagnetic impulse (LEMP).

References

1. PN-EN 62305 Part 2, Protection against lightning Risk Management. (2008) [In English]
2. A. Kern, G. Dikta: *29th2014 International Conference on Lightning Protection (ICLP)*, Uppsala, Sweden, 7a-6-1 - 7a-6-14 (2008) [In Polish]
3. Z. Wróbel, R. Ziemba, M. Gamracki, TTS Technika Transportu Szynowego, **11**, pp. 31-34 (2008) [In Polish]
4. Z. Wróbel, Technical Transactions Electrical Engineering ISSUE 1-E (2), **113** pp. 163-176. (Cracov, Poland, 2016) [In English]
5. Materials of PKP Polish Railway Lines JSC ©, [In Polish]
6. PN-EN 62305 Part 3, Protection against lightning. Physical damage to structures and life hazard. (2011) [In English]
7. PN-EN 62305 Part 4, Electrical and electronic systems within structures. (2011) [In English]
8. M. Łoboda, Scientific Research Papers, Rzeszów University of Technology, Electrotechnics, No. **33**, (2013), [In Polish]
9. DEHNsupport Toolbox. *Help for the designer.* [In Polish]
10. PN-EN 62305- Part 1, Protection against lightning. General principles. (2012) [In English]
11. Z. Wróbel, XVII National Scientific Conference of Electric Traction, SEMTRAK 2016, Zakopane, pp. 156 ÷ 169. (2016) [In Polish].

Children traffic fatalities in Croatia from 2005 until 2016

E. Missoni[1,*], *T. J.* Mlinarić[1], and *B.* Božić[2]

[1]Faculty of Transport and Traffic Sciences, University of Zagreb, Vukelićeva 4, 10000, Zagreb, Croatia
[2]Faculty of Medicine in Zagreb, Neurosurgical Department of "Sestre Milosrdnice" Hospital in Zagreb, Croatia

Abstract. The goal of this paper is to investigate and demonstrate the problem of fatalities of Croatian children under the age of 13 near suburban railways. Specific attention will be payed to the manner and conditions in which they are being instructed and accustomed to use this particular means of transport. Concluding remarks indicate that children should be taught from an early age that although using the suburban railways is very effective and has numerous advantages, in can nevertheless pose a safety risk if they do not behave cautiously (especially when getting on and off the train). In this research the authors examine the traffic injury death rates among children in the Republic of Croatia aged 13 and less over a twelve-year period. From 2005 until 2016 a total of 203 children were fatally injured. The most fatalities occurred in 2006 when a total of 26 children were killed, whereupon in 2016 that figure decreased to 2. Although the number of fatally injured children is gradually decreasing, it still gives cause for concern. In addition, over a ten-year period, from 2006 until 2015, a total of 515 children were seriously injured. Most of these injuries occurred in 2006 (98 children injured), while in 2015 there were only 30 injured children. The overall rate of fatal traffic accidents in Croatia of 8.3 (2012) per 100,000 inhabitants is even now at a high level compared to other European countries.

1 Introduction

We are witnessing an increase in motorisation, growth in the number of vehicles, and ever hastier mobility of traffic participants on a daily basis. Unfortunately, beside all its positive effects, there is also a negative side to increased traffic [1].

Every day thousands of people get killed and injured on roads worldwide. Children and youth go to kindergarten and school, as pedestrians, cyclists, or they drive by car and – sadly never come home, leaving behind grief and pain. Millions of people spend a number of weeks in hospital after serious traffic accidents, after which they are no longer able to lead a normal life, to work and to rejoice, because they remain permanently disabled, becoming a burden on their loved ones. Active participants involved in road traffic make for major factors in accidents [12].

Children in road traffic are defined here as persons aged from 0 to 13 years. Those belonging to age group between 0 and 4 years are considered as babies and toddlers, those between 5 and 10 years are in pre-puberty phase, and those between 11 and 14 are in puberty or early adolescence phase [11].

Each phase within childhood is characterised, and consequently defined, by distinctive levels of physical and psychological skill. Children as road users cannot, therefore, be considered a homogenous group as their abilities and their mode of transport choice differs considerably. Data about the mobility patterns of children as road users in general is rather scarce. In fact, little is known about children under the age of 6 because most of the statistical data starts with children at this age and older.

Within this paper we aim to examine the trends in injury risks for child pedestrians, cyclists and car occupants in Croatia over a twelve-year period, from 2005 until 2016. Motor vehicles crashes are one of the leading causes of death and acquired disability in children from birth to 14 years of age worldwide [1, 2]. For instance, while road traffic crashes account for about 170 child deaths and close to 4 000 serious injuries each year in the UK [3], that number astonishingly rises to over 12,000 children fatalities in China [4, 5].

2 Methods – Data collection and analysis

All the data presented in this paper have been gathered from police reports of the Traffic Police Department, Ministry of Internal Affairs, Croatia, which were registered on the respective scenes of the accidents.

* Corresponding author: missoni@fpz.hr

3 Results

The most child fatalities in Croatia occurred in 2006 when a total of 26 children were killed, whereupon in 2016 that figure had decreased to 2, thus corresponding with an overall declining trend in the number of people killed in road traffic crashes, which had been reduced from the total of 597 in 2005 to 307 in 2016. Although the number of fatally injured children is gradually decreasing, it still gives cause for concern.

Table 1. Crash-related child fatalities in traffic in Croatia from 2005 until 2016 [8].

	YEAR												TOTAL
Age	2005	2006	2007	2008	2009	2010	2011	2012	2013	2014	2015	2016	
0 – 6 yrs.	10	13	13	8	12	6	4	1	5	3	4	1	**96**
7 – 13 yrs.	8	13	12	9	9	4	9	6	3	5	5	1	**97**
Total	**18**	**26**	**25**	**17**	**21**	**10**	**13**	**7**	**8**	**8**	**9**	**2**	**203**

Table 2. Pedestrian injuries causing death.

	2005	2006	2007	2008	2009	2010	2011	2012	2013	2014	2015	2016	TOTAL
0-13 yrs	**6**	**13**	**8**	**5**	**7**	**5**	**5**	**4**	**3**	**4**	**2**	**2**	**64**

Table 3. Vehicle passengers, injuries causing death.

	2005	2006	2007	2008	2009	2010	2011	2012	2013	2014	2015	2016	TOTAL
0-13 yrs	**11**	**10**	**15**	**11**	**13**	**4**	**5**	**0**	**5**	**3**	**4**	**0**	**81**

4 Discussion

The rate of fatal traffic accidents in Croatia of 8.3 (2012) per 100,000 inhabitants is even now at a high level relative to other European countries. Seeing as children, pedestrians and cyclists are the most endangered and vulnerable group of traffic participants, their safety deserves special attention. It has been established that when motor vehicle collisions occur, children are best protected when properly using the appropriate restraint and sitting in a rear row [9].

Of the 32,719 traffic fatalities in 2013 in the United States, 1,149 (4%) were children 14 and younger (National Centre for Statistics and Analysis (NCSA), an office of the United States National Highway Traffic Safety Administration (NHTSA)). In 2013, the 1,149 child traffic fatalities were a 2% decrease from 1,173 in 2012. In 2013, the estimated 172,000 injured children in traffic crashes were a 2% increase from 169,000 in 2012. In 2013, motor vehicle traffic crashes were the leading cause of death for children age 4 and the second leading cause of death for children aged 3 and from 5 to 14 years of age (NCSA).

Furthermore, road danger is a strong disincentive to active transport (i.e., walking and cycling) and thus contributes to increasing levels of childhood obesity. Fear of pedestrian injury can create a vicious circle if an increasingly dangerous pedestrian environment encourages greater car use, leading to higher motorized traffic volumes and greater risks to pedestrians [6]. Reducing injury risks for child pedestrians and cyclists would be an important part of any strategy to promote walking and cycling, and might also reduce social class gradients in child injury mortality. Another method of protecting children from crash-related injuries is the use of child passenger safety seats; when correctly installed and used, they can significantly reduce the risk of crash-related fatalities by up to 71%, and the risk of serious injury by up to 67% [7].

5 Conclusion

From the presented data it becomes apparent that most traffic accident involving children occur when they are pedestrians or car passengers. Therefore we recommend and strongly encourage popularisation and further utilisation of suburban railroad as means of transport suitable for children, ensuring their safety. In view of these facts, it becomes clear that there is a pressing need to develop a nationwide strategy in order to reduce traffic injury death rates among children, which should be primarily directed towards professional training of school and preschool teachers.

Promoting correct behaviour among all traffic participants so that they could continuously acquire appropriate traffic knowledge and traffic culture, should continue to be implemented throughout high-school education, and especially at an university level where, for example, students can gain such knowledge during the course of "Traffic Medicine" lectures delivered at the Faculty of Transport and Traffic Sciences, University of Zagreb, Croatia [10, 11].

References

1. X. Chen, J. Yang, C. Peek-Asa and Liping Li, *Parents' experience with child safety restraint in China*, BMC Public Health, **14**, 318-325 (2014).
2. E. W. Kelli and E. S. Geller, *Increasing the safety of children's vehicle travel: From effective risk communication to behavior change*, J Safety Res, **35**, 263-274 (2004).
3. B. Sonkin, P. Edwards, I. Roberts and J. Green, *Walking, cycling and transport safety: An analysis of child road deaths*, J R Soc Med, **99**, 402-405 (2006).
4. S. Y. Wang, Y. H. Li, G. B. Chi, S. Y. Xiao, J. Ozanne-Smith, M. Stevenson and M. R. Phillips, *Injury-related fatalities in China: An under-recognized public-health problem*, Lancet, **372**, 1765-1773 (2008).
5. X. Zhang, H. Xiang, R. Jing and Z. Tu, *Road traffic injuries in the People's Republic of China, 1951–2008*, Traffic Inj Prev, **12**, 614-620 (2011).
6. P. L. Jacobsen, *Safety in numbers: More walkers and bicyclists, safer walking and bicycling*, Injury Prevention, **9**, 205-209 (2003).
7. F. K. Winston, D. R. Durbin and M. Kallan, *The danger of premature graduation to seat belts for younger children*, Pediatrics, **105**, 1179-1183 (2000).
8. Ministarstvo Unutarnjih Poslova, Sluţba Za Strateško Planiranje, *Analitiku i Razvoj, Statistički Preged. Temeljnih Sigurnosnih Pokazatelja i Rezultata Rada U 2012. Godini,* https://www.mup.hr/UserDocsImages/statistika/2013/statistika2012.pdf
9. M. M. Macy and G. L. Freed, *Child Passenger Safety Practices in the U.S., Disparities in Light of Updated Recommendations*, American Journal of Preventive Medicine, **43(3)**, 272-281 (2012).
10. E. Missoni and I. Missoni, *Analiza smrtnih slučajeva djece u prometu u Hrvatskoj*, Paediatr Croat, **51**, 216-220 (2007).
11. D. Posavec and E. Missoni, *Cestovna sigurnost djece u zemljama Europske unije i u Republici Hrvatskoj*, Suvremeni promet: časopis za pitanja teorije i prakse prometa, 33, 239-242 (2013).
12. E. Missoni, *Ljudski čimbenk u prometu, Sveučilišni udžbenik* (FPZ, Zagreb, 2017).

Comprehensive approach to risk assessment and evaluation regarding constructioning of the first 25 kV 50 Hz AC traction power supply sections in Poland

Marek Pawlik[1,*]

[1]Railway Research Institute, ul. Chłopickiego 50, 04-275, Warsaw, Poland

Abstract. Railway lines in Poland are either equipped with 3 kV DC traction system or not electrified for traction purpose. Presently maximum line speed is 200 km/h while maximum train speed is 250 km/h. Speeding up limit for 3 kV DC traction is estimated to be around 220-230 km/h. Already present in Poland trains for 250 km/h are equipped with three electric power supply systems: 3 kV DC used in Poland, 15 kV 16,7 Hz AC used in Germany, and 25 kV 50 Hz AC foreseen to be used in Poland on high speed lines. Introducing 25 kV 50 Hz AC traction power supply will be associated with safety challenges, which have to be taken into account already during constructioning, starting even at the predesigned phase. Two key questions arise. First of all question regarding methodology, which should be used for safety level acceptance and secondly how to ensure that all safety aspects will be taken into account. Answering first question for railway safety experts seems to be easy – let's apply risk assessment and evaluation methodology described by EU regulation under Railway Safety Directive. The challenge is however to define risk acceptance criteria which will be required, and that is analysed in the paper. The second challenge is even more challenging. Author proposes using 10-by-10 safety matrix which was defined by the author in previous publications [9, 10]. Its concept as well as principles for its application for new type of power supply is also presented in the paper.

1 Introduction

Polish railway network is composed of 19.000 kilometres of railway lines mostly double track. Yearly reports being published by the PKP PLK S.A. (Polish Railway Lines) show, that with station tracks infrastructure is composed by 38.000 kilometres of tracks. Nearly 12.000 kilometres of railway lines (nearly 25.000 kilometres of tracks) are electrified. Electrified lines are double track and electrified together with station tracks. All presently electrified lines in Poland are using 3 kV DC traction system.

For many years the maximum line speed on the Polish railway lines as well as the maximum running speed of the Polish traction units which are running on those lines was 160 km/h. Presently maximum line speed is increased up to 200 km/h. At the same time maximum running speed of some new trains has reached 250 km/h. This shows, that from economical point of view line speeds at least on some lines should be increased minimum to 250 km/h to ensure sensible use of already existing trains and justifying purchase of such high speed trains.

In that respect the challenge is however the traction system [1]. Used 3 kV DC system is a bottleneck for speeding. Speeding up limit for 3 kV DC traction is estimated to be around 220-230 km/h. The trains for 250 km/h already purchased by PKP Intercity are therefore prepared to run under different traction supply systems. The vehicles for 250 km/h are already equipped with three above cited electric power supply systems.

2 Safety challenges associated with introduction of the 25 kV 50 Hz AC traction on railway lines

Introducing 25 kV 50 Hz AC traction power supply for the first time in Poland is a significant change possibly influencing safety of the railway system as a whole. This has to be analysed in details by the infrastructure manager, which will govern such infrastructure (not necessarily PKP Polish Railway Lines). The safety challenges should be taken into account already during construction, starting even at the predesign phase. Therefore identification of potential risks and defining safety requirements and necessary safety measures is a question to be discussed already now.

Defining safety requirements is directly associated with identification of risks. Defining risks requires defining 25 kV 50 Hz AC power supply system together with its borders and interfaces. It seems, that defining system is not a challenge itself. In that respect using definition of the subsystem "Energy" from the railway

* Corresponding author: mpawlik@ikolej.pl

 https://doi.org/10.1051/matecconf/201818006002

interoperability directive [2] seems to be appropriate and quite enough. Subsystems are defined in annex II of the directive. The "Energy" subsystem is defined as "*The electrification system, including overhead lines and on-board parts of the electric consumptions measuring Equipment*". This shows on one side, that power supply to substations are outside railway system. Respective risks however should be taken into account, but it is proposed to take them into account as characteristics of the interface between national power system and subsystem "Energy". On the other side pantographs are not included in "Energy". They belong to the "Rolling stock" subsystem, which is defined as "*Structure, command and control system for all train equipment, current-collection devices, traction and energy conversion units, braking, coupling and running gear (bogies, axles, etc.) and suspension, doors, man/machine interfaces (driver, on-board staff and passengers, including the needs of persons with reduced mobility), passive or active safety devices and requisites for the health of passengers and on-board staff.*", and therefore includes not only current collection but also energy conversion systems and traction engines up to connection with insulated return to substation, which belongs to subsystem "Energy". Risks associated with that interface and with insulated return itself have to be taken into account.

On the basis of the experience gained by the railway companies all over Europe "Energy" subsystem is defined more precisely by a dedicated Technical Specification for Interoperability - TSI known as TSI ENE [3]. The 25 kV 50 Hz AC system's associated risks are minimised in the European countries using such power supply system by applying European technical standards dedicated to traction stationary equipment [4] and traction mobile equipment [5] as well as by applying dedicated European technical standard dedicated for coordination of electrical protection of stationary and mobile traction equipment [6].

As trains are passing state borders since a long time safety related challenges were also discussed by dedicated groups working within International Union of Railways (UIC). As a result UIC has published some leaflets dedicated to traction power supply [7].

Moreover railway tunnels' specific safety aspects are covered by another dedicated Technical Specification for Interoperability - TSI known as TSI SRT [8]. This TSI includes many different requirements e.g. for escape routes in railway tunnels, for ensuring places for rescue services in emergency, for communication, lighting, operational rules, as well as traction power supply both in relation to stationary traction equipment especially regarding reliability of traction system and its power supply in emergency situations and in relation to mobile traction equipment put in rolling stock especially regarding ability to run with fire on board of the train. All those detailed requirements [3-8] minimize risks by imposing defined solutions, however they are not linked with identification of risks.

2.1 Identification of the 25 kV 50 Hz AC power supply risks

For the 25 kV 50 Hz AC power supply system defined together with its borders and interfaces detailed list of risks have to be elaborated. This creates two detailed challenges. First it has to be pointed, that risk is associated with probability of a hazard and with foreseeable hazard consequences. This shows, that on one side some hazards can be assumed as not important as their probabilities are negligible and on the other side some hazards can be assumed as not important as their foreseeable consequences are negligible. However such decisions can only be taken after identification of hazards.

2.2 Identification of the 25 kV 50 Hz AC power supply hazards

The key challenge is to ensure taking into account all hazards associated with 25 kV 50 Hz AC system and its interfaces.

For this purpose it is proposed to use 10-by-10 safety matrix which was defined by author in previous publications [9-10]. Alternative overall safety models which exist are focusing on risk approach to safety un-appreciating technical approach to safety covering railway specific technical challenges. The most important railway specific safety model based on risk approach is a Safety Management System "SMS Wheel" tool prepared and disclosed by the Railway Agency of the European Union [11]. Unlike such models the 10-by-10 matrix is based on two directives. The Railway Interoperability Directive [2], which defines essential requirement "safety" for railway system as a whole and for separate subsystems including "Energy" subsystem defined as above, and the Railway Safety Directive [12], which defines "safety management" scope for railway undertakings and infrastructure managers. Matrix takes into account also security aspects and safety aspects associated with security supporting systems like access monitoring, LPR (Licence Plate Recognition) systems, fire detection, video screening, etc. as well as cybersecurity dimension of all systems, which are using electronic and programmable technical solutions. Security and cybersecurity aspects were defined on the basis of author experience and already discussed with experts dealing with safety in nuclear power plants and air transport which are keen in system approach to hazard identification and risk acceptance. The safety 10-by-10 matrix with security dimension is shown on Fig. 1.

The technical safety aspects are following:

- T1 - malfunctioning safety – what can happen when systems and/or devices are not working properly;
- T2 - construction safety – what can happen when systems and/or devices are affected by exceptional stress, fire, or construction or natural disaster;
- T3 - electrical safety – what can happen in case of overvoltage, short-circuits, presence of electric arch between pantograph and overhead line, lightning, etc;

 https://doi.org/10.1051/matecconf/201818006002

- T4 - preventing unauthorized access and fire – how to ensure appropriate protection against unauthorised access and fire and how to ensure that cases when such protection is broken are notified immediately to respective operational, security and emergency staff;
- T5 - wheel/rail interaction safety – how to ensure protection against derailment in different circumstances and how to ensure appropriate relationships between lengths of the braking distances and spacing of trains imposed by localisations of signals and signs along the tracks;
- T6 - control command and signalling – active safety systems ensuring operational safety and their inherent safety based on fail-safe principle and safety integrity level SIL4 ensuring route setting on stations, train spacing on sections and protection against switching points under running trains;
- T7 - power supply/signalling interaction safety – the ways which are used to ensure minimisation of risk of negative influence from power supply on signalling and control command especially via insulated return and conversions of the energy;
- T8 - safety of the operational rules and staff competences – covering all operational circumstances especially present in degraded modes of operation over lifecycle of technical systems as well as building, verifying and keeping staff competences and health;
- T9 - preventing panic – staff communication in emergency and public address systems, evacuation ways and doors, emergency lighting, etc.;
- T10 - IT support for safety – ensuring collection of relevant data acquisition, storage, transmission and processing for improving safety on the basis of accidents and incidents analyses and reporting.

Two of them are marked with grey background as they are associated with safety aspects not taken into account in the Railway Interoperability Directive [2].

The operational safety management is composed of aspects pointed and described in the Railway Safety Directive [12] and are following:

- O1 - safety Policy;
- O2 - maintenance safety;
- O3 - operational safety;
- O4 - risk assessment;
- O5 - risk assessment monitoring;
- O6 - exchange of safety relevant information;
- O7 - activities in emergency situations;
- O8 - analysing accidents & incidents;
- O9 - internal monitoring of the safety system;
- O10 - safety improvement plans.

Moreover it is important to take into account cyber security aspects. Presently in Poland they are taken into account only for some kinds of technical solutions. Present situation on the Polish railway network is shown on Fig. 2 on the basis of the 10-by-10 matrix. Here there are two different intensities of grey. For some types of solutions cybersecurity is legally required, namely for signalling and control command systems and devices e.g. for electronic interlockings and block systems. This is marked with dark grey not only in relations to control command and signalling safety but also in relations to safety aspects which apply to such systems and devices like: malfunctioning safety, safety of the control command systems inherent operational rules, and IT support for control command and signalling. Infrastructure managers and railway undertakings are already requesting safety proves for other types of systems especially those supporting security although it is legally not required. It is marked grey on the 10-by-10 matrix. Base on that it can be assumed, that generally cybersecurity of electronic and programmable safety technical means (marked dark grey) is and should be required, while cybersecurity of security technical means (marked grey) is requested by users but not required legally as shown on Fig. 2. below.

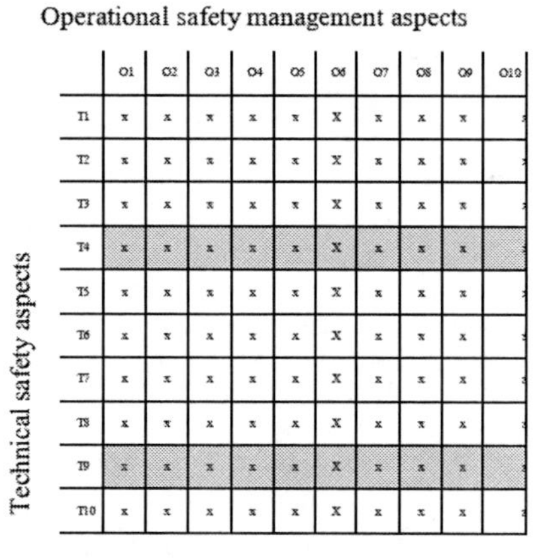

Fig. 1. Safety matrix 10-by-10 with security dimension marked in grey. (own elaboration).

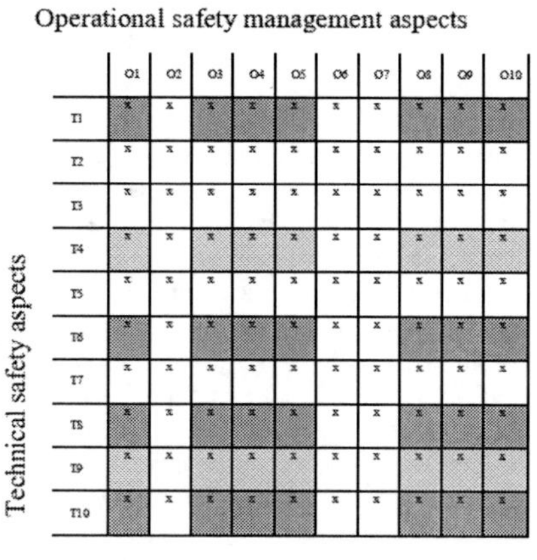

Fig. 2. Safety matrix 10-by-10 with cybersecurity dimension marked in grey. (own elaboration).

This shows, that for safety aspects of the technical means influencing safety, and not only security, cybersecurity have to be seen as a must although it is not clearly stated in binding acts and regulations e.g. for traction control.

As a result following technical safety aspects can be shown as the ones to be analysed in details when introducing 25 kV 50 Hz AC power supply:

- T1 - malfunctioning safety;
- T2 - construction safety;
- T3 - electrical safety;
- T4 - preventing unauthorized access and fire;
- T7 - power supply/signalling interaction safety;
- T10 - IT support for safety;

with focus on following safety management aspects:

- O2 - maintenance safety;
- O3 - operational safety;
- O7 - activities in emergency situations;

as shown on Fig. 3 below.

The first four technical aspects are rather obvious. There is no doubt, that in case of malfunctioning or exceptional stress safety has to be ensured. The electrical safety aspects are also obvious when analysing power supply system as well as prevention against unauthorized access and fire.

Two technical safety aspects are not obvious – power supply/signalling interaction safety and IT support for safety.

The interaction with signalling starts with influence on track occupancy systems disregarding whether track circuits or axle counters are used on stations and along the railway sections between stations.

The IT support for safety starts with remote control and remote monitoring of devices in power supply substations and sectioning points and of different kinds of electrical switches used for maintenance and for protection e.g. against overvoltage due to lightning.

Operational safety management aspects

Technical safety aspects	O1	O2	O3	O4	O5	O6	O7	O8	O9	O10
T1	x	x	x	x	x	x	x	x	x	x
T2	x	x	x	x	x	x	x	x	x	x
T3	x	x	x	x	x	x	x	x	x	x
T4	x	x	x	x	x	x	x	x	x	x
T5	x	x	x	x	x	x	x	x	x	x
T6	x	x	x	x	x	x	x	x	x	x
T7	x	x	x	x	x	x	x	x	x	x
T8	x	x	x	x	x	x	x	x	x	x
T9	x	x	x	x	x	x	x	x	x	x
T10	x	x	x	x	x	x	x	x	x	x

Fig. 3. Safety matrix 10-by-10 safety aspects marked grey are to be taken into account before introduction of the 25 kV 50 Hz AC power systems. (own elaboration).

For all marked areas both safety and security needs to be analysed as well as cybersecurity in case of technical solutions, which are using electronic and/or programmable technical means.

On the basis of those areas and different kinds of incorrect technical states of systems and devices constituting "Energy" subsystem as well as malfunctioning of interfaced systems and wrong functioning of interfaces as well as on the basis of possible external circumstances detail list of hazards have to be prepared for the defined technical solution before it is installed. Preferably before it is designed as already at design phase additional technical safety requirements and/or safety means probably need to be taken into account.

2.3 The 25 kV 50 Hz AC power supply hazard log

Identification of hazards is a first step in system approach to safety. Usually all identified hazards are grouped and put into big tables reflecting their step by step analyse up to the statements, which are influencing everyday operation as well as ever and again maintenance and repair. Proposed step by step approach is shown as a table on Fig. 4.

	T1/O3	T1/O2	T1/O7	T2/O3
Identified hazards				
Associated functions				
Associated device				
Hazard description				
Hazard reasons				
Remarks				
Potential consequences				
Existing safety means				
Frequency level *)				
Severity category *)				
Risk acceptance *)				
Additional safety means				
Responsibility split **)				
Frequency level *)				
Severity category *)				
Risk acceptance *)				

Fig. 4. Step by step approach to hazards(own elaboration):
*) as defined in EN 50126 [13]
**) as defined in regulation 2013/402 [14]
and where T1/O3 means hazards which can occur due to malfunctioning of systems and/or devices and influence ensuring safety during operation; T1/O2 means hazards which can occur due to malfunctioning of systems and/or devices and influence safety during maintenance and operation after maintenance; T1/O7 means - hazards which can occur due to malfunctioning of systems and/or devices and influence activities in case of emergency. And T2/O3 and following areas have to be understood in a similar way.

Identification of hazards have to take into account safety related hazards, security related hazards and cybersecurity hazards for systems and devices within "Energy" subsystem and its interfaces. Each hazard has to be named and put into a table which will form a hazard log. Each identified hazard has to be well understood and therefore it is necessary to point all functions and all systems and devices which are associated with individual hazards. It is important to see not only main functions and devices but also other ones which are affected because risk analyses on the basis of the RAMS standard [13] cannot be applied independently of hazards which are dependent from each other.

Each hazard has to be described. As a result tables describing hazard logs are usually big and difficult to be printed. For instance for constructioning of the four track coverage between Gdańsk Główny and Gdańsk Śródmieście the hazard log can be printed on A0 format paper with font 10 on 12 pages. This makes such analyses quite complex, but as a result of such analyses a number of design and technical changes are being introduced e.g. for quoted coverage as a result of risk analyse even concrete has been changed.

For each hazard primary and secondary reasons have to be pointed. The reasons are important for the identification of the hazards with common reasons. Hazards which occur due to the same reasons have to be analysed together especially from the point of view of safety measures and risk level acceptance. Appropriate remarks have to be put in hazard log.

Identification of each hazard has to be concluded by pointing and describing potential hazard consequences as well as existing safety means which protect against those consequences. This closes identification of hazards and forms a basis for risk analyse.

As risk is associated with frequency and severity of occurrence of a hazard respective analyses have to be done for each hazard and reflected in hazard log.

Frequency levels defined in RAMS Standard [13] are following: *frequent (likely to occur frequently), probable (will occur several times), occasional (likely to occur several times), rare (likely to occur sometime in the system life-cycle), improbable (unlikely to occur but possible) and highly improbable (extremely unlikely to occur).* Pointing frequency level has to be based on expert judgement. As such judgement is important for risk acceptance choosing experts for risk analyses is a tricky decision which has to be based on judgement of experts' experiences. Therefore it is usually treated as setting "council of elders".

As risk is based on frequency and severity the next step is to define severity category for each hazard. Severity categories defined in RAMS Standard [13] are following: *catastrophic (affecting a large number of people and resulting in multiple fatalities, and/or extreme damage to the environment), critical (affecting a very small number of people and resulting in at least one fatality, and/or large damage to the environment), marginal (no possibility of fatality, severe or minor injuries only, and/or minor damage to the environment) and insignificant (possible minor injury).*

The frequency levels and severity categories form a basis for risk acceptance, which is on this basis defined as intolerable, undesirable, tolerable or negligible and associated respectively with following activities: *intolerable - risk shall be eliminated; undesirable - risk shall only be accepted if its reduction is impracticable and with the agreement of the railway duty holders or the responsible Safety Regulatory Authority; tolerable - risk can be tolerated and accepted with adequate control (e.g. maintenance procedures or rules) and with the agreement of the responsible railway duty holders; negligible - risk is acceptable without the agreement of the railway duty holders.*

Risk acceptance judgement is usually marked by making respective cell red for intolerable risk, yellow for undesirable and tolerable risk and green for negligible risk. Colours allow quick understanding where additional analyses are necessary. For all hazards for which risks is red additional safety measures have to be defined. Such measures may be used to lower probability of hazard occurrence and/or hazard potential consequences. After adding such measures, e.g. after re-design or adding extra filters or redundancies, frequency levels and severity categories have to be defined for such hazard once more. As re-design may change frequency and consequences for other hazards such second step is due to cover not only hazards with red risk but all of them.

Moreover second step is used also to verify the yellow risks. They are acceptable under certain circumstances which are associated mainly with split of responsibility. Usually they are based of shifting responsibility from constructioning entity to user. A good example is to point firefighting equipment and backup power supply e.g. for control equipment. Of course they have to be provided during construction, but during years of operation they have to be checked and maintained in appropriate way and in appropriate moments. Shifting responsibility is very important as such information in hazard log gives certainty, that further user includes such activities in his internal regulations, which have to be supervised by infrastructure manager safety management system required as a basis of the safety authorisation without which offering railway infrastructure for railway transport is legally forbidden.

3 Conclusions

Introduction of the 25 kV 50 Hz AC power supply have to be carefully prepared and cannot be based on foreign technical regulations only as it has to work in a different environment e.g. different signalling equipment which work cannot be affected by power supply insulated returns. As a result detail analyses should be undertaken already now. It is proposed to start from an overall safety picture defined by 10-by-10 matrix defined by author in previous publications. Based on that it has been shown that: malfunctioning, construction, and electrical safety as well as preventing unauthorized access and fire, ensuring power supply/signalling interaction safety and ensuring IT support for safety have to be taken into account and analysed in relation to all possible hazards which may occur during 25 kV 50 Hz AC power supply operation and maintenance as well as during activities in emergency situations. It is proposed to create a hazard log and use the frequency and severity concept for accepting individual hazards on the basis of risk acceptance and shifting some well-defined responsibilities to the future user of the new type of power supply based on 25 kV 50 Hz AC traction solutions. As a result effectively paper proposes use of a scientific comprehensive approach to risk assessment and evaluation for the 25 kV 50 Hz traction system introduction.

References

1. M. Lewandowski, *Traction solutions for high speed railways [Trakcje w kolejach dużych prędkości]*, Logistyka, **3**, 1301-1305 (2012) [in Polish]
2. Directive 2008/57/EC of the European Parliament and of the Council of 17 June 2008 on the interoperability of the rail system within the Community (EU OJ L 191/1 of 18 July 2008) with further changes (which is to be superseded by Directive (EU) 797/2016)
3. COMMISSION REGULATION (EU) No 1301/2014 of 18 November 2014 on the technical specifications for interoperability relating to the 'energy' subsystem of the rail system in the Union
4. European Standards dedicated to traction stationary equipment: PN-EN 50119:2009/A1:2014-01 (overhead contact line); PN-EN 50122-1:2011/A1:2011 (electrical safety); PN-EN 50149:2012 (contact lines); PN-EN 50163:2006 (traction power supply voltages)
5. European Standards dedicated to traction mobile equipment: PN-EN 50317:2012 (dynamics between pantographs and contact lines); PN-EN 50318:2003 (validation of simulations); PN-EN 50367:2012 (pantograph and contact line interoperability)
6. European Standard PN-EN 50388:2012 dedicated for coordination of electrical protection of stationary and mobile traction equipment
7. Leaflets of the International Union of Railways (UIC) dedicated to traction power supply: UIC 600 - Electric traction with aerial contact line; UIC 606-2 - Installation of 25 kV and 50 or 60 Hz overhead contact lines; UIC 616 - Rules for electric traction equipment; UIC 618 - Rules for traction transformers and reactors; UIC 642 - Special provisions concerning fire precautions and fire-fighting measures on motive power units and driving trailers in international traffic
8. COMMISSION REGULATION (EU) No 1303/2014 of 18 November 2014 concerning the technical specification for interoperability relating to 'safety in railway tunnels' of the rail system of the European Union
9. M. Pawlik, *Bezpieczeństwo kolei w kontekście powiązań pomiędzy dyrektywami o bezpieczeństwie kolei i o interoperacyjności kolei, analiza z punktu widzenia zarządzania bezpieczeństwem [Railway safety in the context of relationships between railway interoperability directive and railway safety directive, analyse from the point of view of safety management]*, Przegląd Komunikacyjny [Transport Review], **9**, 11-20 (2016) [in Polish] ISSN 0033-22-32
10. M. Pawlik, *Safety, security and cybersecurity in railway operation*, Safety and Reliability – Theory and Applications (Taylor & Francis Group, London, 2017) ISBN 978-1-138-62937-0
11. Wheel – description of the "SMS Wheel" tool prepared for verification of the Safety Management Systems of the railway companies, and disclosed by the Railway Agency of the European Union http://www.era.europa.eu/tools/sms/Pages/SMS.aspx#SMS
12. Directive 2004/49/EC of the European Parliament and of the Council of 29 April 2004 on safety on the Community's railways and amending Council Directive 95/18/EC on the licensing of railway undertakings and Directive 2001/14/EC on the allocation of railway infrastructure capacity and the levying of charges for the use of railway infrastructure and safety certification (Railway Safety Directive) (EU OJ L 164/44 of 30 April 2004) with further changes (which is to be superseded by Directive (EU) 798/2016)
13. EN 50126-1:1999 Railway Applications - The Specification and Demonstration of Reliability, Availability, Maintainability and Safety (RAMS) (which is to be superseded by FpprEN 50126-1:2017)
14. Commission Implementing Regulation (EU) No 402/2013 of 30 April 2013 on the common safety method for risk evaluation and assessment and repealing Regulation (EC) No 352/2009 (EU OJ L 121/8 of 3 May 2013) with further changes 3.5.2013

MATEC Web of Conferences 180, 06003 (2018) https://doi.org/10.1051/matecconf/201818006003
MET'2017

Evaluation of tramway overhead line system in city of Osijek

Ivo Haladin[1*], *Stjepan* Lakušić[1], and *Ivan* Duvnjak[2]

[1]University of Zagreb, Faculty of Civil Engineering, Department for Transportation, Croatia
[2]University of Zagreb, Faculty of Civil Engineering, Department for Engineering Mechanics, Croatia

Abstract. Overhead line system is a vital element of any electrified railway infrastructure. Its performance reflects to all vital parameters such as traction, reliability, availability and safety of railway infrastructure. Operator GPP Osijek runs a 27.5 km long tramway network which plays a key role in transit system of the city of Osijek, Croatia. In 2016 GPP Osijek applied to a tender for acquisition of new rolling stock with 85% EU investment. Mayor requirement prior to rolling stock acquisition was to document, analyse and upgrade current infrastructure (including track structure, electric substations and catenary) to an optimal level. Task of measurement and evaluation of tram track structure and catenary has been appointed to University of Zagreb Faculty of Civil Engineering. Overhead line analysis comprised from several measuring procedures, including supporting columns evaluation, catenary stagger, height, shocks and wear. Such extensive analysis included GPS positioning and on-site evaluation of 1214 supporting columns of overhead line, as well as overhead line measurements using an instrumented tram vehicle on 18 km of track. For conducting overhead catenary wire measurements, a tram pantograph has been fitted with equipment for measuring catenary height, accelerometers for shock measurement, camera, ruler and GPS for stagger measurement. Catenary wire wear was checked manually from catenary inspection vehicle. To conduct measurements, power had to be switched off on all electric substation along the route and tram vehicle had to be towed by a catenary inspection vehicle. All the data has been analysed and presented to end user in a user-friendly and intuitive GIS environment capable of further updates and detailed analyses. Evaluation of overhead line based on direct measurement results pointed out all the defects and weak spots on the system. It resulted in a series of recommendations for reconstruction and upgrade of the catenary system to fit the need of existing state and further development of tramway network.

1 Introduction

Urban transit systems, such as one in Osijek, Croatia often rely on tramway networks as a backbone for delivering quality service to its passengers. Electric energy for powering the rolling stock is delivered from electric substations through the overhead line system. In Osijek, such a system delivers DC 660 V to the rolling stock. Maintenance of such a system is crucial for safe and reliable running of tramway rolling stock. The overhead wire is attached to hangers which are mounted on supporting columns or supporting anchors on building facades.

In 2016, tramway operator GPP Osijek has appointed a task of evaluation of current overhead line system to University of Zagreb Faculty of Civil Engineering, to prepare the documentation for future tenders for reconstruction of tramway infrastructure. This task was a part of integral evaluation of tramway infrastructure in the city of Osijek which also included evaluation of tramway tracks, electrical substations and a feasibility study on 27.5 km of tramway track [1, 2].

The analysis of overhead line consisted of several measurement procedures including supporting columns evaluation, catenary stagger, height, shocks and wear, with a goal of determining the structural health of the system.

2 Measurements and analysis of overhead line

Measurement of the data required to evaluate the state of the overhead line required substantial effort. Measurements were divided into discrete mapping and visual inspection of supporting columns and anchors, and continuous measurement of overhead wire parameters (height, shocks, stagger) using an instrumented T3 tram towed by a catenary inspection vehicle at ~10 km/h. The whole tram pass by over 18 km of track has been divided into 11 subsections.

2.1 Supporting columns and anchors evaluation

Supporting structure of overhead wire plays a significant role in the electrification system. Its stability and position ensure correct position of overhead wire and the contact with vehicle pantograph. Deteriorated supports can have different stiffness, foundation conditions, shape,

* Corresponding author: ihaladin@grad.hr

inclination, and structural damage, which leads to difficulties in ensuring the correct position of overhead wire. In this evaluation all supporting columns and anchoring points have been examined. Total of 1214 measuring points have been acquired and recorded to a GIS application for further analysis. The data set comprises of 1072 columns, 112 anchoring points on building façade and 30 anchor points to other structures such as viaducts. For each measurement, a set of visual inspection data variables (such as overall state, coating state, general comment, column diameter) have been acquired and stored to the GIS database, along with photos and anchoring point position.

Acquisition and evaluation of the state of columns and anchors has been processed using GIS software specifically designed for this purpose [3]. The software allows data acquisition via Android or iOS mobile devices by building custom acquisition forms including with GPS positioning, photos, reports in PDF and different textual and numerical attributes assigned to each point. Such approach allowed a sophisticated and easy to preform analysis of the overhead line supports on such a large dataset. This approach also allows interactive reporting, updating of the database used for maintenance planning. Based on state of the columns, different parameters can be graphically represented for further maintenance planning, such as overall columns state represented in Figure 1.

Fig. 1. Acquired GIS dataset of 1214 analysed supports.

According to analysis, 20 % of columns were characterized as substantially damaged (bent out of vertical axis, corrosion damage, shrapnel damage etc.), as presented in Figure 2; 32 % endangered (minor corrosion or dents); 48 % in good state (not bent, no corrosion or dents).

Fig. 2. Example of substantially damaged columns.

2.2 Overhead wire measurements

Overhead wire had to be checked along 18 km of track. On these sections, a tram T3 has been instrumented with measurement equipment and towed by a towing truck to conduct measurements off the power grid with a speed of around 10 km/h, Figure 3. Towing truck has also been used as an inspection vehicle for overhead wire wear assessment.

Fig. 3. Towing truck and tram T3 during overhead line measurements.

All measurements described in subsections have been synced with GPS data acquired from a GPS receiver mounted on a tram roof to assign position data to all records and further analyse the data using GIS software. HD video recording of the tram pass by has also been synced to all the recorded data. In this way all the continuous measurements can be simultaneously displayed and analysed, and certain events and defects properly located, described and documented.

2.2.1 Overhead wire height

The overhead wire height has been measured using a digital measuring strip attached to pantograph base and pantograph top, recording the height variation, Figure 4. By adding the fixed value of top of rail to pantograph base height, total overhead wire height is calculated. Measurements are recorded continuously at 100 Hz. Such procedure allowed recording of overhead wire along the whole 18 km of tram track.

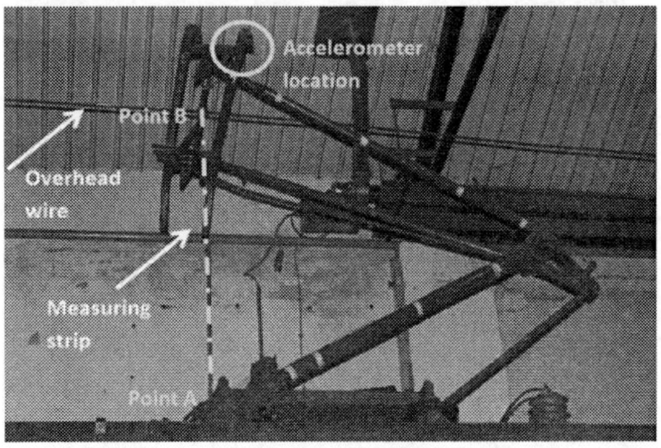

Fig. 4. Overhead wire height and shock measurement setup.

The height of overhead wire has been displayed in a histogram form to analyse the height distribution and occurrence along each of 11 subsections. A minimum height of overhead wire has been carefully examined and documented due to a requirement of free cross section of road vehicles that require 460 mm of height of overhead wire [4], Table 1. Lower level of overhead wire can endanger road users and pedestrians on sections of tram line where road vehicles share the same lane or in an intersection. In such case road vehicles can come in contact with the overhead wire that can result in wire damage, power outage, costly repairs and casualties.

Table 1. Height of overhead wire analysis.

Sub-section	Max / Min [cm]	Height difference [cm]	Below 460 cm [%]
1	523 / 442	81	14.2
2	525 / 441	84	5.1
3	519 / 438	81	17.0
4	523 / 451	73	0.2
5	523 / 430	93	6.7
6	539 / 438	101	9.6
7	527 / 400	127	19.3
8	542 / 438	104	10.4
9	526 / 448	78	4.8
10	543 / 409	133	12.2
11	517 / 402	116	17.1

Height difference in respect to travelled distance has also been analysed because it can lead to loss of contact between the pantograph and the wire, Figure 5.

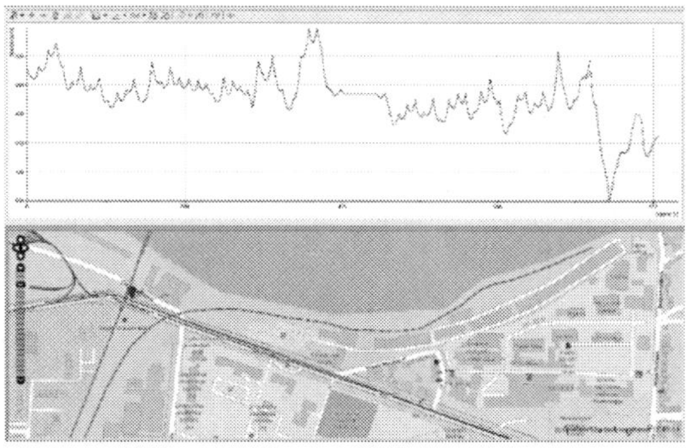

Fig. 5. Substantial height difference of overhead wire (right).

2.2.2 Pantograph shocks and vibrations

Shocks on the pantograph are usually caused by irregularities of contact surface between the pantograph and overhead wire. The discontinuities that result in pantograph vibrations are often caused by joints due to wire break. They can also be caused by cracks in the contact wire that could result in wire breakage. It is therefore important to measure and evaluate shocks and vibrations on the pantograph. To measure vibrations of the pantograph, accelerometers were attached to pantograph head and support rod (Figure 4) in vertical direction. Vibrations were recorded with a frequency of 100 Hz.

Through vibration analysis different average level of vibrations have been recorded on different track segments as well as individual shock events on the catenary, Figure 6. Such data is important for evaluation of overhead line wear, joints, overlaps and potential weak spots.

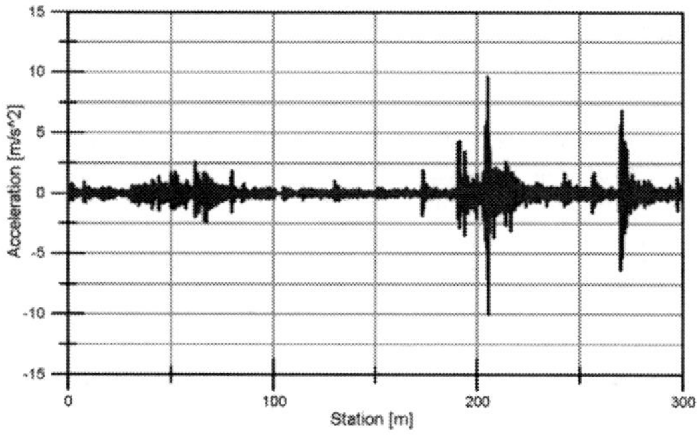

Fig. 6. Overhead wire shock time signal sample.

2.2.3 Overhead wire stagger

During a tram pass by over 18 km of track, a video camera was used to record the pantograph head with a measuring strip attached to it, to record the stagger of the overhead wire in lateral direction, Figure 7.

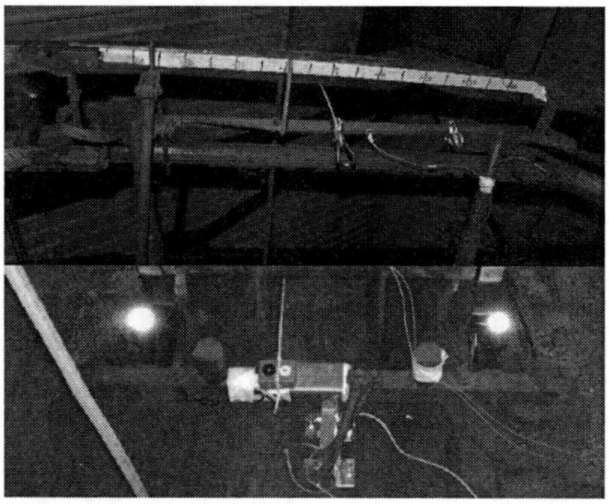

Fig. 7. Overhead wire stagger measurements using camera and measuring strip on the head and camera in the base.

This data has further been processed to create a polygonal curve base on the maximum values the wire has reached in lateral direction, traveling over the pantograph head, Figure 8. Using a GPS signal recorded simultaneously, such data could be geolocated to the tram network and track station.

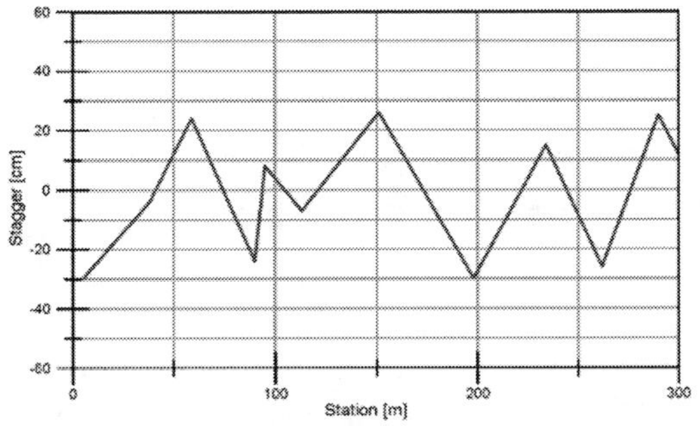

Fig. 8. Pantograph stagger recording.

Depending on pantograph specifications there is an optimal stagger width that leads to even wear of the photograph contact surface. In case of Osijek tram network, it is +/- 30 cm from pantograph centre line [4]. According to that, by histogram analysis, conclusions can be made on wear of pantograph contact surface. Analysis can also pinpoint locations of undesired wire placement (outside of +/- 30 cm range).

2.2.4 Overhead wire wear inspection

Using an overhead line inspection vehicle, the overhead wire has been checked for wear on different cross-section (Figure 9) as well as for wire defects and joints (Figure 10). Such measurements were conducted to asses the overall state of the contact wire on different subsections of the network.

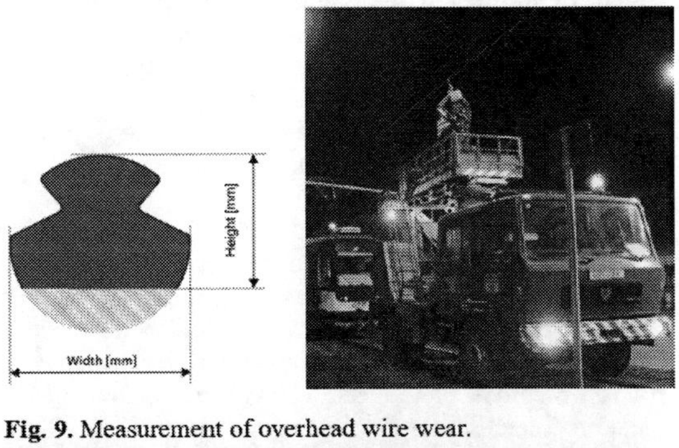

Fig. 9. Measurement of overhead wire wear.

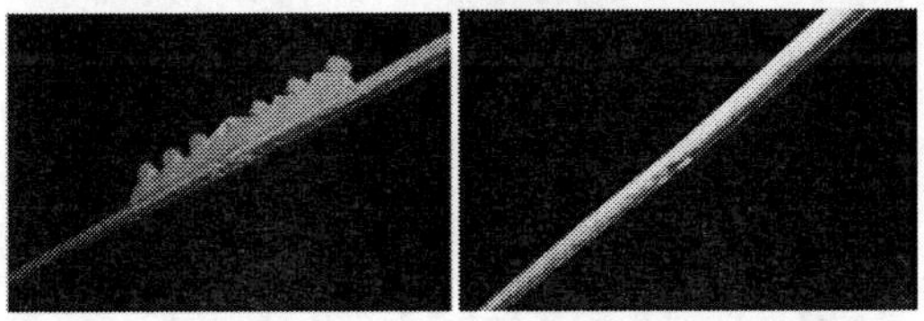

Fig. 10. Inspection of wire joints and potential weak spots.

Such inspection, together with historic data of GPP Osijek maintenance division, revealed 79 joints due to wire breakage, 46 substantially worn spots and other potentially critical points, Figure 10.

3 Conclusions

Overhead wire on tram networks needs to meet high standards of safety and reliability for proper functioning of a transit system. This is can be a challenge in densely populated and narrow urban landscape, with overhead line often anchored to building facades, complex alignment of a tram centreline. Additionally, since tram and road vehicles often intersect or share the same running surface, substantial clearance for road vehicles must be ensured to avoid potential line breakage.

To evaluate the state of overhead line network in whole, different measurement and analysis approaches need to be implemented, as often used on conventional railway tracks [5]. Only the collective knowledge of all components of the overhead line system can lead to correct conclusions.

Overhead line measurements have been divided into discrete mapping and evaluation of supporting columns and anchoring points, discrete measurements of wire wear along the track and continuous recording of wire height, stagger and shocks. To integrate all the measured components, GPS location of each dataset has been stored in a GIS database for further analysis [3].

Conducted measurements and further analysis revealed weak spots and sections in overhead line system that are to be replaced or reconstructed due to weakened or damaged support columns (around 20%), worn overhead wire and elevated level of vibrations and shocks on vehicle pantograph. Furthermore recommendations for correction of overhead line alignment in terms of height and stagger have been given for more than 2000 m of track. Priority levels for reconstruction have also been determined.

References

1. S. Lakušić, M. Ahac, M. Bogut, I. Duvnjak, V. Grgić, I. Haladin, J. Košćak, K. Vranešić, Tramway track condition analysis on GPP Osijek tram network, *technical report* (Univ. of Zag. Fac. of Civ. Eng., 2016) [In Cro]
2. S. Lakušić, M. Ahac, I. Duvnjak, V. Grgić, I. Haladin, J. Košćak, K. Vranešić, Tram overhead line condition analysis on GPP Osijek tram network, *technical report*, (Univ. of Zag. Fac. of Civ. Eng., 2016) [In Cro]
3. I. Haladin, S. Lakušić, Application of GIS software in maintenance management of railway infrastructure, *12th days of Croatian Chamber of Civil Engineers, Opatija, Croatia*, 50-51 (2017)
4. *Regulation on technical conditions and elements for design, construction and reconstruction of urban railway tracks*, Official Gazzette of Croatia, NN 55/1982, 11/1983, 27/1983, 42/1990 [In Cro]
5. H. Neumann, M. Deutzer, Measurement of the Catenary, *technical report* (Deutzer Technische Kohle, 2014)

Inductive influence of 25 kV, 50 Hz electrified single rail traction system

Bosko Milesevic[1,*], and *Ivo* Uglesic[2]

[1]Energy Institute Hrvoje Pozar, Transmission and Distribution Department, Savska 163, 10000, Zagreb, Croatia
[2]University of Zagreb, Faculty of Electrical Engineering and Computing, Unska 3, 10000, Zagreb, Croatia

Abstract. The paper presents the methodology for calculation and measurement of induced voltage caused by railway system operation on the sensitive metallic structures. The arrangement and characteristics of railway system 25 kV, 50 Hz are provided. The procedures of short-circuit current calculation and supply current measurement in the railway system are described. The basic theoretic background of inductive interference is explained. The reduction factors of the communication cable, railway system and environment are mentioned and clarified. On a case study, the results of the induced voltage calculation and measurement is compared. It is proved that induced voltage directly depends of supply current or short-circuit value.

1 Introduction

Electrified traction system enables the transport of people and goods using clean electric energy. The electric energy is transmitted from traction substation to the vehicles through contact network. The current on the contact wire and catenary conductor varies in dependence of the vehicle number and types. Those current is limited by wire's characteristics.

The paper presents the arrangement of 25 kV, 50 Hz single rail traction system supplied from traction substation. The basic characteristics of the system are presented including the electric parameters of traction substation, contact network and the rails [1]. In order to determine the current and voltage conditions on the contact network in the nominal operation conditions, it is necessary to know the types of traction vehicles operating on observed system. The short-circuit current value is calculated from the values of line-to-ground short-circuit current and three-phases short-circuit current at the busses of the power substation which supplies the railway system.

The railway network may also have a return conductor installed on the masts of the contact network and connected to the rails. Its function is important for the current which flow from traction vehicle to the traction substation. The railway system 25 kV, 50 Hz is a single-phase system and it is asymmetric comparing with the power grid. This can affect to the power quality parameters in the high voltage transmission network, but also to the values of power frequency electric and magnetic fields in the vicinity of railway system. The current which returns from traction vehicle to the traction substation has an opposite direction than supply current, so the increase of return current value can affect to the reduction of the magnetic field [2], [3].

Electric railway, as an asymmetric system, has a considerable influence on the sensitive metallic structures located in the same corridor [4]. While the capacitive influence due to 25 kV voltage on the contact network can be neglected, the inductive influence is very important and it should be considered during the normal operation conditions and in the case of line-to-ground short-circuit in contact network for a new lines and reconstructed lines [5], [6].

This paper presents a procedure for the calculation of short-circuit current in the contact network with the known short-circuit currents of the power grid. The curve of the short-circuit current value is determined depending on the distance from the traction substation.

The methods of induced voltage calculation on the sensitive metallic structures in the electrified railway corridor is shown in the case of short-circuit in contact network. The calculation is performed for the communication line in the accordance with international standards. The obtained induced voltage is compared with the limits [5], [6].

2 Railway system 25 kV, 50 Hz arrangement and characteristics

The operation of a single-phase 25 kV (50 Hz) electric traction system is significantly different from the three-phase electric power system which supplies it [1].

Electric traction system is supplied from power transformers located at traction power substation (TPS). These transformers are connected to two phases of the

*Corresponding author: bosko.milesevic@fer.hr

three-phase power transmission system. Traction power supply network is separated by neutral section in two parts which are supplied from different traction substations. Fig. 1 shows the 25 kV 50 Hz electric traction system [7].

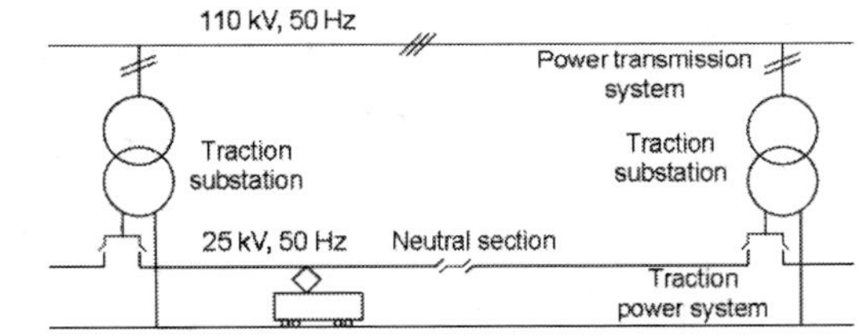

Fig. 1. Electric traction system.

2.1 Current flow in the railway system

Locomotives are supplied with electrical energy through power transformers 110/25 kV. Traction supply network consists of conductors placed above the railway [8]. Conductors are mounted on the masts next to the railway. Locomotives are supplied with electrical energy over the pantograph and the current flows through the rails and ground back in the traction substation [9]. Currents directions are depicted in Fig. 2.

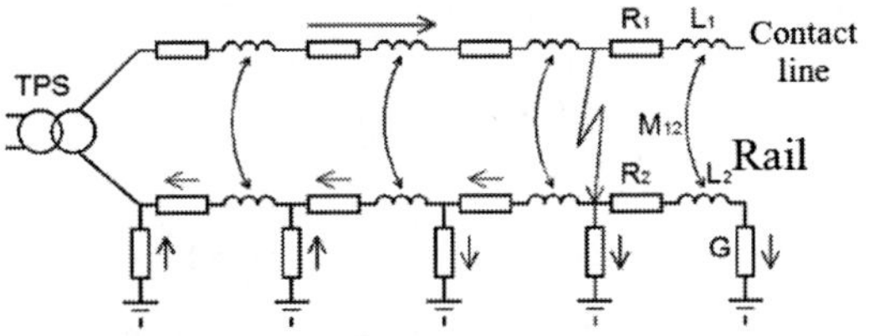

Fig. 2. Current directions in AC railway system.

The values of each current component are crucial for magnetic field distribution around the railway system and indirectly for magnetic coupling. The reduction of magnetic coupling between the railway system and sensitive structures is used into account by reduction factors what is explained in next paragraph.

2.2 Current calculation and distribution in railway system

The calculations of the maximal short-circuit current in the contact network should be performed to determine induce voltage. The impact of the electric power grid is modelled based on the results of three-phase and single-phase short-circuit current calculation on the 110 kV busbars of the traction power substation. Since the short-circuit in traction system is represented as double-phase short-circuit in 110 kV power network (Fig. 1) it is necessary to find the value of positive and negative sequence of equivalent impedance of power grid. Assuming that the negative sequence impedance and positive sequence impedance have the same values, it can be shown that the values of equivalent impedances calculated from double-phase short-circuit current and from three-phase short-circuit current have the same value. The equivalent impedance of power grid can be calculated from equations (1) and (2) using the values of three-phase short-circuit current value and the values of double-phase short-circuit current value respectively [10]:

$$Z_{m3} = \frac{1.1 \cdot U_n}{\sqrt{3} \cdot I_{K3}} \tag{1}$$

$$Z_{m2} = \frac{1.1 \cdot U_n}{2 \cdot I_{K2}} \tag{2}$$

where the U_n is the nominal voltage 110 kV, I_{K3} is the three-phase short-circuit current and I_{K2} is the double-phase short-circuit current.

According to well-known ratio of three-phase and double-phase short-circuit values [11]:

$$\frac{I_{K3}}{I_{K2}} = \frac{2}{\sqrt{3}} \tag{3}$$

and incorporating this ratio to equations (1) and (2), it can be shown that the values of equivalent impedances Z_{m3} and Z_{m2} are equal so the value of equivalent impedance of power grid impedance can be calculated directly from the value of three-phase short-circuit current. The impedances Z_T and Z_{KM} represents traction transformer and contact network. The impedance of contact network is a complex number given in [Ω/m]. The impedance Z_{KM} depends on network geometry, conductor materials and soil resistivity.

Equivalent electrical scheme for short-circuit current value calculation in contact network is presented in Fig. 3. Contact network is connected to power grid through two traction transformers 110/27.5 ± 10 x 1,5 % kV, u_k=10%.

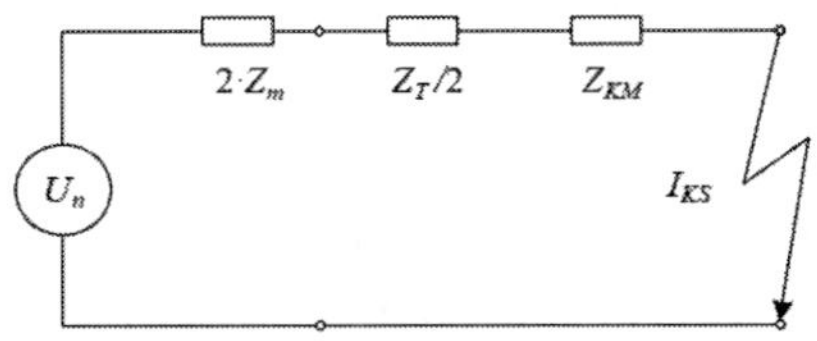

Fig. 3. Equivalent electrical scheme for short-circuit current value calculation.

The impedance of the traction transformers has been determined by its nominal parameters i.e. total power (S_n), voltage level (U_n) and short-circuit voltage (u_k). The impedance of usual traction transformer S_n = 7.5 MVA, u_k=10% is j9.363 Ω.

Now, it is possible to calculate short-circuit current for each point of contact network:

$$I_{KS} = \frac{U_n}{2 \cdot Z_m + 0.5 \cdot Z_T + Z_{KM} \cdot l} \tag{4}$$

where the l is length from traction substation and observed point on contact network.

To determine the maximum short-circuit current, it is assumed that two traction transformers are in operation.

The part of return current that flows through rails depends on parameters such as train distance from substation, rail-to-earth conductance, number of rails which conduct the return current, single or double track line, soil resistivity, etc. In the middle part between the traction vehicle and substation, the return current is 45%-50% of supply current (Fig. 4). The calculations are based on model shown in Fig. 2 and performed in electromagnetic transient program [12], [13].

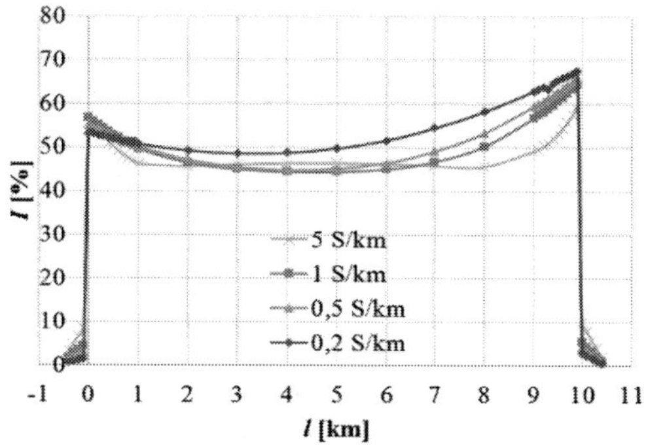

Fig. 4. Return current in rails depending on rails conductance.

3 Inductive interference

The calculations of induced voltages caused by inductive coupling between traction system and sensitive metallic structures should be performed in railway corridor 1 km wide.

For those calculations, the magnetic field distribution in case of short-circuit is a crucial parameter. The inductive coupling between metallic structures appears always when the magnetic field exist in the space. Only the AC magnetic field can induce voltage on the sensitive structure [14].

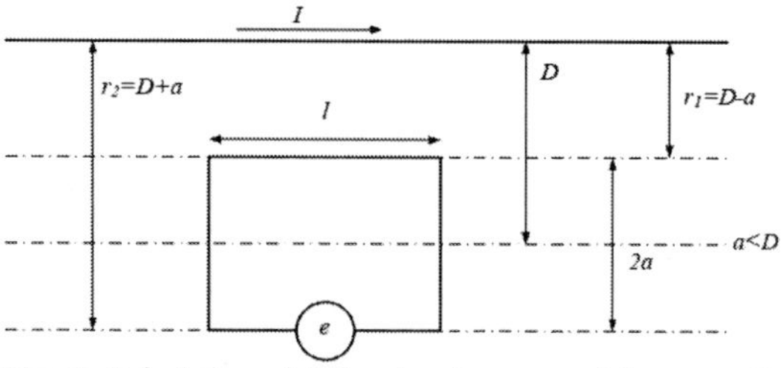

Fig. 5. Calculation of induced voltage caused by magnetic coupling.

The magnetic field that appears in the vicinity of the infinite long conductor is:

$$H = \frac{I}{2\pi r} \tag{5}$$

where the $I = I\,sin(\omega t)$ is the current through the conductor and r is a distance from conductor.

Magnetic flux through the surface S (Fig. 5), caused by magnetic field H, is:

$$\Phi = \iint_S \mu_0 \mu_r H \cdot \vec{n} \cdot \overrightarrow{dS} \tag{6}$$

where the μ_0 is permeability constant of vacuum, μ_r is the relative permeability of the space, $\vec{n}$ and $d\vec{S}$ are the unit length vectors on the surface.

It can be written:

$$\Phi = \iint_S \mu_0 \mu_r H \cdot \vec{n} \cdot \overrightarrow{dS} =$$

$$= \mu_0 \mu_r l \int_{D-a}^{D+a} \frac{I}{2\pi r} dr = \frac{\mu_0 \mu_r l \cdot I_0 \sin(\omega t)}{2\pi} \ln \frac{D+a}{D-a} \tag{7}$$

So, the induced voltage is:

$$e = -\frac{d\Phi}{dt} = \frac{\mu_0 \mu_r l \cdot I_0 \omega \cos(\omega t)}{2\pi} \ln \frac{D+a}{D-a} =$$

$$= -\omega \cdot M \cdot I \cos(\omega t) \tag{8}$$

where $M = \frac{\mu_0 \mu_r l}{2\pi} \cdot \ln \frac{D+a}{D-a}$ is a mutual inductance between metallic structures.

So, the calculation of induced voltage can be performed using the equation:

$$U_i = 2\pi f \cdot M \cdot I_{KS} \cdot l \cdot r_x \tag{9}$$

where the f is nominal frequency, M mutual impedance, I_{KS} short-circuit current and r_x reduction factor.

3.1. Mutual impedance and reduction factor

The contact network and communication line can spread in parallel or can be approaching (Fig. 6). The parallel spread implies that distance change in 5% of the mean arithmetic value of the minimum and the maximum distance on each section (Fig. 7).

Aslant approaching is depicted in Fig. 6 [6]. The mean arithmetic distance is used in calculation with the condition that maximum distance do not excide three times of the minimum i.e. $1/3 \leq a_1/a_2 \leq 3$. If the ratio excides 3, more section should be used.

The mutual inductance of two electrical circuits with the ground as return conductor is very complex to be exactly mathematically expressed. Therefore, approximate equations based on measurements are used. The soil specific resistance has a significant impact on the mutual inductance value because the return current flows through the ground.

The mutual inductance for the frequency 50 Hz depending on the distance between the two circuits and soil specific resistance is given in [5], [7].

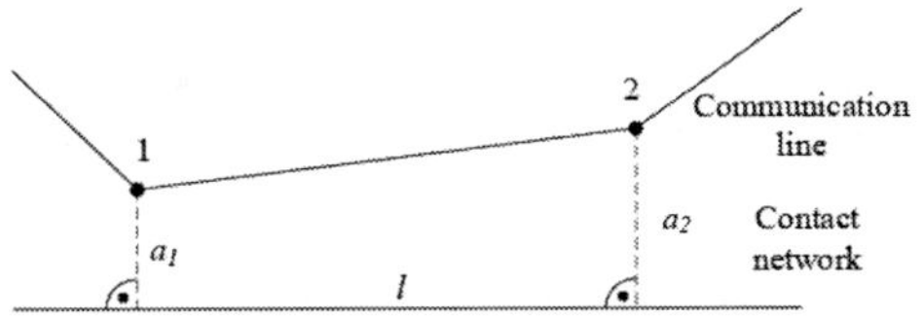

Fig. 6. Determination of the equivalent distance of the communication line from the contact network.

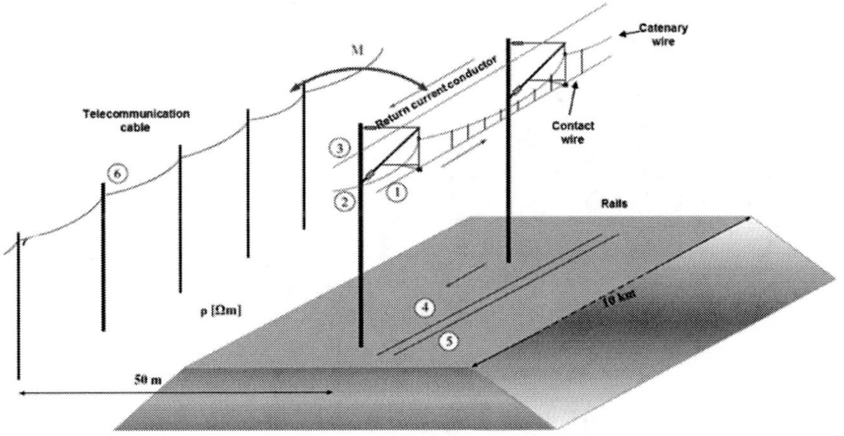

Fig. 7. Inductive coupling in railway system.

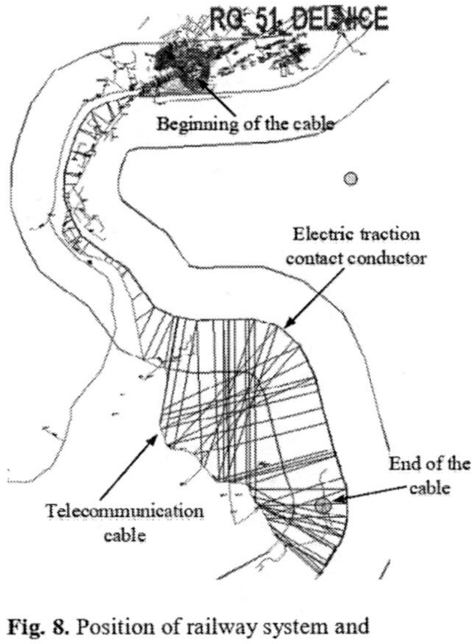

Fig. 8. Position of railway system and communication line.

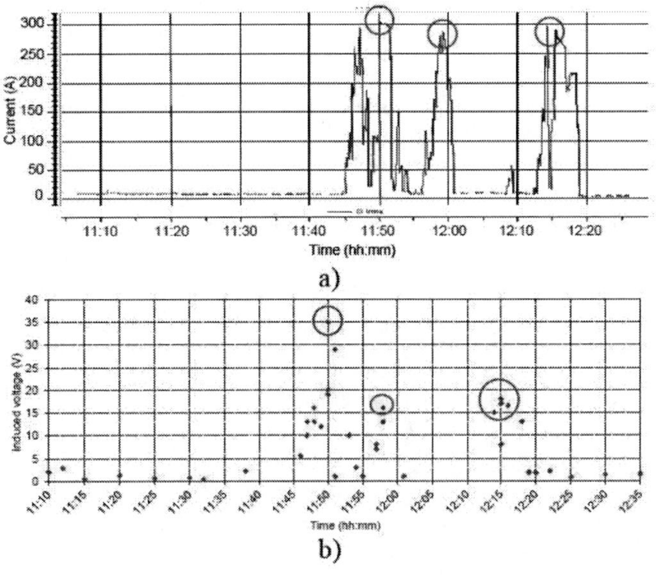

Fig. 9. The current in traction contact conductor and corresponding induced voltage

https://doi.org/10.1051/matecconf/201818006004

The resulting induced voltage reduction factor consists of the reduction factor of rails r_T, the reduction factor of return conductor r_p, the reduction factor of communication cable r_K and shielding factor r_Q:

$$r_x = r_T \cdot r_p \cdot r_K \cdot r_Q \quad (10)$$

If it is possible, the reduction of the environment including threes, grass, other metallic structures etc. can be used.

The reduction factor of the rails r_T is dependent on the ratio of distance from contact wire to rails, the number of the rails, the return current value in the rails, the construction of the rails, the distance from traction substation, the soil specific resistance and other parameters. In the calculations, for the single rail system the r_T factor is used as 0.5 which means that about 50% of the return current flow through the rails.

According to the [5], the reduction factor of return conductor can be estimated with value 0.65.

The reduction factor of the communication cable is given in cable specifications. Diagrams and tables with this reduction is usually provided by the cable manufacturer.

4 Calculation and measurement of the induced voltage

The calculations are made according to the methodology and equations presented in previous paragraph. The measurements include determination of supply current and corresponding induce voltage. In Fig. 9 a) the effective value of current in contact network and in b) the induced voltage on sensitive structure is depicted. The measurements are performed by oscilloscope using the electrical scheme depicted in Fig. 10.

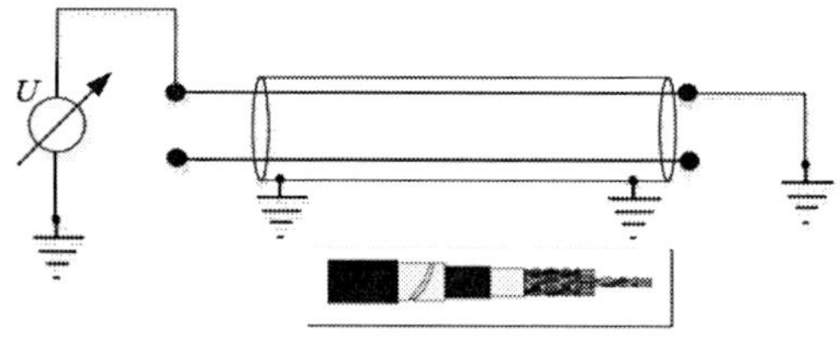

Fig. 10. The electrical setup for induced voltage measurement.

One of the cable wires is connected to the oscilloscope, while the others are idle. The cable sheath, if it exists, should be grounded. It can be easily concluded that the value of induced voltage directly depends on current as a source of perturbation.

The communication cable without armour (r_K=1) is laid in the north-south direction (Fig. 8). It is divided to the 75 sections and orthogonal projection to the rail is founded. The current in contact network was 300 A during the measurements so the same current value is used for calculations.

The induced voltage obtained by calculations was 47,3 V and by the measurements was 35 V. The graph of the induced voltage increase is shown in Fig. 11. The measured value is smaller than the equivalent calculated voltage. This fact is expected because the reduction factor of cable and environment is not considered.

It is recommended to measure the induced voltage on all cables where the calculated voltage exceeds the limit values.

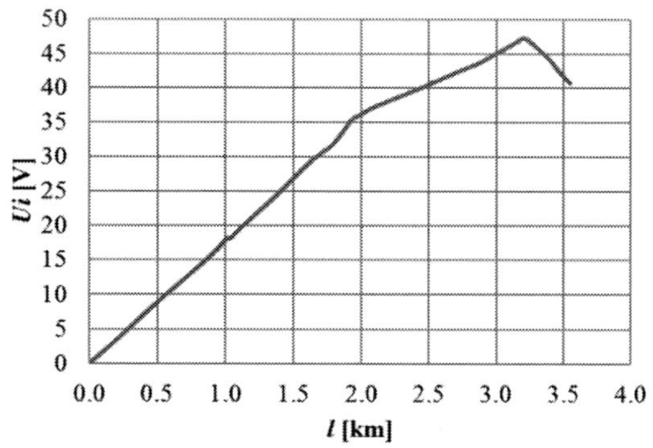

Fig. 11. Induced voltage rises on observed cable.

5 Conclusions

Electrical railway system 25 kV, 50 Hz is a single-phase asymmetric consumer of electrical energy that can significantly affect to sensitive metal structures by inducing voltage. In case of reconstruction of existing railway system or construction a new one it is necessary to check the impact of this system to communication lines.

The short-circuit current in railway system is changeable depending on the location of short-circuit point on the contact network. To determinate value of this current which affect to specific sensitive structures the value of short-circuit current in transmission system and the equivalent electrical scheme of railway system should be provided.

The coupling between the systems is dominantly inductive both in case of short-circuit and normal operation. International standards recommend the procedure for induced voltage calculation. Special attention should be devoted to reducing factors that are usually unknown and changeable. For all communication lines where the value of induced voltage exceeds limit values the measurement tests are recommended.

This work has been supported in part by the Croatian Science Foundation under the project "Development of advanced high voltage systems by application of new information and communication technologies" (DAHVAT).

References

1. Y. Oura, Y. Mochinaga, H. Nagasawa, *Railway Electric Power Feeding System*, Japan Railway & Transport Review, **16**, (June 1998)
2. M.A. Mora, N. Haddad, M. Ney, *Evaluation of the Current in the Rail by 2D and 3D Methods: Influence of the Railway Platform*, IEEE Int. Symposium on Electromagnetic Compatibility, pp. 167 – 172, Pittsburgh, PA, USA, (August 2012)
3. A. Mariscotti, *Induced Voltage Calculation in Electric Traction Systems: Simplified Methods, Screening Factors, and Accuracy*, IEEE Tran. On Intelligent Transportation System, **12**, No. 1, (March 2011)
4. G.C. Christoforidis, D.P. Labridis, P.S. Dokopoulos, *Inductive Interference of Power Lines on Buried Irrigation Pipelines*, IEEE Power Tech Conference Proceedings, Bologna, (2003)
5. ITU Directives concerning the protection of telecommunication lines against harmful effects from electric power and electrified railway lines – Volume II *Inducing Currents and Voltages in Electrified Railway Systems*, ITU, Geneva, (1989)
6. Railway fixed equipment and rolling stock – *Electromagnetic disturbances in electrification 25 kV, 50 Hz – Low frequencies phenomenons* – Calculation tutor, Normalisation Français F 07-11, (October, 1993)
7. B. Milesevic, *Electromagnetic influence of electric railway system on metallic structurec*, (PhD Thesis, University of Zagreb, Croatia, 2014)
8. EN 50149, *Railway applications – Fixed installations – Electric traction – Copper and copper alloy grooved contact wires*, CENELEC, (March 2001)
9. W. Zhang, N. Zhou, R. Li, G. Mei, D. Song, *Pantograph and centenary system with double pantographs for high-speed trains at 350 km/h and higher, Journal of modern Transportation*, **19**, No. 1, 7-11, (March 2011)
10. M. Mandic, I. Uglesic, *Napajanje električne vuče (Railway system supply)*, (Graphis, ISBN 978-953-279-034-4, 2011)
11. J.Schlabbach, *Sort-circuit Currents*, (IET Power and Energy Series 51, ISBN 978-0-86341-514-2, 2008)
12. J. Mahseredjian, C. Dewhurst, *EMTP-RV User Manual, version 2.4*, Powersys, Le Puy-Sainte-Réparade, France, (2012)
13. B. Milesevic, B. Filipovic-Grcic, I. Uglesic, *Estimation of Current Distribution in the Electric Railway System in the EMTP-RV*, International Conference on Power Systems Transients, Seoul, (2017)
14. G.G. Champiot, *Compatibilité électromagnétique Modes de transmission , Technique de l'ingénieur – Génie électrique*, vol. D2, No. D1305, p. 1305.1-1305.20, Paris, (1988)

Operational criteria in the justification of electrification of railway lines

Andrzej Massel[1,*]

[1]Railway Research Institute, ul. Chlopickiego 50, 04-275, Warsaw, Poland

Abstract. The process of electrification of the Polish railways took place from 1936 till early 1990s. Despite the political constraints of the post-war times, the decisions on electrification of particular sections of the railway network were usually based on economic factors. In 1970s and 1980s so called electrification thresholds were used to justify the need for electrification of railway line in question. They took into the number of tracks (single-track lines, double-track lines), terrain conditions (flat terrain, upland terrain) and the traffic structure (the share of passenger traffic) into account. In the framework of National Railway Program till 2023 (KPK) the realization of electrification projects have been undertaken again. The present paper deals with importance of the operational criteria in the decision-making concerning justification of electrification of railway lines. These criteria include the shortening of the journey times on particular sections, shortening of stops on the stations with the change of the type of traction, the savings of train operating companies resulting from the traction unification (change of demand for rolling stock and for workshops for its servicing), improvement of reliability and punctuality of services.

1 Introduction

Electrification of the Polish mainline railways started in the years 1936-1937 with the wiring of three line sections in the Warsaw Railway Node. The process of electrification was continued, although with changing intensity, till early 1990s. The largest growth rate has been observed in late 1980s. The total length of railway lines electrified between 1986 and 1990 has achieved 2485 km, i.e. 497 km per year. The latest large electrification projects in Poland were finalized in 1994 and covered Klodzko – Miedzylesie (36 km) and Olsztyn – Elblag (99 km) sections. Further electrifications covered very short sections (usually newly built connections in agglomerations, for example airport links).

Despite the political constraints of the times after World War II, the decisions on electrification of particular sections of the railway network were usually based on economic factors. In 1970s and 1980s the economic evaluation was performed for one or a few representative sections characterised with typical traffic structure and average line profile, average distances between stops, average train weights and train speeds. So called electrification thresholds were used to justify the need for electrification of railway line in question. They took into the number of tracks (single-track lines, double-track lines), terrain conditions (flat terrain, upland terrain) and the traffic structure (the share of passenger traffic) into account. It is noteworthy, that these methods have been elaborated in COBiRTK (Central Unit for Railway Research & Development), predecessor of Instytut Kolejnictwa (Railway Research Institute).

The European railways have recently been significantly transformed through implementation of the EU legislation imposing "unbundling", i.e. strict separation of infrastructure management and operation of passenger and freight services. Consequently, the approach for railway infrastructure investment appraisal has also been changed. It is also the case of electrification schemes. Relevant costs and benefits of such investment have to be allocated to particular stakeholders: infrastructure managers, train operating companies, competent authorities responsible for contracting public services.

The paper deals with the justification of electrification projects from the point of view of train operating companies (rail operators).

Fig. 1. Rail link in Czerwiensk completed in 2013.

* Corresponding author: amassel@ikolej.pl

2 Rail electrification projects in Poland

According to the Annual Report of PKP PLK for the year 2016, the length of electrified railway lines is 11826 km, while the total length of railways in operation is 18427 km [1]. This means that share of electrified railway lines in Poland equals to 64%. The current shape of electrification in Poland cannot be treated as optimal, mainly due to the missing electrification of some parts of routes, important for passenger and/or freight transport. These missing links are "bottlenecks" in the railway network. They have been identified in the Report of ZNPK (Association of Independent Rail Operators), published in 2012 as major obstacles, inhibiting development of rail services across the country [2].

Table 1. Electrification projects in KPK till 2023.

Line Nr.	Section	Length [km]
25	Tomaszów Maz. - Opoczno	26
38	Ełk – Korsze	99
68	Lublin – Stalowa Wola Rozwadów	103
71	Ocice – Rzeszów	66
201	Maksymilianowo - Gdynia	171
214	Somonino – Kartuzy	8
229	Glińcz - Kartuzy	11
274	Lubań Śląski - Zgorzelec	24
278	Węgliniec - Zgorzelec	26
409	Szczecin Gumieńce (- Tantow)	10

Fig. 2. Electrification projects included in KPK till 2023 (red) and at the planning stage (green).

The process of electrification of railways in Poland, has recently been started again. There are several electrification projects underway, included in KPK (National Railway Program) till 2023 (Table 1) [3]. The total length of existing lines covered with these projects exceeds 500 km. Moreover several kilometres of new tracks will be electrified. Several railway upgrading projects include double-tracking or construction of new tracks, for example between Gdynia Chylonia and Słupsk on line nr 202 and between Kraków and Kraków Bieżanów in Cracow railway node. Electrification investments are also planned for coming years. They are in the planning process, including elaboration of feasibility studies for particular projects (Table 2 and Fig.2).

Table 2. Major electrification projects under study.

Line Nr.	Section	Length [km]
14	Krotoszyn – Leszno - Głogów	116
39,41	Ełk - Suwałki	71
51	Suwałki - Trakiszki	29
68	Lublin – Stalowa Wola Rozwadów	103
137	Kędzierzyn Koźle – Legnica	219
203	Piła – Krzyż – Kostrzyn	161
203	Tczew – Czersk - Chojnice	97
208	Wierzchucin – Chojnice	44
215	Śliwice - Szlachta - Czersk	15
402	Kołobrzeg – Goleniów	100

3 Related works

Methods for analysis of costs and benefits of switching to electrical traction can be found in several publications. First of all it is necessary to mention the Polish works from 1970s and 1980s, in which analyses were based on (so called) electrification thresholds. These thresholds were used to establish the order, in which particular railway lines should be electrified and to define the rational scope of electrification of the national railway network [4],[5].

Various aspects of railroad productivity that might be influenced by the adoption of electrified operation of railways were evaluated by Ditmeyer et al. Productivity was considered from the viewpoint of motive power, transportation economics, signaling and train control, and railway operations [6]. Good example of cost-benefit analysis for railway electrification is presented in the paper by Al-Tony et al. [7]. The paper examines the economic and financial viability of a proposed electrification scheme for the Cairo–Alexandria railway line. The framework of appraisal developed identifies the potential direct and indirect benefits of the scheme, and its costs.

The research team from Gdansk University of Technology proposed complex model for calculating economical threshold for railway line electrification taking the factors regarding constant and variable cost, such as a dynamics of the train or energy prices into account [8].

The paper by Kirkwood et al deals with the typical problems Overhead Line Equipment projects face, when analysing the best option for overbridges. The three competing options (bridge rebuild, track lowering, reduced clearance) have very different capital expenditure (CAPEX) and operating expenditure (OPEX) costs. A model was elaborated to predict these costs over the anticipated assessment period [9].

In the study of rail freight electrification for Southern California, the three key electrification technologies were analysed: straight-electric (with catenary on the entire line); dual-mode (with catenary on some sections); and a linear synchronous motor (LSM) system. These three technology alternatives and each geographic option within each alternative are evaluated based on technology readiness, railroad operations impacts, total capital cost, energy cost savings and total emissions reduction in the region [10].

4 Impact on operations

Very important factor from operational point of view seems to be the flexibility of rolling stock utilisation. Therefore every point of the network, at which changeover from electric traction to diesel traction (or vice versa) occurs, forms significant problem for railway operators due to:

• need for change of the locomotive (passenger and freight trains), or
• need for change of the train (passenger trains).

Every change of rolling stock (electric locomotive versus diesel locomotive, EMU versus DMU) usually generates additional cost for the train operating company, due to need for dedicated infrastructure for servicing traction units (workshops) and for personal at these locations.

The change of locomotive from electric to diesel one (or vice versa) imposes extended stopping time at involved station. Typical time for change of the locomotive (in case of the passenger train) was traditionally around 8-10 minutes. However in case of the new generation of locomotives this time can be extended up to 20 minutes or even more, due to the need for turning off the diesel engine, switching off all onboard computer systems, protecting against electric shock, etc. (Fig.3). The 20 minutes of time waste for the change of the locomotive can be compared with journey time difference, resulting from modernisation of 160 km-long railway line with speed increase from 120 km/h to 160 km/h.

Typical example of problems resulting from changing locomotives is the situation that occurred after diverting Warsaw – Lublin train services (closed for modernisation) from original route to alternative route via Łuków in June 2017. The stopping times for switching the locomotive from electric to diesel (12-14 minutes) have proved to be inadequate and caused significant delays of trains. Therefore starting from September 3 these times were extended to 20-34 minutes to guarantee timekeeping.

Fig. 3. The change of the locomotive on Warsaw – Lublin route.

Another possibility at traction change location is to arrange connection from electric train (operated with EMU) to the diesel one (DMU). In that case some time (usually a few minutes, depending on the layout of the platforms on the station) is necessary for passengers to move from one train to another one (Fig.4). Apart from wasting of time, such change can form inconvenience for passengers, particularly for elderly people, as well as for passengers with reduced mobility (PRM).

Fig. 4. The change of the train at Węgliniec station (Wrocław – Zgorzelec route).

The example of station, where change of train is necessary due to lack of electrification of the final section of the line is Węgliniec in Lower Silesia. Passengers travelling from Wrocław to Zgorzelec change there from electric train (operated with EMU) to the diesel one. Such transfer connections are offered 8 times a day in each direction and a typical time for change is 5-

16 minutes (according to timetable valid from June 2017).

As an alternative for electrification, if travel times are very important, more expensive locomotives with dual mode engines can be used. However from operational point of view, the most practical solution would be uniform traction on entire network (100% of electric traction, 100% of diesel traction). Time losses related to the change of traction can be accepted only in cases of routes with very small traffic. Therefore it is quite understandable, that in several cases a main reason for electrification of particular section is compatibility of the network. When electrifying a line, the connections with other railway lines should be considered. It is noteworthy, that those connections are particularly important in case of traffic disruptions on railway lines and they allow for diverting trains to alternative routes without need for other traction unit.

The good example for importance of alternative electrified routes is the case of traffic breakdown on the Karlsruhe - Basel main line in southwest Germany, one of Europe's key rail freight arteries. The line had to be closed from August 12 till October 2 following a major earth movement caused by groundwater entering the construction site of the new tunnel being built under Rastatt [11]. The double-track section between Rastatt and Baden Baden is already a major bottleneck on this very busy main line, which is used by up to 200 freight trains a day plus over 100 regional and international passenger services. Due to lack of alternative electrified route, huge disruption occurred, involving re-routing of all freight trains through France and substitution of passenger trains (including long-distance trains) with buses.

5 Proposal for operational criteria

In order to assess the effectiveness of the infrastructure investment it is necessary to carry economic analysis. The principles for such analysis are given for example by Dyr et al [12]. However in assessing the need for electrification of particular line segment it seems absolutely necessary to cover essential operational factors in CBA analysis. The following factors can be taken into account:

• the number of trains operated on the segment,
• the number of trains running through from (or to) electrified part of the network,
• the passenger flows on the segment,
• the number of transfer connections from (or to) electrified part of the network,
• the typical time necessary for changing locomotive from electric to diesel one,
• the average time for passenger to change from diesel-operated train to the electric one (and v.v.),
• journey time difference (diesel traction versus electric traction) for all types of trains operated on the line segment and for all stopping patterns),
• existence of alternative routes (electrified, not-electrified).

The electrification brings several benefits for passengers, especially due to reduction of stops at the stations with change of traction and due to shortening of journey times thanks to better traction characteristics of electric locomotives (in comparison to diesel ones) and of EMUs (in comparison to DMUs). The scale of benefits for passengers can be illustrated with the following example.

Let us assume, that the traction change is necessary on given long-distance route. The characteristics of the service is as follows:

• 8 trains per day in each direction,
• 200 passengers per train,
• structure of motivation of journey: 20% duty journeys, 20% journeys to/from work, 60% other reasons,
• stopping time for change of the locomotive (diesel-electric, electric-diesel): 20 minutes
• stopping time necessary for passengers: 2 minutes,
• journey time difference (electric vs. diesel): 10 minutes.

To calculate economic benefits resulting from shortening of the train journey it is necessary to assume unit values of time. According to Jaspers Blue Book for Public Transport project appraisal those values for Poland are as follows:

• 68,46 PLN/h for duty journeys,
• 33,73 PLN/h for journeys to or from work,
• 28,30 PLN/h for other journeys.

Estimated benefit resulting only from shortening of the journey for analysed train service is around 16,7 million PLN per year.

The savings of train operating companies resulting from the traction unification are:

• the change of demand for rolling stock, due its better utilisation (more trainkilometres per vehicle),
• the change of demand for workshops for rolling stock servicing,
• the change of demand for train personnel.

Another benefit (for passengers and for train operators) is the improvement of reliability and punctuality of services.

6 Conclusions

The mainline electrification projects have recently been undertaken in Poland after around 20 years period of stagnation. In the meantime the railway environment has significantly been changed. According to the EU directive, the railway undertakings have been separated from infrastructure managers. The needs of the train operating companies must be taken into account at the stage of planning of railway investment. It is necessary to enhance methods of economic assessment of railway electrification projects.

CBA analyses should not be restricted to short sections in question only. They should rather take costs and benefits of operators, resulting from switching to electric traction on larger area (entire regions) into account.

References

1. PKP PLK, Annual Report 2016, (Warsaw, 2017)
2. ZNPK, *Bottlenecks on the Polish railway network.* (ZNPK, Warszawa, 2012) [in Polish]
3. Krajowy Program Kolejowy do 2023 roku (National Railway Program till the year 2023), (Warszawa, 2015)
4. S. Kuczborski, Z. Polak, W. Wasilewicz, *Criteria for the order of electrification of railway lines*, Przegląd Komunikacyjny, **5**, 3-11 (1974) [in Polish].
5. K. Kożuchowski, Z. Polak, W. Wasilewicz, *Necessary scope of railway electrification in the current situation in fuel and power supply*), Przegląd Komunikacyny, **1**, 8-16 (1980) [in Polish]
6. S. Ditmeyer, J. Martin, P. Olson, M. Rister, B. Ross, J. Schmidt, E. Sjokvist, *Railway electrification and productivity: A Study report*, Transportation Research Record **1029**, 23-30 (1985)
7. F. Al-Tony, A. Lashine, Cost-benefit analysis of railway electrification: case study for Cairo-Alexandria railway line, *Impact Assessment and Project Appraisal*, **18:4**, 323-333 (2000)
8. M. Tomczyk, M. Pacholski, K. Karwowski, *Comprehensive modelling of the cost effectiveness of railway line electrification.* Zeszyty Naukowe Wydziału Elektrotechniki i Automatyki PG, **51**, 197-200 (2016).
9. I. Kirkwood, L. Giuntini, E. Shebab, P. Baguley, *Development of a Whole Life Cycle cost model for electrification options on the UK rail system.* Procedia CIRP, **47**, 1-5, (Elsevier, 2016).
10. Comprehensive Regional Goods Movement Plan and Implementation Strategy. Task 8.3: *Analysis of Freight Rail Electrification in the SCAG Region.* Southern California Association of Governments, (Cambridge Systematics, 2012)
11. A. Green, *DB and SBB seek to minimize Rastatt closure disruption.* International Railway Journal, http://www.railjournal.com/index.php/europe/db-and-sbb-seek-to-minimise-rastatt-closure-disruption.html?channel=537
12. T. Dyr, P. Kazubek, Assessment of transport infratsructure investment co-financed from EU funds, (Spatium, Radom, 2011) [in Polish]

Parameters of DC high-speed circuit-breakers

Artur Rojek[1,*]

[1]Railway Research Institute, ul. Chłopickiego 50, 04-275, Warsaw, Poland

Abstract. DC high-speed circuit-breakers (HSCB) are the basic protection to cut off short-circuit and overload currents. People's and railway infrastructure devices' safety depend on their reliability and parameters. High-speed circuit-breakers, like any other electrical devices, display several electrical and mechanical characteristic parameters, which in many cases are interconnected. The article features these parameters of DC high-speed circuit-breakers that affect the operation safety of electric traction power supply systems. Moreover, it presents ways of their determination basing on laboratory and operational tests. Furthermore, requirements to be met directly and indirectly by DC high-speed circuit-breakers imposed by legal and standardization documents are described. Correlations between the breakers' construction and their parameters are outlined. Basing on tests of leading European producers' breakers, the article presents the impact of parameters of circuit where they are tested as regards HSCB parameters. It also deals with the problem of coordination of short-circuit protection and breaking currents of low values with respect to high-speed circuit-breakers used in Poland.

1 Introduction

High-speed circuit-breakers sorted in traction substations are subject to several requirements regarding their electrical and mechanical parameters as well as their durability and reliability. The malfunctioning of the circuit-breakers may cause railway infrastructure damage or failure, rail traffic disruptions and in extreme cases may even endanger human health or life.

Prior to high-speed circuit-breakers placing into operation, they undergo numerous examinations and laboratory tests. The basic documents relating to high speed circuit-breakers are EN 50123-1 [3] and EN 50123-2 [4] standards for breakers intended for traction substations and sectioning locations (cabins) as well as EN 60077-3 [8] standard, whose provisions need to be read together with EN 60077-1 [6] and EN 60077-2 [7] standards, defining rolling stock breakers' requirements and tests.

Many provisions of these standards are general and permit a large range of parameters. Consequently, this results in the fact that the interpretation of standard provisions and test results is not unambiguous. It is the target operator who decides what type of breaker to be used in a particular application. Nevertheless, laboratory test results do not always translate directly into HSCB parameters in operation. The scope of tests and related issues are described in [13].

2 Breaking off short-circuit currents

High-speed circuit-breakers reducing short-circuit current are used in traction substations and sectioning locations (cabins). They should also be installed in traction units. It means that cutting off the current should take place before it reaches maximum value which depending on the character of short-circuit current flow is a fixed or peak value.

Figures 1 and 2 present short-circuit current flows of different characters and cut off current flows switched off by HSCB. The values shown in the pictures, according to EN 50123-1 [3] standard mean:

- I_{SS} – prospective sustained short circuit current;
- $\hat{I}_{SS}$ – peak of I_{SS};
- $I_{\text{cut off}}$ – cut off current;
- t_c – time-constant of the circuit.

Establishing the time-constant of the short-circuit current t_c in the way described by EN 50123-1 [3] standard is possible only when the short-circuit current reaches its value determined exponentially, as presented in Fig. 1. In case when the short-circuit current flows as presented in Fig. 2, establishing the time constant of the circuit is not possible. The set short-circuit current I_{SS} in Fig. 1 and Fig. 2 has a similar value. However, the steepness of its rise is definitely different.

* Corresponding author: arojek@ikolej.pl

 https://doi.org/10.1051/matecconf/201818006006

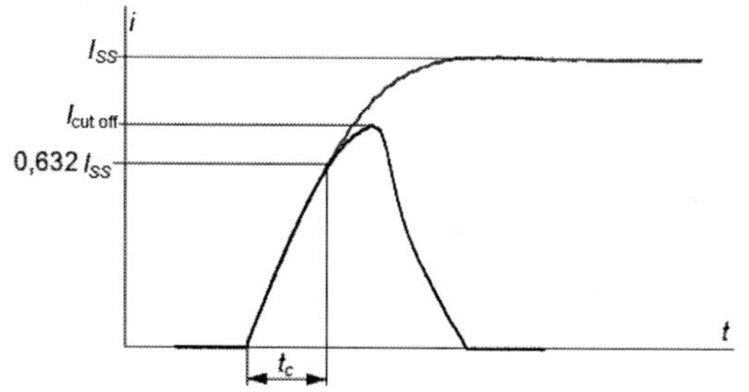

Fig. 1. Short-circuit current oscillogram prior to reaching the sustained value (red) and cut off by HSCB (black)[1] .

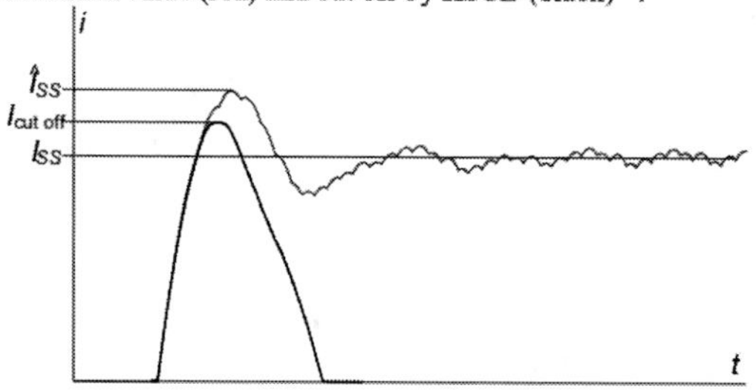

Fig. 2. Short-circuit current flow of value higher than the sustained value (red) and cut off by HSCB (black).

In the event of short circuit on HSCB tripping circuit, the increasing current causes triggering of the breaker when the current value exceeds the level of the trigger setting. Triggering of the breaker should result in its immediate process of limiting and cutting off the current. The opening time t_i of currently operating HSCBs and the level of cut off current $I_{cut\ off}$ depend on the steepness of short-circuit current rise and the peak value, regardless if it flows exponentially or as presented in Fig. 2.

Due to the fact that the short-circuit current flow has a different character, the provision in Clause 11 of EN 50388 standard [5], which is cited in TSI Energy [1] and TSI Loc&Pas [2], seems problematic. This point stipulates that breaking capacity of the circuit breaker in the 3 kV DC system should amount 50 kA of the sustained short circuit current I_{SS}. As HSCB operation should be fast enough to limit the current value as much as possible, a fortiori, to avoid its setting, without specifying the current peak value, the provision in Clause 11 of EN 50388 standard [5] is not precise enough and does not include all possible cases. A provision stating short- circuit current maximum values and the steepness of its rise would be more precise.

On the basis of the above, in order to define HSCB parameters, maximum value of cut off current and its maximum steepness of rise should be provided.

3 HSCB characteristics

High-speed circuit-breakers intended for operation in traction substations and section cabins are divided, inter alia, due to the speed of their operation and short-circuit current limiting. EN 50123-1 standard [3] (points 3.4.7 and 3.4.8) defines three HSCB types: H, V and S. H type breakers are being currently used in railway supply equipment. According to the standard, HSCB type H should prevent reaching the peak value by short-circuit current, the opening time t_i of no more than 5 ms, the total break time t_b should not exceed 20 ms when the value of expected sustained current is at least 7 times higher than the level of the trigger setting and the initial steepness of current has a minimum value of 5 kA/ms.

HSCB characteristics is defined while examining the breaking capacity of maximum short-circuit currents (e.g. duty f). In EN 50123-2 [4] standard, point 5.3.4.2 Table 2, it is defined the tests should be carried out for the time constant resulting from the circuit parameters, however, point 8.3.8.8 in Table 5 stipulates that time constant $t_c = 0$. Meeting this condition is not possible as there always appears inductance in real circuits, thus $t_c > 0$. Moreover, EN 50123-2 standard [4] (Table 5) cites provisions of points 3.4.7 and 3.4.8 of EN 50123-1 standard [3], where conditions regarding time constant are not given, only the steepness of short-circuit current rise is stipulated.

Taking into consideration provisions relating to test parameters for HSCB type H included in the EN 50123-1 [4] and EN 50123-2 [4] standards, the test of maximum currents switching capacity can be carried out in circuits of different parameters, e.g. di/dt = 9,5 kA/ms or di/dt = 5,1 kA/ms. Both values of steepness rise meet the standard provisions – they are bigger than 5 kA/ms, however, the test results are different, which is illustrated in Fig. 3 and 4. Oscillograms presented in these figures were recorded during the short-circuit test of the same HSCB type. Values of prospective sustained short circuit current I_{SS} were similar. The tests were conducted in different circuits – one in the short-circuit laboratory in the circuit without a reactor and the second in the traction substation with the resultant inductance value of reactors 0,8 mH.

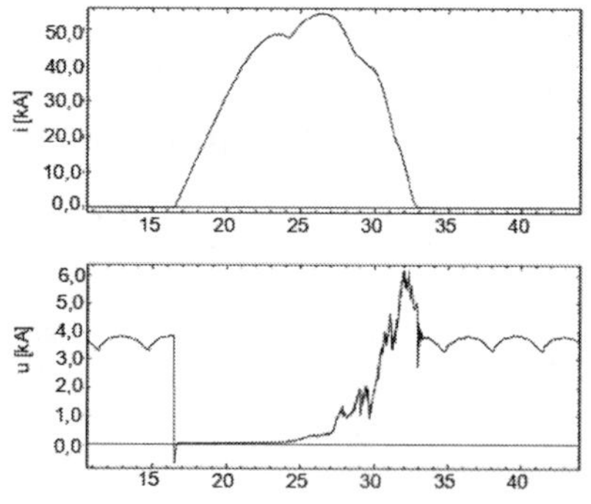

Fig. 3. Short-circuit current and HSCB voltage oscillogram at di/dt = 9,5 kA/ms; t_b = 15,6 ms, arc energy 401 kJ [14].

[1] If not marked differently, figures of recorded measurement data were elaborated for the need of this article during tests of HSCB and DC switchgears carried out for Polish or foreign producers.

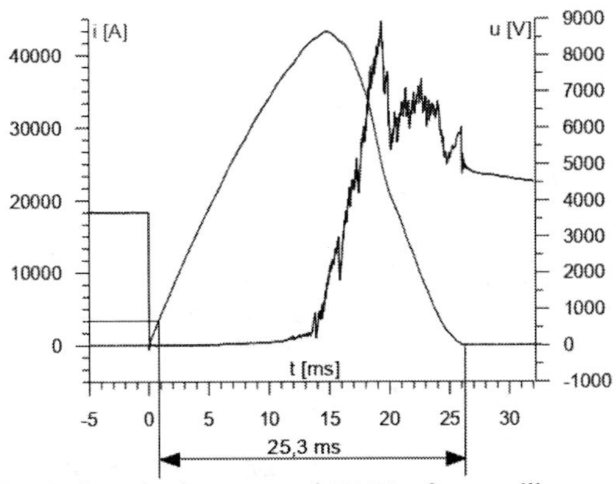

Fig. 4. Short-circuit current and HSCB voltage oscillogram at di/dt = 5,1 kA/ms; t_b = 25,3 ms, arc energy 1268 kJ.

The total break time t_b is the sum of opening time t_i and arcing time t_a. The values of these times depend on the steepness of current changes [11], which result from the parameters of HSCB circuit and construction. In the above-mentioned example, the construction of the breaker did not affect the results (the test were conducted for the same HSCB type) but only the circuit parameters.

Despite the fact that both measurement circuits met the requirements of EN50123-1 [3] and EN 50123-2 [4] standards, the test carried out at di/dt = 9,5 kA/ms confirmed that examined HSCB fulfills the requirements for type H breakers, however, at di/dt = 5,1 kA/ms the total break time t_b is too long. Moreover, attention should be paid to the arc energy value whose most energy must be absorbed by arc chute and arc voltage value. In the event of a circuit with a reactor (Fig. 4), the arc energy is over three times bigger than for the circuit without the reactor (Fig.3) and the arc voltage is approx. 50 % bigger.

The above-mentioned example shows that the parameters of HSCB, which positively underwent laboratory tests in a circuit allowing high steepness of current changes, may not be met in a traction substation real circuit. Stating that HSCB has H, S or V characteristics without defining the steepness of short-circuit current rise is not sufficient for the future user of the breaker.

4 Breaking times

The previous point dealt with the impact of the circuit parameters on the time of cutting off the current by HSCB. This time is also dependent on HSCB parameters, including operating time t_i and arc time t_a.

The operating time largely depends on HSCB construction – the way of its release and contacts opening mechanisms. The arc time also depends on HSCB construction. Its time depends on contacts construction, arc chamber parameters, the arc extinguishing method, magnetic blowout method.

After opening of the breaker's contacts, the arc ignition between them takes place. This arc is short and has a large diameter which causes that its voltage is low. This arc moves to the arc chamber where it significantly lengthens, its diameter decreases and deionization takes place. Since the contacts opening to the moment of arc entering the arc chamber, the arc voltage is low, it does not exceed 10% of voltage source. Such a low arc voltage practically does not cause any short-circuit current limiting. The time when the low voltage arc occurs is defined as contact arc time t_s [15]. EN 50123-1 [3], EN 50123-2 [4] and EN 60077-3 [8] standards do not characterize and do not cover this parameter, despite the fact that it is an essential factor affecting the speed of cutting off the current and the level of its limiting.

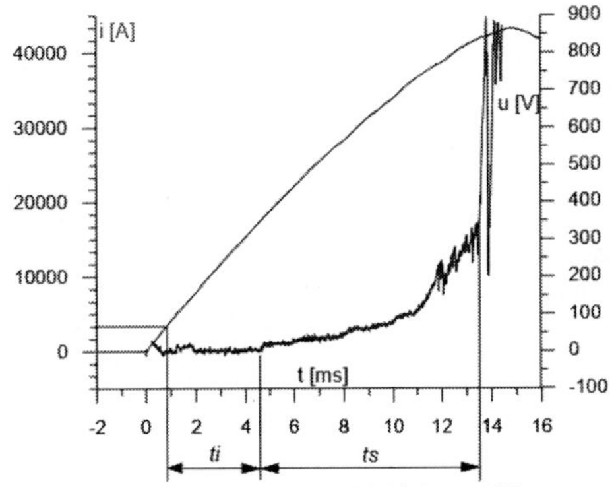

a): t_i = 3,3 ms; t_s = 8,9 ms; $I_{cut\ off}$ = 43,2 kA, t_b = 26 ms

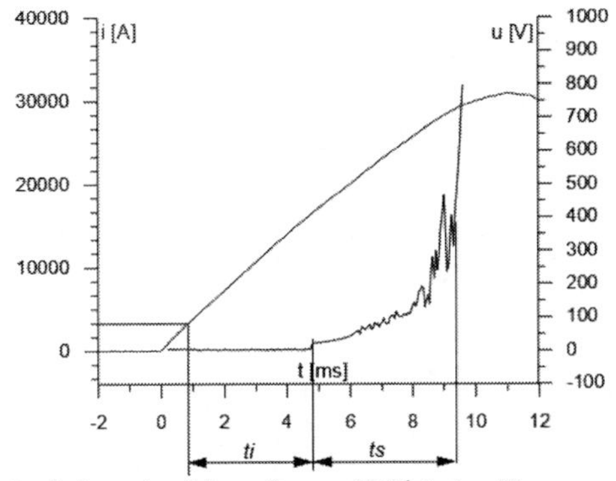

b): t_i = 3,4 ms; t_s = 4,5 ms; $I_{cut\ off}$ = 30,9 kA, t_b = 23 ms

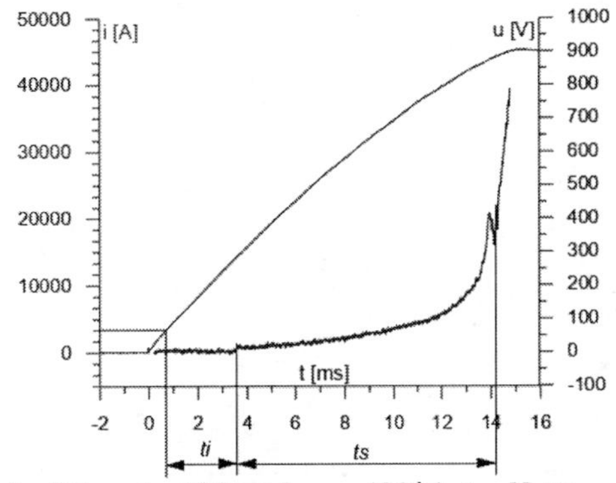

c): t_i = 2,5 ms; t_s = 10,5 ms; $I_{cut\ off}$ = 45,3 kA, t_b = 39 ms

Fig. 5. Short-circuit current and voltage oscillogram, opening time and arc contact time for 3 HSCB types.

https://doi.org/10.1051/matecconf/201818006006

Figure 5 presents the oscillogram of short-circuit current and voltage on three types of HSCB terminals. All flows were recorded during the tests on the same short-circuit and tested breakers had the same levels of setting. The breakers differed in the construction of main contacts and arcing contacts, arc chamber construction and electromagnetic blowout systems.

The analysis of oscillogram presented in Fig.5 shows that the contact arc time is longer than operating time t_i. During the t_s , the current is still rising. The breaker, whose oscillogram are presented in Fig.5c), despite the shortest opening time has the longest t_s time, consequently, the longest time of short-circuit breaking and the highest value of cut off current. This breaker has a gap between the main contacts and arcing contacts, the biggest geometrical dimensions of arcing horns, and its arc chamber does not contain electric conductive elements. This results in the fact that the time needed for the arc to relocate from opened contacts inside the arc chamber is the longest.

The breaker which is characterized by the shortest t_s time and the lowest $I_{cut\ off}$ current (Fig. 5b) has a compact construction of main and arcing contacts, short arcing horns and is equipped with highly efficient magnetic blowout system.

The parameters discussed in this point have a crucial influence on the coordination of short-circuit protections in the vehicle – substation setting, what indirectly affects the reliability of the railway traffic [9]. In order to provide the biggest coordination, the rolling stock breaker should limit the short-circuit current to the lowest level in shortest time. It means that apart from the opening time, also the contact arc time should possible have low values. The bigger the steepness of short-circuit current rise the higher demands relating to t_i and t_s times are placed for rolling stock breakers.

Selecting breakers taking into account only their opening time does not guarantee the protection coordination. In case of long t_s times, the short-circuit current is able to rise to the value when the traction substation's breaker is released. In the protection coordination there is no correlation between requirements placed for substation and rolling stock HSCB. EN 50123-2 [4] and EN 60077-3 [8] standards are not connected which may cause problems while selecting HSCB parameters.

5 Critical currents

Cutting off small currents, called critical currents is a difficult task for HSCBs. Low value of current flowing through magnetic blowout systems results in low value of magnetic field affecting the arc in order of its relocation to the arc chamber, lengthening and extinguishing. Consequently, there are long arc times occurring during critical currents cutting off. The arc time rises then the cutting off current drops.

EN 5012-2 [4] and EN 60077-3 [8] standards do not define permitted arc time t_a during critical currents cutting off. Due to safety reasons and maintaining HSCB durability, there has been introduced a provision into the Polish law stating that critical current is defined as the value of current for which the arc time does not exceed 500 ms [16].

The value and time of critical currents' cutting off depends on HSCB construction, in particular on blowout system. Fig.6 shows the process of critical current cutting off by HSCB equipped only with blowout coil. After equipping this breaker with an additional blowout system [11] (Fig.7) the critical current value decreased about 10 times.

Many HSCB types are bipolar. It means that they can provide switching operation in both current directions. However, the HSCB construction is not symmetrical, consequently the blowout system interaction is not identical for both current directions.

This phenomenon is particularly visible during critical currents breaking. Fig. 8 and 9 show flows recorded during critical currents breaking by the same breaker for both polarizations – forward (the top terminal (+) is connected from the power supply side) and reverse.

Tests of the above-mentioned HDCB showed that critical current's value is 150 A for reverse polarization, i.e. three times bigger than for forward polarization.

The knowledge of the phenomena described above is essential in order to select HSCB for particular applications.

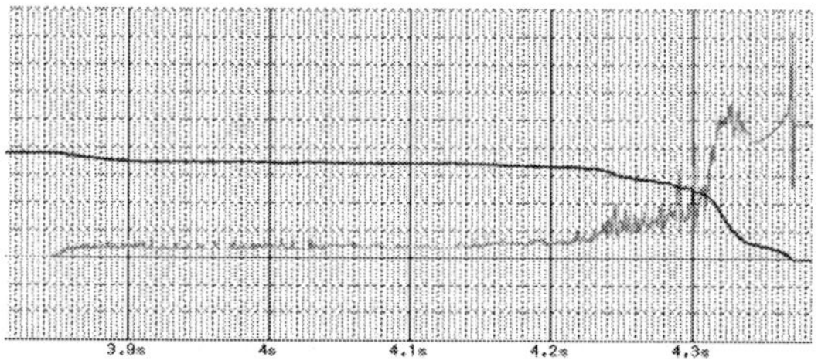

Fig. 6. Critical current cutting off (blue) by HSCB equipped only with blowout coil. $I = 250$ A, $t_a = 535$ ms.

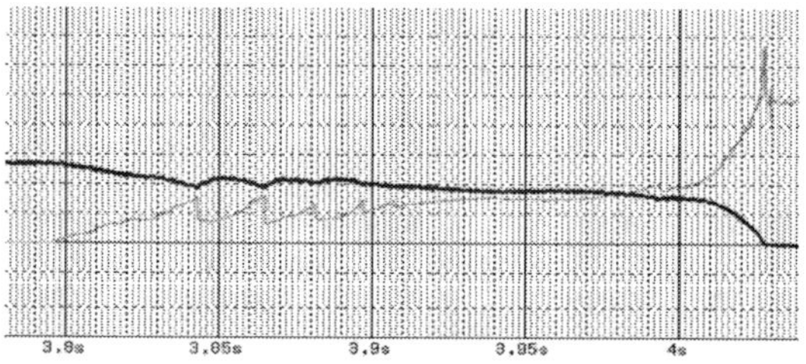

Fig. 7. Critical current cutting off (blue) by HSCB equipped with blowout coil with permanent magnets; $I = 25$ A, $t_a = 231$ ms.

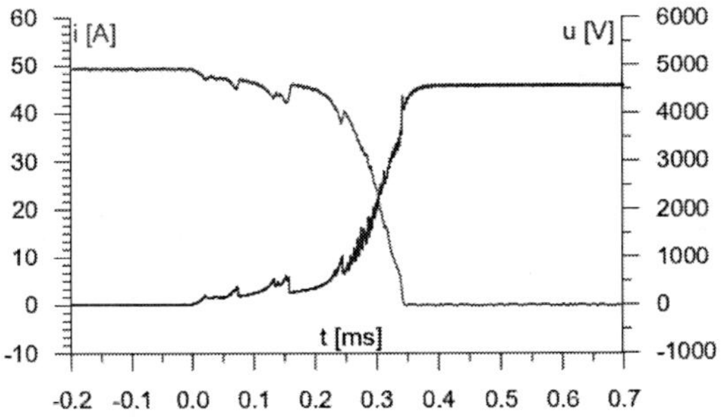

Fig. 8. Current breaking 50 A – forward polarization; t_a = 346 ms

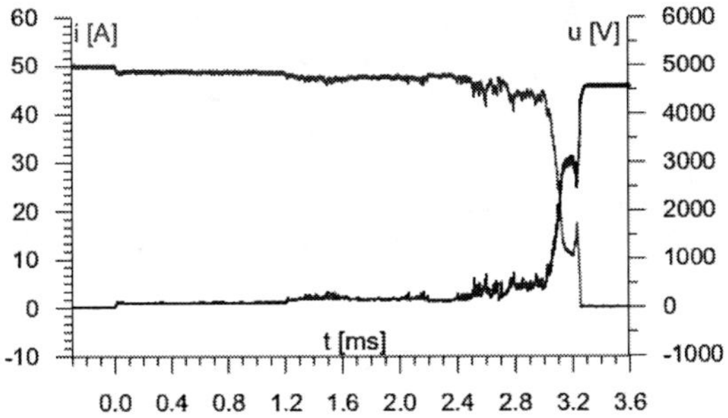

Fig. 9. Current breaking 50 A – reverse polarization; t_a = 3,26 s

6 Conclusions

A proper HSCB choice has a profound significance for reliability and safety of the electric traction supply system functioning. Standards regarding HSCB tests contain many general clauses or such ones that create a possibility of test results' different interpretation. Being uncritically directed only by test results, without the knowledge of the tests conditions does not provide an opportunity of choosing a proper HSCB. Importantly, laboratory tests do not allow defining all HSCB parameters that HSCB will show in real traction power supply systems.

Apart from parameters defined in the standards, the knowledge of HSCB construction is vital for their selection as breakers' construction often determines its characteristics.

In order to make requirements, ways of carrying out tests and their results interpretation precise, a standard amendment relating to HSCB would be useful. Until the standards are amended, due to their current provisions, the entities using HSCB should develop detailed requirements, specifying standards as regards expected parameters and methods of their verification.

References

1. Commission Regulation (EU) No 1301/2014 of 18 November 2014 on the technical specifications for interoperability relating to the 'energy' subsystem of the rail system in the Union.
2. Commission Regulation (EU) No 1302/2014 of 18 November 2014 concerning a technical specification for interoperability relating to the 'rolling stock – locomotives and passenger rolling stock' subsystem of the rail system in the European Union.
3. EN 50123-1:2003. *Railway applications. Fixed installations. D.C. switchgear. Part 1: General.*
4. EN 50123-2:2003. *Railway applications. Fixed installations. D.C. switchgear. Part 2: D.C. circuit breakers.*
5. EN 50388:2012. *Railway Applications – Power supply and rolling stock – Technical criteria for the coordination between power supply (substation) and rolling stock to achieve interoperability.*
6. EN 60077-1:2002. *Railway applications – Electric equipment for rolling stock. Part 1: General service condition and general rules.*
7. EN 60077-2:2002. *Railway applications – Electric equipment for rolling stock. Part 2: Electrotechnical components – General rules.*
8. EN 60077-3:2002. *Railway applications – Electric equipment for rolling stock. Part 3: Electrotechnical components – Rules for D.C. circuit-breakers.*
9. S. Klemba, *Reliability factor in passenger transport modelling.* Railway Reports, 165, 53 – 62 (2014).
10. L. Mierzejewski, A. Szeląg, Research of short circuits and overvoltages in DC electric traction circuits with single-stage transformations110/3 kV. *IVth International Conference MET'99,* 58 – 64 (1999).
11. Ch. Nazeri, *Electromechanical Circuit Breaker.* Patent US 8513558, (2013).
12. A. Rojek, *Switching-off time of electromagnetic blow-out high-speed DC circuit breaker.* Railway Reports, 169, 53 – 60 (2015).
13. A. Rojek, M. Sidorowicz. *Researches and tests of high-speed circuit breakers for rolling stock and substations in 3 kV DC traction power system.* Railway Reports, 159, 7 – 26 (2013).
14. Type Test Report No. 04157-16-0324. IPH Berlin. Berlin, 14 June 2016.
15. F. Wójcik, F. *Theoretical analysis of possibilities for improvement in the effective operation of DC magnetic blow-out circuit breakers.* Electrical Review, 5/2011, 186 – 192 (2011).
16. *List of the President of the Office of Rail Transport on the relevant national technical specifications and standardization documents, whose application makes it possible to meet the essential requirements for the interoperability of the railway system* (Warsaw, 19 January 2017)

Technical parameters of high speed lines as the determinant for selection of rolling stock

Jan Raczyński[1,*]

[1]Research Railway Institute, ul. Chłopickiego 50, 04-275, Warsaw, Poland

Abstract. Choosing a high-speed rail vehicle depends on many factors. On the one hand, there are requirements for ensuring the quality of service for passengers, on the other hand, there are constraints resulting from the parameters of available infrastructure. Also a relation of the benefit and financial costs associated with the purchase and the operation of rolling stock is essential. Technical characteristics of vehicles selected for operating a particular system is a compromise between the three groups of requirements. In this article technical parameters of railway infrastructure and rolling stock are classified and then analysed from the TSI requirements point of view.

1 Introduction

Designing high-speed rail systems requires selection of technical parameters of new or planned for modernization of railway lines. Decisions on investing in railway infrastructure are determined by economic and social needs. These needs can be divided into two groups:

- getting the required travel time between specific regions;
- ensuring adequate transport capacity of a given line in scope of passenger and goods transport.

Technical parameters of high-speed lines result from many conditions: operational requirements, design assumptions as well as access to design solutions and technologies.

Train operation is associated with limitations resulting from the adoption of specific technical parameters for them. Rolling stock parameters are thus determined at the stage of line design.

Although in Europe both railway infrastructure managers and train operating companies are de facto independent entities, choosing new lines technical parameters should take into account the synergy effect. Therefore, the question arises how to make optimal decisions on both infrastructure and rolling stock investments from the operational and economic point of view.

2 Travel time role

In high-speed rail systems design the possibility of obtaining the shortest travel times is of particular importance.

The travel time assumed in the construction or modernization project results from forecasting assumptions for future transport. From the traveler's point of view, it is not important to know how fast he will travel, but at what time he will reach his destination. The volume of interurban travel is strongly dependent on travel time between certain cities or regions.

The influence of the journey time on the performance of passenger transport expressed both in number passengers and passenger-kilometres is much greater than ticket prices. In consequence in spite of higher ticket prices in high speed trains, the demand for transports is increasing along with shortening the journey time (Fig. 1). It is leading railways to the improvement in the competitiveness on the transport market, contributing from fulfilling purposes of the European transport politics and the strategy of the sustainable development of the transport [1]. It is also important factor of the improvement in the competitiveness of states and regions of the European Union [5].

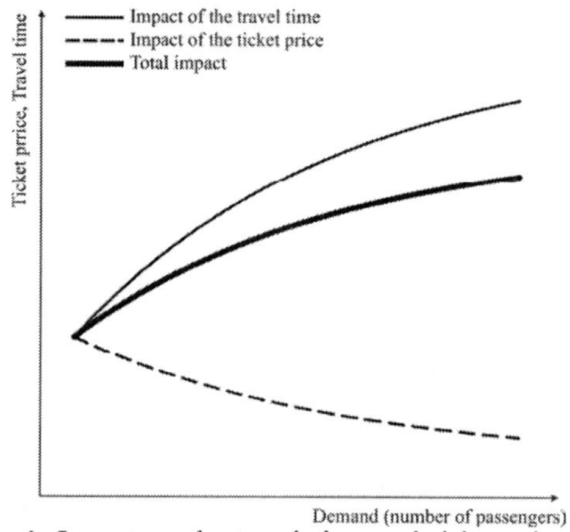

Fig. 1. Impact on the travel time and ticket price on demand in passenger transport. [2]

* Corresponding author: j.raczynski@infotransport.pl

The optimal value of the travel time adoption results in the required technical parameters for a given line and, above all, the maximum speed that can be attained on it. It is obvious that the journey time should be the shortest. The choice of its value is a compromise between the forecasted effects and the costs incurred for the construction of the line.

Factors determining the travel time choice are as follows:

- Ensuring high transport accessibility of concerned regions, including travel time to the railway station. It is therefore a socio-economic factor and it may result from the internal state policy regarding the resources consolidation by obtaining synergy effects and ensuring the harmonious development of regions. In external policy, this objective may be used to strengthen the international position of a country linked to other countries by effective economic ties in which transport plays a key role.
- Provision of large passengers flows who decide to use rail transport as long as it is attractive in terms of speed and travel time. It is important not only to ensure high competitiveness of the railways in relation to other means of transport but also to generate new demand that would not have been created without a radical shortening of travel time in comparison to the one before the new line construction [6].
  The longer is the travel time, the lower is the willingness of people to travel. Shortening the travel time increases the number of passengers. From the passenger's point of view, it is possible to separate certain time intervals determining his willingness to travel. It is assumed in transport forecasts that if travel times are no higher than 90 minutes, the railway shares in the transport market are very high and even exceeding 90%. Shares decrease gradually until they drop significantly after exceeding 3 hours. Air transport is an important competitor in this period of time [7].

Table 1. Travel time and socio-economic effects and railway competitiveness [7].

Travel time	Socio-economic effects	Passengers flows	Railway competitiveness
Up to 1,5 hour	An inter-agglomeration job and services market is created generating economic growth	Relatively high and even passenger flows during the whole day – possible business travel on different day time.	Very high railway competitiveness, in some relations unrivaled
1,5 – about 3 hours	Active, multilateral economic connections are created	Possible one-day-trip travel particularly important for business.	High railway competitiveness.
Above 3 hours	Depending on economic and social relations, friendly conditions for strengthening socio-economic links are created	With the time above 3 hours, the possibility of generated demand drops	The competitiveness of railways decreasing after exceeding 3 hours to very small market shares

- The possibility of reducing unit passenger transport costs calculated per passenger-kilometer. It is the lower as the shorter is the travel time, which is related to better use of production means, i.e. rolling stock and personnel. The only cost element that is increasing with the train speed is the electricity cost. The acceptability of higher energy costs results from compensating them by shortening the travel time, which is also an economic category and has a certain value for the traveler who is able to incur additional travel expenses for a certain service quality level.

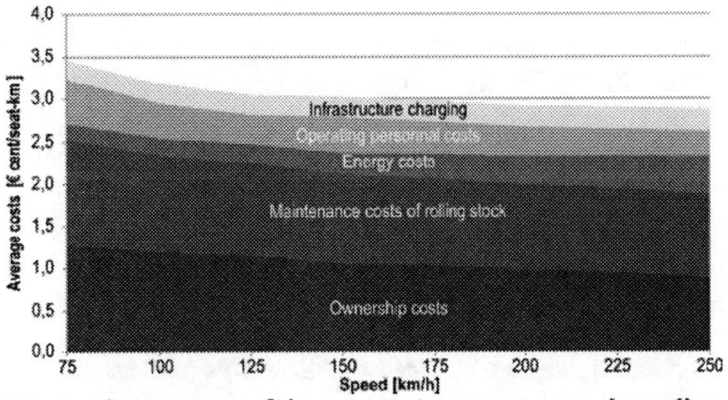

Fig. 2. The structure of the operator's average costs depending on the commercial speed of the train [3].

3 Railway lines categories

Classification of essential parameters for infrastructure and railway lines categorization is the subject of the EU TSI Infrastructure and other TSIs related to the design of railway lines:

- power supply (TSI ENE),
- safety in tunnels (SRT TSI),
- facilities for people with reduced mobility (PRS TSI),
- rail traffic control (CCS TSI) [4].

The TSI Infrastructure specification introduces the both passenger and freight traffic lines classification. Lines with mixed traffic are described by two categories - for freight and passenger traffic.

In terms of requirements for travel time, the railway line maximum speed is a crucial parameter for achieving the goal. This especially applies to long-distance traffic.

The railway line capacity depends on the load and length of the trains. The three most important parameters that characterize them are the gauge, the permissible axle load and the station tracks length, or in the case of passenger transport - the length of the platforms.

The generally accepted criterion for railway lines classification as a high-speed is the possibility of its operation at a speed of at least 250 km/h. EU legislation has introduced a category of lines with a speed of less than 250 km/h for high-speed supplementary lines i.e. feeders or modernized lines, for which reaching a speed of at least 250 km/h due to various objective factors was not possible. The basic parameters for these two lines categories are shown in Table 1.

Table 2. Operating parameters for passenger lines with speeds above 200 km/h [4].

Traffic code	Gauge	Axle load [t]	Speed [km/hour]	Platform length [m]
P1	GC	17	250–350	400
P2	GB	20	200–250	200–400

The parameters „gauge" and „axle load" are treated as minimum requirements. The „line speed", „platform length" and „train length" columns indicate the range of values that are typically used for different traffic types and do not directly impose restrictions on traffic on a given line. It is also acceptable to design new and modernized lines, which will be adapted to larger gauges, higher axle loads, higher speeds, longer platform lengths and longer trains than specified above.

The railway line is also characterized by parameters resulting from the vehicle power supply system. The requirements are described in the ENI TSI. For high speed lines, in Europe are used two AC systems: 25 kV 50 Hz and 15 kV 16.6 Hz. For P2 lines category, a 3 kV DC power supply system is also acceptable, but the economic and energy efficiency of which is lower than for AC systems. The power system determines the power consumed by the trains and the same maximum speed and trains mass.

4 Travel time and maximum speed

From the point of view of railway infrastructure the shorter travel times is as the higher maximum speed is. Her choice is determined by several factors, the most important of which are:

- expected journey time;
- terrain conditions;
- line construction cost;
- technical solutions availability.

The line maximum speed is strongly determined by topographical conditions. High maximum speed requires increasing the railway track curves radius. It is not difficult to implement in a flat and thin area. For such conditions, differences between costs of the 160 to 350 km/h line are small. In difficult topographic conditions, i.e. in mountain areas, the costs of increasing speed are higher. For this reason, in such specific cases lines are designed for up to 250 km/h speed. (Switzerland, Basque Country, Belgian Ardennes region).

Currently in Europe, new lines are built or designed at maximum speeds in the range of 300–350 km/h, less often at 250 km/h. A summary of maximum speeds for Europe is presented in table 3. Larger maximum speeds are used or planned in China, mainly due to the large distances between agglomerations.

When assessing the impact on the travel time of the maximum speed increase a simplified analysis can be used. It compares time differences for different maximum speeds on 100, 200 and 300 km line sections without train stops.

For actual operating conditions in analyzes should be used a commercial speed, depending on many factors, including:

- the number of stops at intermediate stations
- time needed for departures from railway junctions, usually at a lower speed

Table 3. Maximum new line speeds in Europe.

Country	Maximum design speed for new lines [km/h]	Length of a railway line operated, built or planned [km]
Austria	250	266
Czechia	300–350	890
Denmark	250	206
Estonia Latvia Lithuania	250	870
France	320–350	4562
Germany	250–300	2167
Italy	250–300	1269
Poland	300–350	1305
Portugal	300	596
Spain	250–350	4903
Sweden	320	740
UK	400	656
Switzerland	250	159
Russia	400	2978

Source: UIC Statistics (2017)

Table 4. Travel time as a function of the maximum speed.

Maximum speed [km]	Section length [km]		
	100	200	300
200	30	60	90
250	24	48	72
300	20	40	60
350	17	34	51
400	15	30	45

Own work.

For this reason, already at the design phase should be chosen solutions, like the maximum speed. Travel time depends on the track system, the way of servicing passenger stations or the existence of local speed limits. The travel time is also dependent on the solutions of track systems, The ratio of commercial speed to the maximum one on existing and planned lines can reach values even above 0.9, but it is usually lower.

5 Rolling stock parameters

Selected rolling stock should allow maximum use of line parameters both in terms of ensuring the shortest travel time and capacity. Given in table 4 travel times are possible to get under theoretical conditions when the train goes at the line maximum speed and its acceleration and braking are equal to zero.

The essential rolling stock parameters that help to achieve the shortest transit time are as follows:

1. Maximum speed

Usually chosen to make use of the highest maximum speed sections. Currently, on the European market there are available vehicles with maximum speed from 320 to 350 km/h but up to 400 km/h speed is also possible.

2. Acceleration and deceleration

These are parameters that are important when the distances between stations are small equal to several dozen kilometres. High acceleration and deceleration values are achievable with a large number of drive axles at the vehicle construction stage. However, this requires higher power supply efficiency.

3. Passenger flows

In order to ensure the shortest possible train stop time at the station, the interior layout and vehicle doors (their width and number) should allow passengers moving quickly and getting in and out.

4. Power supply

High speed lines of maximum speeds above 200 - 250 km/h are electrified with 25 kV, less often 15 kV AC. AC systems are cheaper to build and are characterized by lower operating costs, primarily due to higher efficiency and lower energy losses on energy transmission. In addition, only they are capable of delivering high power to the vehicle, which at the highest speeds and long train sets reach values up to 20 MW.

Rolling stock capacity is defined by passenger seats number in vehicle and depends on the following parameters:

1. Train length

Due to the infrastructure characteristic his parameter should not exceed 400 m. For P2 lines and lower platform lengths, it may be even smaller. For this reason, the standard length of the high-speed train is 200 m. If the technical conditions allow it, trains can be combined in multiple traction.

2. Vehicle gauge

Larger wagons dimensions allow you to fit more seats in one row. In European conditions, due to historical requirements, it is not possible to use rolling stock with a gauge larger than that included in the GC infrastructure gauge.

3. Axle load

Increasing the capacity of trains is also possible through the use of double-deck structures. However, this causes an increase in the weight of the train and thus an increase in the weight per axle. Small axle loads are required due to the reduction of the dynamic impact of the vehicle on the track. The higher is the maximum speed, the lower the permissible axle loads. The trains with highest maximum speeds have axle loads of 17 tonnes and in the P2 category up to 20 tonnes per axle. This excludes the possibility of using locomotives for trains of a maximum speed above 200 km/h.

6 Summary

Operational requirements for high-speed rail systems determine the selection of technical parameters of railway lines and rolling stock for its service. In case of new line design, the correlation of these parameters can be made at the stage of feasibility studies. When designing railway line, forecasted future new technologies development should also be taken into account. Development trends in the rolling stock construction should be considering while choosing the maximum speed of the railway line. In general, it should be as high as possible. Current technologies allow the track, power supply and traffic safety systems construction adapted for the speed up to 400 km/h. For technical and cost reasons in difficult mountain conditions the line maximum speed can be limited to 250 km/h.

Even if currently maximum European rolling stock speed vary between 250 and 320 km/h it should be taken into account that intensive work on new vehicle designs capable of reaching speeds of up to 400 km/h is underway. In 2017, units with 350 km/h maximum speed were put into operation in China and, ultimately, the construction of them enables running up to 400 km/h.

The choice of the maximum speed should also take into account the effects that can be obtained along with shortening the travel time, i.e. increasing the number of passengers and reducing unit costs for train operating companies, calculated per passenger kilometre.

Analysing European experiences, i.e. average distances between stations, it can be concluded that the optimal maximum velocity values are in the range of 300 - 350 km/h.

In case of choosing rolling stock for existing lines there are restrictions related to infrastructure parameters. In order to get better travel time for a defined maximum speed it is possible to choose rolling stock of higher acceleration value. In case of the necessity to improve train capacity infrastructural constraints are a bigger barrier, e.g. in terms of platforms length, which can be bypassed by the use of a double-deck rolling stock.

Table 5. Relation between infrastructure and high-speed rolling stock requirements.

Operation requirements	Line parameters	Rolling stock parameters
Short travel time	Maximum speed	Maximum speed Acceleration Deceleration Passenger flows
	Power supply efficiency	Power drawn by the train
Sufficiently high transport capacity	Platform length	Train length
	Structure gauge	Rolling stock length
	Axle load	Vehicle mass

References

[1] T. Dyr, and K. Ziółkowska, *Economic infrastructure as factor of the region's competitiveness.* Central European Review of Economics & Finance, **6**(3), (2014)

[2] T. Dyr, and K. Ziółkowska, Efektywność kolei dużych prędkości [w:] Siergiejczyk M. (red.), *Koleje Dużych Prędkości w Polsce* (Wydawnictwo Instytutu Kolejnictwa, Warszawa, 2015)

[3] T. Dyr, and K. Ziolkowska, *Financial effectiveness of high speed railway in Poland,* Technology Transport Systems [Technika Transportu Szynowego], **6** (2015)

[4] *Dziennik Urzędowy Unii Europejskiej* L 356, Tom 57, 17 grudnia 2014.
[5] V. Maráková, T. Dyr, and A. Wolak-Tuzimek, *Factors of tourism's competitiveness in the European union countries*, E a M: Ekonomie a Management, **19**(3) (2016) doi: 10.15240/tul/001/2016-3-007.
[6] Pedro Miguel Cardoso, Martins Network Manager, *Eurocontrol statfor high-speed train model recalibration forecasting air transport workshop,* (7th October, Cologne 2016)
[7] A. Szarata, and J. Raczyński, *Analiza symulacyjna wielkości przewozów dla kolei dużych prędkości w Polsce.* Technology Transport Systems [Technika Transportu Szynowego], **6** (2017)

AUTHOR INDEX